Alternative Energy Sourcebook

7th Edition

P9-CJY-384

7th Edition

Alternative Energy Sourcebook

Edited by John Schaeffer
with contributions from
the Friends and Staff of
Real Goods Trading Corporation

Published by

REAL GOODS

Trading Corporation
Ukiah, California

Credits

Editor & Publisher: John Schaeffer
Authors: Douglas Bath, Gary Beckwith, Mike Bergey, Michael Hackleman, Stephen Morris, Jeff Oldham, Richard Perez, Doug Pratt, John Schaeffer, Joe Stevenson, Steven Strong, Jon Vara and Randy Wimer.
Publication Coordinator: Linda Malone
Design Director: Bob Steiner
Computer Wizardry and Index: Michael Potts
Technical Assistance: Mike Gellerman, Klaus Halbach, Jeff Oldham, Doug Pratt, Douglas Bath, & Gary Beckwith
Proofreading: Andrew Alden, Eileen Husted & the Real Goods Staff
Photography: Sean Sprague & Tom Liden
Illustrations: Tom Jarvis & Kathy Shearn
Cover Design: Linda Lane
Printing: Consolidated Printing, Berkeley, CA

Special thanks & acknowledgments to: Debbie Robertson, Brenda Bianchi, and Jill Andregg.

© **1992 Real Goods Trading Corporation**™. The logo and name Real Goods™ are the trademark of Real Goods Trading Corporation. Nothing in this book may be reproduced in any manner, either wholly or in part for any use whatsoever, without written permission from Real Goods Trading Corporation. Prices, availability, and design of products in this catalog are subject to change without notice. We reserve the right to correct typographical, pricing & specification errors. All merchandise is FOB Ukiah, CA, unless otherwise noted. The ■ symbol indicates that the item is shipped directly from the manufacturer.

© Amory Lovins, forward beginning on page 1
© Michael Hackleman, articles beginning on page 367
© Steven Strong, article beginning on page 119
© Richard Perez, article beginning on page 481

Real Goods Trading Corporation has checked all of the systems described in this catalog to the best of its ability; however, we can take no responsibility whatsoever regarding the suitability of systems chosen or installed by the reader.

Prices Subject To Change Without Notice.

ISBN 0-916571-02-5 - $16
Printed in U.S.A., on recycled paper
Revised regularly - Seventh Edition - April 1992 - 35,000 copies
Over 150,000 copies in print

Real Goods Trading Corporation
966 Mazzoni St., Ukiah, CA 95482-3471 Business office: (707) 468-9292
To order: call 1-800-762-7325 or fax (707) 468-0301

Table of Contents

Appendix

Foreword

Amory B. Lovins
Rocky Mountain Institute
Old Snowmass, Colorado 81654-9199

From 1979 through 1986, the United States got more than seven times as much new energy from savings as from all net increases in supply. Even more astoundingly, of those increases in supply, more came from sun, wind, water, and wood than from oil, gas, coal and uranium. Even as glossy magazine ads were dismissing renewable energy as unripe to contribute much of anything in this century, renewables came to provide some 11-12% of the nation's total primary energy (about twice as much as nuclear power), and the fastest-growing part. The only energy source growing faster was efficiency. Just the *increase* in renewable energy supplies during those years came to provide each year more energy than all the oil we bought from the Arabs. And efficiency, during 1973-86, came to represent an annual energy source two-fifths bigger than the entire domestic oil industry, which had taken a century to build; yet oil had rising costs, falling output, and dwindling reserves, while efficiency had falling costs, rising output and expanding reserves.

To be sure, during about 1986-88—the later years of what, in the telling phrase used in their own country by Soviet commentators, may be fairly called "the period of stagnation"—this momentum declined and in some respects stalled. The impressive successes of efficiency and renewables often fell victim to official hostility and collapsing oil prices (which were largely driven by the very success of energy efficiency). The Reagan Administration's roll-back of efficiency standards for light vehicles immediately doubled oil imports from the Persian Gulf, effectively wasting exactly as much oil as the government hoped could be extracted each year from beneath the Arctic National Wildlife Refuge. While some electric utilities pressed ahead with good efficiency programs, additional electrical usage spurred by deliberate power-marketing efforts was officially projected, by the year 2000, to wipe out about two-thirds of the resulting baseload savings. The same Administration that touted the virtues of the free market pressed home its strenuous efforts to deny citizens the information they needed to make intelligent choices. And Federal tax credits meant to help offset the generally much larger subsidies—totaling at least $50 billion per year—given to renewable's competitors were generally abolished, while most of the subsidies to depletable and harmful energy technologies were maintained or increased, tilting the unlevel playing field even further. These and other distortions of fair competition gravely harmed many sectors of the renewable energy industries, often drying up distribution channels so that even sound, cost-effective options could no longer reach their customers.

Efficiency came to represent an annual energy source two-fifths bigger than the entire domestic oil industry

---

Leadership in some key R&D areas passed from America to Japan and Germany. Cynics began writing premature obituaries of the latest solar flash-in-the-pan. By 1990, the savings achieved since 1979 were no longer seven, but only four and a half times as big as all the net increases in energy supply, and the fraction of that new supply coming from renewables was no longer a bit over half, but only one-third. There were bright spots—Maine, for example, raised the privately generated, almost all renewable, fraction of its electricity from 2% in 1984 to 35% in 1991—but many other efforts and firms faltered or even went under.

Yet throughout that decade's rise, leveling, and sometimes stumbling of the keys to a safe, sane, and least-cost energy future—high energy productivity and appropriate renewable sources—a band of pioneers in Northern California sustained their vision of a way to give everyone fair access to a diverse tool kit for energy self-sufficiency. Through their dedication, thousands of people have had the privilege of discovering that the energy problem, far from being too complex and technical for ordinary people to understand, is perhaps on the contrary too simple and political for some technical experts to understand. The Real Goods team gave, and gives today, an equal opportunity to solve your piece of the energy problem from the bottom up.

Many did exactly that. They discovered that solar showers feel better, because you're not stealing anything from your kids. They found how to get greater security and high-quality energy services from a judicious blend of efficiency and renewables. Having a high do-to-talk ratio, they worked through trial and error, celebrated their inevitable mistakes, and found in Real Goods an effective way to share their experience with a wide audience. (There is, after all, no point repeating someone else's same old dumb mistakes when you can make interesting new mistakes instead.) Piece by piece, with that uniquely American blend of idealism and intense pragmatism, they quietly built, and continue to build a grassroots energy revolution.

The tools of elegant frugality, the technical options that let you demystify energy and live lightly within your energy income, have long been available to anyone in principle. (There's an old Russian joke about the guy who asks whether he can buy various hard-to-find goods. Tired of being always told, "In principle, yes," he exclaims, "Sure, but where's this [Principle] shop that you keep talking about?") Real Goods turns the principle into practice: a useful selection of the things you were looking for, with essential background information, at fair prices, guaranteed, anywhere you want them delivered.

It's especially gratifying to see this catalog's

Lovins (1976) *vs.* Actual Use (1975–90)

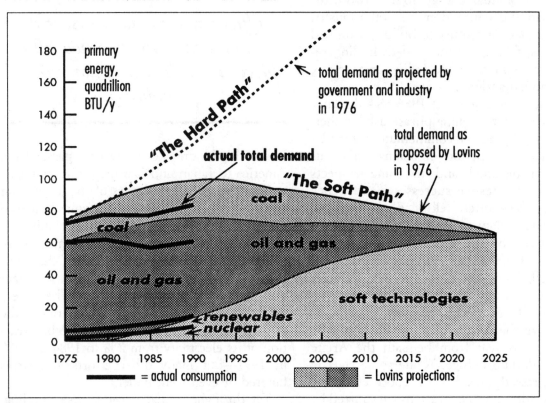

This graph illustrates the pattern of U.S. Gross Primary Energy Use. It demonstrates how much closer Amory Lovins' predictions have come than the government projections.

nice blend of renewable energy supply options with increasingly efficient ways to *use* the energy. In most cases, the best buy is to get the most efficient end-use device you can, *then* just enough renewable energy supply to meet that greatly reduced demand. Our own house/indoor farm/research center of nearly 4,000 square feet, for example, by harnessing some of the efficiency options in this catalog, uses so little electricity for the household lights and appliances—about $5 worth per month—that only about 400-500 peak watts of photovoltaics, a couple of thousand bucks' worth, could entirely meet those needs. Indeed, superinsulation and superwindows, by providing 99% of our space-heating passively in a climate that can get as cold as -47°F, raised our construction costs less than they *saved* us up front by eliminating the furnace and ductwork. All together, saving 99+ percent of our space and water heating load (the latter is virtually all passive and active solar, again made much cheaper by strong efficiency improvements), 90-odd percent of our electricity, and half our water together raised total construction costs only one percent, and paid back in the first ten months. It *looks* nice, too: we tramp in and out of a blizzard to be greeted by a jungleful of jasmine, bougainvillea, and a big iguana offering advanced lizarding lessons under the banana tree—then we remember, as we wipe the steam off our glasses, that there's no heating system; nothing's being used up. And that's all with nine-year-old technology. Now you can do even better.

Today, in fact, the best technologies on the market can save about three-quarters of all electricity now used in the United States, while providing unchanged or improved services. The cost of that quadrupled efficiency: about 0.6 cents/kWh, far cheaper than just *running* a coal or nuclear plant (let alone building it). Thus the global warming, acid rain, and other side-effects of that power plant can be abated not at an extra cost but at a *profit*: it's cheaper to save fuel than to burn it. Similarly, saving four-fifths of U.S. oil by using the best technologies now demonstrated (about half of which are already on the market) costs only about $3 per barrel, less than just drilling to look for more. Such technologies obviously beat any kind of energy supply hands down. But what's the *next* best

buy—the supply choices you need to make in partnership with efficiency in order to live happily ever after? Usually, if you buy the efficiency first, the next best buy will be the appropriate renewable sources. And if you count—as our children surely will—the environmental damage and insecurity that fossil and nuclear fuels cause, then well-designed renewables look even better.

---

*A **single** compact fluorescent lamp will, over its life, keep out of the air a **ton** of carbon dioxide, twenty pounds of sulfur oxides, and various other nasty things.*

That's not to say that every renewable technology makes sense anywhere. Wind machines and small hydro only work well in good sites. Photovoltaics can look competitive with grid electricity (which, by the way, receives tens of billions of dollars' annual direct subsidies and doesn't bear many of its costs to the earth) only if you're upwards of a certain distance (where we are, about a quarter-mile) from a power line —or closer, right down to zero, if you count such benefits as greater reliability and higher power quality. (The Federal Aviation Administration is switching hundreds of ground avionics stations to photovoltaics for exactly those reasons.) Every renewable application is site-specific and user-specific. No *single* renewable source can solve every energy problem. That's part of their strength and their charm.

But wherever you are, if you use energy in a way that saves you money, it's quite likely that *some* kind of renewable energy can further cut your bills, work as well or better, give you a good feeling, and set an example. What more can you ask?

Well, maybe something for the earth—and you get that too. A *single* compact-fluorescent lamp (see lighting section) replacing, say, 75 watts of incandescent lighting with 14-18 watts (but yielding the same light and lasting about 13 times as long) will, over its life, keep out of the air a *ton* of carbon dioxide (a major cause of global warming), twenty pounds of sulfur oxides (which cause acid rain), and various other nasty things. And far from costing extra, that lamp will make you tens of dollars *richer*, by saving more than it costs for utility fuel, replacement lamps, and installation labor. Now put that saving into changing over to renewable energy supplies and you're doing even better. Then take the time you used to put into working to pay your electric bill and put it instead into your garden, your compost pile, a walk, a fishing trip. Take the time you used to work to pay your medical bills and build a greenhouse. Invite your neighbors over to help munch your fresh tomatoes in February and tell them how you did it all. Then...

You get the idea. Here's a treasure-house of things *you* can use to improve both your own life and everyone else's. Read, enjoy, use, learn, and tell: implementation is left as an exercise for the reader.

Amory Lovins is currently director of research for Rocky Mountain Institute. He is a consulting experimental physicist educated at Harvard and Oxford. Mr. Lovins has been active in energy policy in more than 20 countries and has briefed eight heads of state. He has published a dozen books and hundreds of papers. He is probably the most articulate writer on energy policy and strategy today. Newsweek *called him "one of the western world's most influential energy thinkers."*

He has our vote for the next U.S. Energy Secretary!

You can reach the Rocky Mountain Institute at 1739 Snowmass Creek Road, Snowmass CO 81654-9199, phone: 303/927-3128.

Introduction

Welcome to the Seventh Edition of our *Alternative Energy Sourcebook*. As we continue through the final decade of the twentieth century, we contemplate the painful lessons of the last decade—Three Mile Island, Chernobyl, the Exxon Valdez oil spill, and the war in Iraq. As it becomes increasingly obvious that environmentalism must be the preoccupation of the nineties (if we are to survive into the next millenium), we are proud that we have been working on the right stuff.

There is a time for liberal optimism when we pat ourselves on the back for what we have accomplished. There is also a time for radical pessimism which we believe has arrived. It is time to shock the world into action with a heavy dose of reality, however pessimistic, so that we can appreciate the positive steps needed for environmental improvement. The truth is that change is not happening nearly fast enough and the clock continues to tick.

There are only eight short years to the new millenium. We like to think about what the "end of the year" stories will be like in late December 1999. Will the media review the past year? The decade of the '90s? The twentieth century? Or the last millennium since 1000 A.D.? Will they look back in sorrow at all the missed chances to save the planet? If we as a species are to survive another millenium, or even another century, nothing short of an *environmental perestroika* will be required.

While we are cheered to see that the proliferation of sensible energy use that was once envisioned as an "environmentalist's fantasy" in earlier Sourcebooks is now happening routinely, there is still much work to do. Nearly half a century after solar electricity was first developed for the space program of the 1950s, we are becoming ever more conscious that we each must do more to reduce resource consumption to sustainable levels for the benefit of the generations to come. Photovoltaic (PV) power is a prime example of a good technology: permanent, sustainable, benign, modular, maintenance-free, quiet, non-polluting, and clean. It is *in tune* with our planet, whereas other energies and technologies devour the planet. The typical alternative energy home (there are now over 50,000) uses between a third and a tenth of the energy consumed by a conventional home. *Only 14 years ago we sold our first PV panel!* We can only hope that our society will learn the simple lessons of nature less painfully in the twenty-first century.

As you read this, one acre per second of rainforest is being destroyed. Species are disappearing at the rate of 10 per hour, 100,000 per year. Ninety-nine percent of all species that ever existed are now extinct. Entire islands of rainforest in Indonesia are being clearcut to produce plywood· for Japanese concrete building forms that are thrown out, slightly cement-stained, after just one use. And we are cutting and burning trees to create grazing lands for cows to make hamburgers for Americans who have the highest rate of heart disease in history.

> *The proliferation of sensible energy use that was once envisioned as an "environmentalist's fantasy" is now happening routinely.*

It just doesn't make sense. While we, the stewards of the world, play shuffleboard and drink martinis on the deck of the Good Ship Earth, the planks beneath us are rotting and burning. We marvel at the spectacular sunset colored by yet another ecological disaster, while beneath the water's surface, species degradation is going unchecked. Sea turtles and salmon have been depleted by 90%. Taiwan and Japan have put out 27,000 miles of drift nets *every night*, a portion of which break loose and ensnare unsuspecting turtles and dolphins. Drift nets have now been outlawed internationally. Time will tell how effective this new law will be.

Closer to home there is an argument in the

"Most of the great environmental struggles will be either won or lost in the 1990s and ...by the next century it will be too late to act." - Thomas E. Lovejoy, Smithsonian Institution

The Natural
Resources
Defense
Council
estimates that it
takes 3.6
billion gallons
of oil to heat
American
businesses every
year.

Pacific Northwest about spotted owls vs. jobs in the timber industry. Ninety-six percent of the United States is now deforested. The real argument is not about jobs vs. owls, but rather about whether or not our fragile ecosystem can survive the loss of the last 4% of our forests. Lumber companies euphemistically call trees a *renewable resource*. What they create with their tree replantings, however, is not a forest, but a farm that supports almost no animal life—not even birds; *there is no real ecosystem*. The loggers must realize that there are times when we have to look beyond our economic self-interest. Their sons will not be loggers, just as the Japanese whalers' sons will not be whalers.

In southern Chile, life under the ozone hole has reached new and grotesque proportions. There are blind salmon swimming in the sea off Tierra del Fuego, and packs of rabbits so myopic that hunters can pick them up by their ears. Thousands of sheep have been blinded by temporary cataracts. NASA recently revealed that the ozone hole is growing alarmingly faster than had been anticipated. It is now four times larger than the United States, and its outer edge already covers the tip of South America. In the south Chilean port of Punta Arenas, the largest town (population 115,000) located on the cusp on an ozone hole, Chilean scientists estimate that levels of the carcinogenic ultraviolet-B radiation jump to more than 1,000 percent of normal on peak days. Dermatologists have documented melanomic skin cancer appearing at four times the average rate. Dark glasses, hats, and full clothing in midsummer are commonplace in southern Chile. We have sold out our grandchildren's rights to go out in the sunshine and enjoy the beach. Even George Bush has finally reluctantly acknowledged the severity of the problem as new data in 1992 showed the ozone hole revealing itself in North America.

Global warming is the most potentially devastating and life-threatening disaster that we face. Global emissions of carbon dioxide have grown from 1 gigaton (billion metric tons) in 1930 to six gigatons today. Despite the mounting evidence of global warming's devastating effects, carbon dioxide emissions are the one major pollutant that is totally unregulated in America. In Europe a $12 per ton tax on CO_2 emissions has been proposed, meaning a 50 mega-watt coal plant putting out 2 million tons of CO_2 per year would be taxed $24 million.

OK, so what about some "liberal optimism"? American business is finally beginning to see the light, literally. The utility companies have discovered, thanks to the prodding of the Natural Resources Defense Council (NRDC), that energy conservation pays. Pacific Gas & Electric (PG&E) made $15 million in 1990 by saving 280 million kWh of electricity and eliminated the production of 200,000 pounds of CO_2. Many businesses worldwide are understanding that environmental protection costs are not liabilities, but investments in the future. Just as industry learned in the 1930s that safety upgrades may cost in the short run but pay back positively over time, so it goes with environmental improvements in the '90s. Unfortunately, America appears to be seriously lagging behind foreign competition. What can we expect from a corporate culture run by companies like Exxon, the largest corporation in the world, that stonewalls its devastation of Prince William Sound in Alaska.

Japan, after the first oil crisis in 1973, invested 25% of its GNP in improving efficiency for industrial production. This is what has made them the model for economic success in the '70s, '80s, and early '90s. Germany, the other great economic success of the era, was equally aggressive about energy conservation. Eco-products in Germany today are the equivalent of the electronics and microchip industries in America in the '80s. The German model of environmental consciousness will be widened to include the entire European Economic Community in 1992–93. What will happen to our economy if we do not maintain pace?

We've endeavored to demonstrate by example that appropriate technology is not a mystical mirage of the future, but a technology whose time is here. PV is everywhere—from the solar panel that runs our calculators and watches to the PV system pumping water from wells 1,000 feet deep with small efficient pumps so that flowers *do* bloom in the desert. Alternative energy opens a world of new possibilities: radio repeaters, livestock watering, electric fencing, ocean signal buoys, highway billboard lighting, and solar irrigation controllers, to name just a few. PV is fashionable now; you can see it on

BMW sunroofs running the cars' instrumentation, and you can see it on RVs rolling down the highway in search of warmer climates. Even the commercial heavyweights (PG&E, Bechtel Corporation, etc.) have finally figured out that it's often cheaper to get power from an array of PVs than to drop a lengthy transmission line or to run a noisy generator. Many PV manufacturers are now manufacturing photovoltaic modules around the clock. Our customers' questions are no longer "will they work?" but "how many do I need?" and "when can I get them?" Ironic, isn't it that the Three Mile Island, Chernobyl, the *Valdez* disaster, and the Iraqi war have consistently presaged an upswing in alternative energy sales. How quickly we tend to forget about the environmental disasters and go back to our energy-wasteful lifestyles.

Conservation is contagious. It has led us to discover hundreds of new ways to save energy. PV is just the beginning. In this Seventh Edition *Sourcebook* we preview lots of new ideas and novel products, many of which you will find only here. They will save money, energy, and resources up front, and will make you feel better all the while that you use them! For example, did you know that the typical house refrigerator uses 3,000 watts and isn't even as efficient as refrigerators designed in the 1940s? Did you know that those refrigerators hyped with efficiency stickers on the appliance showroom floor are designed with the compressor on the bottom so that heat rises into the very box that it is supposed to cool? Contrast this with our Sun Frost refrigerators (compressor on top), which use only 300 watts—a tenth the power! Tell that to your local appliance dealer. If every American home had a Sun Frost, we could immediately shut down 19 large nuclear power plants!

In this *Sourcebook* you will find new and revolutionary technologies in lighting, both DC and AC. New PL compact fluorescent technology puts out 4 times the light of the old incandescent lighting and lasts 10 times as long. It's been estimated that if the U.S. government outfitted every home in the nation with these revolutionary lights, we would be an energy-exporting nation within 5 years!

Japan and Europe have used instantaneous tankless water heating technology for decades while Mr. and Mrs. Average American hoard (at great expense to themselves and the environment) 40 gallons of hot water through dark of night, over vacations and long weekends, just in case someone might want to wash a glass. Isn't it more reasonable to heat the water as it's needed? *When will our country come to its senses?*

Is it rational to speed the pathogens and bacteria in our toilet wastes into the biosphere on wings of water? Why do we spend thousands of dollars on myriads of septic tanks and leach fields, thousands of gallons of precious potable water, ultimately to pollute our ground water systems with our own bacteria? Over 1,000 Real Goods customers have just said "No!" and now recycle their own waste using the new technologies of composting toilets, many approved by the National Sanitation Foundation.

Alternative energy opens a world of new possibilities

Our strength at Real Goods lies in the diversity of our interests and contacts. We intend to remain the most thorough clearinghouse for all forms of energy-sensible products on the planet. We relentlessly pursue new technologies and endeavor to keep you informed of our finds. We know we can always do better, and we welcome your feedback. We hope to make your lives simpler and happier, to provide dramatic savings to our fragile environment, and to provide a thorough and valuable educational source along the path.

In 1991, we initiated an "Eco Desk" to audit everything we do from an environmental perspective. We try to drive the message home to all our employees, suppliers, customers, and everyone with whom we do business. This year we are involved in innovative and exciting new ways to spread our message of energy independence even further. We have initiated a strategic alliance with an upscale retailer of environmentally sensitive products in New York and Los Angeles to make an ever increasing amount of people aware. We have found a way to involve the citizens who have the most to gain from a healthy environmental future—our children—with our "Real Goods for Real Kids on Real Planets" program, where kids will bring the message of energy sanity into the living rooms of America. Our Institute for Independent Living will bring forth educational programs to teach by example the benefits of independent living and independent energy production. We need to sound our voices loudly for energy and environmental sanity. Let's stop the waste and destruction and start the reparation before the next millennium. Thanks to you all for making it all work.

John Schaeffer

P.S. – We're ahead of schedule by over 50% in our mission to eliminate the production of one billion pounds of carbon dioxide in the 1990s (see article on page 12).

Pictured here is most of the Real Goods crew on a fine spring day in Ukiah.

Global Warming: Environmental Enemy Number One

Global warming is the number one problem facing our world today. *In 50 years, from 1937 to 1988, the world's annual energy consumption quintupled from 60 to 321 quads (quadrillion BTUs), and global emission rates for carbon mushroomed from 1 gigaton (one billion metric tons) to 6 gigatons per year.* Atmospheric carbon dioxide levels have increased since the 1800s from 280 ppm (parts per million) to 350 ppm. Studies on people working in greenhouses containing elevated levels of carbon dioxide have shown that at 500 ppm, long term cardiovascular effects may occur. As the concentration of greenhouse gases in the atmosphere increases, the overall warming effects continue to foul the environment, alter the climate (average temperatures will rise as much as 2 degrees Fahrenheit in the next 30 years), and threaten the habitability of our planet. At the current 6 gigaton carbon emission rate, the potentially devastating level of 500 ppm of environmental carbon dioxide will be reached by the year 2100. Our grandchildren will not thank us. If our current gluttony for energy doubles (if the second and third world match America's irresponsibility), these changes may happen soon enough for us to experience, within 30 years. The endangered ozone layer is a change we already live with: sunbathing at the beach has become an extinct pleasure we recollect for our children.

In 1988 the United Nations convened a panel of scientists called the Intergovernmental Panel on Climate Change (IPCC) to study global warming. These scientists concluded that "rates [of warming] are likely to be faster than [the rates at which] ecosystems can respond, possibly leading to substantial reductions in biological diversity." Today's global carbon emission rate must drop by at least two-thirds to stabilize the atmospheric concentration of carbon as carbon dioxide. This can only be accomplished through a drastic reduction in the use of fossil fuels for energy production.

Armed with the IPCC's discoveries, Sweden plans to phase out nuclear power and substantially reduce carbon dioxide emissions. Seven European nations are organizing a regional global warming treaty. Sadly, the U.S. government (previously led by John Sununu) persists in stonewalling the report, whining that "some species would 'thrive' in a warming world!" (*maybe cockroaches, slime mold, and politicians!*) The world's heavyweight champion producers of carbon dioxide, the Soviet Union (18%) and the United States (20%), are the only two governments advocating no action.

There is really only one answer to the global warming riddle —energy efficiency!

How do global warming and oil wars tie together, and why are we discussing them in our alternative energy catalog? There is really only one answer to the global warming riddle, and it also answers the question of how to avoid reckless oil wars—energy efficiency! Full practical use of the electricity-saving technologies now available would save about 75% of all power used. According to Amory Lovins, of the Rocky Mountain Institute, energy efficiency at European and Japanese levels would save America about $200 billion per year, enough to pay off the national debt by the year 2000. As a bonus, lowering our energy consumption would stabilize global warming and give the atmosphere and biosphere a chance to heal. *If we used oil as efficiently as Japan we would save 7 million barrels of oil per day.*

If the government won't do it, we must. We can begin to reverse these negative trends by making energy-conscious personal choices in the products we buy. In this sourcebook you will find energy-efficient light capsules, water-saving showerheads, Sun Frost refrigerators, solar-electric modules, and many other products which help to curb our energy appetite and can go a long way toward making a dent in reversing global warming and our dependence on fossil fuel based energy systems. In 1990–91 alone our

If your car is going to be idling for more than one or two minutes, it is more efficient to turn it off.

global warming and our dependence on fossil fuel based energy systems. In 1990–91 alone our customers, through their purchases, kept over 309.6 million pounds of carbon dioxide out of the air. It's a small dent (it would take 6,000 purchases like that to reduce 1 gigaton) but a very positive step in the right direction that sets an example to the rest of the world. (As energy costs explode, we get a dividend: energy efficiency saves *us* money.)

As individuals we do as much as we can. The major culprits in global warming are automobiles, which produce from a third to half of the damaging emissions. Real Goods supports and sells affordable electric vehicle conversions, attacking the problem at its source. (See the "*Mobility*" chapter on electric vehicles.)

The global warming disaster won't end until you, our readers, take action and spread the word about the short-sighted and self-destructive path our government is taking. Beyond that, we all need to focus on education, helping everyone we come in contact with to understand the gravity of this problem. Our children and grandchildren depend on us.

–John Schaeffer

Real Goods Products Environmental Impact Report

In order to truly evaluate the impact of your purchases, we asked for the help of the Natural Resources Defense Council (NRDC) in San Francisco and the Rocky Mountain Institute in Snowmass, Colorado to help us quantify the energy impact of what you bought. Our thanks to Chris Calwell (NRDC) and Rick Heede (RMI) for helping us with the formulas and number-crunching. Here's a list of the highlights of what you purchased in 1990 and 1991.

- 358,643 watts of solar electricity (approximately 9,000 solar panels)
- 21,120 watts of wind electricity (66 wind generators)
- 18,000 watts of hydroelectricity (18 hydro plants)
- 138 Sun Frost refrigerators
- 843,200 watts of compact fluorescent lighting
- 1,017 tankless water heaters
- 43 solar hot water collectors
- 13,465 low-flow water-saving shower heads
- 12,157 water-saving faucet aerators
- 311 low-flush toilets
- 372 composting (no-flush) toilets
- 8,381 toilet flush savers
- 232,825 rolls of recycled toilet paper
- 26,070 rolls of recycled paper towels
- 46,423 boxes of recycled facial tissue

Our goal was to analyze the impact these purchases had on the environment for the life of the appliance or the light bulb. Here are some of the energy conversion data that we used to make these calculations:

It takes 11,000 BTUs to generate 1 kilowatt-hour (kWh)
Carbon dioxide emissions per kWh = 1.5 lb
Coal required to produce 1kWh = 1 lb
57% of U.S. electricity comes from coal-fired plants
6.25 million BTUs = 1 barrel of oil
Average U.S. cost of 1 kWh = $0.08
1 car = 500 gallons gasoline per year
1 low-flow shower head = 466 kWh/year in heat savings
1 low-flow shower head = 14,000 gal/year water savings
1 faucet aerator = 4,000 gal/year water savings
1 set toilet dams = 5,475 gal/year water savings
135 trees = 1 ton of pulp

The following are the results of actual energy and environmental savings directly brought about by you, our customer network, by your purchases in 1990–91:

- **You have saved 206.4 million kWh of energy!**
     This represents:
     - 206.4 million pounds of coal or
     - 382,000 barrels of oil or
     - 16,044,000 gallons of gasoline (*enough to keep 32,088 cars off the road!*)
- **You have kept 309.6 million pounds of carbon dioxide out of the air (a major cause of global warming).**
- **You have kept 44.2 million pounds of sulfur oxides (which cause acid rain) out of the air.**
- **You have saved 5 billion, 762 million gallons of water.**
- **You have saved 305,000 trees from destruction.**
- **You have saved $30,960,000 on energy expenses.**

CONGRATULATIONS TO YOU ALL!

We're 50% Ahead of Schedule!

Our Goal for the Nineties: Eliminate 1 Billion Pounds of Carbon Dioxide

In 1990 we took on our most ambitious goal yet. We declared that with your help we planned to eliminate the production of 1 billion pounds of carbon dioxide from our atmosphere by the turn of the century. This will have a major impact on global warming, the preservation of natural habitats, and indeed on life on earth itself. We believe that global warming is the number one environmental problem facing our planet today. (See editorial on pg. 9) The only logical and possible way to solve the problem is for the world to drastically slow down its consumption of fossil fuel. Americans have an insatiable appetite for fossil fuel—coal, oil, and natural gas power our cars and homes and produce most of our electricity. Naturally occurring carbon dioxide (CO_2) and other gases trap heat and keep the earth warm. The burning of fossil fuels is releasing trillions of pounds of extra carbon dioxide into the atmosphere each year, which will likely raise temperatures, causing extensive flooding, more frequent droughts, heightened sea levels, and disrupted farming. Habitats will shrink as plants and animals struggle to adapt to the rising temperatures.

We can stop wasting energy. The way to do this is to drive cars that use less gas, recycle, insulate our homes, install compact fluorescent light bulbs, and purchase energy-efficient appliances. In the first year of the 1990s we were 10% toward our goal. We announced in last year's *AE Sourcebook* that you had already saved over 100 million pounds of carbon dioxide from being spewed into our atmosphere. For our part, we are focusing on energy-saving products.

In order to monitor our progress we have been printing our billion-pound CO_2 meter in each successive *AE Sourcebook* and *Real Goods News Catalog*.

We're proud to report that we are more than 50% ahead of our goal—after two short years you have already eliminated 309.6 million pounds of CO_2 from being generated into the atmosphere! Keep up the good work.

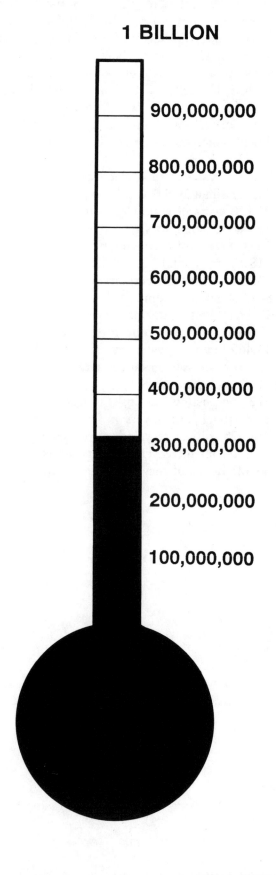

1 BILLION

900,000,000

800,000,000

700,000,000

600,000,000

500,000,000

400,000,000

300,000,000

200,000,000

100,000,000

Make Fuel Efficiency Our Gulf Strategy

The following piece by Amory and Hunter Lovins is reprinted with permission from The New York Times, *December 3, 1990.*

Are we putting our kids in tanks because we didn't put them in efficient cars? Yes: We wouldn't have needed any oil from the Persian Gulf after 1985 if we'd simply kept on saving oil at the rate we did from 1977 through 1985.

Even now we could still roll back the oil dependence that perpetually holds our foreign policy hostage and distorts other U.S. priorities in the Middle East. Just by aiming at greater efficiency, we could eliminate all gulf imports by using only an eighth less oil.

A great place to start would be personal vehicles. Improving America's 19-mile-per-gallon household vehicle fleet by three miles per gallon could replace U.S. imports of oil from Iraq and Kuwait. Another nine miles per gallon would end the need for any oil from the Persian Gulf and, according to the Department of Energy, would cut the cost of driving to well below pre-crisis levels without sacrificing performance.

The Reagan Administration doubled 1985 oil imports from the Gulf when it rolled back efficiency standards. Today's new cars average 29 miles per gallon; the fleet, only 20. Yet 10 manufacturers have built and tested attractive, low-pollution prototype cars that get 67 to 120+ miles per gallon. Better design and stronger materials make some of these safer than today's cars, as well as more nimble and peppy.

And efficiency needn't mean smallness: only 4 percent of past car-efficiency gains came from downsizing. Some of the prototype cars comfortably hold four or five passengers, and two of them would cost nothing extra to mass-produce.

Many other oil savings can help. Boeing's new 777 jet will use about half the fuel per seat of a 727. Technical refinements can save most of the fuel used by heavy trucks, buses, ships and industry. Insulation, draft-proofing and simple hot-water savings can displace most of the oil used in buildings. Superwindows that retain heat in the winter and reject it in summer could save each year up to twice as much fuel as we get from Alaska.

In all, we know how to run the present U.S. economy on one-fifth the oil we are now using, and the cost of saving each barrel would be less than $5. Even achieving just 15 percent of that potential oil savings would displace all the oil we've been importing from the Gulf. Doing that requires only a small additional step. Since 1973, we've reduced our oil use per dollar of gross national product four and a half times as much as we'd need to reduce it today in order to eliminate all Gulf imports.

How can we promote fuel efficiency? Higher gasoline taxes are a weak incentive to buy an efficient car, because gasoline costs six times less than the non-fuel costs of owning and running a car. And since the often higher purchase price of an efficient car about cancels out the lower gasoline bills, the total cost per mile for 20- and 60-mile-per-gallon cars is about the same.

But the 40-mile-per-gallon difference, for cars and light trucks, represents more than twice America's imports from the Gulf. If the security and environmental costs of inefficient cars had to be paid up front, buyers would choose more wisely. The best way is "feebates": When you register a new car, you pay a fee or get a rebate, depending on its efficiency. The fees would pay for the rebates.

Rebates for efficient cars should be based on the difference in efficiency between your new car and the old one—which you'd scrap, thus getting the most inefficient, dirtiest cars off the road first. That's good for Detroit, for the poor, for the environment, and for displacing Gulf oil sooner.

The California legislature recently approved car feebates by a margin of 7 to 1 (they were then vetoed); Connecticut, Iowa, and Massachusetts are weighing them. Feebates are also being considered for new buildings in California, Massachusetts and the four northwestern states, and could be applied to trucks, aircraft, appliances and other energy-consuming goods. Unlike miles-per-gallon standards, feebates reward maximum performance and encourage businesses to bring superefficient models to market.

Energy efficiency is also the key to the decades-long transition to nondepleting, uninterruptible energy sources. Government studies confirm that sun, wind, water, geothermal heat, and farm and forestry wastes can cost-effectively provide, within 40 years, half as much energy as America uses today. Efficiency would raise that share and buy the time need for graceful conversion.

The military alternative to energy efficiency isn't cheap. Gulf jitters have added more than $40 billion a year to U.S. oil imports. Counting military costs, Gulf oil now costs in excess of $100 a barrel.

The more than $20 billion net cost of U.S. forces in the Gulf just from August through December 1990, if spent instead on efficient use of oil, could displace all the oil now imported from the Gulf. It could also create jobs and wealth, improve America's trade balance, stretch domestic reserves, clean urban air, cut acid rain and global warming, and help the poor at home and abroad.

In 1989, the Pentagon used about 38 percent as much oil as the U.S. imported from Saudi Arabia, and estimated that its consumption could readily double or triple in a war. An M-1 tank get 0.56 miles per gallon. An oil-fired aircraft carrier gets 17 feet per gallon. And no good outcome—in dollars, oil, or blood—is in sight.

But from inside an efficient car, the Gulf looks very different. From inside enough of them, its oil becomes irrelevant. National security, peacetime jobs in a competitive economy, and the environment demand immediate mobilization—not of tanks but of efficient cars, not of B-52s but of 777s, and not of naval guns but of caulk guns.

Amory B. & Hunter Lovins
Rocky Mountain Institute

Our Policies

The Real Goods Eco-Desk

The Real Goods Eco-Desk is not made of special material or anything fancy. To tell the truth, it's a regular ol' desk with a phone and a computer on it that is staffed by one of our senior customer service representatives. The Eco-Desk is the clearinghouse for many of the special events, projects, and services that Real Goods becomes involved with. Our motto is: "When in doubt, call the Eco-Desk."

It is the responsibility of the Eco-Desk to make sure that as a company we practice what we preach, whether it's the way we use paper on the copier or the packaging used in our shipments. Many of the special programs are centralized in this function, so remember, "When in doubt..."

Subscribe to Real Goods

We have always tried to find ways to reward, and even glorify, the dedicated customers who read our publications from cover to cover. We think we've come up with a simple, cost-effective way to make sure that these folks will get every precious word that appears on our pages—by subscribing.

Subscribers are the lifeblood of our business. They take energy independence seriously and are anxious to share experiences with others who have made a similar commitment. Subscribers will receive the full roster of Real Goods publications. This consists of the following:

• *The Real Goods Color Catalog.* Published four times a year, this bright, colorful piece has energy-saving information and merchandise for everyone.

• *The Real Goods News,* published three times each year is crammed with articles, customer profiles, unclassified ads, and hard-core information for the person who lives Off the Grid, or might someday. The *News* contains our ever-popular *Readers' Forum* where we turn the soap box over to some of the most interesting people we know—you.

• Back by popular demand is the *Real Stuff* newsletter for *Subscribers Only.* This newsletter is published quarterly and will feature specialized technical information as well as discount merchandise offers made to *subscribers only.* Featured will be special buys on solar panels, Off-the-Grid hardware, overstocks, and discontinued items.

• The *Alternative Energy Sourcebook* is also sent to new subscribers. We can't say much more about the *Sourcebook* than, if you need something for independent living, you'll find it in the *Sourcebook.* We update the *Sourcebook* more or less annually or whenever we find it is outdated.

Perhaps the best feature of our subscriber program is that once you sign up by paying $25 for a domestic subscription, your subscription privileges will continue indefinitely. By sending in a no charge annual renewal form we will keep all the news from Real Goods coming. Even the *Sourcebook,* when new editions appear, will be made available to subscribers on a no-charge basis. You pay shipping and handling only (plus, you have to recycle your old *Sourcebook* by giving it to someone who wants to learn about alternative energy).

Due to the high cost of foreign postage, we must charge $50 per year for all foreign subscriptions. For this annual amount, you will receive each catalog and *Real Goods News* via priority mail, and the current *AE Sourcebook.* *Sourcebook* updates are not included in foreign subscriptions and need to be ordered and paid for as they become available.

The subscriber program is our way of providing our best values to our best customers. We hope you'll take your first important step towards energy independence—**subscribe.**

How To Buy Stock in Real Goods

Real Goods made its first public offering of 200,000 shares of common stock in October 1991. We were over-subscribed in just four months. If you wish to buy or sell stock in our company please contact Ted Prescott at Mutual Securities, Inc. / Cowles, Sabol & Co., Inc., a registered broker-dealer, who executes orders to buy and sell our stock upon request. Mr. Prescott's address is: 415 Talmage Road, Ukiah, CA 95482. His phone number is 800/922-4337 or 707/468-8646. *This is not a solicitation of orders to buy or to sell, and Mr. Prescott is not soliciting orders to buy or to sell.*

Quantity Discounts

We've spent a great deal of time trying to sell in quantity to other retailers for resale, and have come to the simple conclusion that we are not well-prepared to handle such business. Servicing other retailers takes personnel, equipment, and trading terms that we don't have. We are in the process of developing an exclusive line of *Real Goods* products that we will be able to truly wholesale—stay tuned for future updates in our catalogs.

There are several cases when we are able to currently offer discounts:

a. To investors. Real Goods stockholders of record with at least 400 shares are entitled to discounts on all purchases. They should identify themselves at the time of order placement.

b. Multiple quantities of single items. Volume purchases of single items can be discounted—the size of the discount varies with the item. Qualifying quantities vary, so call for a specific quote. We can generally beat just about any price out there for quantity buys.

c. Whole house systems. In cases where an individual is willing to centralize purchasing for a remote home or energy-efficient renovation, Real Goods is willing to negotiate a competitive discount. Available discount will depend on merchandise selected and total volume of purchases.

d. Institutional purchasing. Schools, business, government organizations, and other institu-

tional purchasers are encouraged to contact Real Goods regarding technical advice and bulk purchasing. We generally accept bonafide purchase orders from schools and government agencies. We are uniquely qualified to advise you on energy independence and to provide the necessary products to achieve savings.

At Real Goods we cannot be all things to all people, but we can try!

AE Sourcebook Updates

The AE Sourcebook is updated periodically with new products and new information. (Because prices change frequently, the *Real Goods Color Catalogs* and the *Real Goods News* serve as the pricing update for our products.)

Purchasers of the *AE Sourcebook* will be able to receive subsequent editions by submitting proof of original purchase, $4.50 shipping and handling charge, and the name and address of the person to whom the old Sourcebook has been passed along. In this way, by purchasing the Sourcebook once, you can be assured of always being up to date with what is happening in the world of energy independence.

Discounts are available for schools, libraries, government, or non-profits wishing to purchase Sourcebooks in quantity.

Technical Services

Our technical services are world class, and we find ourselves doing increasing amounts of design work. Newcomers to PV want to be sure they select and install the right system. Solar sophisticates know that we have the latest data, and the best access to solar information.

Technical services are available by phone from 9am to 5pm, Pacific Time, Monday through Friday. We try very hard to answer technical mail within two weeks.

We welcome the opportunity to design and plan entire systems. Here's how our technical services work:

1. To assess your needs, capabilities, limitations, and working budget, we ask you to complete a specially-created worksheet (See appendix). The information required includes a complete list of your energy needs, an inventory of desired appliances, site information, and potential for hydro-electric and wind development.

2. A member of our technical staff will determine your needs, and design an appropriate system with you.

3. At this point we will begin tracking time helping you plan the details of your installation. Your personal tech rep will work with you on an unlimited time basis until your system has been completely designed and refined. He will order parts and assist with assembly, assuring you that you get precisely what you need. He will also interface with your licensed contractor, if need be.

The first hour is free and part of our service. Beyond this initial consultation, time will be tracked in ten minute intervals, and billed at the rate of $60 per hour. If the recommended parts and equipment are purchased from Real Goods, however, this time will be provided **at no charge.**

Our technical department can provide complete working drawings and installation instructions, in addition to phone assistance. We can also provide a professional and experienced installation.

If you prefer to handle installation locally we are happy to refer you to one of our *Local Pros.* These are installers who offer service locally and have provided information about their specialties to Real Goods.

Real Goods has provided no training to the individuals who are registered as "local pros," and our referral is in no way intended to recommend or document any level of qualifications. Referrals are provided purely as a convenience to our customers. We recommend that you check references before using *any* installer.

Our customers assure us that our technical support saves money, time, and hassle. Please feel free to call us for more details about our technical consulting services, or send SASE for answers to specific questions.

An Open Letter from the Real Goods Technical Staff

Our techs, left to right, Doug Pratt, Gary Beckwith, & Douglas Bath.

As Real Goods technicians, we would like to tell you a little about ourselves and our purpose.

Most of us have made a positive choice to be Real Goods Technicians, even though it has meant leaving or refusing more lucrative jobs in the installation field. We feel that here we can be of the most help to the largest number of people who need information to attain energy independence. By contributing to the distribution of this information, we hope to have the greatest impact on promoting positive change.

We strive for truth and accuracy. One great advantage here at Real Goods is the size and experience of our technical staff. Combined, we have a total of over 100 years experience in the field, so there are few situations that we haven't encountered. If an unusual circumstance arises, we can discuss it with one another, and use our extensive library and resources to find a solution. We are strongly committed to standing behind our products and providing any technical support that is necessary. We can also recommend useful books and other publications to help. We now have an extensive list of "Local Pros" if on-site assistance is needed. Some are willing to go anywhere in the world!

We understand that no two customers are alike and each has specific and individual needs. We try to first clearly identify these needs and then determine how to satisfy them in the most sensible and cost-effective way.

While *we* contribute by providing knowledge (this company's most important product), it is *you* the customers, who are enacting the changes. We would like to thank you for doing your part. Without you, the change that has only begun would be a dream. We know that as it becomes a reality, it makes our jobs worthwhile. You can help us by taking the time to clarify your own needs, expressing them clearly (and legibly) to us, and we promise to do our best to help.

Together, we can change the world.

Sincerely,
The Real Goods Technical Staff

Local Pros

As proud as we are of our technical services, they cannot provide the on-the-site service that is sometimes required for equipment installation. To assist customers in linking up with local service professionals we offer a no-charge referral program, called the Local Pro.

Professional solar installers are encouraged to contact Real Goods to get a copy of our Local Pro Application. We will make the information you give us available to anyone who requests it.

This is not an endorsement program, nor does it connote any degree of expertise provided via association with Real Goods. If you are looking for help, and are about to let your fingers do the walking, just let them walk to Real Goods.

Why Don't They Make A...

Real Goods Pays Big Bucks for Brainstorms!

How many times have you wished that a certain product existed? How many times have you said "Why didn't I think of that?" Now is your chance to transform your latent genius invention into a genuine Real Goods product. Our talented technical product development staff, in conjunction with our worldwide manufacturing connections, is capable of producing almost anything. If you can think of it, we can make it happen. Just send your product development idea to Real Goods: Attn: Product Ideas, 966 Mazzoni St., Ukiah, CA 95482. We'll pay you $500 in credit for any idea that makes it into our Alternative Energy Sourcebook. Keep a dated copy of your letter on file — only the first idea received is eligible. *(No Rube Goldbergs, Perpetual Motion Machines, Auto Mileage Extension Devices or Crystal Powered Pet Rocks please. Remember, It needs to be a Real Good.)*

An Invitation to Our Customers

New Product Suggestions

Several years ago we instituted the *New Product Suggestion Program*, figuring that who could tell us what customers want better than customers themselves? We think of it as 300,000 investigative buyers working to broaden our product mix, seeking new products to enhance the quality of energy independent life.

The program works like this: **Send us ideas for new products** that you think we should carry in our catalog—include manufacturer's name, address, and phone number, along with any personal experience you have with the product. If we add your suggested product to our catalog, you will receive a $25 merchandise credit. It's a great way to turn all our creative minds to the benefit of the entire Real Goods family. **Thanks!** Send all your new product ideas to our regular address **Attn: Eco-Desk/New Products.**

Wanted: A Few Good Writers

We try to make all the Real Goods publications readable from cover to cover. We know there are thousands of talented writers out there with vast storehouses of knowledge relating to all aspects of alternative energy, and we're eager to inspire you to pull those keyboards and pens out of the closet.

Gift certificates valued from $25 to $100 to any customer who sends in **an article or photograph** that we use in a future edition of the *Real Goods News* or in the *AE Sourcebook*. Send your articles to our regular address **Attn: Eco-Desk/Article Submissions.**

Come Visit Our Showroom

We invite you all to visit our beautifully renovated showroom in Ukiah next time you're passing through Northern California. We're only two hours north of the Golden Gate Bridge, at the north edge of the wine country, and smack dab in the middle of the solar ghetto (aka: the Solar Capitol of the World!) Follow the map; take the North State Street off-ramp off of U.S. Highway 101.

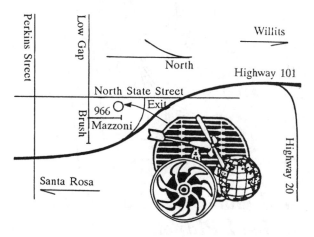

In our showroom, you'll find almost everything shown inside this Sourcebook, many in working applications to demonstrate their use. We try to make our showroom an Exploratorium of Alternative Energy so you can understand and compare.

Showroom hours are 9am–5pm Monday through Friday and 10am–4pm on Saturday. Our staff will help you design an alternative energy system or answer questions you have about our products. If you require extensive technical assistance, please call beforehand to make an appointment.

Real Goods Institute of Independent Living

In 1992 Real Goods will be offering instructional seminars in subjects of relevance to independent living. These sessions will be conducted by members of our technical staff, using the facilities here in Ukiah and the Mendocino County area to provide a more interactive educational environment than is possible through our publications.

Full information on schedule and curriculum is available by calling or writing our **Eco-Desk/Institute of Independent Living.**

Real Goods in Your Home Town

Again, there are many energetic organizations who request our presence at local events, energy fairs, Earth Day celebrations, etc. We love you, we support you, we wish you every success....... but we're a small company, and we don't have the staff or resources to participate in as many events as we would like to.

The best way for Real Goods to help you locally is via our ability to communicate. We are glad to provide literature and sign-up sheets for people who want to learn more about energy independence. In this way people at your event can share in our most important product—knowledge—and we can concentrate on what we do best, i.e. providing that knowledge. If you would like literature and sign-up sheets, write or call our Eco-Desk.

August is Customer Appreciation Month

August is a great time to come and visit Real Goods. The phone activity is not quite so chaotic, so we have more time to spend with you personally. The Solar Energy Expo and Rally (SEER) takes place in nearby Willits, we hold our stockholders' meeting, the Electrathon race is held next door at the fairgrounds, and there are lots of special buys in our showroom.

Plan your vacation around a visit. We promise sunshine and hospitality.

The Real Goods Bridal Registry

This is one of our more outlandish ideas. But, it works! If your dream of nuptial bliss includes independent living, then you can register with Real Goods so that well-wishing friends can honor you with the gift of a much-needed inverter rather than the sterling tea service that will end up in the closet. To register, call or write our **Eco-Desk**.

Off the Grid Day

The first-ever Off the Grid Day was held in 1991 as we encouraged people from coast to coast to flip their breakers for one day to experience life without power. We will be celebrating Off the Grid Day in August at Real Goods, but the good news is, you can celebrate it in your local community any ol' day you choose.

Use the Off the Grid concept to promote energy awareness in conjunction with fundraising, political activism, a fun event—whatever you feel is most appropriate. We will help by providing information and access to materials to help make your event a smashing success. **Call our Eco-Desk** for information.

Real Goods for Real Kids on Real Planets

We receive many requests to support non-profit organizations. There are just too many good causes for a small company like ours to support in a meaningful way. Because of this, we have decided that the best financial contribution we can make to a non-profit is to actively assist you in your local fund-raising efforts by making available our logistical support.

The program, called "Real Goods for Real Kids on Real Planets," is designed to harness kid power in the most productive way- by helping the community and the planet. In doing so we can provide your qualifying organization with a fund-raising vehicle that sells earth-friendly products in your community. Orders will be selected from a modified Real Goods catalog provided at no charge to your organization. Resulting orders are processed and fulfilled by Real Goods, with your organization receiving a rebate for each order placed.

Full details and an application form are available by writing Real Goods, **"Attention: Eco-Desk/Real Kids."** Help your community, help your organization, and help the planet. And let Real Goods help you do it!

The Real Goods Story

The Ride to Boonville

As did many of his contemporaries in the 1960s and early '70s, John Schaeffer, founder of Real Goods, experimented with an alternative lifestyle. After graduating from Berkeley, where he was exposed to every strand of the lunatic fringe, he moved to a commune outside of Boonville, California, a mountain community as picturesque as it is isolated. There he lived a life in pursuit of self-sufficiency.

Despite idyllic surroundings, needed elements of life were missing—specifically money and power. (We're talking about a boy who was raised in the exurbs of Los Angeles!) Not that he needed a lot of money and power, but he wanted some small amount to balance what he had grown up with and complete deprivation. Self-sufficiency, in other words, was a more appealing concept than reality.

He discovered power from the sun. A photovoltaic panel hooked up to a storage battery could power lights, a radio, and even a television. Despite his departure from a purist's position, he became the most popular person on the commune to visit whenever it was time for "Saturday Night Live."

The money came from a job. Unfortunately, the only job he could find was as a computer operator in Ukiah, some 35 twisty miles away.

Self-sufficiency, in other words, was a more appealing concept than reality.

Knowing that he would be making the daily trek over the mountain to the big city, it was not unusual for his friends to ask them to buy supplies needed on the commune. As a conscientious, thrifty person Schaeffer spent many hours plying the hardware stores and home centers of Ukiah, searching for the best deals on fertilizer, bone meal, tools, and other goods related to the communards' close-to-the-earth lifestyle.

One day, while driving his Volvo (what else?) back to Boonville after a particularly vexing shopping, the thought occurred to him. "Wouldn't it be great if there was *one* store that sold all the products needed for independent living, and sold them at fair prices."

The idea of Real Goods was born.

In recent years the focus of Real Goods has been energy. The company now can claim to be the oldest and largest catalog firm devoted to the sale and service of alternative energy products. Do not think of the company as either old or large, however. They are devoted to the same principles that guided their founding way back in 1978—quality products for fair prices and customer service with courtesy and dignity.

Many of the Real Goods staff, like John Schaeffer, are veterans of a movement that had the right inclinations, but wrong technology. Now, whether you are a full-time off-the-gridder or have simply added a note of energy sanity to your life, the Real Goods staff is committed to helping you achieve the independence you desire.

Will the Real Doctor Doug... Please Stand Up?

Scandal in Ukiah?

Can it be that the company that prides itself on giving customers the straight-from-the-shoulder, no-holds-barred view of the world be perpetrating a fraud? Has the last of our institutions gone the way of hype and hucksterism? Say it ain't so!

Let's just say that we've become a little trapped by our own white lie.

There really is a *Doug*. And he really can cure things, although he is much better at fixing mechanical things, such as engines and pumps, than people. You would not want him operating on you with a scalpel, but if your question involves which wire to connect to which terminal (a delicate operation that can be just as important to your health), you could not be in better hands.

The original Doug is Doug Pratt, and while he holds the honorary degree of Doctor of Electrons from the Institute for Independent Living, he is not a real doctor.

That's not the scandal. The truth is that there is more than one Dr. Doug. In fact, there is a whole team of Dougs. One is named Douglas, another Gary, then there is Terry and Ross and Randy and Mike.

At some point the person of Doctor Doug became a concept. Maybe it was related to the fact that Ross knew a little more on radiation safety, or Gary was particularly teched-up on solar tracking, but we found that the job of being the technical expert was too big for any one person. Also, if limited to one Dr. Doug (who possesses, despite local rumors, only two ears and one mouth), only one Real Goods customer at a time could receive technical advice. Thus, it came to pass that a whole team of "techies" were bestowed with the honorary doctorate.

Technical service is the heart of Real Goods. While we sell books, toys, and gadgets, each item of merchandise is connected to a body of knowledge that enables you to achieve an independent lifestyle. This service is available for a fee (a very reasonable $40 per hour), but it also comes **free** when you purchase your products from Real Goods.

Whether your Doctor is Gary, Mike, Randy, or even Doug, you will receive friendly, experienced advice from someone whose experience encompasses years of personal experience. He will talk to you on the telephone, answer your letters, and even meet with you in person if you come to Ukiah. Attend a session at the Institute for Independent Living, and you will likely find a Doctor Doug as your instructor, too. And if you can stump him with a question, chances are one of the other Doctor Dougs on staff will be able to come to the rescue.

The little man climbing out of the mailbox in the *Sourcebook* looks a lot like Doug Pratt, but in reality, he stands for many others who will answer to the same name—Doctor Doug.

The Real Goods Curmudgeon

The Real Goods Curmudgeon sprang to life one dismal day when I had to review one too many greener-than-thou books. Dang! I thought, don't these folks get the irony in harvesting trees to publish one more impassioned plea to stop deforestation?

There's a time warp at play here. Even the american president (the tall belligerent one; there may be a new one by the time you read this), a relic from less pressing times, has lately admitted that the ozone hole may be an immediate problem in kennebunkport (I'm an equal opportunity uncapitalizer). Presumably he and his ilk will belatedly reach the same conclusion about greenhouse gases. Being kinder and gentler, we remember that he (and even many of us) formed our idea of the world at a time when conventional wisdom agreed that certain resources—the oceans, the atmosphere, fossil fuels, the carrying capacity of the planet—were inexhaustible. We have undergone an enormous sea-change since we were in grade school; our training and operating principles are at odds with present challenges. We may drown in our own garbage.

Curmudgeonhood comes to me honestly. My grandfather was a classically crusty Missourian who believed nothing undemonstrable. A civil engineer, he flew exactly one time, as a young man, when a barnstormer took him aloft and scared him thoroughly, and he would never fly again. My grandmother, his wife, was born and raised before telephones and therefore declined to acknowledge their invasive clamor; nor did she drive; flying, however, tickled her. Some of the most interesting people I've studied with and worked for have been curmudgeons of a high order: Miss Samon, Robert Frost, John Kenneth Galbraith, Hanley Norins, and Margaret Hamilton Clark among others. "Show me!" they all said. Is everyone in the state of Missouri a curmudgeon?

My grandmother formulated her operating system under the tutelage of a strict New England protestant churchman, who borrowed commentary from Emerson and Thoreau. As a youth I remember rolling my eyes when she brought forth some Thoreauvian utterance, but lately I have come to admit we could find worse folks to borrow from. To wit:

When we consider how soon some plants which spread rapidly, by seeds or roots, would cover an area equal to the surface of the globe, ...how soon some fishes would fill the ocean if all their ova became full-grown fishes, we are tempted to say that every organism, whether animal or vegetable, is contending for the possession of the planet... Nature opposes to this many obstacles, as climate, myriads of brute and also human foes, and of competitors which may preoccupy the ground. Each suggests an immense and wonderful greediness and tenacity of life ... as if bent on taking entire possession of the globe wherever the climate and soil will permit. And each prevails as much as it does, because of the ample preparations it has made for the contest,—it has secured a myriad chances,—because it never depends on

spontaneous generation to save it. (Journal, 22 March 1861)

Companies like Real Goods have lives of their own: callow youth, concupiscent teenage years supposedly giving way to mellow middle age and rewarding golden years. Since the last edition of the *Sourcebook*, maturity has come to Real Goods, leading me to conclude that "dog years" (in this case at least) are useful units for understanding the relationship between companies and humans; the golden years should start three years from Thursday. Until then...

Many of us bemoan the loss of the "Old Real Goods," which was into newsprint, crowded pages, guerilla marketing, countercultural values, and a pragmatic attitude toward the ways and standards of "modren amuricun biznis." At Real Goods we wonder every day if we can keep our values and continue to grow, growth seeming inevitable given the task at hand. We are trying to "preoccupy the ground" in an ethical and sustainable way, before less scrupulous competitors—the kudzu of the corporate world—prevail. Sometimes we falter because of hastiness, inattention, incomplete understanding, or greed; we catch ourselves or are caught by you, faithful readers, and get things put right again; a spirit of perfectionism, critical scrutiny, and dogged tenacity pervades the Real Goods organism—call it a Curmudgeonly spirit, if you will—and I am honored to be asked to give it voice.

As you can see, they've given me my own typeface, too. We will meet again in the marginalia. Until then: review all assumptions.

— Michael Potts

How To Use This Book

We've worked hard to make this Seventh Edition of the Sourcebook more organized and user-friendly than ever. You'll find cross referencing comments in the margins directing you to other places in the book for more information. We have fine tuned our index to make the location of specific products and systems much easier. Over the past year we have sorted out our product mix into eight well defined categories — you might call it a *Dewey Decimal System* for appropriate technology. The eight icons (graphic representations) that you see on these two pages will appear frequently throughout our publications. Tab down the right margin to see just how far each section goes. Whether you're a casual user of energy-efficient systems or a hard-core off-the-grid devotee, you'll be able to rapidly find what you're seeking. We've kept all the intense technical information and charts in the appendix at the back of the book.

Remember, the Sourcebook's primary goal is education. While it indeed carries the backbone of our product offerings, with a fast-changing industry, you must rely on our periodic catalog updates to keep abreast of recent developments. Here is a summary of the eight sections that lie before you:

The Efficient Home

Everything to maximize energy efficiency & environmental sensibility in your home, with an emphasis on the conventional AC, electrical grid-connected home. This section includes super-efficient lighting, outdoor solar lighting, refrigeration, water heating, water conservation, water & air purification, radiation safety, recycling, nicad battery charging, non-toxic household cleaners, and other energy saving products.

Shelter

For many, the creation of a dream house is the most important (and expensive) tangible accomplishment of a lifetime. The state of shelter is really the state of Humankind's dominance over, and mastery of, the environment. We must think about shelter from the drawing board stage. Locating local, efficient, intelligent, and natural resources is something of a lost art. This section deals with innovative alternative housing: Geodesic Domes, Yurts, Underground Housing and more.

Power

Our Power section is the heart of the Sourcebook, and is the place from which our business was born. It includes the sources of sustainable power systems: the sun, wind, & water. You'll learn about the origin and workings of a photovoltaic module (solar panel), and ways to figure out whether solar-electricity makes sense for you in your location. Also included are wind generation, hydro-electric generation, large storage batteries (everything you need to know), controllers, meters, and safety concerns for independently powered home. Converting the power you create from low-voltage DC to usable household-AC power with an inverter is covered in depth.

Off-The-Grid Living

This section includes the tools and technology that make energy independence a reality and allow the independently powered homeowner to live with all the creature comforts of any city dweller. It covers low-voltage lighting, low-voltage appliances, cooling systems, refrigeration, water pumping, instantaneous water heaters, and composting toilets. Also included are some of the electronic marvels that you can enjoy without being connected to the power grid like TVs, video players, and telephones.

Tools

This section is a bit of a catch-all for miscellaneous tools. It includes tools and technology that make energy independence a reality. From hand-powered lawn mowers, garden composters and kitchen tools, to binoculars, hand tools, and even wood cookstoves.

Mobility

This section introduces Electric Vehicles that will begin to sound the death knell for the internal combustion engine. Although it is a little short of product at this time, we include several conversion kits to change that gas-guzzler into an electric efficiency machine. There is lots of meaty information here about the history, theory, and workings of electric vehicles.

Just-For-Fun

For young and old, this section deals with products that teach, delight, and make the planet a more pleasant place. The emphasis is on educational toys and gifts. Included are lots of solar-powered wonders for children and adults, T-shirts, maps and charts and other fun stuff.

Knowledge

Knowledge is the most important product of Real Goods. This section contains a thorough library of books on all aspects of renewable energy, independent living, and healthy households. There are also videos on fascinating subjects. We hope it will be a jumping off point to educate yourself to the joys of independent living.

The Efficient Home

Our mission at Real Goods has broadened over time. Originally, we just provided off-the-grid power to people living in remote home-sites. Now we are trying to provide the means to energy emancipation for everyone.

As our task and business grew through the late 1980s, Real Goods became a household word in the estimated 50,000 solar-powered homes in America. In terms of real change, however, the company's impact has been minimal. If we are ever to effect larger scale change, we know, it will need to be through the nearly 100 million American households that are on-the-grid.

The good news is that there is plenty of improvement to be made. By implementing simple conservation measures, assisted by using some of the many energy-saving devices that have reached the market in recent years, one can reduce energy consumption by over 75% in the average AC-powered home.

In other words, you don't have to go off-the-grid to strike a blow for energy independence, and you don't have to go off-the-grid to use Real Goods!

But we warn you—energy independence can be addictive! Once you've converted energy-hogging incandescent lights to super-efficient compact fluorescents; once you've installed solar-powered garden and driveway lights; once you've changed to rechargeable batteries; once you've installed water-saving devices; once

you've changed to nontoxic household cleaners, insulated your attic, and degaussed your home from electromagetic radiation, you just might find yourself saying, "What's Next?"

You might even be one of the bold ones who someday clips the umbilical cord of the power line that brings you electrons (at considerable expense) from your friendly neighborhood utility. As you progress in your journey to independent living, you will be amazed to see how easy energy independence can be. The important thing is to take the trip one step at a time, and keep making forward progress.

The horizons of energy sanity have risen dramatically in recent years, fueled by images of burning oil wells, oil-covered coastlines, and belching smokestacks. Again, there is a good news side to the story, as many innovative products have appeared on the market. Our Real Goods product development team is moving as fast as possible to make solar products more accessible to conventional households. As this edition of the *Sourcebook* goes to press, we are actively at work on a backyard solar retreat, a complete PV (photovoltaic) starter system that will give homeowners the flexibility to convert to independent power gradually as utility rates increase while solar prices remain constant. (You can keep abreast of market innovations by subscribing to the "Real Goods News," our internal publication that introduces new products and tells the latest of what's happening in the world of independent living.)

We live in a society of instant gratification. Our energy addiction did not develop overnight, nor will it be undone instantly. The first steps, however, can be taken today, and you can learn about them in this chapter.   – J.S.

The Kitchen Dreadnought

Not long ago, on a Sunday afternoon, our electric popcorn maker had a meltdown. It returned to the plastic slime from whence it came, nearly burning down our house in the process. We had no microwave popcorn on hand. To be brutally honest, we faced the bleak prospect of an afternoon with no junk food.

Suddenly, from the cobwebby recesses of my mind, an idea emerged like the Creature from the Black Lagoon. Uncontrollably, it gurgled forth: we could cook the popcorn the old-fashioned way, in a pan on the woodstove.

"It'll never work," said my wife.

"You can't cook popcorn in a pan," said my ten-year-old.

"Get a grip," said the teenager.

But it did work, and it tasted good. Even the teenager asked for seconds.

Imagine a new product, to be called the Real Goods Dreadnought 2001. It is an energy system for the most used room in the house —the kitchen. It is designed to use a renewable fuel that is locally produced—wood. It heats the living space, and does all cooking.

We mean *all* cooking. The Dreadnought (the name comes from a class of British battleships circa World War I that supposedly made all other war ships obsolete) can replace the cooktop, oven, toaster, and microwave in the modern home. Moreover, it warms your dishes, dries your mittens, and can even provide hot water for domestic use. To top everything off, this device uses no electricity (not even a digital clock!) and is, to some extent, self-cleaning. It

will last for 50 years, and it is—guaranteed—the best-looking piece of furniture in your house.

With all our talk of appropriate technologies and sustainable lifestyles, isn't it discouraging to find that society had a perfect answer to something, and we simply forgot about it?

The Real Goods Dreadnought 2001 would surely sell like hotcakes (which it then could cook), while revolutionizing American life. To state it simply, it sounds too good to be true.

What has just been described in the guise of a new fangled gizmo is a kitchen cookstove, similar to the one that was probably in your grandmother's house. These were staples of the country kitchen at the turn of the century, compact—even at 600 pounds—powerhouses that were the energy focal point for Victorian life. From this one item (it does not seem right to call it an "appliance") the family received comfort, sustenance, and luxury.

The demise of the kitchen cookstove came with the increased use of fossil fuels. "Clean, efficient" electricity, produced at a remote plant and piped through landscape-disfiguring lines, became a popular replacement fuel, along with "modern" gas. The ornate, shiny nickel of the cookstove was replaced with chrome strips, the exposed cast-iron sides with sleek, but flimsy, enameled sheet metal.

The heart of the cookstove's efficiency revolved around the fact that it was always on. Grandma could play that sucker like a harp, alternately rough and gentle, slamming in the wood or delicately raking the coals for just the desired effect. She could produce more special effects on her stove than Stevie Wonder can on a synthesizer! The kitchen was her environment, and nothing gave her control of it like the Dreadnought.

The cookstove was meant to be used winter and summer. In the warmer season the firebox size was reduced, resulting in less heat output, and from time to time the stove would get a few days' rest. Sometimes it was even moved out to a summer kitchen, a room that was closed off when the family "closed ranks" during the cold weather.

"You can't cook popcorn in a pan,"
said my ten-year-old.

With the coming of central heat, and the inferno in the basement that gobbled black gold and gave us uniform comfort at fingertip command, the warmth from the cookstove became redundant.

So did Grandma's skills.

We entered the age of instant-on, with power to suit our every whimsy. Want a gourmet dinner at a moment's notice? No need for advance preparation or any cooking skill. Just pull a frozen, prepared Truite Almondine from the freezer and pop it in the microwave. Fresh biscuits? Ever heard of Pop 'N' Fresh Dough? Maybe some Sara Lee cheesecake for dessert.

The cookstove, which was always on, gave way to instant-on. Now appliances with digital readouts and colorful monitor lights can launch into service in a nanosecond, grilling, zapping, or nuking our food into shape in time for the six o'clock news. So what if there is no smell of baking bread in the house anymore?

The cookstove was removed to provide addi

tional floor space. Having an energy center in the kitchen was no longer necessary when there was a veritable power plant in the basement, not to mention a nuke up the road. Grandma, or rather her grand-daughter, was now freed of the drudgery of sustaining life and could devote her time to commuting and joining the rat race. If the cookstove did not wind up at the landfill, long before its useful days had past, perhaps it was a decorative plant holder at a local antique shop.

I suppose all this is excusable under the guise of progress, but there has been a price to pay, and the price is independence. It is not feasible for all of us to trade in our GEs and Westinghouses and Jenn-Aires and Vikings and Garlands for the plant holder at the antique shop. But at least if we can learn from the past we can move forward.

The Kitchen Dreadnought was more than a good idea; for a generation of Americans it was a way of life, and a product that bonded the family as much as the minivan does today. In our rush to progress we have become dependent on a power supply that carries with it very high costs, ranging from nuclear disaster to global warming. Perhaps the greatest threat has been to our independence. Deprived of our thermostats, we no longer know how to keep warm. Without our freezers and microwaves, we can barely feed ourselves.

The kitchen cookstove is a product whose time has passed, or has it? You can still get a kitchen cookstove, made from the patterns originally developed years ago and improved over time. The new cookstove has made concessions to technology and efficiency, but it still looks great and works better than ever. Most importantly, it still does what the Jenn-Aire will never do; it provides the kitchen with a hearth. There is a new logic taking hold, and something very old is suddenly making a lot of sense.

— Stephen Morris

Notes and Workspace

AC Lights

A revolution has taken place in commercial and residential lighting. Until recently, we have been bound by the ingenious but crude discoveries that Thomas Edison made over 100 years ago.

Mr. Edison was never concerned with how much energy his incandescent bulb consumed. Today, economical, environmental, and resource issues force us to look at energy consumption much more critically. Incandescent literally means "to give off light as a result of being heated." With standard light bulbs, as much as 95% of the energy consumed is given off as heat; the light is only a small by-product.

New technologies are available today that improve on Edison's invention in reliability and energy savings. Although the initial price is higher than standard light bulbs, cost-effectiveness is far superior in the long run. When businesses understand how much can be saved with compact fluorescents, they often replace even bulbs that haven't burned out yet.

The types of efficient lighting on the market include: new-design incandescent lights, quartz-halogen, standard fluorescents, and compact fluorescents.

New Incandescent Lights

Inside an incandescent light is a filament that is heated, giving off light in the process. The filament is delicate and eventually burns out. Some incandescents incorporate heavy-duty filaments or introduce special gases into the bulb to increase life. While this does not increase efficiency, it increases longevity by up to 4 times.

Quartz Halogen Lights

Tungsten-halogen (or quartz) lamps are turbocharged incandescents. Compared to standard incandescents, these produce a brighter, whiter light and are more energy-efficient because they operate their tungsten filaments at higher temperatures. In addition, unlike the standard incandescent light bulb which loses approximately 25% of its light output before it

burns out, halogen lights' output depreciates very little over their life, typically less than 10%.

To make these gains, lamp manufacturers enclose the tungsten filament in a relatively small, quartz-glass envelope filled with halogen gas. During normal operation, the particles that evaporate from the filament combine with the halogen gas and are eventually redeposited back on the filament, minimizing bulb blackening and extending the lamp life. Where halogen lamps are used on dimmers, they need to be occasionally operated at full output to allow this regenerative process to take place.

Tungsten-halogen lamps produce about 40-50% more light per watt input than standard incandescents and last longer, having useful service lives ranging from 2,000 to 2,500 hours, depending on the model. Tungsten-halogen lamps are very sensitive to operating voltage.

There are many styles of halogens available for both 12V and 120V applications. The most popular low-voltage models for residential applications include the miniature multifaceted reflector lamp (MR 16) and the halogen version of the conventional A-type incandescent lamp, which incorporates a small halogen light capsule within a protective outer glass globe. The reflector lamp offers a light source having very precise light beam control, allowing you to distribute light exactly where you need it without wasteful spillover, while the latter option is intended for general ambient-lighting applications. A new family of 120VAC voltage, sealed-beam reflector halogen lamps are now coming on the market to replace standard incandescent reflectors. You'll hear more about these in forthcoming issues of the *Real Goods News*.

The higher operating temperatures used in halogen lamps produce a whiter light, which eliminates the yellow-reddish tinge associated with standard incandescents. This makes them an excellent light source for applications where good color rendition is important or fine-detail work is performed. Because tungsten-halogen lamps are relatively expensive compared to standard incandescents, they are best suited for applications where the optical precision possible with the compact reflector models can be effectively utilized.

Never touch the quartz-glass envelope of a halogen lamp with your bare hands. The natural oils in your skin will react with the quartz glass and cause it to fail prematurely. Because of this phenomenon, and for safety reasons, many manufacturers incorporate the halogen lamp capsule (generally about the size of a large flashlight bulb) within a larger outer globe.

Fluorescent Lights

Fluorescent lights are still trying to overcome a bad reputation. For many, the term connotes a long tube light that emits a blue-white light with an annoying flicker. These limitations have been overcome with current technology.

The tube, no matter how it is shaped, contains a special gas at low pressure. When an arc is struck between the lamp's electrodes, electrons collide with atoms in the gas to produce ultraviolet radiation. This, in turn, excites the white phosphors (the white powder coating the inside of the tube), which emit light at visible wavelengths. The quality of the light fluorescents produce depends largely on the blend of chemical ingredients used in making the phosphors; there are dozens of different phosphor blends available. The most common and least expensive are "cool white" and "warm white". These, however, provide a light of relatively poor color rendering capabilities, making colors appear washed out, lacking luster and richness.

The tube may be long and straight, as with standard fluorescents, or there may be a series of smaller tubes, in a configuration that may be screwed into a common fixture. These are called **compact fluorescent** (cf) lights.

There are two popular styles of compact fluorescent lamps being used with screw-in adapters: the "PL" type (as designated by Philips Lighting Corp.) and the "quad" type.

The PL lamp consists of two straight, parallel, interconnected glass fluorescent tubes mounted side-by-side on a plug-in base. These are most readily available in 5, 7, 9, and 13-watt models. The 9 and 13-watt models are more than 6" in length, making them too long to fit inside some fixtures or reflectors. To reduce the overall length of these lamps and of higher wattage models, lamp manufacturers developed

A single compact fluorescent lamp will, over its life, keep out of the air a ton of carbon dioxide, twenty pounds of sulfur oxides, and various other nasty things.
- Amory Lovins, in the Foreword

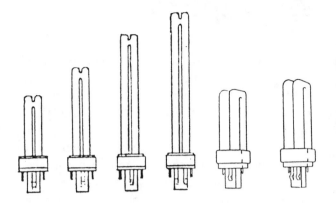

"quad" lamps, which incorporate four short, interconnected glass fluorescent tubes, rather than the two longer tubes characteristic of the PL lamps. At roughly half of the overall length of PLs, "quad" lamps provide approximately the same light output (as PLs of equivalent wattages) and are available in a variety of models from 9 to 28 watts, the most common being the 9 and 13-watt models.

Color Quality

The color of the light emitted from a fluorescent bulb is affected by the phosphors that coat the inside surface. Industry has adopted the terms "color temperature" and "color rendering" to describe this light. Color temperature, or the color of the light that is emitted is measured in degrees Kelvin (K) or absolute temperature, ranging from 9,000°K (which appears blue) down to 1,500°K (which appears orange-red). The color rendering index (CRI) of a lamp, on a scale of 0–100, rates the ability of a light to render an object's true color when compared to sunlight. The following chart will demystify the quality of light debate.

Type of Light	CRI	Deg.K
Incandscent	90-95	2,700
Cool white fluorescents	62	4,100
Warm white fluorescents	51	3,000
Compact fluorescents	82	2,700

Phosphor blends are available that not only render colors better, but also produce light more efficiently. Most notable of these are the fluorescent lamps using tri-stimulus phosphors, which have CRIs in the eighties. These incorporate relatively expensive phosphors that peak out in the blue, green, and red portions of the visible spectrum (those which people are most sensitive to), and produce about 15% more visible light than standard phosphors. Wherever people spend much time around fluorescent lighting, specify lamps with higher (80+ CRI) color rendering ratings.

Ballast Comparisons

All fluorescent lights require a **ballast** to operate, in addition to the bulb.

The ballast regulates the voltage and current delivered to a fluorescent lamp and is essential for proper lamp operation. The electrical input requirements for each compact fluorescent lamp varies and hence, each type/wattage requires a ballast specifically designed to drive it. There are two types of ballasts which operate on AC: **core-coil** and **electronic.** The **core-coil ballast**, the standard since fluorescent lighting was first developed, uses **electromagnetic** technology. The **electronic ballast**, only recently developed, uses **solid-state** technology. All DC ballasts are electronic devices.

Relative to the core-coil ballast, electronic ballasts weigh less; operate lamps at a higher frequency (20,000+ cycles per second vs. 60 cycles); are silent; generate less heat; and are more energy-efficient. However, electronic ballasts cost significantly more, particularly DC units because they are manufactured only in very small numbers. With few exceptions, electronic ballasts for AC lamps are presently only available as a part of integral units. This is because the bulbs last the same amount of time (about 10,000 hours) as their ballasts. Magnetic ballasts, on the other hand, can last up to 50,000 hours, and often incorporate replaceable bulbs. (Don't buy extra bulbs when you buy the light though, because you'll never remember where you put the replacement four years later!)

All of the compact fluorescent lamp assemblies and pre-wired ballasts featured in the AE Sourcebook are equipped with standard medium, screw-in bases (like normal household incandescent lamps.) The ballast portion, however, is wider than an incandescent light bulb *just above the screw-in base.* Therefore, fixtures having constricted necks or deeply recessed sockets may require a socket "extender" (to extend the lamp beyond the constrictions). These are readily available in several sizes at most hardware stores.

Savings

The main justification for buying fluorescent lights is to save money. The new compact fluorescents provide opportunities for tremendous savings without any inconveniences. Simple payback calculations prove cost effectiveness. **Fluorescent lights typically last 10 times longer and use one-fourth the energy of standard lights.**

Lifetimes of lights (hrs)

Standard incandescents	1,000
Long life incandescents	3,000
Quartz-halogen	2,250
Compact fluorescents	10,000

Efficacy of Lights (approx. lumens per watt)

Incandescents	16
Compact fluorescents	60

The next question, of course, is "How much?" The more a light is used, the more that can be saved by replacing it with a compact fluorescent. The following calculations assume that the light is on for an average of 6 hours per day and power costs $.10 per kWh (approximate national average).

	Panasonic 27W	Incandescent 100 watt bulb
Cost of bulb	$29	$.50
Product life	4.5 years	167 days
Watts used	27	100
Annual Energy Cost	$5.91	$21.90
Bulbs replaced in 4.5 yrs	0	10
Total Cost	$55.60	$103.55
Savings over life	$47.95	-------

Health Effects

Many people complain about fluorescent light flicker. Migraine headaches, loss of concentration, and general irritation have all been blamed on fluorescent lights. The **frequency** of a light describes the speed of the flicker. Different ballasts will run the same bulb at different frequencies. If a flicker is perceptible to humans, the light is operated by a **core-coil**, or **magnetic** ballast, which flickers 60 times per second. **Electronic** ballasts, which operate at around 20,000 cycles per second, have totally eliminated perceptible flicker and the ensuing complaints.

The other issue concerning health and fluorescent lighting is the miniscule amount of radiation contained *only* in core-coil ballasts. The amount is so small that we don't consider it a health hazard. Several other appliances such as smoke alarms and some watches have the same relative amounts. For those concerned, the electronically ballasted fluorescent lights contain no radioactive materials whatsoever.

AC Fluorescents & Remote Energy Systems

For a discussion of power factor, and choosing AC or DC appliances, see the article, "AC or DC, That is the Question" on page 144.

When planning a remote energy system, one must make numerous decisions as to whether AC or DC power is appropriate for specific appliances. Fluorescent lights are available in both AC and DC. Most people using an inverter choose AC lights, because of the wider selection, a slight quality advantage, and lower price. However, DC models may be a little more efficient, due to something called **power factor**. There are some older inverters that have problems with magnetic ballasts.

Please note that when using *AC ballasted fluorescents*, there are some losses associated with the "power factor" of the ballast that occur in the generation and distribution systems (house wiring, generators, inverters, and transformers) which are often not accounted for. Power factor (pf) is the ratio of watts ("real power") to volt-amperes (total "apparent" power), and measures how much of the power drawn by the load is "real" (in phase with voltage) and thus able to do work. Simply divide the real watts drawn by a fluorescent lighting system by the ballast's power factor to determine the volt-amps. Unless you properly allow for this demand, your generator or inverter may not have ample capacity to power all of your lighting loads, or your wiring may be overloaded. Because most compact fluorescents generally draw such small amounts of current at 120V,

the line losses will be minimal in most residential applications (assuming standard wire gauge sizes are used, #12 or #14), even with low-pf ballasts. Most manufacturers unfortunately do not publicize the power factors of their lights, mainly because it is not an issue for those receiving their power from the utility company (you are only charged for the power you use *before* considering power factor).

This concern does not affect DC fluorescent ballasts (current and voltage are always in phase in DC systems) or any incandescents (they have a pf of 1). Power factor is, however, a concern with some electronic devices, and *all* AC *inductive* loads, including not only ballasts but also transformers and any appliance or tool equipped with an inductive motor.

Most inverters will operate compact fluorescent lamps satisfactorily. However, because all but a few specialized inverters produce an alternating current having a modified sine wave (versus a pure sinusoidal waveform), they will not drive compact fluorescents which use electromagnetic, core-coil ballasts as efficiently or "cleanly" as possible and may emit an annoying buzz. Electronic ballasted compact fluorescents, on the other hand, are more tolerant of the modified sine wave input, and will provide better performance, silently.

Compact Fluorescent Applications

Unfortunately, a compact fluorescent light bulb is shaped differently than an "Edison" incandescent. This is the biggest obstacle in retrofitting light fixtures. Compact fluorescents are longer and sometimes wider. Additionally, the ballast, the widest part, is located at the base, right above the screw-in adapter. We recommend using a ruler to measure the available space, and referring to the sizing chart before purchasing or use the life-size cut-out drawings in the Appendix.

Compact fluorescents have many household applications—table lamps, recessed cans, desk lamps, bathroom vanities, and more. As manufacturers become tuned in to this relatively new market, light fixtures suited for compact fluorescents are bound to become increasingly available, but at the moment options are somewhat limited.

With **table lamps**, the "harp" which holds the lampshade is sometimes not long or wide enough to accommodate these bulbs. An inexpensive replacement harp can solve this problem. Table lamps are an excellent application for compact fluorescents.

Recessed cans are limited by diameter, and often cannot accept the wide ballast at the base. There are now complete kits available for cans less than 4½" in diameter, and include the bulb, ballast, can, and reflector. Desk lamps can be difficult to retrofit, but complete desk lights which incorporate compact fluorescents are readily available, as are bathroom vanities.

Hanging fixtures are one of the easiest applications for compact fluorescent lights. One of the best applications is directly over a kitchen or dining room table. Usually the shade is so wide that it couldn't get in the way. It may even be possible to use a Y-shaped two-socket adapter (available at hardware stores) and screw in two lights if more light is desired.

Track lighting is one of the most common forms of lighting today. The Real Goods showroom and office building is filled with track lights that use compact fluorescent bulbs. When selecting your tracking system, choose the fixture that will not interfere with the ballasts, which is right above the screw in base. It is best to use one of the reflector lamps such as the SL18R40 or the Dulux reflector series.

Reflectors

Several manufacturers offer a variety of reflectors for modular and integral compact fluorescent lamps to help you satisfy many lighting needs (see Figure 4). These plastic or metal units clip on (e.g., the PL reflector), or may be built into the housings (e.g., the SL 18 R40) of compact fluorescent lamps.

If a directional light source is needed, such as in recessed downlights, tracklights, and "bullet"/floodlight fixtures, a lamp with a reflector should be used to direct the light where it is desired. In wall- and ceiling-mounted fixtures having only white-painted surfaces serving as reflectors, a clip-on reflector which fits most

"PL" and "quad" compact fluorescent lamps can effectively double the light output from the fixture.

Dimming

Compact fluorescent lights should *never* be dimmed. Using these lights on a dimmer switch may be a hazard. If you have a dimmer switch you wish to retrofit, the best option is retrofit the light and the switch. This is a simple, inexpensive procedure. Don't forget to turn off the power first!

Starting Time

The start-up time for compact fluorescent lamps varies. It is normal for most core-coil compact fluorescents to flicker for up to several seconds when first turned on while they attempt to strike an arc. Grounding the fixture greatly eases the starting process. If your compact fluorescent lamp flickers for more than 3 seconds before coming on, it is probably not adequately grounded. One way to overcome this is to touch or stroke the bare bulb while it is attempting to come on. If this is frequently the case, we suggest you loosely install a copper grounding piece (available from us) between the lamp's tubes and wire it to a grounded point.

Most electronically ballasted units start their lamps instantly, though, depending on the ambient temperature, it may take from several seconds to several minutes for the lamps to come to full brightness. The start-up times for lamps driven by core-coil ballasts vary from almost instantaneous to several seconds (they flicker during this brief period). These, too, may take from several seconds to several minutes to "warm up" and attain full brightness, depending on the ambient temperature.

In our opinion, the very brief start-up time, which is only apparent in some of the fluorescent lights, is a very small price to pay for the energy savings and the subsequent good feeling about doing less harm to our fragile environment.

Caution: Several vendors are offering customized PL lamps which turn on instantly. They do this by disconnecting part of the circuit built into the lamp base. This voids any lamp and ballast warranty and may cause premature failure of the ballast.

Cold Weather/Outdoor Applications

It is important to realize that fluorescent lighting systems are sensitive to temperature. Manufacturers rate the ability of lamps and ballasts to both start and operate at various temperatures. Light output and system efficiency both fall off significantly when lamps are operated above or below the temperatures for which they are designed to operate. The temperature at which any given fluorescent system will effectively work varies greatly, depending on the specific lamp and ballast combination. The optimum operating temperature for most fluorescents is between 60° and 70°F, though most standard double-ended types will operate satisfactorily between 50° and 120°F. Special lamps and ballasts must be specified for applications outside of these temperature ranges such as low-ambient ballasts for cold applications.

Temperatures below freezing inhibit compact fluorescent lamps in starting and in attaining full brightness (normally reached between 50°-70°F). In general, the lower the ambient temperature, the greater the difficulty in starting and attaining full brightness.

We have found that most lamps will start at temperatures 10°–20°F lower than those stated by manufacturers. The lower wattage PL models (5/7/9 watts) function better in colder temperatures than the 13 watt units, and PL9 and PL13 lamps have lower starting temperatures than their "quad" lamp equivalents. At present, all electronically ballasted lamps (including the SL 18) are rated to start in temperatures down to 0°F.

To help improve the operation of compact fluorescents used in cold temperatures, we recommend that they be installed in enclosed fixtures. This insulates them, improving their ability to both start and attain full brightness. It is also very important that the fixture be well grounded; a grounded metal fixture or reflector near the lamp, ideally within 1/4" of the lamp,

helps it to start (i.e., strike an arc of electrons between the electrodes). If this is not possible or if the lamp flickers for more than several seconds before coming on, a thin strip of copper can be loosely inserted between the tubes of the PL lamp and attached to a grounded point.

Most compact fluorescent lamps are not designed for use in wet applications (e.g., in showers or in open outdoor fixtures). In such environments, the lamp should be installed in a fixture rated for wet use.

Service Life

Most manufacturers rate lamp life based on a three-hour duty cycle, meaning that the lamps are tested by turning them off-and-on once every three hours until they burn out. *Turning lamps off-and-on more frequently will decrease lamp life while keeping them on longer will increase lamp life.* Generally, it is more cost-effective to turn fluorescent lights off whenever you leave a room for more than ten minutes, since the greatest cost associated with operating a light is for the electricity it uses versus the cost to replace it. Please note that the rated lamp life represents the *average life* of a lamp; as a result, some will have longer lives and some shorter. As illustrated below (Figure 5), a typical compact fluorescent lamp will last as long as (or longer than) 10 standard incandescent AC light bulbs or 5 standard incandescent AC floodlights, saving you the cost of numerous bulbs, in addition to much electricity from the greater efficiency.

Three-Way Sockets

Compact fluorescents can be screwed into any three-way light socket, but they will only operate on full light output.

Full Spectrum Fluorescents

Many customers have inquired as to why we don't sell full spectrum fluorescents. First of all, we're not convinced that they perform as promised. The Ultraviolet (UV) end of the lighting spectrum is the first part of the lamp's spectrum to disappear lasting only a fraction of the overall lamp life. Therefore the purported benefits of full spectrum lighting are extremely short lived and thus overrated and do not justify the high price of these lamps. Further it's very difficult for us to ship anything as fragile as fluorescent tubes through the mail. That's also why we recommend that you purchase *all* your lamp tubes locally—there is no difference between a 120V tube and a 12V tube—only the ballast is different.

Dulux EL Compact Fluorescents

These AC lamps are one of the most efficient and lightweight compact fluorescent units available to screw into standard 120VAC light fixtures. They incorporate a built-in electronic ballast (absolutely no hum on inverters!), which operates the lamp at high frequency (35 kHz), starts it instantly, works in temperatures from 0 to 140°F, and are completely silent. Many of our customers prefer these bulbs to the popular Panasonic light capsules as the Osrams contain absolutely no radioactivity of any kind. These lights are best installed inside table lamps or other light fixtures, which will diffuse their bright light and keep dust from settling on them. The Dulux EL's rated lifetime is 10,000 hours. Four wattages are available; the 11- watt and 15-watt are our best sellers. All Dulux lamps will start down to 0° F.

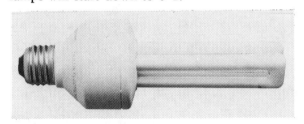

36-111	Dulux EL 7	$25
36-112	Dulux EL 11	$25
36-113	Dulux EL 15	$26
36-114	Dulux EL 20	$26

Dulux EL Reflector Lamps

The Dulux EL Reflector combines the convenience and efficiency of the Dulux EL lamp with a high-performance parabolic reflector. With 10 times longer lamp life and up to 75% less power consumption, the Dulux EL Reflector is the energy-saving alternative for R30 and R40 incandescent lamps. These lamps give a warm, pleasant incandescent-like light with excellent color rendering. They are ideal for track lighting and down lighting in offices or homes. Two models are available: the Dulux EL-R-11 runs on 11 watts and is equivalent to a 50 watt incandescent, and the Dulux EL-R-15 is 15 watts and equivalent to a 75 watt incandescent. The EL-R-11 has a 70° beam spread, puts out 600 lumens, is 5-7/8" high by 4-7/8" in diameter and weighs 4.1 oz. The EL-R-15 has an 80° beam spread, puts out 900 lumens, and is 7-1/4" high by 4-7/8" in diameter. Life expectancy is 10,000 hours.

| 36-131 | Dulux 11 Watt w/Reflector | $29 |
| 36-132 | Dulux 15 Watt w/Reflector | $29 |

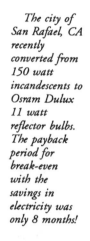

The city of San Rafael, CA recently converted from 150 watt incandescents to Osram Dulux 11 watt reflector bulbs. The payback period for break-even with the savings in electricity was only 8 months!

See the full size cut-outs in the appendix, pages 494-497 and the lamp comparison charts on page 48. Extended lamp harps on page 42.

Panasonic Electronic Light Capsules

These brand new electronic light capsules represent a technological breakthrough in energy-efficient lighting. The 18-watt compact fluorescent light is equivalent to a 75-watt standard incandescent light bulb and is the brightest of any equivalent compact fluorescent presently available with *frosted diffuser/globe*. It lasts, on average, 10 times longer than standard incandescent light bulbs, minimizing the need to purchase 10 replacement lamps. Its solid-state circuitry allows it to start instantaneously and operate completely flicker free. The diffuser provides a *soft* even illumination, which is particularly desirable in bare bulb applications. The light fits most table lamps, when installed with our specially-matched extended lampshade harp (the metal framework supporting the lampshade), as well as in many ceiling, wall, hanging pendant, or pole mounted fixtures. 7.4" long by 3.0" in diameter, 1,100 lumens. *No hum on inverters and no radioactivity!* **May not be used with dimmers.**

Full size cutouts for the most popular bulbs are on pages 494-497.

Must we always demand instant gratification? Some of these lamps - the Core/Coil ballasted ones - take a second or two to start. Put them where instant light will not be required. Can't we slow down just a bit to save a few barrels of oil?

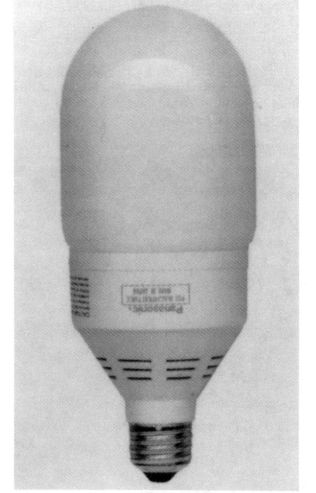

36-141 18W Electronic Light Capsule $25

Electronic Twin Light Capsules

Panasonic has recently come out with this brand new energy-saving twin-tube light capsule. Just like other compact fluorescents, these lamps last up to 10 times longer than a conventional incandescent bulb. Because they are electronically ballasted they have instant start, they produce more lumens, at less weight. They fit into standard incandescent sockets. These are twin-tube bulbs and do not include a diffuser like the standard light capsules or the other Panasonic electronic light capsules listed above. The 27-watt bulb is the *brightest compact fluorescent on the market and puts out the equivalent of a full 100-watt incandescent*. It is 7.8" long by 2.1" in diameter and puts out 1,550 lumens. **May not be used with dimmers.**

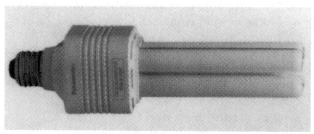

36-143   27-Watt Twin-Tube (100w equiv.) $29

To order any of these products
or for more technical information
call us toll-free at
1-800-762-7325

Panasonic Light Capsules

These are one of the most efficient 120V bulbs on the market and a best seller for us. The technology is still young — first there was the Mitsubishi Marathon bulb, but they canceled their marketing agreement with the USA. Next came GE, which we discontinued in early 1989 because of the boycott. Now we have Panasonics, which are working out as well as or better than any previous product.

These attractive, core-coil ballasted, compact fluorescent lamps offer a high-quality, economical option for providing light from 120VAC fixtures. Available in either globe- or tubular-shaped, white-frosted glass bulbs, the 15-watt Light Capsule produces a "soft" light comparable to that from a standard 60-watt incandescent. But a 60-watt incandescent typically lasts only 500-1,000 hours while these light capsules last up to 10,000 hours! And only use 1/4 of the electricity! These one-piece units operate at line frequency (60 Hz) and screw into standard light bulb sockets. While they at first glance seem expensive, they pay for themselves in less than three years and over their life save more than $40 when compared to standard incandescent light bulbs.

Note: Because these units are relatively heavy, they may make some fixtures top-heavy. When installed in unsheltered outdoor areas, use only in weatherproof fixtures. *These units may not be used with dimmers.* See the appendix for actual size drawing of our compact fluorescents to be sure they will fit your fixtures. Not suitable for use inside globed or sealed fixtures.

The ballast is included inside the bulb, and a standard edison screw base makes it simple to use with any fixture. The lights come in two configurations: the G15 is the globe-shaped bulb which is 3.7" in diameter and 5.7" long, and the T15 is the tube-shaped bulb which is 3.1" in diameter and 6.5" long and able to fit in narrower fixtures. Minimum starting temperature is 32°F. Both bulbs use only 15 watts but put out the equivalent of 60 watts of incandescent lighting. Their efficacy is 48 lumens per watt.

36-101	Light Capsule — Globe (one)	$17
36-103	Light Capsule — Globe (ten)	$155
36-102	Light Capsule — Tube (one)	$17
36-104	Light Capsule — Tube (ten)	$155

Socket Extender

Sometimes your compact fluorescent lamps may have a neck that is too wide for your fixture. This socket extender increases the length of the base. The extender is also useful if a bulb sits too deeply inside a track lighting or can fixture, loosing its lighting effectiveness. Extends lamp 1 3/16".

36-405	Socket Extender	$3

Radioactivity of Some Compact Fluorescents: Panasonic Light Capsules employ a rare-earth element called promethium that in its gaseous form aids in starting the lamp. Promethium puts out very small quantities of ionizing radiation. Chemically, promethium is almost identical to the very elements which are used in tristimulus phosphors. This extremely small quantity is of even less chemical toxicity concern than the phosphor itself. We believe that the minuscule amount of radioactivity involved is of little concern considering the tremendous energy savings brought about by this wonderful technology. If the radioactivity concerns you, we recommend you consider the electronically ballasted lamps, which contains absolutely no radioactivity.

Replacement
lamp for
Reflect-A-Star
is sold on page
253. Use the
13-watt Quad
bulb with pin
base.

*A word of
caution: "Don't
eat a compact
fluorescent
lamp." CFLs,
like all
fluorescents,
contain small
amounts of
mercury vapor
to help carry
the current.
This volatile,
toxic metal will
escape if a
fluorescent
lamp is broken.
If you do break
one, it's best to
leave the room
until it's had
time to
ventilate...
Some CFLs also
contain a very
small amount
of krypton-85
or promethium-
147,
radioactive
isotopes which
help start the
bulb at low
temperatures.
The amount of
radiation that
you'll encounter
from working
in close
proximity to
such bulbs is
less than one
millionth of
one percent of
the typical
background
radiation. We
don't consider
these levels of
radioactivity a
concern, but if
you do, get
electronically
ballasted
integral CFLs,
which contain
no radioactive
elements.
Rocky Mtn.
Institute*

SL18 Compact Fluorescents

Phillips Lighting pioneered the PL compact fluorescent. They are now partners with us in our *Real Kids* fundraiser program.

Introduced to the U.S. market in 1983 and since much refined, the SL18 was the first electronic-ballasted 120VAC compact fluorescent, and it produces the same amount of light as a 75W standard incandescent, drawing only 18 watts; the reflector version — SL18R40 — will replace a 75W incandescent floodlight. These screw-in, one-piece units operate at high frequency, work in temperatures from 0 to 140°F, are silent, and screw right into standard light bulb sockets. The SL18s are rated to last for 10,000 hours. Their lightweight polycarbonate housing and prismatic diffusor/lens keep dirt from settling on the bulb, making them an excellent choice for use in applications where a bare bulb is presently used on most regular fixtures. It measures 7 3/16" x 2 1/2". The Reflector version, measuring 7 1/4" x 5", is designed for use anywhere a directional light source is desired and may be installed in recessed-can or track fixtures or simple surface mounted, porcelain-type lampholders. The SL18s are one of the most energy-efficient compact fluorescent lamps presently available.

Note: When installed in unsheltered outdoor areas, use only in weatherproof fixtures. **Caution: May not be used with dimmers.** Check SL18's dimensions to make sure it will fit your fixture.

36-121 SL18 $25
36-122 SL18R40 $27

Reflect-A-Star

This is the most durable, high quality, modular compact fluorescent floodlight available. Its 13-watt replaceable bulb produces the same amount of light as a 75-watt incandescent floodlight and will typically last the life of at least five of them. The ballast is engineered to provide 50,000 hours of reliable service, this product incorporates a coil ballast, ALZAK finished aluminum reflector, and clear acrylic prismatic lens into its design. It screws into most recessed downlight can fixtures and track fixtures which presently accommodate 4.5" and larger diameter incandescent floodlights. The fixture is rated for use in damp locations and may be used outdoors in locations protected from the weather. Operates down to 0° F. Choose between a 5.25" diameter reflector and 4.5"—the larger one produces more light. Both units are 6" long and 2" wide at screw-in base end. Available in 120VAC only, and it uses 16 watts. Use Quad 13 bulb with pin base (31-122) for replacement of bulb.

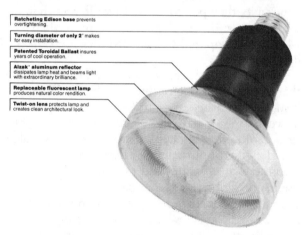

Ratcheting Edison base prevents overtightening.

Turning diameter of only 2" makes for easy installation.

Patented Toroidal Ballast insures years of cool operation.

Alzak® aluminum reflector dissipates lamp heat and beams light with extraordinary brilliance.

Replaceable fluorescent lamp produces natural color rendition.

Twist-on lens protects lamp and creates clean architectural look.

36-313   Reflect-A-Star (5.25") $49
36-314   Reflect-A-Star (4.5") $49

Fluorever Lamp

These lamps are highly recommended for remote power systems because of their extremely high Power Factor (pf) of .955. Power Factor is not a big concern for utility based consumers a low pf lamp essentially sneaks extra power past your meter without being recorded (some folks might feel that's even a benefit). But in remote systems you're paying the bill for that hidden power use. The Fluorever lamps have the highest pf, and therefore the most efficient operation of any compact fluorescent now available. They have can be used to replace a 60W incandescent bulb. These lamps also feature a built-in thermal protection switch. If lamp temperature exceeds 140°F the switch will automatically shut the lamp off to prevent overheating and shortening of lamp/ballast life. Once cooled, approximately 15 minutes, it will reset and resume operation. This allows mounting the lamp in enclosed fixtures. *Note: these lamps will "buzz" very slightly with all modified sine wave inverters.*

36-105	Fluorever 11 Watt	$25
36-106	Fluorever 16 Watt	$25

Circular Fluorescent

Designed to last 9,000 hours, our new electronically ballasted circular fluorescent, made by Lights of America, is designed for situations where maximum illumination is required. The Tri Phosphor tube provides a 10% brighter and warmer light that is far closer to sunlight than standard cool white circular fluorescents. 9" diameter by 3" tall. Ideal for table lamps or hanging lamps. Turn it into a fixture with the optional circular diffuser. *May not be used with dimmers.*

36-145	30W Circular Light	$25
36-146	Circular Diffuser	$8

Ultra-Short Compact Fluorescent System

This new versatile, modular system is great for illuminating areas where lower light levels are desired. The system is available in four versions: (1) as a bare bulb package, (2) with glass PAR 38 (4.75" diameter) floodlight reflector, (3) with white frosted glass (3.75" diameter) decorative globe diffuser, and (4) with ornate frosted glass candle flame diffuser. All use the same ballast adapter base and replaceable PL7 lamp. The screw-on reflector and diffuser options offer additional utility for a variety of lighting applications. The low-power factor core-coil ballast and PL lamp draw 10.8 watts and replace comparable 40 to 60- watt incandescents. Operates down to -20°F and can be installed in outdoor locations protected from the weather. Adapter base diameter is 2-5/8". Screws into conventional light fixture socket.
Available in 120V AC only.

36-151	Bare Bulb Package	$25
36-152	w/PAR 38 Reflector	$39
36-153	w/White Frosted Glass	$37
36-154	w/Ornate Frosted Diffuser	$44

See the full size cut-outs in the appendix, pages 494-497 and the lamp comparison charts on page 48. Extended lamp harps on page 42.

All our AC compact fluorescents work great with inverters.

Circular Fluorescent for Recessed Can

Our brand-new Panasonic fluorescent is perfect for downlighting situations to install into recessed ceiling cans. It emits the equivalent of 40 – 50 watts but uses just 15 watts of electricity and lasts at least 10 times longer. It features a core-coil ballast, has a rated life of 7,500, hours and measures 4.5" tall by 5" in diameter. *May not be used with dimmers.*

36-144 Panasonic C-15 Circular Fluorescent $19

Lamp Harp Retrofits

PL-13 lamp with pin base sold on page 253 can be used as a replacement bulb for the Dazor lamp.

Some compact fluorescent lamps are larger than standard incandescent bulbs and will not fit ordinary floor or table lamps unless you change the harp. The harp is the rigid piece that holds the shade over the bulb. We offer 10" & 12" harps to satisfy most lamp conversion needs. They are very easy to install, requiring no tools.

36-403	**10" Lampshade Harp**	**$3**
36-404	**12" Lampshade Harp**	**$3**

Dazor Asymmetria PL Task Lamp

This Finnish-made, 120VAC, swing-arm lamp is one of the most versatile and energy-efficient of its kind. Its sleek, contemporary design, articulating arm, and adjustable reflector head enable you to provide good-quality task lighting where you need it. The built-in specular aluminum reflector is asymmetrically shaped to bias light distribution in one direction, lowering reflective glare on work surfaces and computer screens. The Asymmetria has a 34" arm with three swiveling joints, each with its own tension control, eliminating the temperamental behavior of spring-tensioned lamps. The fixture's 13-watt compact fluorescent lamp delivers a soft, warm colored light, without the excessive heat that makes working by comparable incandescent desk lights uncomfortable. Available in both black and white with weighted base and clamp. Total fixture power consumption, including ballast, is 15 watts.

Note: The plug-in ballast, located at the end of the power cord, sometimes emits a slight audible hum which can be easily muffled by plugging it into an outlet situated underneath a desk or behind furniture. Also beware, there are numerous imitations of the Dazor now on the market which, though lesser priced, are also not nearly so well made.

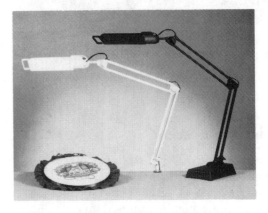

36-302-B Task lamp w/weight base (black) $69
36-302-W Task lamp w/weight base (white) $69

Modular Compact Fluorescents with Replaceable Bulbs

We have been swamped with requests for a compact fluorescent lamp (like the Panasonic T-15 & G-15) with a replaceable bulb. Our new Q-Lux lights look like the Panasonics, but the Lexan diffuser unscrews, allowing easy bulb replacement, and you are off and running for another 10,000 hours. Ballast is a core-type (magnetic) with a life of at least 50,000 hours. The Q-Lux is 13 watt and replaces a 50-watt incandescent. Tube light is 6.75" long by 2.875" diameter; globe light is 6.875" long by 4.5" diameter.

36-501	Q-Lux Tube Light	$25
36-502	Q-Lux Globe Light	$25
31-122	Pin Base Replacement Lamp	$12

Halogen Par 38 AC Lights

These efficient halogens supersede standard incandescent AC PAR 38 spot and flood lights. They produce a whiter, brighter light while consuming 40% less power. And they last two to three times as long; average life is 2,000 hours. Refit your store with them, or use them in and around your home, and watch your power bills drop. The 45 and 90-watt lamps replace the standard 75-watt and 150-watt PAR 38's respectively. They're diode free, nonflickering, and dimmable. The efficient halogen burner produces a white tungsten light, 3,000°K, without lamp blackening.

36-123	Spot Light (45-watt)	$11
36-124	Spot Light (90-watt)	$11
36-125	Flood Light (45-watt)	$11
36-126	Flood Light (90-watt)	$11

All our AC compact fluorescents work great with inverters.

Light Fantastic Lamp Retrofit

Our new lamp conversion kit is the answer to a lot of compact fluorescent retrofit problems. The installation can't get much easier — unplug lamp cord, take out the energy hog incandescent, and put in the PL adaptor and lamp. Next, plug the ballast/wall cube into the wall plug and the lamp cord into the ballast — you're done! This kit will work with any light fixture that uses a cord and wall plug (except lamps that have the shade clipped onto the lamp itself). Available in most sizes; 13 watt replaces 75 watt, 9 watt replaces 60 watt, and 7 watt replaces 40 watt. All have the same color and light output as the incandescent they replace.

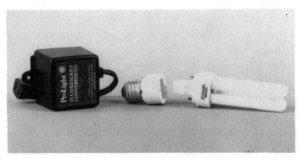

36-601	Light Fantastic — 7 Watt	$29
36-602	Light Fantastic — 9 Watt	$29
36-603	Light Fantastic — 13 Watt	$29
36-604	Light Fantastic — Quad 9	$33
36-605	Light Fantastic — Quad 13	$33
36-606	Light Fantastic — 22 Watt	$36

See page 253 for replacement lamps to fit the Light Fantastic.

Elba Electronic Ballast

Since the introduction of energy-efficient ballasts, we have been searching for the most satisfactory model. In particular, we wanted a high-quality ballast that would not produce a hum when powered by an inverter. We have found it, manufactured by the Elba corporation. Months of testing proved the Elba superior to all competitors. This exceptional ballast is the recipient of many design awards, including a gold medal in 1982 from the New York International Inventors Expo. But in spite of its excellence, we are offering the Elba at a substantial savings over the Etta ballast. It is designed to operate 40-watt, 34- watt, and "U" fluorescent tubes in parallel. This means that it will power either one or two lamps; and if one lamp burns out, the other will remain lit. Ultra-efficient, the unit draws 2/3 the power of a conventional ballast, with the same light output (.99 power factor; 0-140°F. It's CEC approved and UL listed. Magnetic radiation is a super-low 1 milligauss at 8". Install the Elba in any 120VAC fluorescent fixture. It has a 15-year expected life.

36-205 Elba Electronic Ballast   **$49**

Retrofit Recessed Downlights

You may find yourself stumped by the project of replacing the existing recessed incandescent downlights in your home or office ceiling with compact fluorescents. This kit is ingeniously designed to solve the problem. The system installs into your existing mounting frame, J-box, and mounting hardware. The trim flange completely covers the old ceiling cutout. Kit components include a precisely engineered reflector with 30° cutoff (replaces R and PAR lamps); lampholder with socket, flex, cap and wire; and ballast attached to J-box door. Ballast is enclosed F-can, Class P, high-power-factor type, rated for 120V. The unit fits Marco MX7, Halo H7, Capri R10, and Progress P-7. Fits most 7" cans. Included is a 13-watt, pin-base fluorescent lamp that puts out as much light as a 75-watt incandescent. Cannot be used with dimmers.

EcoWorks Ecological Light Bulbs

Even if compact fluorescents are too big for your light fixtures, you can take a step forward in energy savings and economy. EcoWorks ecological light bulbs use less energy and last three times longer than standard incandescent bulbs. At $0.08 per kWh, you'll save $2.80 to $4.00 per twin pack depending on the wattage. Use them wherever you can't use compact fluorescents: EcoWorks incandescent bulbs fit in any fixture and *will work with dimmer switches.* They are an energy-saving and nuclear-free alternative to standard incandescent bulbs made by nuclear weapons contractors such as GE, GTE Sylvania, and Philips. The brass base won't freeze in the socket, and six filament supports protect against vibration. Made in the USA by a nuclear-free company. 10% of profits go to nuclear-free organizations. Bulbs last 3,000 hours on average. Unlike all fluorescents, these bulbs contain no mercury. Available in 54 watts and 90 watts. Good choices for seldom-used lights in closets, outbuildings, etc.

38-110 Ecoworks-54W (8 bulbs)   **$15**
38-111 Ecoworks-90W (8 bulbs)   **$15**

36-133 Downlight Retro Kit   **$89**

Compact Fluorescent Vanity Lights

These vanity light bars look exactly like their energy-hogging counterparts, the incandescent watt-gobblers. The big difference is that you can feel good about having the light you need, instead of guilty about using so many lamps. Each bulb is a 5-watt compact fluorescent, instead of the typical 25-watt incandescent. Just add up the number of lamps you plan to use, and see the enormous energy savings! Lamp life is 10,000 hours. The color is 2700° K (warm white), very close to incandescent; your reflection won't have that "Night of the Living Dead" look that cool white fluorescents give. The attractive vanity light bars are available in polished brass or chrome, and in three different sizes: the 3-light, 18-1/16" long; 4-light, 24" long; and 6-light, 36" long. All are 4-1/4" wide and project out from the wall 6-3/4", including the lamps. (Lamps must be ordered separately and you must use the vanity lamps listed below. Please note: these lamps cannot be used with dimmers and the bulbs have plug-in PL, not screw in, bases.)

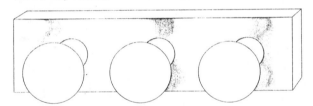

36-137-B	3 Lamps—Brass	$59
36-137-C	3 Lamps—Chrome	$59
36-138-B	4 Lamps—Brass	$69
36-138-C	4 Lamp—Chrome	$69
36-139-B	6 Lamps—Brass	$89
36-139-C	6 Lamps—Chrome	$89
36-140	Vanity Lamp	$16

Rechargeable Security Light

Plugged into a standard 120V wall outlet, this emergency light comes on automatically during a power failure or it can be unplugged for use as a mini-flashlight if needed. The light will operate continuously for about 50 minutes. When fully discharged and plugged in again, it recharges in 36 hours. But it's made for more than just urgent situations; it also functions as a night light. The built-in sensor activates the unit when the room is darkened, and turns it off when daylight arrives or the main light is turned on. The emergency light measures 2.4" by 3.25" by 1.38".

Motion Switch

This motion-activated light control will light stairs, back halls, garages, and basements even when your hands are full and can't reach the light switch. The secret is an infrared sensor that reacts to heat in motion. When a person enters a room, the light goes on. When the control senses no heat in motion for 3-30 minutes (it contains a built-in adjustable timer), the light goes off. Replaces standard wall switch and comes with manual bypass. Rated up to 300 watts. **Incandescent lights only!**

25-404	Motion Switch	$39

25-407	Rechargable Security Light	$23

Flicker Free Night Light

The FlickerFree night light automatically turns on at dark and shuts off at dawn without the anemic start-up flicker of other lights. The secret is two photocells instead of one. It uses standard night light bulbs (7 watt), and costs only pennies a month to operate.

25-406 Flicker Free Night Light   **$8**

Home Guard Emergency Light

The super-bright fluorescent bulb on this rechargeable emergency light illuminates up to a 10' by 10' room. It turns on automatically when there is a power failure so that your room will never be without light. The slim space-saving profile plugs into a regular 120V outlet, adding an extra light to your room. Unplug it whenever you need a portable light. The five-watt replaceable bulb lasts three hours on a charge. The fold-out stand on the back lets it stand up wherever you need it.

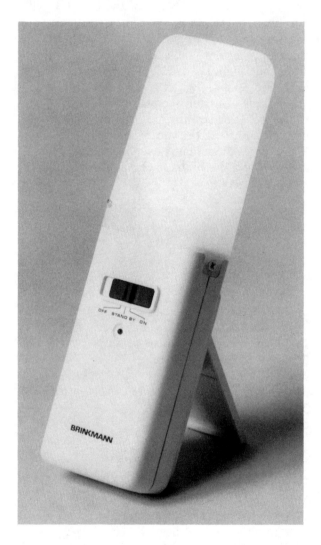

25-409 Home Guard Light   **$35**

To order any of these products
or for more technical information
call us toll-free at
1-800-762-7325

Automatic Off Light Disc

This amazing little disk contains a wizard: a circuit chip that is a real money saver. Simply stick the self-adhesive disk onto the bottom of an incandescent bulb and screw it back into the socket. You now have a light that will turn off automatically after 10 minutes — no more burning on and on, fruitlessly illuminating an empty room. As if that were not enough, the light dims, as a warning, just before turning off. That's not all: just flip the switch once to override the 10-minute timer, and the light will stay on for 2 hours. Automatic Off contains Soft Start™ electronics, which protect the light bulb filament from the initial power surge each time the light is turned on. This makes the bulb last 4 times longer. Automatic Off is 99% efficient and is UL listed.

25-410 Automatic Off   **$10**

Nightlight

The Nightlight is one of the best products for a child's room that we've ever seen. You attach the self-adhesive disk to the end of a bulb (incandescent only) and return it to the socket. The first time you turn on the switch, the light comes on at 100% intensity and dims to 5% over a 20-minute period, then remains at 5% until you turn off the light. Your child will drift off into slumberland as the light slowly dims —perhaps while you tell a story or sing a lullaby. If you turn the switch on and off twice, the light operates at 5% continuously. If the switch is flipped on and off three times, the light stays on at 100%.

25-411 Nightlight   **$10**

Sleeplight

The Sleeplight works much like the Nightlight, except that it dims all the way to off after 20 minutes. The rest of the features are the same. Both devices have Soft Start™ circuitry for initial power surge protection, quadrupling bulb life. They are 99% efficient and UL listed. Since the gradual waning of light is more like the way it happens in nature, it can be very soothing. Try it!

25-412 Sleeplight   **$10**

Electronically Ballasted lamps

Item Code	Description	Incandescent Equivalent	Wattage	Power Factor	Lumens	Size h x w	Shaded lamp	Recess Can	Track light	Operating Temperature
36-145	30w Circular Fluorescent	150	30	.65	2240	9x3½	YES	NO	NO	32-140
36-143	Panasonic 27w twin tube	100	27	.60	1550	7¾x2¼	YES	YES	OK	32-95
36-114	Osram Dulux EL 20	75	20	.56	1200	7¾x2¼	YES	YES	NO	0-140
36-121	Phillips SL 18	75	18	.60	1100	7¼x2½	OK	YES	OK	0-140
36-122	Phillips SL 18w Reflector	75	18	.60	1100	7¼x5	NO	YES	OK	0-140
36-141	Panasonic 18w	75	18	.59	1100	7½x3	OK	NO	NO	32-95
36-113	Osram Dulux EL 15	60	15	.52	900	6¾x2¼	YES	YES	NO	0-140
36-132	Osram Dulux EL 15 Reflector	60	15	.52	900	7x5	NO	YES	YES	0-140
36-112	Osram Dulux EL 11	50	11	.54	600	5½x2¼	YES	YES	OK	0-140
36-131	Osram Dulux EL 11 Reflector	50	11	.54	600	6x5	NO	YES	YES	0-140
36-111	Osram Dulux EL 7	40	7	.42	400	5½x2¼	YES	YES	OK	0-140

Core / Coil Ballasted lamps

Item Code	Description	Incandescent Equivalent	Wattage	Power Factor	Lumens	Size h x w	Shaded lamp	Recess Can	Track light	Operating Temp	Replace bulb
36-101	Panasonic G-15	60	15	.52	720	5¾x3¾	NO	YES	OK	32-95	NO
36-102	Panasonic T-15	60	15	.52	720	6¼x3	OK	OK	OK	32-95	NO
36-106	Fluorever 16	60	16	>.9	900	6¼x2¼	YES	YES	YES	32-140	NO
36-313	Reflect-A-Star 5¼	60	14	.54	860	7¼x5¼	NO	YES	YES	0-140	YES
36-314	Reflect-A-Star 4½	60	14	.54	860	7¼x4½	NO	YES	YES	0-140	YES
36-603	Light Fantastic 13w Twin	60	15	.55	600	8x1¼	YES	NO	NO	0-100	YES
36-605	Light Fantastic 13w Quad	60	15	.55	860	5½x1½	YES	NO	NO	0-100	YES
36-501	Q-Lux Globe	58	16	.55	860	7x5	NO	YES	OK	0-100	YES
36-502	Q-Lux Tube	58	16	.55	860	6¾x3	OK	OK	OK	0-100	YES
36-602	Light Fantastic 9w Twin	50	11	.55	600	7¼x1¼	YES	NO	NO	0-100	YES
36-604	Light Fantastic 9w Quad	50	11	.55	575	5¼x1½	YES	NO	NO	0-100	YES
36-105	Fluorever 11	40	11	>.9	575	5¾x2¼	YES	YES	YES	32-140	NO
36-144	C15 Recess Can Fluorescent	40	15	.52	450	4¾x5¼	NO	YES	YES	41-95	NO
36-151 through 36-154	Ultra Short Fluorescent System	40	11	.42	400	5½ x various	OK	YES	YES	-20 to 95	YES
36-311	Janmar J-910	40	10	.42	400	8x5	NO	NO	NO	-20 to 140	YES
36-312	Janmar J-910-65	40	10	.42	400	9x6	NO	NO	NO		YES
36-601	Light Fantastic 7w Twin	40	7	.55	400	6¼x1¼	YES	NO	NO	0-100	YES

LAMP.TBL 16 April 1992 chart by Michael Potts based on information developed by Jeff Oldham, Doug Pratt, Douglas Bath & Gary Beckwith

The fixture applications in these charts are generalities. There are always exceptions to various applications. Use the full size cut-outs in the appendix for exact fit. Usually, if it fits and it won't get wet – use it!

Differences between Core/Coil & Electronic Ballasts:

Core coil ballasts generally last about 50,000 hours and many have replaceable bulbs, which last about 10,000 hours. Exceptions are the C15 Recessed Can lamp, the Panasonic G & T-15 lamps, and the Fluorever lamps with about a 10,000 hour life expectancy. Core/coil ballasts flicker when starting and take a few seconds to get going. They also run the lamp at 60 cycles per second, some people are affected in a negative way by this flicker.

Electronic ballasts last about 10,000 hours, the same as the bulb, and do not have replaceable bulbs. Electronic ballasts start almost instantly with no flickering. They also run the lamp at about 30,000 cycles per second. For the many people that suffer from "60 cycle blues" this is the energy efficient lamp you've dreamed of.

Solar Outdoor Lighting

Lighting for driveways, yards, porches, decks, or walkways can keep burglars or animals away, and prevent guests from stumbling in the dark. The thought of digging trenches to bury wires, and having to remember to turn the lights on in the evening and off in the morning is often enough to discourage us from brightening up our property effectively.

Solar energy is one of the most useful and cost-effective solutions to outdoor lighting. There is an increasing selection of available products, and at Real Goods we offer what we believe are the highest quality, most reliable, and most economical lights available. Most of the units are self-contained, meaning there is no need to run any wires from the house. Some offer a photo sensor, which automatically turns on the light when it gets dark. "Security" lights often have a motion sensor, great for scaring away thieves and predators.

The typical unit includes a small solar panel, a rechargeable battery, and some form of efficient light. The number of hours that they operate each night depends on the size of the solar panel and the type of light. Some have switches that let you control the brightness and therefore, the running time per night. A few are capable of running all night. Almost all have replaceable batteries and bulbs.

Dear Dr. Doug:
Can I use one of these lights to light up my billboard?,

Answer: The lights that appear in this section are far too dim for lighting a billboard. If you have a need to light up a sign or billboard, call a Real Goods technician and we will design a system for you.

Siemens Solar Lights

All Siemens solar lights employ single-crystalline solar cells, the same that are used in the M55 industrial solar modules—these new outdoor solar lights are strictly top-of-the-line. The solar cells are highly efficient. These cells provide brighter and longer running solar lights. Siemens employs an exclusive type of nicad battery that is extremely tolerant of high temperatures. Unlike conventional nicad batteries, these high-temperature nicads will continue to accept a full charge from the solar cell even in hot temperatures in excess of 50ºC. While most competitive solar lights use either butyl tape or an EVA backing to encapsulate their cells, Siemens uses an exclusive silicon based compound designed to correctly encapsulate the cells. All lights have a 1-year warranty except the Prime Light and the Sensor Light, which have a 2-year warranty. All units will need direct sun for a few hours a day.

The Pathmarker

The Pathmarker is used to outline or define driveways, paths, walkways, patios, or steps. It comes with an 8" stake for easy mounting. A large textured acrylic lens emits a soft red glow by way of a specially designed high-brightness LED that will never burn out. The LED is driven by premium quality AA nicad batteries through environmentally sealed circuitry. The batteries are charged by a Siemens Thin Film Silicon (TFS) industrial solar panel which is fully protected by a polycarbonate lens cover. The Pathmarker's body and stake are injection molded from premium quality black ABS treated with UV inhibitors to prevent the fading and brittleness associated with untreated plastics in long-term sunlight exposure. It operates up to 12 hours every night because of the low amp-draw of the red light. The soft red glow promotes a feeling of safety. One-year warranty and 3-year battery life. (Battery is replaceable.)

Appropriate technology is not a mystical mirage of the future, but a technology whose time is here.

34-321 Pathmarker Solar Light (set of 2) $45

Mini Coach Light

The Mini Coach Light is an economical lantern-style light that mounts atop a two-piece 12" stake. An incandescent bulb provides a soft glow within a unique diffraction grated lens. The Mini Coach Light will run up to six hours per night. It decoratively accents driveways, walkways, and patios and comes with a full one-year warranty. The nicad batteries have a three-year life and the bulb has a nine-month life.

34-322 Mini Coach Light $49

Pathway Light

The Pathway Light is an economical pagoda-style light that mounts atop a two-piece 12" stake. An incandescent bulb provides a soft glow within a unique prismatic lens. The bulb is driven by high temperature Nicad batteries through environmentally-sealed circuitry. The batteries are charged by a powerful single crystalline solar cell which is encapsulated in a waterproof silicone compound. The Pathway Light decoratively accents driveways, walkways, patios, and gardens. Up to six hours nightly run time. Full one-year warranty. Three-year battery life. Nine-month bulb life.

34-323 Pathway Light $59

Prime Light

This is our brightest, longest running solar light. It will accent your driveway, patio, and walkways and provide twice the run time of ordinary incandescent solar lights. Prime Light is up to seven times brighter than the competition, with a three-year bulb life. The hi-low bulb gives a choice of run times and brightness level: 10 hours on low, 5 hours on high. A 12" adjustable stake is included. Full 2-year warranty.

34-327 Prime Light   **$89**

Siemens Sensor Light

The Sensor Light eliminates the need to wastefully burn an outside light while you're away and will keep you from stumbling in the dark. The solar-charged Sensor Light's built-in detector automatically turns on the light when triggered by heat or motion, then turns off when you leave the area. This can make it effective in deterring prowlers, too. The energy-efficient Dulux fluorescent bulb is up to 4 times brighter than competitive solar lights and will last over 5 years. There are no timers and only one switch to set (off, charge only, on). An adjustable sensitivity control reduces false triggers by your cat, a raccoon, tiny UFOs, etc. The unit mounts easily to wall, fascia, soffit, or roof eaves. No wiring or electrician required. Extra battery capacity allows up to 2 weeks of operation without sun. Comes with mounting bracket hardware and bulb included. Full 2-year warranty.

34-303 Sensor Light   **$195**

Be sure to see our Nicad section for your battery needs.

To order any of these products or for more technical information call us toll-free at
1-800-762-7325

Notes and Workspace

Flashlights

Navigator Lite

Our rechargeable Navigator Lite is a high-tech, pure bright red light source that greatly reduces night blindness. When the light is turned off, the eyes rapidly return to their full nighttime abilities. This is a must for sailors, astronomers, and pilots, and also very handy for campers and outdoorspeople. It will produce continuous illumination from its rechargeable batteries for over 24 hours. The recharger plugs into a standard cigarette lighter plug or 12V socket, where it will recharge in 6 hours. The light measures 4" by 2" by 1" thick and comes with a 4" velcro strip for attachment to clothing, table, or any cloth. The light will last in excess of 100,000 hours.

37-307 Navigator Lite $45

Solar Torch

When you need light in the darkest hour, what do you find? Dim glow, dying batteries. Solara's high-performance solar flashlight does not need battery replacement. Simply hang in a south-facing window, on the provided bracket, and it's always ready when you reach for it. 24 hours (3 days) of solar charging gives a full two hours run time. If you need a quicker recharge, the included AC charger will do the job in 16 hours. Its dual-beam parabolic reflector allows you the choice of a 5° or 45° beam spread. The Solara's ergonomic design makes it comfortable to hold and use. Two-year guarantee.

37-311 Solar Torch $45

Solar Lantern

"Your Solar Lantern just arrived and me and the boys are blown away! Life is tough here in the Saudi desert. It's bad enough not getting to drink beer or take regular showers, but until you guys came along it was even getting hard to see at night! It seems there is a shortage of D-cell batteries and the ones we can get don't last very long. Now that we have your lantern that's charged by the sun (no shortage of that over here!) we can even do reconnaissance at night — kind of amazing that our government didn't think of that, huh?" PFC Larry Harbiter, Operation Desert Shield, APO NY.

If you're like us, you're tired of all the Hong Kong–produced throwaway flashlight lanterns. We've finally located the state-of-the-art rechargeable solar lantern, made by Hoxan. This lantern is the epitome of quality. It provides up to 5 hours of continuous lighting on a single charge with its bright 6-watt fluorescent bulb and broad 180° field of illumination. It is lightweight (3.3 lb) and very durable (made of corrosion-resistant, high-impact plastic). The solar cells consist of 14 single crystal individual cells with a maximum output of 2.5 watts. The energy storage unit is a compact lead-acid-sealed battery with 4 volts and 6.0 amphours and will last year after year (minimum 1,000 hours life) with solar recharging. The lantern measures 5.9" long by 2.2" wide by 12.1" high. Although the price seems high, we find this lantern to be well worth it.

37-302 Solar Lantern $109

The Forever Light

Here is a miniature sized (1-7/8") rechargeable light that will give you years of use and you'll never have to change a battery or a bulb. It is energized by solar cells and entirely self-contained. Simply expose it to sunlight for 2–3 hours and it's ready to give weeks of intermittent use. The semiconductor LED emits a brilliant yellow light, turned on by a simple squeeze. It's great for key chains.

37-306 Forever Light (FL-100) $16

All-in-One Solar Flashlight and Battery Charger

Our solar flashlight/battery charger serves two very useful functions. It recharges and operates on multiple battery sizes: AA, C, and D. It incorporates a heavy-duty, high-quality case that is water- and weather-proof. The built-in encapsulated solar panel is made of high-efficiency solar cells. The unit provides two sets of "battery size adapters" that allow you to operate and recharge multiple battery sizes. An LED charge indicator light shows when a solar charge is occurring. In addition to being extremely useful around the homestead, this product is also ideal for use in disasters, power emergencies, outdoor adventures, and, yes, even wars! *Nicad batteries recommended but not included.*

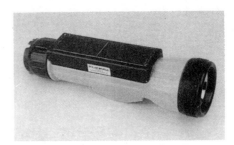

37-309 All-in-One Solar Flashlight $39

Solar Flashlight

Our newest solar flashlight is very compact and lightweight. Place in a sunny spot and its two AA Nicad batteries will charge from the built-in solar panel. Charging time is 7-8 hours in full sun. Running time on a full charge is about 1-1½ hours. The flashlight also doubles as a solar AA Nicad charger, charging two AA batteries at a time. Batteries are included.

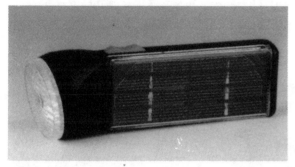

90-469 Solar Flashlight   **$19**

Dynalite Flashlight

We have had numerous requests for a manual flashlight. With continuous squeezing action the Dynalite produces a small steady light, perfect for anyone who does not want to rely on batteries. The light may not be as bright as a conventional flashlight's, but it will not let you down in an emergency. This flashlight is compact, shockproof, and has a shatterproof lens. We have installed a diode in the unit ourselves for further reliablility.

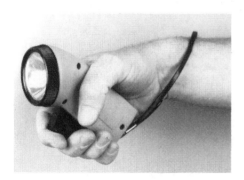

37-308 Dynalite Flashlight   **$9**
37-312 Replacement Bulbs(pack of 2)   **$5**

Tops Light Forehead Lamp

Here is a simple but extremely useful product. It answers the problem of chopping wood in the darkness on cold nights with an awkward flashlight lodged in your mouth. It's a high-tech version of the old coal miner's lamps and has a multitude of uses. It's also ideal for using under the hood or dash of a car, camping and hiking, bicycling, or for hunting for your lost cat at night.

The Tops Light is made of ABS plastic and requires 4 AA batteries to operate, which are not included. The lighting angle is fully adjustable as is the length of the head strap, which is secured down to the proper size with velcro. A spare light bulb is included with the lamp. We used the Tops Light for our nicad battery testing and discovered that it will operate for 4 hours on a set of fully-charged Golden Power nicads.

37-304 Tops Light Forehead Lamp   **$9**

Bright Eyes

This simple invention makes you wonder how you lived without it for so long. It's a hands-free flashlight that can be worn over standard eyeglasses and is so lightweight that you hardly know it's on. The Bright Eyes glasses operate on 4 AAA batteries, which are included! The quality and durability exceeds the simple Tops Light. Bright Eyes operates approximately 1-1/2 hours on a set of batteries.

37-305 Bright Eyes Head Lights   **$14**

Cordless Sensor Light

The Cordless Sensor Light gives you a glowing welcome as you step into your darkened entry hall with an armload of groceries. It lights the way in your garage, workshop or stairway. Or conserves energy in places where people often forget to turn out the light: bathrooms, corridors, etc. In Auto mode, it comes on when the infrared detector senses motion up to 10 feet away. There are On and Off settings, too. Place Sensor Light on a desk, table, or counter; or wall-mount it. Works on 4 C batteries (not included).

25-413 Cordless Sensor Light   **$39**

**To order any of these products
or for more technical information
call us toll-free at
1-800-762-7325**

Rechargeable Batteries and Chargers

Americans use two billion disposable batteries every year. Anyone who uses cameras, flashlights, portable radios and stereos, or battery-operated toys can testify to the high cost of regularly replacing standard alkaline batteries. Now the concern over the proper disposal of spent batteries is mounting. Nickel-cadmium (Nicad) batteries and the new nickel metal hydride (NMH) batteries offer a reuseable solution to the planned obsolescence of zinc-carbon (standard) or alkaline-manganese (alkaline, high capacity) batteries. The use of rechargeable batteries can cut the cost from as much as 10 cents per hour of operation to a fraction of a cent per hour. We therefore find economic merit as well as ecological merit through the reduction of disposable batteries in our waste stream.

There are some important differences in how rechargeable batteries operate, and the consumer should be aware of these. Nicads and NMH batteries have steady operating voltages of 1.2 volts per cell, whereas standard alkaline batteries have an initial voltage of about 1.5 volts which gradually declines as charge is used. While the difference is usually inconsequential, it can affect performance of some devices.

Rechargeable batteries typically have one-third to one-half the energy capacity of standard batteries. Comparisons can be difficult, as standard batteries rarely state the capacities on their label, but simply claim to "keep going, and going, and..." Battery capacity is rated in amphours (Ah) or in the case of small batteries, "milliamphours" (mAh). Nicad and NMH manufacturers typically label how much energy is storable and at what rate the energy should be optimally replaced. Recent advances made to rechargeable batteries have increased their capacity, and these are the batteries that Real Goods carries. When purchasing rechargeables, one should be sure to compare **capacity** as well as cost.

Consider purchasing several sets of rechargeable batteries, one to operate the equipment while the others are recharging. When the appliance shows weakness or fails, just swap batteries with the fresh ones. Having fully charged batteries will avoid the frustration that otherwise might be caused by the lower capacity of the rechargeable batteries.

Solar Battery Chargers

These small solar chargers and the 12V charger are great for charging batteries at the campsite or in the car. They will take longer to charge batteries than a conventional AC charger because the solar panel yields a relatively small output and charging occurs only during periods of direct sun. AC chargers can draw more power for more batteries and charge day and night. If you can afford to charge at a leisurely rate, the solar chargers are great, and the power is *free* and *non polluting*.

Important note on nicad charging: Nicads need to be exercised to give them their full capacity. When you receive them they have been "asleep" on the shelf. They need to be awakened by getting fully charged and discharged for 3 or 4 cycles before their memory is "*stretched*" enough to hold a full charge. They should then be occasionally fully discharged (perhaps once or twice a month). Customers new to nicads should try this simple method before calling us and saying your batteries are defective.

Charging Times for Nicad Battery Chargers[1]
Rated in Hours

	4AA Charger	MSNC	12V Charger[2]
1 @ 240 mAh AAA	2-4	2-4	2-4
2 @ 240 mAh AAA	4-6	4-6	2-4
3 @ 240 mAh AAA	6-8	5-7	3-5
4 @ 240 mAh AAA	8-10	7-9	5-7
1 @ 700 mAh AA	6-8	4-6	4-6
2 @ 700 mAh AA	14-18	8-12	5-7
3 @ 700 mAh AA	20-24	12-16	6-8
4 @ 700 mAh AA	24-30	16-20	7-8
1 @ 1100 mAh AA	11-15	10-12	10-12
2 @ 1100 mAh AA	22-26	20-22	12-14
3 @ 1100 mAh AA	33-40	24-27	13-15
4 @ 1100 mAh AA	44-50	30-34	14-16
1 @ 1800 mAh C		10-13	12-15
2 @ 1800 mAh C		20-25	15-17
1 @ 3500 mAh C		21-26	20-26
2 @ 3500 mAh C		40-45	30-35
1 @ 4000 mAh D		25-30	25-30
2 @ 4000 mAh D		50-60	40-50

[1]Charging rates assume a fully discharged battery, and full intensity sunshine on chargers. A solar charger will receive anywhere from zero to eight hours of sun per day, depending on geographic location and local weather.

[2]The 12V charger will take longer than an AC charger but much less time than a solar charger, as it may charge through the night. The 12V charger **can** overcharge Nicad batteries if they are left in the charger indefinitely. Solar chargers **will not** damage batteries as they have substantial rest (not charging) periods and therefore the batteries may be left in the charger for storage while in the sun.

120V AC chargers can overcharge batteries if left charging indefinitely. To determine proper charging time, take milliAmpere-hour (mAh) capacity of largest battery being charged and divide by AC charger rate (100 mAh for Golden Power 4AA charger or 200 mAh for Golden Power Universal charger).

4 AA Solar Nicad Charger

This is our most popular solar battery charger. We've sold over 15,000 of these in the last two years. We've seen them selling in other catalogs for up to $20! It couldn't be more simple to operate — just leave the unit in the sun for 2-4 days and your dead AA nicads will become fully charged! The most common use is for "Walkman" portable tape players. With a set of AA cells in the Walkman and a set of batteries always in the charger, you'll never be without sounds. Not designed to get wet. (*for charging times, please see chart*)

50-201   4 AA Charger   **$14**

Solar Button Battery Charger

Our Solar Button Battery Charger is designed especially for charging mercury button-type batteries. These batteries are not specifically rechargeable, but most people can squeak a significant amount of use out of batteries they used to toss after one life. Button batteries typically power hearing aids, cameras, alarms, calculators, watches, hand-held electronic games, and many other appliances. Of the 20 million hearing-impaired people in the USA, many use as many as two batteries per day! Charging time is 2–6 hours depending on light intensity, size, and condition of battery. The unit is supplied with a suction cup for window mounting, plastic carrying bag, and a complete instruction booklet. It measures 2-1/4"L x 1-3/8"W x 3/8"D.

Metered Solar Nicad Charger

This great solar battery charger goes one up on the four AA charger. You can use it to charge four AAs, and two Cs or two Ds, at the same time. It will also partially rejuvenate alkaline batteries for a short period of time. It's the only solar charger that comes equipped with a meter that shows battery strength (for AAs only) at a glance before charging. The meter works perfectly to check alkaline battery strength, but with nicads, because of their properties, it will show only if they are totally dead. The MSNC comes highly recommended. Not designed to get wet. (for charging times, please see chart)

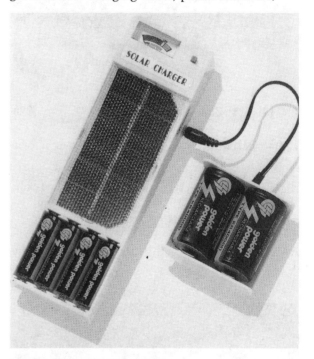

50-203   MSNC Nicad Charger   **$26**

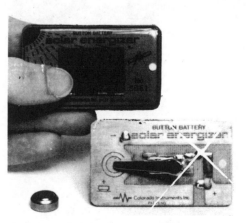

50-205   Solar Button Battery Charger   **$19**

Mercury button batteries will safely recharge to about 90% of previous capacity with the solar button charger, i.e.: they lose 10% capacity on each cycle.

The 12 volt charger coupled with a 10 watt solar panel makes a great high powered battery charger. Use the Solarex unbreakable module (11-513, $139) and the 12 volt socket (26-108, $3), to make it work.

12-Volt Charger

The 12-volt charger will recharge AA, C, and D nickel-cadmium batteries in 10–20 hours from your 12-volt power source. It will charge two AA or two C or two D batteries at one time. It comes with a 12-volt male cigarette lighter-type plug to go into any 12-volt socket. (Make sure the cigarette lighter works without the ignition key on.)

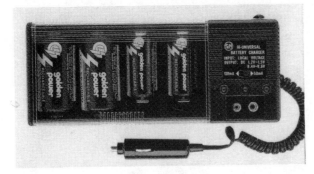

50-214 12V Charger   **$24**

120V Nicad Charger for All Sizes

Our 120V universal chargers provide complete charging from any AC power source. This 4-cell model features LED lights (to indicate charging is in progress) and has reverse polarity protection. It will charge any combination of four AAA, AA, C or D nicad batteries simultaneously or two 9-volt batteries.

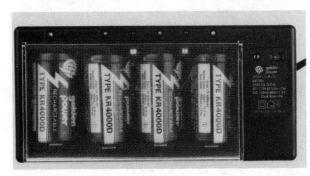

50-223 4-Cell AC Charger   **$15**

Our Golden Power nicads are the highest quality construction and will give 900 to 1000 cycles life expectancy, unlike Panasonics and other brands of nicads which only give 500 cycles or less.

Overcharging will decrease nicads' life expectancy.

High-Powered Nickel Cadmium Batteries

These are the batteries you've been waiting for. Strong enough to make "rechargeable living" more practical and convenient than ever. Designed to keep performing so long that they'll almost become family heirlooms. Yet priced to allow you to fulfill all your power requirements. Our new high-powered AAs have a capacity of 700 mAh (compared to 500 mAh for Panasonics); the C is 1,800 mAh (compared to 1,200 mAh), and the D is a full 4,000 mAh, more than three times greater than Panasonics; now we're offering a AAA with 240 mAh as well. These higher-capacity nicads will last much longer on a charge and more closely approximate the energy storage capacity of disposable alkaline batteries. We strongly recommend that you compare the storage capacity of our Nicads with others on the market. Ours are the most cost-effective Nicads available.

50-109	AAA Nicads ea.(240 mAh)	$2.75
50-106	AA Nicads ea.(700 mAh)	$2.75
50-107	C Nicads ea.(1,800 mAh)	$4.75
50-108	D Nicads ea.(4,000 mAh)	$7.00

Environmentally Friendly Rechargeable AA and C Batteries

Our brand-new *Nickel Metal Hydride* batteries are often referred to as *green* rechargeable batteries, because they don't contain toxic materials such as cadmium, lead, mercury, or lithium. At 1,100 mAh for the AA and 3,500 mAh for the Cs, they store nearly twice the capacity of our our standard rechargeables and will last twice as long. (The nickel metal hydride batteries last roughly 2/3 as long per charge as an alkaline but have an infinitely longer life.) Further, these new batteries don't have the "memory" effect of conventional rechargeables. (NMH AA batteries are a little fatter that AA batteries. They may not fit into the tightest battery compartments. **Sold as individual batteries—not pairs!**

50-105	*Green* AA Batteries	$8
50-104	*Green* C Batteries	$16

Compact Charger for Your AA Nicads

This compact charger is almost the size of a cassette, just 3" by 4.75". It will charge four AA or AAA or a combination of two of each from any standard AC outlet and comes with a convenient selectable charge rate.

50-224 Compact Charger   **$12**

These four companies will accept your old batteries:

Chemical Waste Management
Controlled Waste Division
W. 124 N. 9451 Boundary Rd.
Menomonie Falls, WI 53051
(414) 255-6655.

Inmetco
245 Portersville Rd, Route 488
Elwood City, PA 16117
(412) 758-5515.

Mercury Refining Company
790 Watervlite-Shaker Rd.
Latham, NY 12110
518/785-1703.

Environmental Systems Corporation (ENSCO),
P.O. Box 1957
El Dorado, AK 71731
(501) 863-7173.

We have found that brand new NMH batteries, and to a lesser extent nicad batteries, are very "stiff" and resistant to initial charging. The small solar chargers may not have sufficient power to overcome this resistance until the batteries are broken-in with 2 or 3 cycles. Or try charging just one new battery at a time for the first cycle.

NMH batteries cannot be "quick charged". Attempting to do so will reduce life expectancy. AAs have a maximum charge rate of 150 MA, Cs have a maximum of 350 MA. All our chargers - AC, DC, or solar, are OK for either of the NMH batteries.

"I thought you might enjoy a photograph of my low-tech adaptation of one of your 4AA chargers to give me a nice solar shave".

— *August Salemi*

Energy Conservation

Reflectix Insulation

Reflectix is a wonderful insulating material with dozens of uses. It is lightweight, clean, and requires no gloves, respirators, or protective clothing for installation. It is a 5/16 inch thick reflective insulation that comes in rolls and is made up of seven layers. Two outer layers of aluminum foil reflect most of the heat that hits them. Each layer of foil is bonded to a layer of tough polyethylene for strength. Two inner layers of bubblepack resist heat flow, and a center layer of polyethylene gives Reflectix additional strength.

Reflectix can be used wherever standard fiberglass insulation is used, without the necessity of wearing goggles or a face mask for installation. The standard thickness has an R-value comparable to standard insulation (see chart). Reflectix inhibits or eliminates moisture condensation and provides no nesting qualities for birds, rodents, or insects. Other benefits include reductions in heating and cooling costs that accelerate the payback time of the cost of installation over ordinary insulations. It is Class A Class 1 Fire Rated and nontoxic.

Reflectix BP (bubble pack) is used in retrofit installations. Proper installation *requires* a 3/4 inch air space on both sides of any Reflectix products. The best application for Reflectix products is preventing unwanted heat gain in the summer time. Reflectix is also a great add-on to already insulated walls, providing a radiant barrier, vapor barrier, and R-value.

R-Value Table for Reflectix

R-Value ratings need to be clarified for different applications and more specifically for the **direction of heat flow**. "Up" refers to heat escaping through the roof in the winter or heat infiltration up through the floor in the summer. "Down" refers to preventing solar heat gain through the roof in the summer. "Horizontal" refers to heat transfer through the walls.

Up	Down	Horizontal
8.3	14.3	9.8

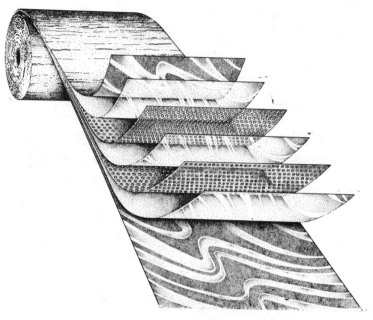

As well as its most common usage as a building insulator, Reflectix has a myriad of other uses: pipe wrap, hot water heater wrap, duct wrap, window coverings, garage doors, as a camping blanket or beach blanket, cooler liner, windshield cover, stadium heating pad, camper shell insulation, behind refrigerator coils, and a camera bag liner.

Reflectix BP (Bubble Pack) comes in 16", 24" and 48" widths and in lengths of 50 feet & 125 feet. (We cannot UPS rolls larger than 125 feet.)

■56-503-BP	16" x 50' (66.66 sq. ft)	$35
■56-502-BP	16" x 125' (166.66 sq. ft)	$85
■56-511-BP	24" x 50' (100 sq. ft)	$49
■56-512-BP	24" x 125' (250 sq. ft)	$125
■56-521-BP	48" x 50' (200 sq. ft)	$109
■56-522-BP	48" x 125' (500 sq. ft)	$259

Attn: Alaska and Hawaii Customers: *Please add $15 additional freight for first class shipment.* For air shipment call for freight quote before ordering.

05-220   Add'l freight to HI & AK $15

Reflectix works best when bouncing heat UP, as in under floors for winter time, or under rafters in summertime. In all cases, reflective insulation needs an air space on the hot side to work.

Bubble Pak Staple-Tab Reflectix

For the most efficient, energy-saving refrigerator available today, see the Sun Frost on page 274-276.

This Reflectix product is the same bubble-pack insulation as the standard Reflectix insulation with the added feature that it's made to be installed between framing members as opposed to on the surface. This product makes installation far easier as it eliminates the need to add the furring strips that are needed to form an air space with standard Reflectix. It comes with easy-to-install staple tabs. **Same pricing as BP insulation above—but be sure to specify Staple-Tab by ending the product number with ST instead of BP.**

- ■ 56-503-ST 16" x 50' (66.66 sq. ft) $35
- ■ 56-502-ST 16" x 125' (166.66 sq. ft) $85
- ■ 56-511-ST 24" x 50' (100 sq. ft) $49
- ■ 56-512-ST 24" x 125' (250 sq. ft) $125
- ■ 56-521-ST 48" x 50' (200 sq. ft) $109
- ■ 56-522-ST 48" x 125' (500 sq. ft) $259

Hawaii & Alaska Customers: *You must add additional freight for first class shipment of $15.* For air shipment call for freight quote before ordering.

05-220 Add'l freight to HI & AK $15

Tape for Reflectix

Reflectix makes a 2-inch aluminum tape that is excellent for bonding two courses of Reflectix together. It works far better than masking tape or duet tape. It also has reflectability and is highly recommended for any Reflectix installation. Two sizes of tape rolls are available: 30 feet and 150 feet.

56-531 30' Roll, 2" Reflectix Tape $3
56-532 150' Roll, 2" Reflectix Tape $9

Cord Caulk

Air infiltration is the number one cause of heat loss in our homes. Cord Caulk is a great draft and weather sealing substance that comes in a 100 foot roll. The cord itself is 3/16 inch in diameter. It stops drafts, window rattles, and moisture damage to sills and floors. This is a very simple product: a soft fine fiber yarn that is saturated with synthetic adhesive wax polymers. It can be applied at below freezing temperatures, and can be removed and reused numerous times without affecting the sealant characteristics required for weatherization.

Cord Caulk adheres well to wood, aluminum, steel, paint, glass, rigid vinyl, rigid plastic sheet, and nylon. It will adhere poorly to plastic foams, poly films, dirty, flaking surfaces, and silicone. The inventor won an award for this energy invention at the Boston Edison Centennial Invention Competition in 1986. Every house utility closet should contain at least one roll of Cord Caulk!

56-501-D Cord Caulk (wood grain brown) $15
56-501-W Cord Caulk (white) $15

Super "R" Radiant Barrier

NASA uses Super "R" Radiant Barrier to protect astronauts from the extreme temperatures of space. It can also save you money on heating and cooling costs for your home or business. Simply rolled out over the existing insulation in your attic or, for the best results, stapled to the underside of the roof rafters, Super "R" reflects 97% of all radiant heat out of your home in summer and back into your home in winter. It improves the efficiency of your insulation, greatly reduces operating time of your air-conditioning and heating equipment, and helps maintain a constant temperature throughout the house. Made of mesh-reinforced aluminum, Super "R" Radiant Barrier is environmentally safe and nontoxic. It is lightweight, easy to install, and Class A/1 fire rated. A truly cost-effective energy saver and comfort enhancer. 4' x 125' roll (500 sq. ft).

56-500 Super "R" Radiant Barrier $99

Draft & Safety Cover

Our new Care Cover is a neat little energy-saving device that doubles as a safety accessory. It's an electrical outlet cover with a draft gasket, and sliding, spring-loaded doors in front of the sockets. When the outlet is not in use, the doors are automatically closed, preventing air infiltration (the leading cause of home heat loss) and discouraging the probing fingers of small children. Care Cover fits all standard three-prong receptacles; it's available in ivory color only.

56-524 Care Cover (set of 3) $9

If you want to cut the power costs on your washing machine, see our Guzzle Buster Kits on page 265.

Programmable Thermostat

Our programmable thermostat, made by Honeywell, features six daily temperature settings, seven day advance programming, digital room temperature readout, armchair programming, and automatic heating/cooling conversion. The Weather Monitor program analyzes the time needed to heat your house and controls the thermostat so that your house will be at a preset comfort level at the programmed time. Accuracy is +/- 2°F; Size: 7" x 3¾" x 1¼". Use for most 24V gas heating, oil heating, and electrical cooling systems. Designed to replace any 24V termostat, which is the most common type of control for forced-hot air systems. Also includes a 9V battery backup (battery not included) to save programing during power outage.

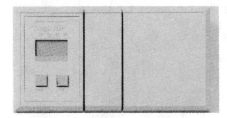

25-405 Thermostat   **$69**

"We bought a fancy analog thermostat from a local supplier, and it never worked. We replaced it with your digital unit and are very pleased. It works exactly the way it should, every time. And it has saved its cost in less than 6 months."
Barry Dalsant, Arcata

**To order any of these products
or for more technical information
call us toll-free at
1-800-762-7325**

Radiation Control

The dangers of extremely low frequency (ELF) electromagnetic radiation have been strongly suspected since the early 1970s. Studies have indicated that long-term exposure to ELF fields emitted by electric blankets, computer monitors, TVs, some bedside clocks, and other appliances may interfere with the body's internal self-regulating systems. ELF energy pulsates, and it can influence human biological functions such as normal cell reproduction. With so many subtle factors affecting our health, it is essential to stay informed about these invisible pollutants and to eliminate or reduce as many of them as we can from our living environment.

Real Goods claims no credentials as medical or scientific experts. Instead, our product offerings and actions are governed by what we hope qualifies as common sense. On the subject of ELF hazards, we are governed by the single thought: *Why take chances with something as important as health?*

Nuclear Radiation

When it comes to issues of health and safety, one of the bitterest lessons that United States citizens have learned is, "Don't depend on government agencies or industry to look out for you." This is doubly true for one of the greatest environmental dangers of all: nuclear power. Time and again, Americans have learned of inadequate inspections, faulty design, improper installation, poor maintenance, human error, coverups, and outright "disinformation." The response has been a rising wave of citizen activism. People are beginning to take charge of their destiny.

At Three Mile Island, the Pilgrim plant south of Plymouth, Massachusetts, and other nuclear power facilities, groups of concerned residents in nearby towns have taken up the task of monitoring ambient radiation levels on a daily basis, and forwarding the data to plant operators, state health officials, and the Nuclear Regulatory Commission. The instrument that most often stands sentinel is the Radalert monitor.

In contrast, government and public utility inspectors generally read their radiation detectors only four times a year. So any releases of concentrated radiation are flattened out when averaged over a three-month period. But a troubling number of these "spikes" of radiation have occurred and continue to occur. Officials tend to dismiss them as insignificant. However, there has been a documented increase in cancer and leukemia in the area surrounding the Pilgrim plant.

Though the near-disaster of Three Mile Island and the Chernobyl devastation got the big headlines, many scientists are greatly concerned about the potential peril of continual small, unreported releases of radiation. We urge our readers who live within the zone of influence of government or private nuclear reactors, or near any industrial source of radioactivity, to set up their own monitoring networks and lobby officials to increase their vigilance and enforcement of regulations. – *James Pendargast*

Copam Low-Radiation VGA Monitor

After carefully examining every low-radiation computer monitor we could get our hands on, we decided that the Copam high-resolution, low-radiation super VGA color monitor provides the best value. It produces a crisp clear 1024 x 768 (8514A) display with less than 2.5 milligauss extremely-low-frequency radiation at a viewing distance of 12 inches! This monitor is designed for use with IBM PS/2 and other IBM and compatible (clone) computers which have standard VGA, super VGA or IBM 8514A functions. The monitor has a large 14 inch screen for easy viewing. The Copam color monitor obtains its extremely low extremely-low-frequency and very-low-frequency radiation levels by creating an equal and opposite field which cancels out the fields normally generated within the monitor. Radiation levels are less than the Swedish VDT standards.

Computer users will appreciate our new low-radiation VGA color monitor, and anyone who uses 48 inch twin tube fluorescent lights at home or at work will benefit by replacing the old ballast with a new Elba electronic ballast, which gives more light, uses less power, and according to our tests produces far less electromagnetic radiation than any other ballast on the market today.

■**57-115 Copam Monitor** $495
■**57-116 VGA Card** $295

Safe Computer Monitor

This is *an extremely-low radiation* computer monitor. Our own tests have found that a 1-milligauss envelope extends a maximum of 2-feet from most monochrome monitors in any direction. Even laptop computers give off magnetic fields (from backplane drives and screen lighting) which travel 4 to 6 feet. These fields penetrate almost everything, *including all anti-glare screens.*

The Safe Computer Monitor consists of a monochrome style shielded liquid crystal display with an 8-1/2" by 5-1/2" screen with 640 x 200 pixel resolution. Onboard electronics are completely shielded so it emits *only .5 milligauss at 6-inches* magnetic radiation. This monitor will work with the IBM PC, XT, AT. Guaranteed for one year on parts and labor.

■**57-110 Safe Monitor** $1,095

Trifield Meter

Low-level electromagnetic radiation may be a hazard to human health; many studies have raised that suspicion. And the dangers of microwaves are well known. The versatile TriField meter will let you discover the calm backwaters of the magnetic swamp in which we live. With it you can test microwave oven door gaskets, appliances, cellular phones, computers, and other electronic equipment. You'll be able to construct a magnetic safety "map," so as to move couches, chairs, beds, and especially, cribs out of "hot spots" (children seem to be more susceptible to radiation damage). The meter has three easy-to-read scales, red-lined to show field levels that may be harmful. It reads magnetic, electric, and microwave pollution quickly and accurately without the need to correct for the influence of your own body. Since the human body absorbs radiation from all directions, a meter has to read on all three possible axes to get an accurate measurement. Other meters require tipping and turning; the Trifield reads the X,Y, and Z axes simultaneously. It also compensates for different magnetic field frequencies, to accurately indicate the electrical currents induced within the human body. It will measure fields through the non-ionizing range from 30 Hz - 3 Billion Hz. *9V battery included.*

57-108 TriField Meter $169

Magnetic Scanner
Alerts You to Potential Hazards

Run safety checks on your home and office with our new, economical Magnetic Scanner. This sophisticated instrument has five LEDs, indicating AC 1, 2, 5, 15, and 30+ milligauss, giving much better resolution. You'll learn not only how far away you need to sit from your TV or video monitor, but also how strong a field it's putting out. The hand-held detector is very compact: at 1-3/4 x 1-1/4 x 1-1/2 inches, it easily fits in a shirt pocket. The sensors are omnidirectional, detecting the same fields your body receives.

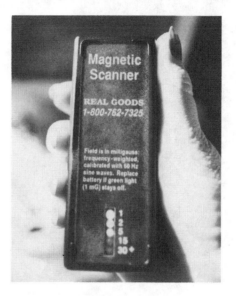

57-101 Magnetic Scanner $79

We've had a lot of customers abuse our return policy with radiation meters by testing their homes, then returning the meter. For this reason we can only fully refund returns on radiation meters when they truly are defective! Otherwise, we must charge a 25% restocking fee.

We offer monitoring devices for two very different types of radiation. Geiger counters only measure hard nuclear radiation, such as emissions from nuclear power plants or radon gases. These meters are represented by the Pocket Geiger Counter and the Radalert. Electromagnetic radiation is a completely different animal. It is generated by AC power grids and appliances, and requires different monitoring equipment. These meters are represented by the Tri-Field meter and the Magnetic Scanner.

The Body Electric

Dr. Robert Becker & Gary Seldon. This is the definitive work on Energy Medicine (electro-medicine), the rapidly emerging new science that promises to unlock the true secrets of healing. The book does an excellent job of getting vitally important information out of the scientific priesthood and into the hands of the general public— who should be making the important environmental decisions! 364 pages.

80-610 Body Electric   **$11**

'innocent until proven guilty' is great when applied to people, but it's bad strategy for carcinogens and other health risks. We energy addicts just don't want to know that the grasses seldom bloom beneath the high tension lines.

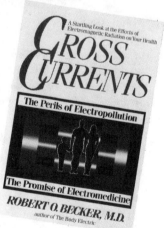

Radalert Digital Radiation Monitor

The Radalert is a high-quality digital Geiger counter sensitive to alpha, beta, gamma, and X-radiation. The product of choice for environmentalists; it measures subtle low-level radiation changes, a must if you live near a nuclear power plant. Digital counting technique and accumulated counts mode allow greater sensitivity in measuring low dose radiation compared to other Geiger counters. Utility is increased with an accumulated counts mode for comparing indoor and outdoor averages and an adjustable alert to signal radiation changes. Operating range is 0 to 19,999 counts per minute (0 to 20 mR/hour). An output jack lets you connect the radiation monitor to a computer for hard copy needs. Runs on 9V battery for months of use. Belt loop carrying case included. *Battery included. 25% restocking fee on all returns.*

57-106 Radalert   **$369**

Cross Currents

Dr. Robert Becker. This book discusses the hazards of EMR from ordinary household appliances, utility power lines, radar, & microwave transmitters. This book is fascinating reading and will give you background information and motivation to clean up the magnetic pollution in your local environment!

80-612 Cross Currents   **$22**

Monitor 4 – Pocket Geiger Counter

The Monitor 4 is the best low-cost nuclear radiation monitor we could find for giving instantaneous readings. Other nuclear monitors cost thousands of dollars, are bulky to carry, and use hard to find 22.5-volt batteries. The Monitor 4 is small enough to fit in a shirt pocket and uses a standard 9-volt battery. It's a must-have item if you live near a nuclear plant. (Plant operators don't usually tell you about spills and accidents until it's too late!) The Monitor 4 reads ranges from 0.5 to 50 mR/hour. It has a switch-selectable audio/visual count indicator. *25% restocking charge on all returns.*

57-105 Geiger Counter   **$319**

**To order any of these products
or for more technical information
call us toll-free at
1-800-762-7325**

Home Safety Tests

Lead Alert Kit

The Frandon™ Lead Alert Kit helps find lead in the household environment—in paint (surfaces, toys,) ceramicware, enamel pots, solder (cans and plumbing) and other possible hazards. Lead poisoning continues to be a concern if you have small children or eating utensils of foreign origin. The impact of lead poisoning is both hard to diagnose and severe. It begins with anorexia, weight loss, weakness, or anemia and can result in irreversible brain damage. If you suspect the presence of lead in your environment, this simple kit will give you much needed answers.

   Kit contains two 1-ounce vials of indicator fluid which can make up to 50 tests. Each vial must be used within three to five days of activation. The test will work only on solid objects, it will not test water.

57-102 Lead Alert Kit   **$29**

Carbon Monoxide Detector

Your home may be invaded by a deadly enemy - carbon monoxide (CO). But you won't know it. This highly poisonous gas, produced by incomplete fuel combustion, is odorless, tasteless, and colorless. Even in small amounts, it can cause headaches, dizziness, shortness of breath, anxiety, and irritability. It changes color at greater than 100 ppm and changes back when fresh air is present. Using the Detector, test for CO in your house, garage, car, or RV. Check for dangerous flue gasses by placing the Backdraft Spillage Indicators on natural-draft water heaters and furnaces. It includes two backdraft detectors and one CO detector. It could be the safest move you ever made.

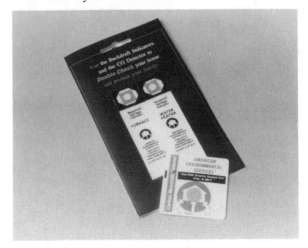

57-119 Carbon Monoxide Detector   **$14**

Watch our catalogs for more Home Safety testing equipment in this fast developing field.

Notes and Workspace

Water Conservation

Toilet Lid Sink

This ingenious water saver supplies clean water for hand washing, then uses it to fill the tank for the next flush. These units have been used in Japan for decades. The unit, constucted of durable plastic, has the appearance of porcelain, fits rectangular tanks (up to 8" wide) and is attached with moisture-resistant Velcro. When the toilet is flushed, incoming fresh water is rerouted through the chrome fixture into the basin, then filtered into the tank and bowl. Water automatically shuts off when it reaches normal fill level in the tank. This sink is a boon to people with limited arm or hand movement, since they need not struggle with faucet handles. And it's so easy that children are more likely to wash their hands. Excellent for small spaces, too; no separate washbasin is needed. Installation is very simple, requiring no tools and only a few minutes of time.

46-120 Toilet Lid Sink (15"–18" long) $35
46-121 Toilet Lid Sink (18"–20.5" long) $35
46-122 Toilet Lid Sink (20.5"–22" long) $35

Low Flow Toilets

Toto LF-16 Toilet

The Toto LF-16 is a 1.6 gallon per flush (gpf) toilet, which has the highest performance results of any 1.6 gpf toilet on the market according to Uniform Plumbing Code requirements. Manufactured in Japan, the Toto is easy to install and uses standard U.S. flush mechanisms, mounting hardware, and rough-in dimensions. Constructed of high-quality vitreous china, it meets all current code standard requirements (IAPMO, BOCA, SBCCI). The water surface area is 8 x 7½ inches, or 60 square inches. *Shipped from the East Coast. Seat not included. Add the additional freight charges to the standard order form shipping charges.*

■44-701-W Toto LF-16 (white) $190
■44-701-B Toto LF-16 (bone) $215
■44-701-G Toto LF-16 (gray) $215
05-207 Western US add'l freight/Toto $45
05-208 Eastern US add'l freight/Toto $20
05-209 HI & AK add'l freight/Toto $110

Ifo Cascade Toilets

Many of our veteran readers will remember the beautiful Ifo toilets that we featured several years ago. We discontinued them because the price went over $400 per toilet with exchange rate fluctuations. Now Ifo has licensed a U.S. factory to manufacture their fine toilets, and we're proud to reintroduce them to you. Ifo has always stood out because of its beautiful European styled tank shape and its fine working mechanisms. Ifo has nearly 50 years of worldwide experience with flush toilets.

Ifo 3180

The 3180 is Ifo's most popular toilet. It has an adjustable flush from 1 to 1½ gallons. It comes with a standard 12 inch rough-in and has a water surface of 5½ x 5¾ inches. It accepts a standard round seat (not included).

Ifo 3190

The 3190 was designed for the American market and is IAPMO listed. It is a 1.5-gallon flush toilet that comes with a standard 12-inch rough-in. The water surface is much larger than the 3180: 8½ x 9 inches, allowing for easier cleaning. It accepts a standard elongated seat (not included).

■44-511   Ifo 3180 - White $269
■44-512   Ifo 3180 - Bone $339
■05-200   Pallet charge $20
■05-210   Air Freight crating charge $50

Add one (only) $20 pallet charge (or air freight crating charge for AK/HI) whether you order one or many toilets. *All toilets shipped freight collect from Northern California*

■44-516   Ifo 3190 - White $329
■44-517   Ifo 3190 - Bone   . $389
■05-200   Pallet charge $20
■05-210   Air Crating charge $50

Magic Flush

Magic Flush is the most impressive water-saving device we've found, and we've had hundreds submitted to us. Easily installed in two minutes in any toilet using a flapper or drop valve, it saves twice as much water as toilet dams! As a bonus the unit includes a new flapper valve, which usually needs replacing anyway. The adjustable flusher (our tank holds 8-1/2 inches of water and now flushes with just 2-1/2 inches) comes with a flow restricter for the bowl rinse cycle, which alone saves half the rinse water.

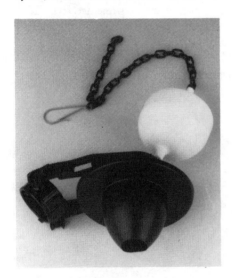

46-203 Magic Flush   **$15**

Water Conservation Kit

An average family of four will save 30,000 gallons of water per year by simply installing one low-flow showerhead, two faucet aerators, and a set of toilet dams. We've packaged all these water-savers together with a toilet leak detection kit, an instruction card, and a 26-page booklet on saving water. This is the ideal kit for California, now in its 6th drought year. This is a $45 retail value but we're offering this kit at a price that will make saving water irresistible.

46-109 Water Conservation Kit   **$19**

Toilet Dams

Each time your toilet is flushed, it uses 3 to 7 gallons of water. Toilet dams are plastic barriers that isolate part of your toilet's tank so that water in this section does not run out with the flush. By installing two toilet dams you can save 2 to 3 gallons per flush. For a family of four that amounts to a savings of up to 12,000 gallons per year! Our toilet dams install easily in seconds without tools into all standard toilets. Pressure is maintained for a forceful flush. Dams are constructed of stainless steel and thermoplastic and do not interfere with a normal flush. If only 10,000 people installed our toilet dams, 120 million gallons of water per year could be saved.

46-201 Set of 2 Toilet Dams   **$11.50**

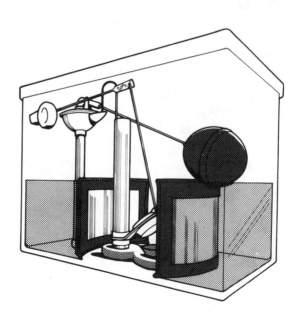

WaterMate

In many drought areas, water is strictly rationed. The penalty for overuse ranges from a painful 10–25% surcharge to virtual disconnection. Water users, in most cases, are aware of exceeding their allowances only after the fact—when the water bill arrives. And even in regions of plentiful water, it behooves us all to conserve this precious natural resource, while benefiting from lower water bills or pumping costs. Water-Mate allows easy monitoring of household water consumption on a fixture-by-fixture basis. It requires no electricity and can be installed behind a showerhead; at a washing machine; or on a hose, to keep tabs on garden watering and car washing, or to check for leaks. The LED readout can be in gallons, cubic feet, or water units, simply by pushing a button. The display also indicates the amount of water use per month for each of the last 13 months, as a comparison to the current one- or two-month totals, to correspond to the billing cycle.

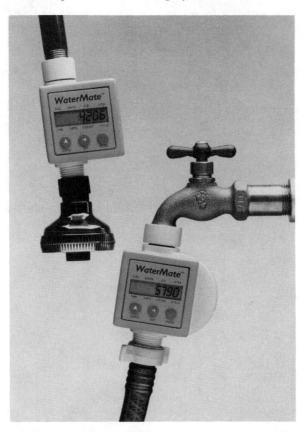

46-160	1/2" WaterMate	$59
46-161	3/4" WaterMate	$69

The Mistic Arch

We discovered this wonderful cooling product at the Solar Energy Expo in Willits. It was the single most popular item there on a 100-degree day. We bought it for our hardworking overheated warehouse staff, and they use it for cooling down on hot days. It puts out a very fine mist that is extremely soothing yet not wet enough to really soak your clothes. It hooks up simply to a garden hose. It consists of 1/2-inch PVC pipe with a brass ball valve shutoff. It has eight 3-gallons-per-hour misters equally spaced and will operate on any water pressure over 20 psi. The easy-to-read blowup diagram allows you to assemble the arch in about 10 minutes! A can of PVC cement is included. This is a major improvement over "running through the sprinklers!"

46-150 The Mistic Arch $59

Water-Saving Showerheads

If a family of four takes 5-minute showers each day, they will use more than 700 gallons of water every week—the equivalent of a three-year supply of drinking water for one person. —50 Simple Things You Can Do to Save the Earth.

Showers typically account for 32% of home water use. A standard shower head uses about 3 to 5 gallons of water per minute, so even a 5-minute shower can consume 25 gallons. According to the Department of Energy, heating water is "the second largest residential energy user." With a low-flow showerhead, energy use and costs for heating hot water for showers may drop as much as 50%. Add one of our instantaneous water heaters for even greater savings. Our low-flow showerheads can easily cut shower water usage by 50%. A recent study showed that changing to a low-flow showerhead saved $0.27 of water per day and $0.51 of electricity per day for a family of four. So, besides from being good for the Earth, a low-flow shower head will pay for itself in about two months!

Save energy too with our more efficient and lifetime life expectancy tankless water heaters. See pages 318-320.

A close-up from our working showerhead display in our showroom. On the right is the "hardware store low-flow" which actually uses close to 3 gpm. On the left is the Spradius Aerator. Next from the left is the Lowest Flow (46-104), then the Plusational (46-107), the Deluxe (46-106), and second from the right is the Spa 2000 (46-101).

The five showerheads listed immediately below all work incredibly well. In our showroom we have a showerhead testing module which holds six heads and measures both pressure (psi) and flow (gallons-per-minute). We can unequivocally recommend all of these showerheads for both performance and styling! We invite you to visit our showroom and judge for yourself.

Lowest Flow Showerhead

This is by far our best selling and finest designed showerhead. It can save up to $250 a year for a family of four by cutting hot and cold water use by up to 70%. At 40 psi it delivers 1.8 gpm, with a maximum at any pressure of 2.4 gpm. Manufactured in the USA of solid brass, chrome plated, it exceeds California Energy Commission standards. It comes standard with a built-in on/off button for soaping up and standard 1/2-inch threads so that a wrench is all that's necessary for installation. Fully guaranteed for 20 years. Specified by the City of Los Angeles.

46-104   Lowest Flow Showerhead $12

For those with high mineral contents that regularly plug your shower head, these 3 heads are incredibly easy to clean. The head unscrews just below the soap-up valve, a stainless steel disc comes out, wipe it clean or drop it in vinegar for really tough cases, and you're back in business.

Plusational Showerhead

This brand new technology showerhead gives an excellent water stream. It features a new minimum/maximum water-saving feature. You can adjust the shower to give 1.0, 1.4, or 1.7 gpm at 40 psi. The stream this head puts out is very close to a pulsating showerhead. It means much greater rinsing efficiency at flows of 4 quarts per minute and less. It is made in the USA of solid brass, and is chrome-plated with a 20-year guarantee. It also includes an on-off valve. Specified by the City of Santa Monica.

Deluxe Low-Flow Showerhead

Made from solid brass and with real 14 kt. gold plating, this Deluxe showerhead is very similar to our Lowest Flow Showerhead but much more elegant looking (if you like gold!). This showerhead is engineered for low-flow and does not use flow restrictors or pressure compensators to reduce flows. This head has the same flow ratings as the lowest-flow and is fully guaranteed for 20 years. Also equipped with an on-off valve. The Deluxe looks like the Plusational showerhead but has a superior full conical spray.

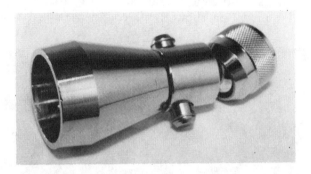

46-106   Deluxe Low-Flow Showerhead $18

46-107   Plusational Showerhead $17

Spa 2000 Showerhead

A good low-flow, high-performance showerhead reduces water consumption and heating costs significantly while still getting the deed done admirably. The Spa 2000 showerhead is one of the best available. Its uniquely engineered venturi design delivers a vigorous, pulsating spray using just 2.5 gpm, and its handy built-in pressure-reducing lever can be used to save even more water in high-pressure areas or simply to reduce flow when you are soaping up. Durably constructed, its head is made from white Celcon plastic, and the shower-arm coupling is chrome-plated brass. Fits standard, ½-inch threaded shower-arms.

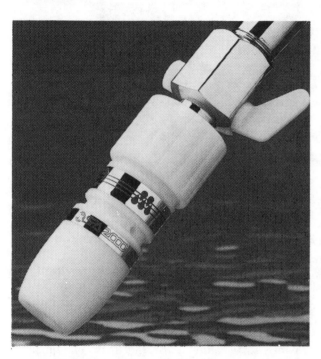

46-101 Spa 2000 Showerhead $24

Hydro-Powered Lighted Shower Head

This lighted shower head is actually a mini-hydro electric power plant that lets you take a romantic shower in the dark. It puts on a great show by actually filling the streams of water with light. Installation is simple. Each Shower-Star comes with a "Mood Pak" including red, blue, green, amber, and two clear bulbs. Available in either chrome or brass.

The Ultimate Shower

We've located the perfect water saving shower-head for comfort. Certified as a Low-Flow fixture, the Ultimate Shower uses only 2.74 gallons per minute, even with its 3-1/2 inch showerhead that has 127 holes. The all-directional showerhead does not sacrifice spray quality as most available water-saving fixtures do. The wide-covering consistent rainstorm effect that emanates from the head produces a sensation never experienced with conventional showers. The Ultimate Shower allows over 20-inches of height flexibility with its universally adjustable, all-directional arm. Constructed of solid brass, chrome plated, the shower includes a wing spanner wrench and a water flow restrictor. Shower arms install to industry standard 1/2-inch pipe thread.

46-102 Ultimate Shower $59

63-312 Chrome ShowerStar $45
63-313 Brass ShowerStar $55

Low-Flow Faucet Aerators

According to Home Energy magazine, we would save over 250 million gallons of water every day if every American home installed faucet aerators. Installing aerators on kitchen and bathroom sink faucets will cut water use by as much as 280 gallons per month for a typical family of four.

Bathroom Faucet Aerator

This is one of the aerators included in our Water Conservation Kit. Installed in the bathroom a typical family of four will save up to 100 gallons every day. Very simple to install and constructed of solid brass, chrome plated. Sold in sets of two. Uses 1.5 gpm at 30 psi. Fits both male and female faucets.

46-108 Bathroom Aerator (set of 2)   **$4.50**

Kitchen Faucet Aerator

This is one of the aerators included in our Water Conservation Kit. Installed in the kitchen sink, a typical family of four will save up to 50 gallons every day. Very simple to install and constructed of solid brass, chrome plated. Sold in sets of two. Uses 2.5 gpm at 30 psi. Fits both male and female faucets.

46-110 Kitchen Aerator (set of 2)   **$4.50**

Deluxe Faucet Aerator

This low-flow faucet aerator with finger tip on-off lever will cut hot and cold water use by up to 60%. It is dual threaded both internally and externally to fit almost all male and female faucets. It limits faucet flow to 2.75 gpm at any water pressure. It's made of solid brass, chrome plated. It installs simply in seconds and can save thousands of gallons of wasted water every year. The on-off fingertip control lever allows the user to temporarily restrict the flow of water to a trickle without readjusting at the hot and cold controls. It's ideal for shaving, brushing, and washing dishes.

46-103 Deluxe Faucet Aerator   **$7.50**

Spradius Kitchen Aerator

Here is a two-position faucet aerator that saves lots of water. It's rated at 2.5 gpm, and it swivels 360 degrees to direct spray or stream to every part of your sink. When pushed up, it functions as an aerator keeping water from splashing out of the sink by injecting air into the water to soften it. When pulled down for spray, it saves water with a great spray pattern that's handy for cleaning or rinsing. It installs quickly and without tools. The double swivel design allows you to rinse the entire sink all the way up to the edges, and eliminates the need for an expensive sprayer hose.

46-131 Spradius Kitchen Aerator   **$11**

Water Purification

Water, Who Needs It

Water is vital to our bodies. Like the surface of the Earth, our bodies are mostly water. The average adult contains 40 to 50 quarts of water which must be renewed every 10 to 15 days. With the intake of fruits and vegetables, you are receiving water, but you still must drink at least six glasses of water daily to enable your body to function properly. Water is the base of all bodily functions.

Your Body Depends Upon What You Drink

The EPA released information on December 14, 1988 that stated there is some toxic substance in our ground water no matter where we live in the U.S. Even materials added to our drinking water to protect us (such as chlorine) are linked to certain cancers and can form toxic compounds (THM's).

The adage "If you want something done, do it yourself" applies to drinking water. The most obvious solution to pollution is a point-of-use water purification device. The tap is the end of the road for water consumed by our families. There are no pipes or conduits to leach undesireable elements into our drinking water beyond this point.

To make the best choice for a water purification system, first examine the problems we can encounter:

Biological Impurities

Bacteria, Virus, and Parasites

Years ago, waterborne diseases accounted for millions of deaths. Even today in underdeveloped countries, an estimated 25,000 people die daily from waterborne disease. Effects of waterborne microorganisms can be immediate and devastating.

Modern municipal supplies are relatively free from harmful organisms because of frequent monitoring and routine disinfection with chlorine or chloramines. This does not mean municipal water is free of all bacteria.

People with private wells or small rural water systems have reason to be concerned about the possibility of microorganism contamination from septic tanks, animal wastes, or other sources. (There is a little community in California, where 4,000,000 gallons of urine hits the ground daily from dairy cows!)

Approximately 4,000 cases of waterborne diseases are reported every year in the U.S. Many of the minor illnesses and gastrointestinal disorders that go unreported can be traced to organisms found in water supplies.

Inorganic Impurities

Dirt and Sediment or Turbidity

Most waters contain suspended particles of fine sand, clay, soil, and precipitated salts. Turbidity is unsightly and can be a source of food and lodging for bacteria, and can interfere with effective disinfection.

Total Dissolved Solids

Total dissolved substances consist of rock and other compounds from the earth. The entire list could fill this page. The presence and amount of total dissolved solids in water represents a point of controversy among water purveyors.

Here are some facts about the consequences of higher levels of TDS in water:

1. High TDS results in undesirable taste which can be salty, bitter, or metallic.

2. High TDS water is less thirst quenching.

3. Some individual mineral salts may pose health hazards. The most problematic are nitrates, sodium, sulfates, barium, copper, and fluoride.

4. The EPA Secondary Regulations advise a maximum level of 500 mg/liter (500 parts per million) for TDS. Numerous water supplies exceed this level. TDS levels in excess of 1000 mg/liter are generally considered unfit for human consumption.

5. High TDS interferes with the taste of foods and beverages.

6. High TDS make ice cubes cloudy, softer, and faster melting.

7. Minerals exist in water mostly as **inorganic** salts. In contrast, minerals having passed through a living system are known as **organic** minerals. They are combined with proteins and sugars. According to many nutritionists, minerals are easier to assimilate when originating from foods. (Imagine going out to your garden for a cup of dirt to eat rather than a nice carrot.)

8. Water with higher TDS is considered by some health advocates to have a poor cleansing effect, because water with low dissolved solids has a greater capacity of absorption than water with higher solids.

Toxic Metals or Heavy Metals

Among the greatest threats to health are the presence of toxic metals in drinking water — Arsenic, Cadmium, Lead, Mercury, and Silver. Maximum limits for each are established by the EPA Primary Drinking Water Regulations. Other metals such as Chromium and Selenium, while essential trace elements in our diets, have limits imposed upon them when in water because the form in which they exist may pose a health hazard. Toxic metals are associated with nerve damage, birth defects, mental retardation, certain cancers, and increased susceptibility to disease.

Asbestos

Asbestos exists as microscopic suspended mineral fibers in water. Its primary source is asbestos-cement pipe which was commonly used after World War II for city water supplies. It has been estimated that some 200,000 miles of this pipe is currently in use to transport drinking water. Because these pipes are wearing, asbestos is showing up with increasing frequency in drinking water. It has been linked with gastrointestinal cancer.

Radioactivity

Even though trace amounts of radioactive elements can be found in almost all drinking water, levels that pose serious health hazards are rare—for now.

Radioactive wastes leach from mining operations into groundwater supplies. The greatest threat is posed by nuclear accidents, nuclear processing plants, and radioactive waste disposal sites. As containers containing these wastes deteriorate over time, the risk of contaminating aquifers becomes a toxic time bomb.

Organic Impurities

Tastes and Odors

If water has a disagreeable taste or odor, the likely cause is one or more organic substances ranging from decaying vegetation to algae; hydrocarbons to phenols.

Pesticides and Herbicides

The increasing use of pesticides and herbicides in agriculture affects water. Rain and irrigation carry these deadly chemicals into groundwater as well as into surface waters. Since more than 100 million people in the US depend upon some groundwater for drinking water, the increasing contamination is a major concern. Our own household use of herbicide and pesticide substances contribute to the contamination. These chemicals can cause circulatory, respiratory and nerve disorders.

Toxic Organic Chemicals

The most pressing and widespread water contamination problem results from the organic chemicals created by industry. The American Chemical Society listed 4,039,907 distinct chemical compounds as of 1977, and they began their list only in 1965. The list can grow by 6,000 chemicals per week!

Every year approximately 115,000 establishments produce more than 1.3 billion pounds annually of many different chemicals in the U.S., a business worth $113 billion. This is a very difficult juggernaut to stop or change.

> *There is some kind of toxic substance in our ground water no matter where we live in the U.S.*

The EPA says there are 77 billion pounds of hazardous waste generated each year in the U.S., 90 percent of which is disposed of improperly. This equals 19,192 pounds of hazardous waste for every square mile of land and water surface in the U.S. (including Alaska and Hawaii.)

Repeat the statistic to yourself slowly.

There are 181,000 manmade lagoons at industrial and municipal sites, 75 percent of which are unlined. Many are within 1 mile of wells or water supplies. Information on the location of these sites, their condition, and containments ranges from sketchy to non-existent. Will this be the horror story of the millenium?

Chemicals enter our drinking water from many sources. The effects of long term exposure to toxic organics, even in minute amounts, are difficult to recognize. Contaminated drinking water may look and taste perfectly normal. Users' symptoms include headache, rash, or fatigue—symptoms so common that they are hard to diagnose as water related. The more serious consequences of drinking tainted water are higher cancer rates, birth defects, growth abnormalities, infertility, and nerve and organ damage. Some of these disorders take decades to manifest themselves.

Chlorine
Trihalomethanes (THM's) are formed when chlorine, used to disinfect water supplies, interacts with natural organic materials (e.g. by-products of decayed vegetation, algae, etc.). This creates toxic organic chemicals such as chloroform, and bromodichloromethane. Chlorinated water has been linked to cancer, high blood pressure and anemia. Anemia is caused by the deleterious effect of chlorine on red blood cells.

Methods for Solving Our Water Problems

Sorting Through the Options
What are the alternatives for the seeker of pure water? There are a mind-boggling number of water systems sold. Following is a brief analysis of each option's strengths and weaknesses.

Centralized Water Treatment
Building hi-tech water treatment plants to remove impurities is not a practical solution for two reasons:

1. Only 2% of water supplied to our homes is used for human consumption.

2. It is not economically feasible to service rural areas with large municipal treatment centers.

Treatment isn't logical for the water we use for our lawns, to flush our toilets, and to fight fires.

There is one additional disadvantage to centralized water treatment; even treated water can pick up contaminants from pipes before reaching the tap.

Boiling Water
Boiling is a useful method for killing bacteria and other living organism during emergencies, but is not recommended for long term use.

Boiling may kill germs, but dirt, sediment, dissolved solids, bad taste, or odor remain, as well as possible chemical contaminants.

Bottled Water
An increasing number of people find solutions for safe drinking water by paying $.80 to $2.00 per gallon to drink water prepared and bottled by someone else. The price reflects the costs of bottling, storage, trucking, fuel expenses, wages, insurances, and advertising, making it extremely cost ineffective, particularly when some water sources have proved inferior to what comes out of the tap.

A point-of-use water system, eliminates middleman costs, and can provide purified water for pennies per gallon.

Point-of-Use Water Treatment

The most efficient and cost effective solution to water purity is to treat *just* the water to be consumed. Devices for point of use water treatment are available in a variety of sizes, designs, and capabilities.

Mechanical Filtration

Mechanical filtration acts much like a fine strainer. Particles of suspended dirt, sand, rust and scale (i.e. turbidity) are trapped and retained, greatly improving the clarity and appeal of water.

When enough of this particulate matter has accumulated, the filter is discarded. This type of filter is called a pre-filter.

Activated Carbon Adsorption

Carbon adsorption is the most widely sold method for home water treatment because of its ability to improve water by removing disagreeable tastes and odors, including objectionable chlorine.

Activated carbon (AC) is processed from a variety of carbon based materials such as coal, petroleum, nut shells, and fruit pits steamed to high temperatures in the absence of oxygen (the activation process). The process leaves millions of microscopic pores, and crevices. One pound of activated carbon provides from 60 to 150 acres of surface area. The pores trap microscopic particles and large organic molecules while the activated surface areas cling to or adsorb the smaller organic molecules

While AC theoretically has the ability to remove numerous organic chemicals like pesticides, THM's, TCE, and PCB, the actual effectiveness is highly dependent on several factors:

1. The type of carbon and the amount used.
2. The design of the filter and the rate of water flow (contact time).
3. Length of service.
4. The types of impurities it has removed.
5. Water conditions (e.g. turbidity, temperature, etc.)

A disadvantage of carbon filters is that they can become a base for the growth of bacteria. When the carbon is fresh, practically all organic impurities and even some bacteria are removed. The accumulated impurities can become food for bacteria, enabling them to multiply within the filter. The high concentration is considered by some to be a health hazard.

After periods of non-use (such as overnight) a quantity of water should be flushed through a carbon filter to minimize the accumulation of bacteria.

Oligodynamic, Silver Impregnated or Bacteriostatic Carbon

A manufacturer who adds (impregnates) silver compounds to the surface of the carbon granules is trying to inhibit bacteria growth within the carbon bed. However, EPA sponsored testing of such filters have shown that they are "neither effective nor dependable in meeting these claims" [EPA Report #EPA/600-D-86/232 October 1986].

Some manufacturers have also made misleading claims that silver impregnated filters will eliminate bacterial contamination from virtually any water source. The low concentration of silver, however, is insufficient to destroy influent waterborne bacteria or to protect from contaminated water under normal flow conditions.

Pyrogens can induce fever (from dead bacteria). Bacteria destroyed in silver impregnated carbon can still end up in your drinking water.

Because silver is toxic to humans, these filters are regulated by the EPA and must be registered. This registration doesn't imply any EPA approval, but certifies that the carbon will not release more than 50 parts per billion of silver—the maximum safe level.

Chemical Recontamination of Carbon Filters

Another weakness of carbon filters is the chemical recontamination which can occur when the carbon surface becomes saturated with the impurities it has adsorbed — a point that is impossible to predict. If use of the carbon continues, the trapped organics can release from the surface and recontaminate the water with more impurities than those contained in the raw tap water.

To maximize the effectiveness of carbon, it should be kept scrupulously clean of sediment and heavy organic impurities such as the by-products of decayed vegetable matter and microorganisms. These impurities prematurely consume the carbon's capacity and prevent it from doing what it does best — adsorbing light weight toxic organic impurities like THM's and TCE, and undesirable gases such as chlorine.

Solid Block Carbon

This is obtained when very fine pulverized carbon is compressed and fused with a binding media (such as a polyethylene plastic) into a solid block. The intricate maze developed within the block insures contact with organic impurities and, therefore, effective removal. The problem of channeling (open paths developing because of the buildup of impurities, and rapid water movement under pressure) in a loose bed of granulated carbon granules is eliminated by solid block filters.

Block filters can be fabricated to have such a fine porous structure that they are capable of mechanically filtering out coliform and associated disease bacteria.

Among the disadvantages of compressed carbon filters is reduced capacity due to the inert binding agent and a tendency to plug up with particulate matter, requiring more frequent replacement. They are also substantially more expensive than conventional carbon filters.

Limitations of Carbon Filters

A properly designed carbon filter is capable of removing many toxic organic contaminants, but falls short of providing protection from the wide spectrum of impurities which have been referred to previously.

1. They are not capable of removing any of the excess Total Dissolved Solids.

2. Only a few solid block or carbon matrix systems have been certified for the removal of lead, asbestos, VOC's, cysts, fecal coliform, and other disease bacteria. Large *suspended* materials will be removed by some filters. Small *dissolved* materials can't be removed by carbon filtration.

3. They have no effect on harmful nitrates, or high sodium and fluoride levels.

4. For any carbon filter to be effective (even for organic removal), water must pass through the carbon (whether it be granular or compressed) slowly enough that complete contact is made. *This all-important factor is referred to in the industry as contact time.* At useful flow rates

of 0.5 - 1 gallon per minute, the flow rate is determined by the amount of carbon.

One must read the Data sheets provided by responsible manufacturers carefully to verify claims. Many companies are certified with the National Sanitation Foundation (NSF), whose circular logo appears on their data sheets.

Minerals in Drinking Water

Manufacturers and sellers of AC filter systems often claim "We need minerals in water — these are essential for good health." This statement has **never been proven** by scientific studies. Therefore, the value of minerals in drinking water remains a moot point. Filter proprents make their point about minerals to obscure the fact that their product will not remove dissolved solids. The same process that keeps dissolved minerals, keeps total dissolved solids, hardness, and some heavy metals, as well.

Carbon Filters in Summary

AC filters are an important piece of the purification process, although only a piece. AC removes chemicals and gasses. AC won't remove total dissolved solids, hardness.

Distillation

Distillation is the process of heating water to steam and recondensing it by cooling. Distillation mimics the hydrologic cycle of nature (the sun causes evaporation over the earth's bodies of water and condensation/precipitation occurs over the land masses).

Distillation will remove impurities such as sediment, dissolved solids, nitrates, sodium, toxic metals, and microorganisms. These are basically left behind when the water is converted to steam.

Some toxic organic chemicals will vaporize with the steam and be carried over into the distillate with the water. To solve this problem, an activated carbon filter must be incorporated either before or after the boiling chamber.

Sophisticated "fractional" distillers will remove these organics by heating water in fractions until the boiling point is reached. The organics are vented out at each step of the heating process.

Even with the problem of organics addressed,

there are still disadvantages with distillers:

1. Distillers are time consuming to maintain and clean. Impurities and total dissolved solids are left behind in the boiling chamber. A hard scale builds up on the heating element and in the boiling chamber. If this scale is not removed, efficiency will be impaired.

2. The product water must be cooled quickly as its elevated temperature encourages the regrowth of airborne bacteria.

3. The process of rapid distillation will drive away free oxygen dissolved in the water. Many scientists and doctors refer to distilled water as dead water. The absence of free oxygen will also give water a flat taste.

4. Distilled water is costly because of the energy required to vaporize all drinking and cooking water (an exception to this is a solar distiller). Every rate increase from the utility company makes distilled water even more expensive.

Deionization

The process of deionization (DI) is worth discussing even though it doesn't seem practical for household water treatment. DI is a chemical process that utilizes minute plastic beads called resins. As untreated water flows over these treated resins, the ions of total dissolved solids are leached from the water.

When the resin beads become saturated they must be removed, and regenerated with acid or caustic chemicals.

DI removes *only* charged particles (total dissolved solids). DI is not capable of removing dirt, rust, sediment, pesticides, organic toxins, asbestos, bacteria, or virus. It is, therefore, best used in conjunction with other water treatment methods.

The resins also will provide an environment that encourages bacteria growth.

Water softeners work by the principle of ion exchange as well. The resin beads in a water softener will give two ions of sodium for an ion of calcium or magnesium. With the removal of the calcium and magnesium ions, the water is no longer hard.

Ultrafiltration/Reverse Osmosis —

Osmosis occurs in living organisms in which there is a piece of tissue or a membrane with fluids on either side. Fluids having a lesser concentration will be drawn through the tissue/membrane to mix with fluids having a greater concentration, equalizing the concentration on both sides.

Osmosis occurs when two fluids of differing concentration are separated by a semi-permeable membrane. The fluid passes through the membrane in the direction of the most concentrated solution.

The value of minerals in drinking water remains an undecided question — no one knows for sure.

Oxygen passing from the lungs into the blood stream, and water and nutrients penetrating the root structure of a tree use osmosis.

When we quench our thirst, a quantity of water enters our stomach. This water is diffused into our system to replenish what is lost during the normal life processes.

In the natural world, surrounding and inside us, there are vast networks of biological membranes. These screening barriers govern the selection and passage of chemicals and fluids, in essence, controlling the traffic of the life processes.

Reverse Osmosis is exactly the opposite of Osmosis. In Reverse Osmosis (RO), water having a lesser concentration of substances is derived from water having a higher concentration of substances. Tapwater with dissolved solids is forced by the water pressure in our pipes against a membrane. The water penetrates the RO membrane, leaving up to 99% of the dissolved solids behind.

The RO membrane is an *ultimate* mechanical filter, with a pore size of two one hundred millionths of an inch in diameter—too small to be seen even by an optical microscope!

By the remarkable phenomenon of RO, particles smaller than water molecules can be removed! The molecules diffuse through the membrane in a purified state, and collect on the opposite side.

Ultrafiltration/RO membranes remove and reject such a wide spectrum of impurities from water using *minimal added energy*—just water pressure. RO gives the best water available for the lowest price expended.

Reverse Osmosis Effectively Reduces the Following:

1. Particulate matter, turbidity, sediment, etc.
2. Colloidal matter.
3. Total dissolved solids (up to 99%).
4. Toxic metals.
5. Radium 226/228
6. Microorganisms (potable water only)
7. Asbestos.
8. Pesticides and herbicides (coupled with AC).

Reverse Osmosis and Activated Carbon Adsorption

Ultrafiltration/RO alone will not remove all of the lighter, low molecular weight volatile organics such as THM's, TCE, vinyl chloride, carbon tetrachloride, etc. They are too small to be removed by the straining action of the RO membrane, and their chemical structure is such that they are not repelled by the membrane surface. Since these are some of the most toxic of the contaminants found in tap water, it is very important that a well-designed carbon filter be used in conjunction with the membrane.

In some applications, AC is used before the membrane. In **all** quality RO systems, there is AC after the membrane. This means that post AC filters don't have to contend with bacteria and all of the other materials which cause fouling and impair performance.

Not all RO systems are created equal, in performance or price. The engineering and experience behind the RO design is critical to overall performance and dependability.

Note: The typical time required to purify one gallon of RO water is three to four hours. RO uses water to purify water. This is the "rate of recovery." RO units will use from three to nine gallons of brine (wastewater) to make one gallon of purified water. Quality units will have automatic shut-off when full.

Brine is necessary to remove excess accumulated materials from the RO membrane. If left in the system these will impair efficiency. One can direct brine water outside as an additional drip line.

The cost of water energy for a fine RO sys-

Little River, Mendocino, California.
Photo by Sean Sprague

tem will be about $1.33 per month if one pays for water at the rate of $1.00 per 100 cubic feet.

You Can Guarantee the Quality of Your Drinking Water

An individual's main decision is how comprehensive a water treatment system to purchase and install. A system which combines technologies will produce better water than a system incorporating just one.

Choose technologies that will meet your needs for a long time. If you are making a commitment to pure water, begin to enjoy the benefits today.

Water Filtration Systems

Have you noticed how much information about drinking water has been appearing in the media over the years?

Periodically, reports surface about violations in safety standards at bottling water facilities. One 1985 example was entitled *Bottled And Vended Water: Are Consumers Getting Their Money's Worth*. It was prepared by the Assembly Office of Research for the State of California. In the cover letter to the report it stated: "Consumers of bottled water who sometimes pay 1000 times the cost of tap water have the right to purchase a safe and wholesome product. Current state law contains no specific requirements to make bottled or vended water more wholesome or safe than tap water." The report then goes on to cite some of the infractions which the study uncovered. More recently, the United States General Accounting Office published report GAO/RCED-91-67 entitled *Food Safety and Quality: Stronger FDA Standards and Oversight Needed for Bottled Water*. (FDA difines bottled water as water that is sealed in bottles or other containers and is intended for human consumption. Bottled water excludes soda, seltzer, flavored, and vended water products). This report states "After temporarily exempting 'mineral water' from bottled water standards in 1973, FDA has not developed alternative standards for it or even defined it [Although FDA has not officially defined mineral water, it is generally considered a type of bottled water that contains various dissolved minerals, such as copper, iron, sulfate, and zinc]. As a result, bottled water, including mineral water, may contain levels of potentially harmful contaminants that are not allowed in public drinking water." The report includes background, sample problems, conclusions, and recommendations.

These are just two examples that illustrate the problems of relying on bottled water. Just because you are paying a fortune for it does not mean that you are fully protected from contamination! A very valid alternative is to purify water coming out of your tap before it gets consumed. You save all the costs of packaging and transportation, while potentially receiving a superior product. Plus, it's more convenient to purify water right at the tap just as it is more convenient to have your own refrigerator than to rely on the ice man to deliver.

Who is a candidate for a water purifier, and how do you decide which system is appropriate? These are the questions we are helping you answer.

The majority of people who purchase a water purifier are previous users of bottled water. If you are still using tap water, chances are you haven't "bought in" to the need to spend money on better taste and quality in water. Once a person changes water habits, the two most popular options are to purchase bottled water, or a point-of-use drinking water system.

By means of our definition, the word "filter" refers to a carbon drinking water system. The carbon system will reduce whatever it has been verified to reduce by independent testing. "Purifier" refers to a slower process, such as reverse osmosis, which also greatly reduces dissolved solids, hardness, and organics. A purifier takes several hours to process each gallon of water. Make sure that the claims of any purifier you consider have been verified by independent testing.

Be aware certain states require third party documentation of the performance of any water system claiming to reduce contaminants known to have adverse health effects. (These contaminants are known as Health-Related Contaminants.)

Here is a list of items which are considered to be health-related contaminants: Total coliform bacteria and turbidity, arsenic, barium, cadmium, chromium, fluoride, lead, mercury, nitrate, selenium, silver, benzene, carbon tetrachloride, 1,2-dichloroethane, 1,1-dichloroethylene, p-dichlorobenzene, endrin, lindane, methoxychlor, total trihalomethanes, toxaphene, trichloroethylene, 1,1,1-trichloroethane, vinyl chloride, 2,4,5-TP (silvex), 2,4,-D, and radium--226 & 228.

Whenever you see a claim that a system will remove any one or more of these contaminants, ask to see the Data sheet.

There are two kinds of bottled water: Purified, and Spring. *Purified water* is usually prepared by Reverse Osmosis, De-Ionization, or a combination of both processes. Distillation can also be considered purified water, but the label will usually specify "distilled." *Spring water* is either acquired from a mountain or artesian spring, or is no more than processed tap water. Spring water will generally have higher total dissolved solids than purified water. (Purified water is better for battery use and steam irons because of lower total dissolved solids.)

Without wanting to over-complicate the decision-making process, here is a simple exercise to help you determine if you will be happier with a filter or a purifier.

Step 1: Look for Total Dissolved Solids, Hardness, and pH written on the label of your favorite bottled water, and write down the results:

Bottled Water Results	
TDS	
Hardness	
pH	

If not listed on the label, call up the bottler of your water (they often have an 800 number or can be found through directory assistance) and ask for the information. Specify that you want both the normal levels as well as *how high they range*.

Next: (you're not quite through) call your own municipal supplier of tap water. Speak with a water chemist at the water company. Remember to ask for the normal amounts as well as how high they range — this is very important. Find out the same information **for these six items:**

Tap Water Results –
Tds:_____
Hardness:_____
pH:_____
Residual chlorine:_____
Iron:_____
Manganese:_____

Step 2: Compare the first three bottled water results with the first three tap water results. How close are the numbers of your tap water to the bottled water? If your bottled water is higher in TDS and Hardness than your tap water, a good filter should satisfy your needs.

Step 3: If your bottled water is lower in Tds, and Hardness, then you should consider a good purifier, because the odds are against you being satisfied with a filter alone because your tap water is "heavier" in TDS and Hardness than your preferred bottled water.

If you are not drinking bottled water at present, then a good filter will provide enough improvement over tap water to make you thrilled with your new purchase.

Important information for you if you have city/municipal water:

For anyone planning to purchase a reverse osmosis water purification system, it is very important to call your water supplier and get the **Tap Water Results** information shown earlier. This will help us choose the proper membrane configuration for your purifier.

Important information for you if you are using your own well water:

Wells can have problems with staining, sediment, rotten egg smell, iron, manganese, tannin, etc. These will often be accompanied by tell-tale signs. These problems are pre-treatment problems, and must be corrected before you purchase any point-of-use water treatment device. Knowing this can save you headaches and money. A water test is advisable if you have pretreatment problems. If you aren't concerned about coliform bacteria, or lead, there is a low priced test available. If you are concerned about bacteria, lead and heavy metals, pesticides, and VOC's, then the 93 item test by NTL is recommended.

If you spend the time to obtain the figures for your tap water as outlined in this article, you'll be able to purchase a water treatment system you'll love and respect. -Randy Wimer

National Testing Laboratories (NTL) More Extensive Water Sample Testing

We are working with National Testing Laboratories (NTL) to provide a thorough analysis of your water. Now that the EPA has determined that the ground water in more than 30 states is seriously polluted, we feel it's essential to thoroughly test your water. NTL has two laboratory facilities, one in Michigan and one in Florida, which perform the full range of drinking water analysis, inorganic, organic, and bacteriological. Their laboratories are certified in 13 states to perform analyses of drinking water, using only US EPA approved methods, and a strict quality assurance program. *We've compared National Testing Laboratories with lots of other testing services and find them to be the most comprehensive, accurate, and reasonably priced.* Their program is simple: you order the test kit from us; we mail it directly to you; you fill the sample bottles and ship the kit to the NTL lab. NTL will analyze the samples according to the test series purchased and will send the full report and explanatory letter back to you within 10 working days. Your test will show if your water contains all of the listed pollutants in amounts higher or lower than EPA limits. A cover letter interprets your test and clearly explains what action (if any) is recommended to ensure that your water is safe to drink.

METALS:
Arsenic
Barium
Cadmium
Chromium
Copper
Iron
Lead
Manganese
Mercury
Nickel
Selenium
Silver
Sodium
Zinc

INORGANICS AND PHYSICAL FACTORS:
Total alkalinity (as $CaCO_3$)
Chloride
Fluoride
Nitrate (as N)
Nitrite
Sulfate
Hardness (as $CaCO_3$)
pH (standard units)
Total dissolved solids
Turbidity (NTU)

VOLATILE ORGANICS (VOCs):
Bromoform
Bromodichloromethane
Chloroform
Dibromochloromethane
Total trihalomethanes

Benzene
Vinyl chloride
Carbon tetrachloride
1,2-Dichloroethane
Trichloroethylene (TCE)
1,4-Dichlorobenzene
1,1-Dichloroethylene
1,1,1-Trichloroethane

Acrolein
Acrylonitrile
Bromobenzene
Bromomethane
Chlorobenzene
Chloroethane
Chloromethane
O-Chlorotoluene
P-Chlorotoluene
Dibromochloropropane (DBCP)
Dibromomethane
1,2-Dichlorobenzene
1,3-Dichlorobenzene
trans-1,2-Dichloroethylene
cis-1,2-Dichloroethylene
Dichloromethane
1,1-Dichloroethane
1,1-Dichloropropene
1,2-Dichloropropane
trans-1,3-Dichloropropane
cis-1,3-Dichloropropane
2,2-Dichloropropane
Ethylenedibromide (EDB)
Ethylbenzene
Styrene
1,1,2-Trichloroethane
1,1,1,2-Tetrachloroethane
1,1,2,2-Tetrachloroethane
Tetrachloroethylene (PCE)
1,2,3-Trichloropropane
Toluene
Xylene
Chloroethylvinyl ether
Dichlorodifluoromethane
cis-1,3-Dichloropropene
Trichlorofluoromethane
Trichlorobenzene(s)

MICROBIOLOGICAL:
Coliform bacteria

and it will check for the following pecticides:

PESTICIDES AND HERBICIDES:
Alachlor
Aldrin
Atrazine
Chlordane
Dichloran
Dieldrin
Endrin
Heptachlor
Heptachlor Epoxide
Hexachlorobenzene
Hexachlorapentadiene
Lindane
Methoxychlor
PCBs
Pentachloronitrobenzene
Simazine
Toxaphene
Trifluralin
Silvex 2,4,5,-TP
2,4-D

NTL Water Check With Pesticide Option

This is the best analysis of your water available in this price range. Your water will be analyzed for 73 items, plus 20 pesticides. If you have any questions about your water's integrity, this is the test to give you peace of mind. The kit comes with five water sample bottles; blue gel refrigerant pack (to keep bacterial samples cool for accurate test results), and easy-to-follow sampling instructions. You'll receive back a two-page report showing 93 contaminant levels, together with explainations of which contaminants, if any, are above allowed values. You'll also receive a follow-up letter with a personalized explanation of your test results, plus knowledgeable, unbiased advice on what action you should take if your drinking water contains contaminants above EPA-allowed levels. The test will check for the following parameters:

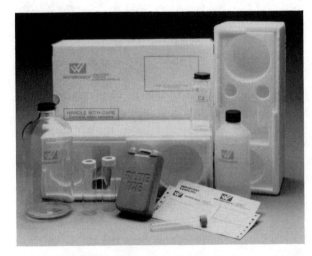

42-003 NTL Check w/Pesticide Option   **$149**

Laboratory Water Test

For Private Water Supplies Only (Private wells, springs, catchment, etc.)

We've found through lots of experience that if you're not on municipal or pretreated water, we can't conscientiously sell you a water purification system without first knowing these specific particulars about the water you have: pH, total dissolved solids (TDS), hardness, iron, manganese, copper, and tannin. This test normally costs $40, but we're providing it to our customers for only $17. This analysis tells us enough about your water so that we can make an informed recommendation to you regarding what system will work best with the water you have.

Upon receipt of your $17 we will send you a questionnaire and a small plastic bottle. Fill out the questionnaire, return the water sample, and you'll have your results back in a few weeks. It's the only way we can both be sure you're getting the right filter for your system! This test does not show contamination problems such as lead, coliform bacteria or chemicals. Refer to our water check with pesticide option for this.

We need to stress that if you're on a municipal or pretreated water system we don't need to test your water to recommend a filtration system for you.

42-000 Water Test (non-city water) $17

The World's Smallest Water Filter is in a Straw

For the first time you can drink fresh, clean water wherever you go. The Clean Sip is small enough to fit in your pocket, yet contains the latest technology in water filtration—a combination of three micro filters, activated charcoal, and a patented, high purity metal alloy. (For use with potable water supplies only.) The result is reduction of chlorine, algae, fungus, scale, and sediment. Life is approximately five 8-ounce glasses of water per day for up to six months, depending on the condition of the water.

42-607 Clean Sip Filter Straw $14

Katadyn Pocket Filter

The Katadyn Pocket Filter is *standard issue* with the International Red Cross and the armed forces of many nations and is essential equipment for survival kits. Manufactured in Switzerland for over half a century, these filters are of the highest quality imaginable, reminiscent of Swiss watches. The Katadyn system uses an extremely fine 0.2-micron ceramic filter that thoroughly blocks pathological organisms from entering your drinking water. A self-contained and very easy to use filter the size of a 2-cell flashlight (10" x 2"), it will produce a quart of ultra-pure drinking water in 90 seconds with the simple built-in hand pump. It weighs only 23 oz. It comes with its own travel case, and a special brush to clean the ceramic filter. The replaceable ceramic filter can be cleaned 400 times, lasting for many years with average use. This filter is indispensible for campers, backpackers, fishermen, mountaineers, river runners, globetrotters, missionaries, geologists, and workers in disaster areas.

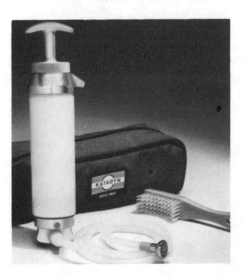

42-608 Katadyn Pocket Filter   **$239**
42-609 Replacement Filter   **$140**

Purwater Reverse Osmosis (RO) Filters

All RO filters will use 4 to 9 gallons of water to produce 1 gallon of clean water. This "waste" water can be used for watering plants or can be cycled back into an unpressurized holding tank.

The process of reverse osmosis was developed at great expense by the U.S. government and is incredibly simple. Tap water is forced against a semipermeable membrane using only water pressure as the power source. No electricity or other energy source is required. When water is applied under pressure, the properties of the membrane allow the pure water molecules, but **not** the pollutants, to pass through.

Reverse osmosis (RO) works like the human kidney. The RO membrane is the ultimate mechanical filter. RO will reduce common impurities, along with salts, metals, total dissolved solids, and hardness (depending on the initial water quality and the water pressure).

Purwater 2

The Purwater 2 is a reverse osmosis system that uses a 2-gallon holding tank with an easy, quick disconnect. It features a built-in flush valve, a carbon post filter, has a 4:1 recovery rate and is compact, fitting nicely on a countertop. It is 12-inches high, 16-1/8 inches wide and 8-5/8 inches deep. There is a 3-year pro-rated warranty on the RO membrane, which is renewed with each new RO membrane installation.

Purwater 4

The Purwater 4 has been tested using NSF standard 58 protocol (documentation available upon request). The Purwater 4 produces up to 9 gallons of purified water daily with a 4:1 recovery rate. It uses a graded density cellulose prefilter to reduce particulate matter. This improves water quality, protects the membrane, and extends its useful life. The Advanced Automatic Shut-Off assures peak efficiency and water conservation. It's totally user-friendly and serviceable—the ultimate in uncompromised purity. It can be permanently installed under your sink easily, so all you see is the chrome faucet on your sink. It also works well with an ice-maker (with the optional ice maker kit). The tank measures 8-inches in diameter by 20-inches. A Purwater 4 with two filters, membrane, head, and shroud measures 12 inches wide by 16 inches high, and mounts on the wall. The Purwater 4 as pictured is using a TFC membrane with a presediment and a precarbon filter, in addition to the postcarbon filter. Please call Randy to determine your filter configuration. There is a 3-year prorated warranty on the RO membrane, which is renewed with each new RO membrane installation. The Purwater 4 includes all you need to hook it up. The costs listed below are out of warranty replacement costs.

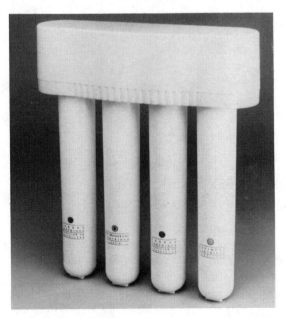

■42-202	Purwater 2	$375
Replacement parts:		
■42-204	CTA Membrane	$75
■42-205	TFC Membrane	$129
■42-206	Carbon Replacement Filter	$18

■42-212	Purwater 4	$639
■42-207	CTA Membrane	$75
■42-208	TFC Membrane	$125
■42-209	Carbon Replacement Filter	$24
■42-210	Presediment Filter	$24
■42-215	Ice Maker Kit	$69

PowerSurvivor 12V Watermaker

Fresh water is a necessity at sea and in remote areas. In many third world countries getting fresh, clean water often poses a problem. The PowerSurvivor is the most compact, most efficient desalinator in the world. Drawing only 4 amps at 12 volts, it will turn sea, brackish, or contaminated water into crystal-pure drinking water at the rate of 1.4 gallons per hour. In an emergency, it can be operated manually with an easy-to-attach handle. It is highly reliable with few moving parts and only takes up one cubic-foot of space. The PowerSurvivor combines reverse osmosis (RO) and energy recovery technology. The Powersurvivor will last approximately 2,500 to 3,000 hours, when it will need an overhaul from the company.

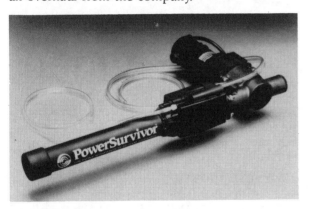

■**42-301 PowerSurvivor 12V Watermaker $1,995**

Survivor 06 Hand Water Purifier

The new Survivor 06 is a hand-operated water purifier that will convert seawater, brackish water, or contaminated water into fresh and pure drinking water. It will produce one cup every 13 minutes at 30 strokes per minute, or six gallons per day. This is the only manually operated desalinator in the world. It uses no batteries, generators, or alternators. It combines reverse osmosis technology with energy recovery technology. A prefilter removes larger particles and debris. The unit measures 5"H x 8"L x 2-1/2"W. It sucks to a maximum suction height of 10 feet and will works under a temperature range of 33° to 120°F. The unit will perform for approximately 1,000 hours.

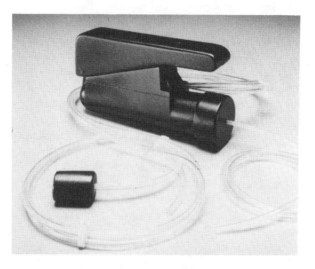

■**42-302 Survivor 06 Watermaker $695**

Floatron Solar Pool Cleaner

The Floatron is a safe and economical alternative to the chemical marinade in which most of us soak in our pools. When floating on the water, the solar panel converts sunlight into harmless low-power electricity. This energizes the mineral electrode, resulting in ionization. Floatron is solid state with no moving parts, no batteries, portable, and cost-effective. The alloyed electrode depletes after two or three seasons (depending upon pool size) and can be replaced in one minute for $50. No more chlorine allergies, red eyes, discolored hair, or bleached bathing suits! Floatron will typically reduce chemical expenses by 80%. The Floatron measures 12-inches in diameter by 6-inches high. A water test kit is included with each purchase as well as a 1-year warranty from the manufacturer. The Floatron is effective for pools up to approximately 40,000 gallons.

The Floatron should not be used in combination with an ozone purification system.
42-801 Floatron $299

Rainshower Shower Filter

The Rainshower is a great shower filter that removes chlorine and other contaminants from your water. It is molded out of ABS plastic and is only 5 inches long, including our finest low-flow showerhead. The new filter will remove 90% or more of residual chlorine from shower water and aids in the control of fungus and mildew through its unique use of nontoxic KDF-55 and its "redox" technology.

Chlorine and residual chlorine is very hazardous to hair, skin, eyes, and lungs. We actually can take in more chlorine from one 15-minute shower (*but don't ever waste that much water!*) than from drinking 8-glasses of the same water in one day. Chlorine plays havoc with our skin and hair, chemically bonding with the protein in our bodies. It makes hair brittle and dry and can make sensitive skin dry, flaky, and itchy.

Rainshower's use of electrochemical oxidation technology makes it effective for thousands of gallons of water, not the hundreds of gallons of the less effective activated-carbon shower systems. The filter converts free chlorine into a harmless water soluable chloride which washes out of the filter. The copper-zinc "redox" mixture can reduce inorganic compounds in the water such as lead and cadmium. Also, zinc is generally regarded as a nutrient for the skin. The product meets FDA standards for copper and zinc in drinking water.

The Rainshower is easy to install and comes with our lowest flow shower head. Savings in water and energy costs can pay for this product in less than 6 months for the average family. A roll of teflon tape is included for a leak-free installation. Each unit comes with a back-flushing adapter. A 1-year warranty is provided.

There is, after all, no point repeating someone else's same old dumb mistakes when you can make interesting new mistakes instead...
- Amory Lovins, in the Foreword

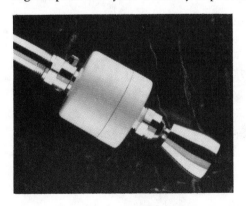

42-701	Rainshower Shower Filter	$79
42-703	Rainshower Replacement Filter	$69

Clearwater Tech Ozone Purifiers

Clearwater Tech makes several different ozone generators for different applications. The ozone is manufactured in the generator by intaking air, which is composed of 20% oxygen (O_2) and bombarding it with a specific light frequency. This frequency causes the oxygen molecules to disassociate and reassemble as ozone (O_3). Ozone is the most powerful oxidizing agent available. When ozone is drawn into the spa or pool water, it will kill bacteria, virus, or mold spores that come in contact with it. Ozone has a life expectancy of approximately 20 minutes. Several short cycles through the day are recommended.

S-1200 Ozone Purifier

The S-1200 features a polished stainless steel reaction chamber, thermally protected self-starting ballast, weathertight (outdoor approved housing), and a 17-inch specially designed high-output ultraviolet lamp. The S-1200 is wired to the pump circuit to be on when your pump is on. The unit has convenient mounting brackets, comes with all necessary fittings, and is easy to install. If the spa is run daily, as recommended by most spa manufacturers, four separate 1-hour cycles in a 24-hour period will generate a sufficient amount of ozone to keep the spa free of biological contamination. The 12-volt S-1200 has the same features, and works the same. It will treat the same capacity of water as the AC model. The S-1200 will purify up to 1,000 gallons for spas and 2,000 gallons for pools.

42-811	S-1200 Ozone Purifier (110V)	$295
42-812	S-1200 Ozone Purifier (12V)	$360

PR-1300 Ozone Purifier

This system comes with its own 24-hour timer and compressor (on the AC system **only**; the 12-volt unit doesn't have the timer). This means that the PR-1300 can run independently of the circulation pump in your spa or pool. It comes with all necessary tubing, check valve, and fittings for installation. There is also a diffuser stone which can be attached to the ozone delivery line and submerged into any vessel of water.

You can treat yourself to a lavish chlorine-free bath by using the system in your bathroom, or anyone else's bathroom since the system is totally portable. You can treat your friends' spa before you use it with this portable water treatment system. For those of you who store your water, whether you have a spring, or a catchment system, treat your water with ozone instead of chlorine, when you are trying to deal with bacteria, iron, or other problems in your storage tank. The PR-1300 features a GFCI (Ground Fault Circuit Interruption) circuit breaker on 110V systems only. This gives state-of-the-art electrical protection. It has a weathertight cabinet, coated with a baked on enamel finish for years of corrosion free service. The UV lamp is encased in a polished stainless steel reaction chamber, and can be replaced by the homeowner in minutes. The compressor rating at 12V is 2.5 psi. Average lamp life is 9,000 hours. Power consumption is 40 watts. The units are rated up to 1,000 gallons for spas and 2,000 gallons for pools. The size is 20" x 9" x 4". Maximum pressure is 20 psi.

42-821 PR-1300 Ozone Purifier (110V) $395
42-822 PR-1300 Ozone Purifier (12V) $435

CS-1400 Ozone Purifier

The CS-1400 is a UV ozone generator designed to be used on swimming pools up to 15,000 gallons or spas up to 2,500 gallons. You can double the output and capacity by adding another CS-1400. It has a polished stainless steel reaction chamber, a thermally protected self-starting ballast, weathertight enclosure (outdoor approved), and a 29-inch specially designed high-output ultraviolet lamp. With no moving parts, the CS-1400 requires virtually no maintenance and will provide years of uninterrupted service. Wire the CS-1400 to the pump circuit, so each will work together to keep your pool or spa perfectly clear. It will also work with a compressor (not included.) The CS-1400 comes with a 1-1/2 inch venturi injector suitable for use with a single-speed pump.

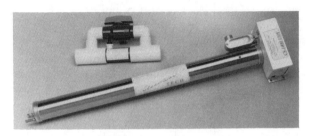

42-831 CS-1400 Ozone Purifier - 110V $545
42-832 CS-1400 Ozone Purifier - 12V $595

Each UV lamp is rated at 9,000 hours. Running the S-1200, the PR-1300, or the CS-1400 for four 1-hour intervals per day equates to 5-years of operation, with a energy consumption worth approximately 85 cents per month.

Sterling Springs Filters By Eco Resource

The Sterling Springs cartridge system has been certified by the Water Quality Association for reduction of chlorine and taste and odor. Utilizing the latest in hydrotechnology, the replaceable cartridge contains four filtration levels and is rated very conservatively for 2500 gallons (but we recommend you change once per year). This filter is easy to install, easy to use, easy to replace, and is available in either the under-counter or countertop models.

Under-Counter (UC) & Countertop (CT) Filters

These units contain a new multifunctional medium called Ecolyte and premium grade activated carbon for water treatment. Proven effective in polishing water supplies to assure good quality drinking water, Ecolyte is an insoluble, inorganic media that employs different modes of action in the removal of contaminants: Ionic adsorption works like a strong cationic ion exchange resin, but with a greater binding capacity. Oxidation/reduction of adsorbed contaminants occurs on the surface of the medium complementing the adsorbtion by GAC. This state-of-the-art technology provides assurance of a safe quality drinking water.

42-615	Sterling Springs CT	$119
42-610	Sterling Springs UC	$149
42-616	SS Replacement Cartridge	$25

Dual Filter

This is an under-the-sink dual filtration system with replaceable Sterling Spring cartridge (as explained above), and a replaceable 5 micron prefilter. It comes with a countertop chrome faucet set. A great system for city water filtration.

42-611	Sterling Springs Dual	$179

Triple Filter

This is an under-the-sink triple-filtration system featuring the replaceable Sterling Spring cartridge, a 5-micron replaceable presediment filter for suspended solids (like dirt and rust), and a recleanable sub-micron ceramic filter. This ceramic filter is being used world-wide to provide quality drinking water. Some of the uses for this filter are for rural encatchment systems and for RVs.

42-625	Sterling Springs Triple	$269
41-132	5 Micron Repl. Filter	$ 7
42-618	Ceramic Repl. Filter	$ 59

Seagull Carbon Filters

To expand your options for the purest, safest water, we are introducing the Seagull IV systems. They will complement the Sterling Springs line, as our high-end cartridge filters. Seagull IV (X-1) products are among the very few qualified for continuous use aboard international airlines. They are manufactured in the U.S.A. from stainless steel and other high grade components. Water is filtered through a unique microfine structured matrix, with a high flow rate of 1-gallon per minute at standard pressures(30-40 psi). The replaceable cartridge should last 9 to 15 months with ordinary use.

Here are some of Seagull IV's unique features:

1. Ultrafine filtration down to 0.1 micron—small enough to remove all visible particles, as well as Giardia, cysts, harmful bacteria, and larger parasites.*

2. Molecular sieving and "broad spectrum" adsorption mechanisms remove chlorine and many organic chemicals such as pesticides, herbicides, solvents, lead, and foul tastes and bad odors.*

3. Electrokinetic attraction removes colloids and other particles even smaller than those removed by microfine filtration.*

Seagull IV X-1F

Under-sink unit, with revolutionary ceramic disc faucet for countertop and accessories for easy installation.

42-650	Seagull IV X-1F	$389
42-657	Replacement Cartridge	$69

Seagull IV X-1D

Countertop unit, faucet-attaching, diverter unit for apartments, cabins, etc.

42-651	Seagull IV X-1D	$359
42-657	Replacement Cartridge	$69

Seagull IV X-1P

Same as X-1D with the addition of a hand-pump high-capacity unit for use on non-pressurized systems.

42-652	Seagull IV X-1P	$389
42-657	Replacement Cartridge	$69

*See performance data sheet for specific contaminants. Certain states may prohibit health claims as a matter of local or state law. Such claims not in compliance are hereby withdrawn. Please check with appropriate officials as necessary.

First Need

The First Need is the best-selling portable water purification system in America. Weighs less than 16 ounces. The same removal effectiveness as the Seagull IV systems. Great for camping, backpacking, and travel.

42-653 First Need $55

Trav-L-Pure

The almost legendary effectiveness of Seagull IV and First Need drinking water systems is now available in Trav-L-Pure, a self contained, ultra-convenient configuration. Incorporating 150-micron rough screening, 15-micron prefiltration, and 0.1 micron filtration into a single, self-sealing container, one-hand operation makes the Trav-Ler the easiest system, ever, to use. Also available with a Cordura traveling case. Dimensions: 6-5/8"x4-1/2"x3-3/8".

42-654	**Trav-L-Pure**	**$99**
42-655	**Trav-L-Pure w/Carrying Case**	**$119**
42-656	**Replacement Canister**	**$29**

Ten-Year Warranted Chemical Removal Filter

We are happy to add the Aqua-Pure under-the-sink Chemical Removal filter to our product line. It comes with a 10-year warranty. It reduces chlorine and chloramines by 99%, and will reduce over 95% of many hazardous volatile organic chemicals including THMS, PCE, and more. It also reduces dirt, rust, taste, and odor. This filter has been tested and listed under NSF (National Sanitation Foundation) Standard 53 for the reduction of volatile organic chemicals (VOCs). Here's how it works: Water is first passed through a 5-micron filter, which screens out microscopic particulate matter. The water is controlled by a flow regulator and then flows through an activated carbon depth bed along an extra long flow-path. This maximizes water contact time in the carbon bed. The flow regulator assures that the water remains in contact with the largest amount of chemical-adsorbing carbon for the maximum amount of time. Adsorption is the most effective method for reducing organic chemicals, and this carbon bed contains the equivalent of many acres of chemical removing surface area. Industrial spills and agricultural spraying can seep into ground water and eventually into water supplies. This system is for private wells as well as city water supplies where the threat of hazardous VOCs might be suspected. The filter cartridge lasts a family of three for 5 months. It is certified for 200 gallons.

■**42-213**	**Chemical Removal System**	**$179**
■**42-214**	**Replacement Filter**	**$29**

Breathe Easy with a Panda

Recent EPA studies have found that air inside a typical home is some 3 to 6 times more polluted than the air outside. It is suggested that indoor air pollution will be the largest emerging environmental health issue in the next 5 to 10 years. Most of the months of the year find our homes with windows closed to keep heat either in or out. In our relaxed comfort we are inhaling formaldehyde from drapes, carpets, pressed wood, and insulation; nitrogen oxides and sulfur dioxide from gas and wood stoves; ammonia from cleaning materials, bathrooms, and soiled diapers or cat boxes; VOC's from plastics and dry-cleaned clothes, as well as molds and mildew, viruses, fungi, and bacteria. We are convinced that allotropic oxygen (ozone) is by far the best way to purify and cleanse indoor air. According to Ed McCabe in his book O_2xygen therapies, "Pure cold plasma ozone stimulates the immune system in low concentrations," and "the use of ozone as a healing agent is being used in clinical practice, because it manifests bactericidal, viricidal and fungicidal actions." Because of the reception of our air purification equipment, we are introducing the Panda home purification device. This less than 10 pound system, with its variable output control, can duplicate outside ambient ozone levels, about 0.03 parts per million, in a home or a single bedroom.

The Panda is for a medium-size home or apartment, and the Panda Plus is for a larger home. Both systems will react with offensive odor-causing molecules in the air and reduce them to their basic, naturally occurring components. After the air is cleaned, these systems will continue working to abate the source of most odors, and the bacteria, whether airborne or settled. The Pandas are almost maintenance-free, and contain a patented self-cleaning capability that will ensure years of like-new operation. They are built to run continually, consuming only about 40 watts of power.

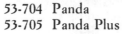

53-704 Panda $475
53-705 Panda Plus $595

Solar Car Air Cleaner

Our Solar Catalytic Car Air Cleaner installs instantly on your dash or back deck. A monocrystal silicon solar cell powers a soundless fan, which pumps fresh air through a catalytic/activated carbon filter. Space age design and technology will run for years, eliminating tobacco and pet odors.

64-240 Solar Catalytic Car Air Cleaner $19

To order any of these products
or for more technical information
call us toll-free at
1-800-762-7325

Inline Sediment Filter

Sometimes your water isn't dirty enough to mess with fancy and expensive filtration systems and all you need is a simple filter. Our inline sediment filters accept standard 10-inch filters with one-inch center holes. They are designed for cold water lines only and meet National Sanitation Foundation (NSF) standards. Easily installed on any new or existing cold water line (don't forget the shutoff valve), they feature a sump head of fatigue-resistant Celcon plastic. This head is equipped with a manually operated pressure release button to relieve internal pressure and simplify cartridge replacement. They're rated for 125 psi maximum and 100°F. They come with a ¾-inch FNPT inlet and outlet and measure 14" high by 4-9/16"in diameter. It accepts a 10-inch cartridge and comes with a 5-micron high-density fiber cartridge.

41-137 Inline Sediment Filter   **$49**

Water Filters

Our rust and dirt cartridge is made of white cellulose fibers with a graduated density. These filters collect particles as small as 5 microns (2 ten-thousands of an inch). These are NSF-listed components that take a maximum flow of 6 gpm. Our taste and odor filters are made with granular activated carbon. These filters effectively remove chlorine, sulfur, and iron taste and odor. Maximum flow is 3 gpm. Note: filters should be replaced every 6 months to prevent bacterial growth or as needed. This is the cartridge to use with the inline sediment filter.

41-138 Rust & Dirt Cartridge (2)   **$14**
41-436 Taste & Odor Filter (2)   **$34**

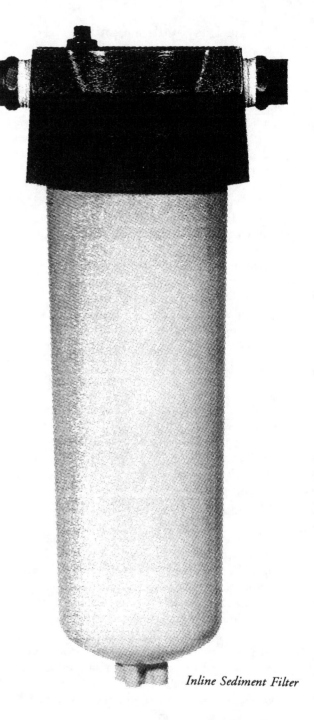

Inline Sediment Filter

Non Toxic Household Products

Ossengal Stick Spot Remover

Ossengal Stick is wonderful stuff. It gets rid of spots and stains better than anything we've ever tried—**and** it's organic, odorless, free of industrial chemicals, and easy to use. It's made from the purified gall of oxen by the Dutch, and when you use it you can understand why they're famous for getting things clean. We asked a chemist why Ossengal works so well. He sent a page of words like "short chain organic acids, phospholipids, enzymes..." The gist of it seems to be that an ox needs tough-acting stuff to break down what it eats, and it acts the same way on spots and stains on fabric.

Ossengal is a pure white stick of soaplike substance, packaged in a small pocket-size pushup tube. To use it you just moisten the spot, rub it well with the stick, brush or knead it in, then rinse. Most stains just disappear.

54-101 Ossengal Stick Spot Remover $7

Citra-Solv Natural Citrus Solvent

Citra-Solv is one of our favorite products—it will handle nearly all your cleaning needs. It dissolves grease, oil, tar, ink, gum, blood, fresh paint, and stains, to name just a few. It replaces carcinogenic solvents, such as lacquer and paint thinner, toxic drain cleaners and caustic oven and grill cleaners—*you won't believe how well it cleans your oven!* Use Citra-Solv in place of soap scum removers, which use bleaching agents. Citra-Solv is composed of natural citrus extracts derived from the peels and pulp of oranges. It is highly concentrated and can be used on almost any fiber or surface in the home or workplace *except plastic.* Citra-Solv is 100% biodegradable, and the packaging for Citra-Solv is 100% recyclable.

"Citra-Solv is the greatest cleaner we have ever used. It removes bubble gum from the carpets, makes old asphalt look like new and we're even using it to clean the jungle gyms." Margaret Foust, Keaau, HI

54-140 Citra-Solv (16 oz. bottle) $8
54-141 Citra-Solv (32 oz. bottle) $13

Oasis Biocompatible Laundry Detergent

Wash water containing Oasis Laundry Detergent can be used in the garden without harming plants. It can even produce better growth than plain water, because it biodegrades into a mix of essential plant nutrients. By running a hose from your washing machine to your garden, you can capture up to 1,000 gallons a month headed right down the drain. Your clothes will come out clean (we tested it) and your fruit trees will grow stronger with Oasis! Concentrated, one gallon does 64+ loads. (1/8 to 1/4 cup per load) *Contains no phosphates.*

For an energy-saving Solar Clothes Dryer, see page 355.

54-151	Laundry Detergent (quart)	$8
54-152	Laundry Detergent (gallon)	$24

Oasis Biocompatible All-Purpose Cleaner

A true superconcentrate, Oasis is excellent for hand dishwashing, hand soap, and general cleaning. Perfect for camping, it can also be used for hand laundry and diluted for body soap or shampoo. Like Oasis Laundry Detergent, it biodegrades entirely into plant nutrients with no plant or soil toxins.

54-155	Oasis All-Purpose Cleaner (quart)	$7
54-156	Oasis All-Purpose Cleaner (gallon)	$21

Greywater Information Booklet

This is an excellent small booklet on just about anything you ever wanted to know about greywater disposal and usage. It contains all the information you need to construct and use a greywater system. Written by the folks at Oasis, it includes the guidelines to the approved use of greywater from the Goleta, California Water District in the appendix.

80-203 Greywater Information Booklet $5

Orange Air Freshener

This Mia Rose fragrant spray will purify your air while you freshen it. This real citrus non-aerosol functions much like an ionizer machine in that each droplet contains millions of active electrical charges (ions) that attract and neutralize offensive odors. It also elimates airborne bacteria and smoke. Contains no artifical ingredients or additives, only essential oils distilled from citrus fruits. The spraying mechanism is a spray pump that uses no fluorocarbons. This is very effective on all household odors. There is no animal testing and Orange Air Therapy is USDA approved. This 16 fluid ounce size will last for at least 6 months under normal usage, making it very economical.

54-143 Orange Air Therapy $15

Citri-Glow Cleaning Concentrate

Throw out those poisonous household products: new Mia Rose Citri-Glow is a total home cleaner that really works. It's been formulated with the health of your family and the preservation of our natural environment in mind. Use it in your kitchen and bathroom with perfect safety: it's nontoxic, noncaustic, color safe, bleach-free, with no animal testing or products used. It's amazingly effective and economical: one quart makes up to 5 gallons. We don't usually go in for hyperbole, but Citri-Glow is truly miraculous.

54-142	Citri-Glow (quart)	$7
54-148	Citri-Glow (gallon)	$16

Aquasave Waterless Carwash

It can take 100 gallons of water to wash a car. It takes only 3 to 5 ounces of Aquasave 2000—and no water at all. Is your water rationed? Then this product is absolutely essential. Hoseless? No problem. And even in a water wonderland, why squander and pollute a precious resource, when Aquasave *does the job better?* Just spray it on and wipe it off. It cleans gently but superbly, emulsifying dirt, bugs, even tar, so that they release easily from the surface and adhere to the towel. The lubricating formula protects the finish, helps cover fine scratches, and adds a brilliant gleam. Use it on your boat, motorcycle, RV, or around the house. Aquasave is successful and safe on paint, vinyl, glass, mirrors, chrome, stainless steel, finished wood, stove tops, and more. It's nonflammable, nontoxic, and contains no phosphates.

54-172	Aquasave (16 oz.)	$9
54-173	Aquasave (64 oz.)	$25

Yellow Jacket Inn

Since so many of our customers live in rural situations, we thought it appropriate to offer some relief from the yellow jacket menace. These reusable traps capture incredibly large numbers of the stinging pests—enough to make outdoor eating a pleasure again. They use natural bait such as tuna (only the dolphin-safe variety!) or chicken (or chicken-scented tofu). They're ideal for taking camping or fishing as they weigh only 6 oz. and can be reused over and over. These traps are made by Seabright Laboratories, a company dedicated to non-toxic devices and humane ways of dealing with pests. Seabright is very environmentally conscientious and ships with popcorn instead of styrofoam!

"We have used numerous products and pest control services without much success. We then set out about three dozen of your roach traps and caught over a thousand roaches in a few days. It is now three months since we began using your traps and I am pleased to say, we are 'roach free' for the first time in memory!" - Sister Jean Marie, St. Vincent De Paul Dining Facility, Oakland, CA.

54-201 Yellow Jacket Inn   **$6**

Humane Mouse Trap

Often, the first interaction that a child has with "wild" animals is a mouse caught and deformed in a standard mouse trap. The Smart Mouse Trap teaches a child the idea of working in harmony with animals and the ecosystem. This is a humane and effective trap, a forgiving answer to unwanted visitors. Why kill? For a little more effort, you can take the mouse to the nearest brush or wooded area and release it to live out its tiny life. It is a joy to feel the shared compassion with a child watching a trapped mouse escape to freedom. This trap is simple to set—the bait is a soda cracker, it will catch mice that standard traps can't and you can use it again and again. The trap is make of Kodar plastic with two stainless steel springs. It measures 2 by 3 by 7 inches.

Cockroach Traps

Rid your home or shop of cockroaches without using poisons. These effective traps are nontoxic and safe for pets, children, and adults. Used in an uninfested dwelling, they help prevent invading insects from establishing colonies. Place them where roaches tend to congregate: under kitchen and bathroom sinks; under the refrigerator and behind the stove; in the pantry; near or under a warm appliance such as a TV, stereo, heater, or air conditioner. The traps contains natural food bait to lure roaches inside where they are stuck in to an adhesive. Discard and replace the traps after 2 months or as they fill with roaches. The trap's active ingredient is 52% boric acid. Twice as effective as the Roach Motel. Awarded the Good Housekeeping Seal of Approval. Three packages of two traps each.

54-205 Roach Traps   **$11**

54-202 Smart Mouse Trap   **$10**

Moth-Proofing Cedar Products

Cedar keeps moths, fleas, ticks, and silverfish out of drawers, suitcases, garment bags, and closets. And it lasts a lifetime—light sanding renews the fresh cedar scent. No more nasty naphtha odor! Nontoxic to your family and pets. The cedar comes from nonendangered trees grown on private lands in the Ozarks. Blocks measure 2½" by 1½" by ¾", balls are 7/8" in diameter; scatter them in dresser drawers. Closet protector hangers, 3" by 9" by 3/16", have a rotating brass-plated hanger.

54-120	Cedar Blocks (12)	$11
54-121	Cedar Balls (50)	$11
54-122	Cedar Hangers (6)	$14

AirSpray Bottle

The average American has 46 aerosol cans around the house, according to the EPA! Many of the gases used in today's aerosol cans are as harmful to the ozone layer as CFCs. Our Airspray Bottle is a nongas spray bottle that totally eliminates the need for gas-filled aerosol spray packs. The simple solution uses compressed air as the propellant that you pump up easily by hand. The bottle is refillable and reusable. It can be used for a wide range of products: household cleaning products, hair spray, deodorant, and dust-off products in computer maintenance or photography. This product won the 1989 best packaging award for Europe. We're impressed at how simple it is to use.

| 54-144 | AirSpray Bottle | $7 |

Be sure to see our Tool Section for enery-saving appliances.

Bat House

Our sturdy, pine, weather-resistant bat houses will comfortably house 30 bats, capable of consuming 30,000 harmful insects in one night. The house (9.5"W by 16.75"H by 8.5"D) contains four internal chambers with perching grooves that are accessed from under the bottom of the house, which easily attaches to a tree or building. The open bottom discourages squirrels and birds from moving in.

America's Neighborhood Bats

This 128-page book by Merlin Tuttle is a wonderful companion to your bat house. It provides a wealth of helpful information on bat behavior, biology, and habitat. It includes range maps, a source list, and easy-to-understand text. Spectacular color photos!

80-830 Bat Book   **$12**

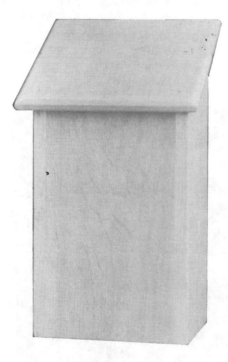

54-210 Bat House   **$29**

To order any of these products
or for more technical information
call us toll-free at
1-800-762-7325

Recycled Products & Recycling Aids

Americans produce 154 million tons of garbage every year—enough to fill the New Orleans Superdome from top to bottom, twice a day, every day. 50% of this trash is recyclable!
—50 Simple Things You Can Do to Save the Earth.

Recycling saves energy, thus reducing acid rain, global warming, and air pollution. Recycling paper uses 60% less energy than manufacturing paper from virgin timber. Recycling a glass jar saves enough energy to light a 100-watt light bulb for 4 hours. 75,000 trees are used for the Sunday edition of the *New York Times* each week, yet only 30% of newspapers are recycled in the United States. If we all recycled our Sunday papers, we could save over 500,000 trees every week.

Recycling cuts down on landfill. The average American throws away 4 pounds of garbage per day. By 1994, half the cities in the U.S. will run out of landfill space. By recycling it is possible to cut our waste stream by 80%.

If you need information on recycling programs in your area call the Environmental Defense Fund at 1-800-CALL-EDF.

Global warming is the number one problem facing our world today. In 50 years, from 1937 to 1988, the world's annual energy consumption quintupled from 60 to 321 quads (quadrillion BTUs), and global emission rates for carbon mushroomed from 1 gigaton (one billion metric tons) to 6 gigatons per year.

Recycled Paper Products

It takes an entire forest—over 500,000 trees—to supply Americans with their Sunday newspapers every week. —50 Simple Things You Can Do to Save the Earth.

Consider a few telling facts: Each ton of recycled paper produced saves approximately 17 trees. Each ton of recycled paper produced saves approximately 4,102 kWh of electricity compared to virgin paper production which is enough to power the average home for 6 months. The manufacture of recycled paper requires 7,000 gallons less water per ton compared to virgin paper. The manufacturing of recycled paper reduces overall emissions of air pollution.

Americans use 50 million tons of paper annually, which means we consume more than 850 million trees, or 580 pounds of paper for each American. We know that waste paper and paper products constitute nearly half of all municipal solid waste. Making new paper from old paper uses 30% to 55% less energy than making paper from trees—and it reduces related air pollution by 95%.

We've researched the market for a manufacturer who uses a maximum of post-consumer waste in making recycled paper products and are happy with our choice of the "Envision" line of toilet paper, paper towels, and facial tissue. These products are made from virtually 100% recovered materials!

Toilet Paper

Made from 100% recycled paper, and made from 100% post-consumer waste, this two-ply toilet paper is unbleached and dioxin-free. It has 500 sheets to the roll compared to 300 for standard virgin toilet paper. Because of excessive weight, we must charge extra for Alaska and Hawaii.

51-102	Toilet Paper Full Case (96 rolls)	$55
51-101	Toilet Paper Half Case (45 rolls)	$29
51-107	Toilet Paper Sampler (12 rolls)	$8
05-221	Add'l freight (case to HI & AK)	$15
05-222	Add'l freight (1/2 case to HI & AK)	$8

Paper Towels

Made from 100% recycled paper, these two-ply paper towels are very strong and extremely absorbent. They come 100 sheets to the roll; each one 11-inches by 9-inches.

51-104	Towels Full Case (30 rolls)	$34
51-103	Towels Half Case (15 rolls)	$18
51-108	Paper Towel Sampler (6 rolls)	$8
05-223	Add'l freight (case to HI & AK)	$10
05-224	Add'l freight (1/2 case to HI & AK)	$6

Facial Tissue

Made from 100% recycled paper, these two-ply facial tissues are extremely soft and gentle. These tissues come 100 sheets to the box.

51-106	Facial Tissue Full Case (30 boxes)	$26
51-105	Facial Tissue Half Case (15 boxes)	$14
51-109	Facial Tissue Sampler (6 boxes)	$6

Cotton String Shopping Bags

Made from 100% cotton and manufactured entirely in the USA, our string bags are the finest on the market, made of a heavier construction than just about all that you've seen. Over the years we have tested over 100 different string bags. To be considered they must withstand the rigorous "JS Rip Test" (John pulls and tries to tear). 95% were rejected for failing this test alone. These will easily hold two to three times the weight of a common grocery store plastic bag and a little more volume. Under stress (dropping, etc.) they will greatly outperform either plastic or paper. The soft cotton handle is comfortable even when fully loaded. They're also ideal for picnics, trips to the beach, and even for storing food in your kitchen. When not in use they fold up easily and can be carried in your pocket.

51-901 String Bag   **$5**

All-Purpose Cotton Garment Bag

The garment tote is made from 100% natural 8-ounce cotton canvas. It holds up to 12 garments on hangers and doubles as a clothes hamper or laundry duffel bag. No more plastic wraps or nylon bag. This easy-to-clean and easy-to-carry bag is all you need.

51-909 Garment Tote   **$26**

Canvas Shopping Bags

It takes one 15 to 20-year old tree to provide 700 paper shopping bags, and in a big super-market 500 paper bags could be used every hour according to "Save A Tree". This means that 8 to 9 trees per store are destroyed every day in the name of convenience. Any item purchased and sacked can go into our canvas bags. We've been deluged since Earth Day 1990 with samples of canvas bags sprouting up from new cottage industries across the country, so we were able to pick the best. Our bags are made of 100% cotton canvas in 10 ounce weight and natural color to avoid chemical dyes. The bags measure 8"D x 13"W x 17"H. The top hem is double sewn to provide the handles with maximum strength for heavy loads. They are designed with a flat bottom which allows them to stand up for easy filling, just like paper.

51-902 Canvas Shopping Bag   **$9**

Reusable Lunchbags

Our nylon lunchbags eliminate the need for paper bags. They're washable and include an easy to seal velcro closure. Each bag is decorated with an attractive earth logo. Cheaper and easier to use than lunch boxes! Green and turquoise background with a black border around the earth. Approximately 12"H x 6½"W x 4"D.
51-221 Reusable Lunchbag   **$7**

110   REAL GOODS

Canvas Lunch Bags

Our new 100% cotton, 10-ounce unbleached canvas lunch bags provide an attractive and ecological alternative to lunch boxes and throwway paper sacks. Complete with a velcro closure, carrying handles, and a forest green trim, these bags measure approximately 12"H by 6"W by 4"D. Seven percent of the proceeds from the sale of each bag is donated to Save the Earth.

51-222 Canvas Lunch Bag   **$7.75**

Reusable Coffee Filters

Our dioxin-free, 100% cotton muslin reusable coffee filters contain absolutely no chlorine. They replace standard throwaway filters and will last approximately 2 years with daily use. They are easy to wash in hot water, they won't tear, and stains can be easily removed by an occasional soaking in boiling water with baking soda. Sold in packages of two. The basket filter is for automatic coffee makers. The #2, #4, and #6 are Melita sizes.

51-211	#2 Filter (2 ea.)	$5.50
51-212	#4 Filter (2 ea.)	$6.50
51-213	#6 Filter (2 ea.)	$7.50
51-214	Basket Filter (2)	$8.75
51-215	#2 Cone	$5.50
51-216	#4 Cone	$6.50
51-217	#6 Cone	$7.50
51-218	Mini Basket	$7.75

To order any of these products
or for more technical information
call us toll-free at
1-800-762-7325

Recycling Bins

Our reinforced space-saving recycling bins are ideal for home recycling. They come with easy-grip handles and the legs lock securely for safe stacking. Each bin comes with a drop front to allow access when stacked. Bins are made from a tough polyethylene that won't rust, corrode or mildew. Get one for aluminum, one for newspaper, and one for glass or plastics. Dimensions: 20-7/8"L x 15-1/8"W x 13-3/8"H.

51-201 Recycling Bins (each) $14
51-203 Recycling Bins (set of 3) $39

Aluminum Can Crusher

Our new can crusher allows more than five times as many flattened cans to fit in the same storage space as regular uncrushed cans. Constructed of heavy gauge steel with a baked enamel finish. It mounts very easily in your kitchen or next to your soda machine at work. It measures 14"H x 15"W x 4-1/2"D . The large compaction chamber accepts 12 oz. or 16 oz. cans without adjustment. Mounting screws not included.

Multi-Crush Can Crusher

Try as you may to limit your purchases of canned drinks, the empties seem to pile up as if they were mating and reproducing. Multi-Crush reduces the size of the problem. It makes recycling aluminum cans faster and easier than any other can compactor. The unit holds up to six cans at once, automatically feeding them into the crushing chamber and ejecting without interruption. Just mount it on the wall and place your can receptacle beneath. Multi-Crush features all-steel construction and a lifetime guarantee.

51-210 Multi-Crush Can Crusher $24

51-209 Can Crusher $19

Paper Boy Newspaper Recycling Aid

Here is an inexpensive, easy to assemble newspaper holder and bundler. It is a durable product made from recycled, biodegradable cardboard material and makes recycling newspaper easy. The Paper Boy has a storage compartment for scissors and twine. Its unique design makes bundling easy. It suspends the stack of newspapers so the bundler can reach under it from four directions to place twine. Then, tie the knots, remove the bundle and Paper Boy is ready for the next stack of newspapers. Does not include twine or scissors.

51-205 Paper Boy   **$6**

The Earth Pencil — From the Blackfoot Tribe

The natural dignity of Native Americans, and their concern with elemental principles, are beautifully expressed in the Earth Pencil, a basic writing instrument made by Blackfoot Indians in the shadow of the Rocky Mountains. Crafted of incense cedar wood, without varnish or paint, even the sawdust is recycled as fuel for the factory heating system. The lead is a combination of natural graphite, clay and wax. Available in sets of four dozen pencils with rawhide binding; or one dozen in a natural cedar box.

51-936 Earth Pencils (48)   **$13**
51-937 Earth Pencils (cedar box)   **$15**

Recycled Map Stationary

What happens to the old topographic maps printed by U.S. and Canadian map companies? Until now, they ended up as solid waste in landfills! We're recycling misprinted and outdated maps into beautiful and unique stationery. Our package includes 50 envelopes and 100 sheets of stationary packed in an attractive 10" x 13" map envelope.

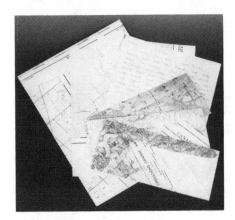

51-904 Map Stationary   **$14**

EnvyLope Envelope Maker

EnvyLope makes perfect envelopes from any kind of paper like recycled gift wrap, kids' drawings, computer paper, and more. Complete with instructions, template, nontoxic glue stick, and address labels. All you need are scissors, paper, and about a minute to make a unique eye-catching envelope, 4 inches tall by any width. Great for a school project or rainy day activity.

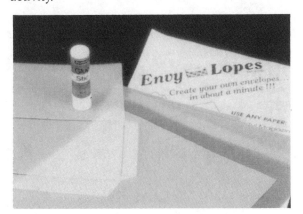

51-206 EnvyLope   **$6**

There's an old Russian joke about the guy who asks whether he can buy various hard-to-find goods. Tired of being always told, "In principle, yes," he exclaims, "Sure, but where's this [Principle] shop that you keep talking about?"
- Amory Lovins, in the Foreword

Notes and Workspace

Shelter

For many, the creation of a dream house is the most important (and expensive) tangible accomplishment of a lifetime. Our children grow and establish their own households, but our home remains to accommodate future generations. Questions of where and how to shelter our families reach deep into the human psyche, and touch both the practical and spiritual sides of our lives. After all, a home is the very essence of being and the wellspring from which mood, comfort, and creativity emanate.

With humans, however, only a few of the species are builders. The rest are nomads, taking refuge in what was built before. Henry David Thoreau put it well in *Walden*:

There is some of the same fitness in a man's building his own house that there is in a bird's building its own nest. Who knows but if men constructed their dwellings, with their own hands, and provided food for themselves and families simply and honestly enough, the poetic faculty would be universally developed, as birds universally sing when they are so engaged? But alas! We do like cowbirds and cuckoos, which lay their eggs in nests which other birds have built.

We have all observed the childlike excitement of owner-builders deep in reflective thought, or watched them stroll through just-framed house fantasizing details of future rooms. Only 20% of the new houses built in America today are constructed by owner-builders. The rest are created by mass-production house factories that speculate on the demand for housing as others speculate on the stock market.

The state of shelter is really the state of humankind's dominance over, and mastery of, the environment. By most accounts we have not fared well in recent years. Our earliest ancestors improvised shelter from available materials or slept under the stars. With ever-growing populations, expanding agriculture, and the advent of metal tools, our shelter needs were inexorably altered. Familial knowledge of building crafts, formerly passed on from generation to generation, were lost to industrialization. Desirable building materials have become scarce, victims of our shortsighted squandering of natural resources.

Our ancestors built with readily available local materials: grasses, leaves, twigs, canes, mud, animal skins, saplings, and turfs. These made excellent, well-insulated structures. Today, however, the products made for the mass market are prefabricated from unpronounceable compounds that had their origins in the sands of the Mideast. Their designs do not accommodate the nuances of individual homesite or local climate, but rather compensate with power plants that can adjust the internal environment as needed with the simple addition of an appropriate number of megawatts.

The statistics associated with inefficient home design are staggering: American houses currently use $40 billion of fuel and $75 billion of electricity every year. Every year, $13 billion escapes through the holes and cracks of our residential buildings (according to the American Council for an Energy Efficient Economy). Had we continued on the path of energy conservation that we set forth on in 1973 in the wake of the first Arab oil embargo, we could have saved enough energy to have completely eliminated the national deficit by the turn of the century.

But we didn't. We went right back to our wasteful energy consumption habits, while economic rivals Japan and Germany rigorously applied available technologies to their homes and manufacturing processes. As a result, we will enter the twenty-first century strangled by debt and as dependent on foreign oil as we were before the embargo. Germany and Japan will increasingly control the world's wealth.

We must think about shelter from the drawing board stage. Locating local, efficient, intelligent, and natural resources is something of a lost art, but is essential to the success of the process. Only those with the luxury of designing structures from the ground up have the option of incorporating local building materials with the wisdom of local stewardship. The trick is to combine these attributes with the technical knowledge and material developments that the late twentieth century has afforded us. It is a process that is intensely complex, but immensely rewarding; one that every one should experience at least once in their lives.

Chapter 2, "The Efficient Home," showed

cost-effective ways to cut energy consumption in an existing home. The person designing a new home from scratch, however, has a once-in-several-generations opportunity to achieve resource efficiency. The technologies are changing fast. There are new structures on the market that weren't around just 14 months ago when we were writing our last *AE Sourcebook*. Advances in underground housing, for instance, are staggering, making that option more cost-effective and elegant than ever.

This is the first time the *Sourcebook* has included a section on shelter. We promise that our emphasis in this area will grow in coming years. We're working on making available shelter-related products as well, from recycled structural building materials to nontoxic carpets and paints. (We've even worked extensively with one customer in Los Angeles to create a 100% energy-independent home built of steel recycled from American cars of the 1950s and '60s!)

This chapter contains the shelter-related products that we currently sell or represent. While we promise to continually increase our offerings, we can never offer the full roster of materials to enable you to create your energy-efficient dream home. In the Knowledge section in this *Sourcebook* you will find many references to help you in the practical application of the most ancient of processes—building a comfortable house.

– J.S.

POWER DOWN

The Stone Foundation

The numbers seem outlandish now. Even having lived with them for most of my life, they require slow repetition. Two hundred sixty acres, eight thousand dollars.

In 1970, my wife and I stared at a map. The world had gotten very crazy, out of control.

Our institutions, now derisively referred to as "the establishment," were taking it on the chin, and constant bombardment about Vietnam, Black Power, Kent State, drugs, and corrupt politicians made things seem worse than they really were. We had to get out.

The goal was to find a place that offered sanctuary—peace, quiet, honest people, and not too many of them. "There," we decided, pointing to a lobsteresque jut of land extending out into the North Atlantic. That's where we should go.

Nova Scotia.

Neither of us had been there. In fact our collective sum of knowledge about the place was close to zero. But someone, a friend of a friend, told us that it was very beautiful—"like Vermont surrounded by the ocean"—and that land was very cheap.

We spent months researching, first with books, but later by time. We learned the valuable lesson that if you want to move someplace, you have to be willing to invest the time. Our investment paid off when we found 260 acre parcel overlooking the valley on one side, the Bay of Fundy on the other. It was mostly wooded, with 18 cleared acres and a farmhouse that looked as if it had been lifted from Andrew Wyeth's famous painting, "Christina's World." The price was $8,000.

The farmhouse had seen better days. Let me be completely honest; the farmhouse was a wreck. The farmer who owned it had stored grain in it, and collapsed the floor. The sills were rotten and the north wall of the foundation had fallen in; the windows were all out and the chimneys were on the verge of collapse. Only a fool would attempt such a massive restoration. Or two fools, in this case.

The first order of business was the foundation. It was made of unmortared fieldstone, in a wide variety of sizes and shapes. Getting started was difficult. I had never built anything of stone. In addition to my lack of experience, I also had no tools and no help and no money.

We lived far, far from the grid. Every morning we would awake in the shelter of our wooden tent, as we called the farmhouse, and after a breakfast that featured luxuries such as hot water, it was down into the hole. I spent the day cursing and wrestling, becoming intimate with inanimate objects that unfailingly obeyed the laws of physics that I was only beginning to understand. An entire day could be spent on fewer than ten rocks. Putting the three-dimensional puzzle together was as stimulating as it was exhausting. At the end of each day I was sore, tired, and dirty. By the next morning I couldn't wait to go back into the hole.

Occasionally, friends would come over to marvel, not at my handiwork, but at the sheer magnitude of our obstinence. There were a thousand "strong back, weak mind" jokes at my expense, many of them made by me. When summer ended, however, and we closed up for the season, I felt an overwhelming sense of pride that the farmhouse, for the first time in years, was weathertight.

The Nova Scotia dream did not materialize as planned. The Canadian government, figuring that our real interest in emigrating was their free medical coverage, got in the way. So did other tentacles of reality, such as earning a living. There wasn't much of a market for one-man fieldstone foundation crews.

The people who knew me were bemused at the turn of events. After all, I spent a summer of back-breaking labor in a damp, dark environment where it was me and the stones, all for the sake of a wreck-of-a-farmhouse that was, despite my efforts, still a wreck-of-a-farmhouse. Now we couldn't even live there.

We sold the place eventually and made a tidy profit that became the down payment on our next house, then the next, then the next. We now live mortgage-free, and it's all because of the growth of that first investment. As for the farmhouse, the subsequent owner had the good sense to tear it down, but I am proud to say that the foundation survived several winters without once giving in to the frost. "It's a pity," said my friends, "that after all that work, you never got to live in that house."

But they are wrong. We experienced the best of that house. We came to appreciate how the structure had been situated on its site, how the trees had been planted for shade and windbreak, and how important the use of the lever and the inclined plane were to the preservation of the human back. And we learned that good shelter needs good design more than it needs unlimited power. The only pity is that it took that summer for me to learn something that everyone should learn much earlier in life—the importance of good shelter.

— Stephen Morris

When we live close to our natural surroundings, we come to know and love them and to build in ways which reflect our sense of joy in being part of them. Our buildings connect us to, rather than isolate us from, the natural forces of the place, and they take their own form from the special spirit of the region which arises from its unique climate, geography and community of living things. Their palette of color is attuned to the space-filling white light of snow country, the pastels of fog country, the green light of the forest, or the golden sunsets of the tropics. They know their world and are fully a part of it.

Tom Bender,
In Context

This western Michigan home is snug and bright through the winter on its stone foundation and with its solar array providing power. Our thanks to Richard Orawiec.

Designing an Environmentally Responsive Home

Solar Design Associates is a leading pioneer in the design of sustainable buildings. Over the past 15 years, they have completed dozens of energy-efficient and energy-independent residences and other buildings across the country, many of which are powered by photovoltaics. We asked Steve Strong, Solar Design's president, to share some thoughts on designing what he calls the "environmentally responsive home". Real Goods has forged a working alliance with Solar Design Associates so that together we can offer you the most professionally designed and engineered system available anywhere.

The methods of housing design and construction in popular use today leave a great deal to be desired. Houses commonly being built seldom respond to the site, most often curtail your spiritual freedom, are woefully inadequate in an energy engineering analysis, consume resources, create pollution, and almost never improve the landscape.

Many understand these shortcomings, for they are often quite obvious. How, then, can we achieve the freedom and environmental harmony that we all seek?

What is needed is an intelligent approach to the overall design process where your home and its support systems are designed together to function together in an integrated manner. In this way, each enhances the other to produce a comfortable, environmentally responsive living environment that consumes a minimum of nonrenewable resources.

Sustainability

Each year, Americans **waste** more energy through inefficiency than two-thirds of the rest of the world uses. As a country, we presently consume the vast majority of the world's resources and produce the vast majority of the world's pollution. This is simply not sustainable. We are rapidly running up against finite supplies of energy and resources while simultaneously realizing the limits of our planet's

Solar-powered, all-electric, environmentally-conscious residence in Carlisle, MA

Building a sustainable home is a major step you can make toward building a sustainable society.

ability to absorb waste and pollution.

The built environment is responsible for a substantial part of this problem, and architects are substantially responsible for the built environment.

Fortunately, the last 15 years have brought many changes to the design profession. Architects with vision have come to understand it is no longer the goal of good design to simply create a building that's aesthetically pleasing - buildings must be environmentally responsive as well.

They have responded by specifying increased levels of thermal insulation, healthier interiors, higher-efficiency lighting, better glazings and HVAC equipment, air-to-air heat exchangers, and heat-recovery ventilation systems. Significant advances have been made and this progress is a very important first step in the right direction.

However, it is not enough for society to continue to enjoy the comforts of the late twentieth century; sustainability must become the cornerstone of our design philosophy. Rather than merely using less nonrenewable fuel and creating less pollution, we must come to design sustainable buildings that rely on renewable resources to produce some or all of their own energy and create no pollution.

Using technology available off the shelf today, it is now possible to design buildings that produce their own thermal and electrical energy on-site from renewable resources. Even in the upper latitudes, energy-independent buildings are now a reliable design goal.

Building a sustainable home and incorporating sustainability into other aspects of your life and work are major steps you can make toward

See our knowledge section for more information on intelligent housing.

the building of a sustainable society.

Sustainability is the cornerstone of our design philosophy. The houses and buildings we design require a minimum of nonrenewable energy. Some are energy-independent, and others actually produce a surplus of energy, which can be exported to the utility grid for the benefit of society as a whole. We employ natural, long-lasting materials indoors and out to create a healthy, high-quality, sustainable living environment with enduring beauty.

Let your house take form naturally in response to yourselves and the land around you.

Form

The needs of the individual family members and the special characteristics of the site they have chosen are the natural determinants of form for a family's home.

A home should be a manifestation of the unique spirit, aspiration, and character of its owners and the land they have chosen

---

Just as each person is an individual and no two sites are exactly the same—unless made so— there is no good reason I can see, unless you are a developer or prefabricator, why our houses should all turn out the same.

In the ideal sense, a home should simply be a manifestation of the unique spirit, aspiration, and character of its owners and the land they have chosen to surround them.

While we have successfully worked with many architectural 'styles' from very traditional to ultra-contemporary, we've found contemporary architecture to provide the greatest design freedom. Free of the rigid constraints that traditional styles impose, we are able to shape and orient the house to the sun and welcome in its life-sustaining energy.

Through the use of natural materials, your house will be beautiful in and of itself.

The design process begins with a visit to the client's land. We walk the land with the client, giving consideration to views, access, outdoor recreation, privacy, weather and wind exposure, solar orientation and all special characteristics of the site. Since the amount of energy your house will require is affected significantly by the characteristics of the building site, the energy

options are carefully assessed with the client before the best homesite is decided upon.

Once the optimum homesite is located, we discuss the client's requirements and work with them to develop a detailed "design program" outlining all that they want their new home to be and what they expect from it. Then we'll revisit the site again before beginning schematic designs.

Materials and furnishings

Fine arts museums are filled with vases, cutlery, housewares, and furnishings from civilizations afar. And visitors continually remark on their basic beauty, their clean simple lines, and their pleasing form, for they are truly beautiful.

What I find most fascinating is that the majority of these objects are just common, everyday furnishings that were designed primarily for their useful service.

Our culture has strayed far from this ideal of functional beauty. Today, everyday objects are designed merely to function—and many don't even work well. Things intended to be beautiful are designed only to look at—and most are not even pleasing. A design that combines both function and beauty is truly of a higher order.

If you want a naturally beautiful house, build with natural materials to encourage beauty from within, making a clean and simple statement while adding integrity and a sense of harmony to the whole.

Through the honest use of natural materials, your home will be beautiful in and of itself. Decoration as appliqué, the superficial substitute for true innate beauty, will be unnecessary.

Where practical, we prefer materials indigenous to the area such as local timber products or stone from a nearby quarry. This tends to better integrate the house with its surrounding environment while supporting the local economy and reducing transportation costs.

We take care to research and select healthy materials, furnishings, and finishes with low-toxic content to create a healthy living environment. We also hold maintenance requirements as an important criterion. Our goal is to create a beautiful, healthy, and efficient home for you that will endure for many years with a minimum of upkeep.

Passive Solar Design

Houses are traditionally designed in **resistance** to nature. Their support systems are specified to meet the tremendous forces of earth's energy head on, attempting to maintain comfort regardless of cost and wastefulness.

A primary goal in our design of an environmentally responsive house is to create a pleasant and comfortable living environment that functions in harmony with nature.

This is achieved by first taking advantage of passive solar gain—the direct utilization of solar energy without processing through an active or energy-consuming mechanical system. Through proper orientation, north-wall shelter, south-facing glass, sun-controlling architecture, and the use of the structure as thermal mass, the sun can play a direct and useful role in your home and its energy support.

By designing to optimize this free solar gain, and using proper insulation, high-performance glazing, internal mass, and other energy-conserving elements, the energy requirements for your home, and hence the operating costs, will be substantially reduced before you even consider any alternative energy hardware.

Energy

During a year of engineering work on remote energy support systems for the 52 gate-valve stations on the Alaskan pipeline, I gained some unique, behind-the-scenes insight into the world's fossil fuel industry.

The $9 billion, 800-mile, trans-Alaskan project is just a small part of the 0verall program envisioned for the world. The known Alaskan oil reserves at Prudhoe Bay could be pumped dry in less than 10 years. If the present rate of world fossil fuel consumption is to continue, costs will escalate and extraction schemes even *more* farfetched will be proposed and (must be) implemented to meet demand.

There is little disagreement, however, that the final end point will come within our lifetime when it simply takes more energy to extract the remaining oil and gas than the extraction process can yield. Long before that, petroleum will become far too precious a commodity to simply burn up for its thermal content.

If you want a naturally beautiful house, build with natural materials to encourage beauty from within

---

All along the way, we increase concentrations of carbon dioxide and other greenhouse pollutants in our atmosphere as we continue to burn fossil fuels. It is physically impossible to remove the carbon dioxide from the waste effluent of carbon-based fuel combustion without expending more energy than the combustion itself can yield. The prospect of irreversible global warming looms ominously on the horizon.

Unfortunately, their massive investments in nonsustainable, "conventional" energy systems have made it very difficult for most major energy suppliers to look with favor on the near-term transition to sustainable sources and systems. Most have wrongly regarded the widespread utilization of energy alternatives not as a common-sense course for a troubled civilization, but as an immediate threat to their bottom line.

It is vitally important that we begin this transition while we still have the relative luxury of time and resources available to make these changes in an orderly manner and avoid their serious disruptions to our way of life that prolonged procrastination will certainly bring. Now, as always, the meaningful decisions come to rest with the individual.

Before we can apply our intelligence and technology on a world scale to make the transi-

Use natural energy and intelligent design to create a joyous place to live.

Open your house to the south— reach out and welcome the sunshine.

Now, as always, the meaningful decisions come to rest with the individual.

tion to a sustainable energy economy, a philosophical and political about-face will be required of many national governments and most utilities and multinational energy giants.

In many areas, this process has already begun, especially in the utility sector where forward-thinking energy planners have demonstrated conclusively that it's far easier, faster, and much less costly to meet new electricity demands by reducing energy waste than by building more power plants. But the pace of world progress will be slow until the near-term advantages and the eventual long-term inevitability become clear to all concerned. Fortunately, as individuals, we don't have to wait that long.

Your Alternatives

For true energy efficiency, design your home and its energy system together to work together.

High-Performance Architecture — There are many cost-effective options you can put to use right now to reduce or eliminate your dependence on nonrenewable fuels. Our approach to the design of a new home begins with an energy-efficient building envelope. By carefully choosing the materials, insulation, infiltration barriers, and finishes for the exterior structural system and incorporating state-of-the-art innovations such as low-emissivity glazing, we create a living environment with high thermal integrity. It requires a minimum amount of energy to maintain warmth and comfort, and its geometry is configured to take maximum advantage of passive solar gain.

With supplemental space heating requirements reduced to a minimum by the high-performance building envelope, there are many options to supply the remaining thermal energy that your new house will need. In fact, the exterior envelope of your new home can be made so energy-efficient that, in many areas of the country, your new home may require no conventional back-up heating system at all.

Solar Domestic Water Heating — All homes require domestic hot water, and we almost always recommend an active solar domestic water heating system. The system is sized to satisfy the domestic hot water requirements calculated for your family, and the solar collector array may be designed as an integrated part of the structure.

This integrated collector design results in

greatly enhanced aesthetics while reducing the loss of thermal energy from the back and sides of the solar collectors. The solar array is assembled using the highest quality selective-surface, black-nickel-chrome plated, all-copper solar absorber plates with a tempered, high-transmission glass cover that can serve as the finished weathering skin of the roof.

Active Solar Space Heating — For residences whose size or geographical location necessitate additional space heating to supplement that contributed by passive solar gain, we often recommend increasing the size of the active solar collector array beyond that required to heat the home's domestic hot water. This option makes thermal energy collected and stored during sunny periods available to provide space heating when the weather turns bad. During the warmer months when space heating demand is low, the active solar system can provide heat to a swimming pool or spa if desired.

For even distribution of thermal energy to the living space, we favor a radiant floor heating system. Such a system is quiet, consistent, takes up no space, delivers an even, upwardly radiated warmth throughout the house, helps to distribute direct passive solar gain, and does not interfere with furniture placement. It is an elegant and most efficient method of heating a space.

Electricity — Other residential energy system configurations are available to attain higher degrees of energy efficiency or to be fully energy-independent. For example, you can generate all of your own electricity with a properly designed solar photovoltaic (PV) system.

Photovoltaics are a truly elegant means of

producing electricity on site, directly from the sun, allowing you to take control of your energy destiny and create your own lifestyle without concern for energy supply or environmental harm. No pollution, no by-products, no depletion of resources—these solid-state devices simply make electricity out of sunlight.

Photovoltaics are the ideal source of power for an environmentally-responsive home whether it's your principal residence or a vacation retreat. With the right design, the sunlight that falls on your homesite will power your home. Your PV power system can also easily be configured to provide on-site solar recharging of electric vehicles—now, or in the future.

We have designed and built many photovoltaic-powered residences across the United States. Some of these homes are utility-interactive, selling surplus power generated by the PV system to the local power company during sunny days and then "buying back" power at night. Others are completely stand-alone with no connection to the utility grid and store electrical energy in a battery bank for use at night and during cloudy weather.

Live lightly on the earth, without having to take a vow of deprivation to do it.

When possible, we prefer to integrate your photovoltaic array with the roof of the structure. This produces superior aesthetic results and can actually improve performance of both your PV system and your house. We also have design experience with wind and water power systems, although these are heavily site-dependent sources and thus are less frequently employed.

Systems Integration

Imagine a home built in harmony with the environment. A house that's healthy inside and out, satisfies its own energy requirements, and creates no pollution. A delightful, sun-filled house designed to suit your lifestyle, free from dependence on scarce, nonrenewable energy resources.

You can have the type of home and lifestyle you desire while setting a shining example of environmental stewardship for others to follow.

People with a high degree of concern for environmental issues who also desire a comfortable, high-quality living environment typify our residential clients. The houses we design for them nicely reconcile the conflict between their environmental commitment and the desire for a comfortable lifestyle. Our clients live lightly on the earth, without having to take a vow of deprivation to do it.

Whatever degree of energy independence you desire, your home and its energy support system should be designed together as an integrated concept. In this way, your home and its energy system will complement each other, optimizing passive solar gain, moderating peak loads, combining pleasure with function in a solar sunspace, collecting and storing thermal energy and generating and storing electricity.

When designed together as an integral system, your home and its energy system will provide a much greater overall level of comfort and efficiency than if planned separately.

Your home represents a substantial investment, both in terms of your resources and of your time and life energy. The care and effort you put into its planning will be returned to you each day for many years to come.

All the photos used in this article were provided by Solar Design Associates of Harvard, MA.

Steve Strong's book, The Solar Electric House, can be found on page 407 of this Sourcebook.

Pacific Yurts

The yurt is an architectural wonder, a legendary dwelling that has been in continuous use for centuries. Invented by the nomadic Mongols of Siberia, this circular, domed abode is as near-perfect a blend of beauty, simplicity, and functionality as any human habitation throughout history.

The Pacific Yurt is ideal as either a recreation retreat or a year-round residence for "living lightly on the Earth." It may be kept unelaborate or include modern amenities such as plumbing, electricity, and multilevel deck systems. Other uses for the yurt include a hot tub/spa enclosure, a workshop/studio, a resort/conference center, a ski hut, a remote base camp, or temporary housing for an owner-builder.

The easily ventilated yurt is well adapted to warmer climates—as well as the Alaskan North Slope. It can be insulated and equipped with a heater or woodstove. Even in extreme cold, it heats efficiently and comfortably. Naturally strong, it can be reinforced to withstand high winds and heavy snow.

The low cost per square foot makes the Pacific Yurt an outstanding value and an economical alternative to higher-priced standard frame structures. You can easily transport your yurt in a small pickup, then set it up quickly, virtually anywhere, for a comfortable stay. You can take it home, or leave it up permanently.

These yurts are surprisingly easy to erect. The 30-foot yurt takes two people less than a day to set up, and the 12-foot model only a few hours. Materials are of the finest quality available, including a center ring of cross-laminated, kiln-dried fir, select fir rafters, galvanized steel tension cable, electronically bonded, vinyl-laminated polyester top cover, and large clear vinyl windows with screens. All top covers come with a 5-year pro-rata warranty, and custom Duro-Last top covers come with a 15-year warranty. Many custom options are available, including extra windows, solar skylight arc, insulation, and extra doors.

Once you've tasted yurt life, you may conclude that the human race made a mistake in abandoning the old nomadic ways.

I have had one of their 20 foot yurts out at my commercial salmon camp on the remote west side of Kodiak Island since 1986 and cannot praise it enough. It comfortably houses three crewmembers and all their stuff, plus a woodstove for the four months we spend harvesting salmon... Another nice thing about our yurt is its' acoustics. Perhaps it is their roundness or maybe it's the dome overhead, but live music in the yurt simply glows and reverberates as if it were a tiny Carnegie Hall of the north."
Toby Sulivan, Kodiak, AK

"The yurt manufactured by Pacific Yurts is one of the most beautiful structures I have ever encountered."
–Joan Halifax, Ph.D., Author

- ■ 12'D 115 Sq.Ft. 8'"H 350# wt. $2,195
- ■ 14'D 155 Sq.Ft. 8'9"H 450# wt. $2,885
- ■ 16'D 200 Sq.Ft. 9'3"H 550# wt. $3,285
- ■ 20'D 314 Sq.Ft. 10'0"H 700# wt. $4,295
- ■ 24'D 452 Sq.Ft. 11'6"H 900# wt. $4,875
- ■ 30'D 706 Sq.Ft. 13'"H 1,200# wt. $6,585

Shipping and packing are extra. Allow 4-6 weeks for delivery. All yurts are shipped freight collect FOB Oregon. *Do not send money to us; we will refer you directly to the manufacturer.* **Send SASE for color brochure and prices on options.**

Econ-O-Dome
Saves Up to 80% in Materials & Time

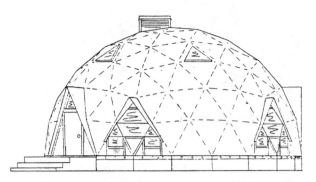

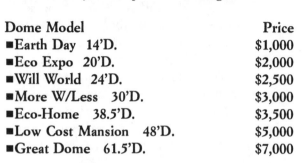

This Econ-O-Dome is one of the classrooms for our Institute for Independent Living.

Far too much wood, money, and precious time has been wasted in the production of conventional housing. The geodesic dome shape is naturally energy and materials efficient, because it has the smallest possible surface area for a given volume. And surface area determines not only the amount of materials used, but also the amount of heat that can escape through your walls. Our customers who are aware of these benefits have continually asked us where to obtain a high-quality dome kit. Our search for a low-cost, reliable, easy-to-build, energy-efficient geodesic dome is over: we are glad to introduce **Econ-O-Domes**. The designers of these structures have been making domes for over 10 years and have worked out all the "bugs." The kits are easy to assemble, since all struts (2 X 4 construction) and hubs (coated metal) are predrilled, and the package comes with a complete and understandable instruction manual. All the materials for a 38½-foot dome can actually fit in the back of a full-sized pickup truck! Prices listed are for the structural frame, which you can put up in only 2 days with just an extension ladder and a drill gun —though scaffolding will help. Screws are included. Plans for cutting the wall boards (very little waste involved) are included; or the precut pieces can be obtained at additional cost. (To create an efficient greenhouse, simply cover the frame with plastic sheathing.) Other options include a double shell design, or 2 X 6 strut dimensions for extra insulation space. Installation is also available at an added cost. Please write or call for additional literature. Shipping is **Freight Collect** FOB California. Allow 6-8 weeks for delivery. *Do not send money to us; we will refer you directly to the manufacturer.*

Dome Model	Price
■**Earth Day** 14'D.	$1,000
■**Eco Expo** 20'D.	$2,000
■**Will World** 24'D.	$2,500
■**More W/Less** 30'D.	$3,000
■**Eco-Home** 38.5'D.	$3,500
■**Low Cost Mansion** 48'D.	$5,000
■**Great Dome** 61.5'D.	$7,000

We set up 38½' Econ-O-Domes at both the Solar Energy Fair held in Willits during the August heat wave and at the Eco Expo in LA. They went up with amazing speed—both cases taking less than 12 hours to erect. Both domes were sold on-the-spot.

**To order any of these products
or for more technical information
call us toll-free at
1-800-762-7325**

Earthship

Here is a professionally produced video by Dennis Weaver of his independent sustainable living space, Earthship. It is narrated with the same enthusiasm and down-home twang he brings to commercials. The video is a solid prospectus for a responsible way to build, using surplus material—tires and cans. The book is a builder's guide to an innovative construction technique integrating recycled waste with conventional building materials. The primary construction material—used tires—are a post-consumer disposal nightmare, and can be recycled into a building, **free**. A single tamped-earth tire weighs 300 pounds, and is stable, immovable, and cheap: the houses in the book cost from $20 per square foot. Interior walls made with adobe-covered cans are strong, light, and can be sculpted to any shape. An exciting concept. 'Just as small problems lead to big problems, so do small solutions lead to big solutions.' Dennis grinds his favorite axe—the greenhouse effect—and emphasizes the 3 Rs —Reduce, Reuse, Recycle.

The companion books, Earthship, Vol.1 (230 pages) and Vol. 2 (260 pages), by Michael Reynolds, are competently home-grown how-to manuals showing a builder how to use these new building materials. They provide detailed information from design to finish with Vol.2 giving interior details and including a chapter about getting permits.

80-117	Earthship Video	$29
80-118	Earthship, Vol. 1	$29
80-131	Earthship, Vol. 2	$29

Beauty, Efficiency and Economy in Underground Homes

Terra-Dome underground homes and commercial buildings dispel the stereotype of underground living as dark and damp. Terra-Dome designs emphasize openness and esthetic flexibility, at a price tag that rivals conventional construction. Modular design is the key to the low price (approximatiely $18 per square foot for the basic structure), while energy savings are extraordinary—up to 90% savings over traditional designs. Modules are available in versatile 24' and 28' configurations, connected by 16' arches. Walls are 10" thick, and the dome itself will support 8' of earth. Architectural styles range from Colonial to Tudor, Contemporary to Mediterranean. Terra-Dome builds homes that are exceptional in every way and at an incredible price. *Call for more data and an informative brochure, or order our 30 minute video showing houses, commercial structures, and building processes.*

80-106 Terra-Dome Video   **$20**

Shelter Systems LightHouses

One of our long-standing customers has a business called Shelter Systems. They make a wonderful portable shelter that is as easy to set up as a tent but far more durable. Portable shelters have a wide variety of uses. They can be used to make a shop or studio, to give children a playroom of their own, or as a greenhouse.

The Lighthouse

The LightHouse sidewalls are a 5½ ounce polyester canvas, which is resistant to sun degradation, watertight, flame retardant, breathable, and will not rot or mildew. Its color is light tan for good reflectivity and interior brightness. The translucent skylight above the sidewalls is constructed of a woven, rip-stop, UV-resistant film. It creates a pleasing interior light and opens to provide an excellent fresh air vent that is rainproof. A sunshade of silver and black rip-stop covers the top of the dome to block and reflect the sun. The frame is PVC tubing for strength and longevity. Patented Lexan clips connect the LightHouse's covering without puncturing or weakening. Each panel of the LightHouse is shingled over the next so that the dome breathes and remains leakproof.

The LightHouse 18 and LightHouse 14, 18 feet diameter and 14 feet in diameter respectively, have four tipi style doors spaced evenly around the dome. The Lighthouse 10 is 10 feet in diameter has one door. There are no zippers to fumble with or break. Clear vinyl windows above the doors let you see out in all directions. In tropical weather the sidewalls can be rolled up to provide unsurpassed ventilation. There's always plenty of light and fresh air in the LightHouse. The LightHouse comes complete with stakes, guylines, vent tubes, spare parts, and an instruction booklet that details floors, site selection, anchoring, cooling, winterizing, and stove installation.

Options for LightHouses

Mosquito Net Doors: No-see-um netting is used in addition to fabric.
Floor: Made of a material similar to the blue rip-stop tarps you may be familiar with but with fire retardant and of a higher grade.
Sun Shade: 6-feet by 12-feet black and silver rip-stop with 4 grip clips.
Liner: A duplicate of the outer covering. It comes with fasteners and cords attached, creating a dead air space between the layers.
Porch: 5-1/2-feet by 5-1/2 feet arched square, it is tied to the main domes with two poles supporting the outer corners.
Net Wall: Side wall panels are opened out functioning as awnings, exposing the panels of netting underneath. These net panels must be installed at the time of purchase.

■92-201	LightHouse 18	$725
■92-256	LH 18 Floor	$99
■92-253	LH 18 Winter Liner	$639
■92-259	LH 18 Net Wall	$99
■92-202	LightHouse 14	$639
■92-255	LH 14 Floor	$79
■92-210	LH 14 Winter Liner	$525
■92-258	LH 14 Net Wall	$79
■92-203	LightHouse 10	$359
■92-254	LH 10 Floor	$49
■92-257	LH 10 Winter Liner	$280
■92-250	LH 10 Net Wall	$25

(Options for all three sizes)

■92-252	Mosquito Netting for doors	$11
■92-251	Porch	$59
■92-204	Sun Shade	$29

Shipping to Alaska & Hawaii:

	1st class mail	UPS 2nd day air
10'	$40	$45
14'	$55	$65
18'	$80	$90

"I have been completely satisfied with the quality and apperance of our dome. So far it has handled 20" of rain and wind storms very well. The dome is my year round home. I have a bright, airy but warm home inexpensively. The tipi style doors are 100% improvement over zippered doors. Living in the dome, one becomes intimate with the sun, clouds, and waning and waxing moon. I love it."
Paul Guree, CA

Shelter Systems Solar-Dome

Both the Solar-Dome and Lighthouse can be used for shelter, storage, or shop spaces.

Many greenhouses are too big and too expensive for the average gardener, but not the Solar-Dome. Vegetables and flowers receive the growing conditions they need: shelter from cold, rain, wind, frost, and birds, with plenty of room for the gardener to work standing up, store tools, build up flats, hang potted plants, or care for mature plants. The Solar Dome sets up in 20 minutes without tools. All hubs are factory attached and all poles are interchangeable. Just insert them into the connectors and it's up. Turn or move the dome into any desired position or lift to wherever it's needed most in the garden plot. If a season occurs where your Solar-Dome isn't being used, just take out the poles, roll the greenhouse up, and store it in a closet or on a shelf.

The Solar-Dome is built of a super-strong woven rip-stop greenhouse covering. It is treated with ultra-violet inhibitors, designed specifically for long sun exposure in greenhouse use. The filtered light that comes through is excellent for plants and will never burn leaves like glass or clear vinyl will. It transmits 90% of available light. The clips that join the PVC poles at each hub have a patented design. They're made of an especially strong plastic called Lexan. They grip the canopy, providing greater strength than sewn seams or other grommets.

The ventilation tubes ensure plenty of fresh air with the doors open or hooked closed. Also included are stakes, hooks for hanging potted plants, and an instruction manual. There are no zippers in the Solar-Dome. Simple hook closures have proved to be the best. Carts and wheelbarrows can be wheeled in and out, as the Solar-Dome's doors open completely along the ground.

▪92-245	Solar-Dome 20 (20'D X 10'H)	$859
▪92-246	Solar-Dome 18 (18'D X 9'H)	$679
▪92-247	Solar-Dome 14 (14'D X 7'H)	$575
▪92-248	Solar-Dome 11 (11'D X 6'4"H)	$325
▪92-249	Solar-Dome 8 (8'D X 7'4"H)	$175

Shipping to Alaska & Hawaii:

	1st class mail	UPS 2nd day air
8',10',11'	$40	$45
14'	$55	$65
18'	$80	$90
20'	$100	$110

Power

Sunlight is energy. Sunlight is permanent, free, nonpolluting, totally quiet, and universally available.

Think of the sun as a giant power plant generating radiant energy at a kilowatt (kW) rate estimated to be a staggering 110 trillion kW. Even though less than one-billionth of this energy is intercepted by the earth, every 10 square feet of the earth's surface facing the sun is estimated to receive about 1,000 watts at midday.

Here's a fun fact for your solar tidbit repertoire: If an area the size of Manhattan Island in New York was covered with photovoltaic modules (solar panels) the entire energy consumption of the United States could be solar-generated. The idea of turning Manhattan into a giant solar farm may strike some as ludicrous, but it dramatically illustrates the potential of solar power.

Why, then, do we still have an oil-based economy, nuclear power plants, and a menacing problem of global warming from the burning of fossil fuels? The answer is, because solar power is not yet cost-effective. Yet, the reason solar power seems so expensive has a lot to do with the fact that the kilowatts from other energy sources are kept artificially low by the oil-friendly policies of our government.

There are myriad hidden costs unattributed to the gallon of gasoline that you buy. You know there has to be some economic sleight of hand at work when a gallon of water purchased at a convenience store costs 20 cents more per gallon than gasoline, refined from imported crude, at the same store's gas pump!

The hidden costs of oil, coal, and nuclear generation of electricity include health care and cleanup costs of air pollution, the government subsidies to the coal, oil, and nuclear industries, and the tremendous costs of keeping the Persian Gulf oil lanes open and safe militarily. While a barrel of West Texas crude may run only $20 on the spot market, popular scientist Carl Sagan estimates the true cost as $80, while the Greenpeace organization says it is as high as $200 to $500 per barrel!

Whomever you believe, it is clear that there are economic forces whose interests are served by keeping solar energy economically out of the general populace's reach.

The 1980s will stand out in American energy history as an era marked by unprecedented greed. While Germany and Japan became economic successes by implementing energy-efficiency plans, we saw Ronald Reagan dismantle the solar panels (installed by Jimmy Carter) on the White House roof and firing the scientists at the Department of Energy involved in solar research. George Bush, meanwhile, cruised the waters off Kennebunkport in a speedboat so powerful that it looked like a waterborne space shuttle, all the while designing plans for drilling oil in the Arctic National Wildlife Refuge.

At Real Goods we saw correspondingly dramatic changes in our photovoltaic suppliers during this period. From 1979 through 1988, we purchased 95% of our solar modules from Arco Solar, a company based in California. Then a curious thing happened. Arco sold out to Siemens, a German conglomerate. Two Japanese companies, Hoxan and Kyocera, began to absorb market share. Today, both Germany and Japan have laid the groundwork for enormous production increases in renewable energy products, including photovoltaics. America, meanwhile, is having a jolly time bouncing over the waves with George Bush's "National Energy Strategy," which promises more oil, coal, and nuclear power.

A barrel of West Texas crude may run only $20 on the spot market, but popular scientist Carl Sagan estimates the true cost as $80.

Meanwhile we degenerate into a nation of Japan-bashers, venting our frustration with a sledgehammer on some poor Toyota (note that no one ever bashes a BMW). Perhaps the energy would be better directed at our own government, demanding leadership in the energy field before our trade position falls to a new worldwide low.

The intention here is not to roil the political waters, but to provide background to the key issue surrounding alternative energy — cost. At Real Goods we try to break everything down to dollars and cents. Our economic analyses may not be the most sophisticated but the grounding principle is unarguable: "If it doesn't save money, it isn't worth doing, period."

Where our philosophy differs, however, is in the way we filter our thinking by time. We try not to be blinded by the short-term costs, which are heavily influenced by the price of technology, but to look at the overall investment. This philosophy works for compact fluorescent lightbulbs, electric vehicles, super-efficient refrigeration, and photovoltaics. Amortization times might be longer, the hidden costs exposed, but if you find a product on our pages, it means that in the long run, it pays.

The "Power" chapter is the heart of our *AE Sourcebook*. We are not so naive as to think people are about to live without power or even to experience inconvenience for the sake of the environment. They will, however, take action if they see an economic benefit. Over the past hundred years or so (a period that future history books will refer to as the Age of Oil), we have become so narrow-minded and ignorant of the fuel options available that we have overlooked some obvious options.

Read this chapter thoroughly, and refer to the appendices in the back of the book as you hone in on the power strategy to govern your own life. The time spent on up-front planning will always pay off in the end. Take the plunge.

First of all, let's make the assumption that an average home's comfortable power load is 4.6 kWh per day using energy-efficient lights, a Sun Frost refrigerator, and a wood or gas cookstove, oven, and space heating. Let's also assume adequate insulation. This power requirement can be fully powered by our Remote Home Kit 4 (cost = $10,495) to power a full-scale home with a comfort level of an efficient city dwelling hooked up to the power line. The power generated by the solar system will power the vacuum cleaner, washing machine, microwave, large color TV, VCR, satellite dish receiver, computer and printer, most all power tools, and some water pumping. Now clearly if the power company charges you $10 per foot to hook into their power grid, the breakeven distance is only 1,050 feet or 1/5 of a mile, to make the alternative energy system instantly more economical. And you'll never have to pay a utility bill, although you will have to replace your batteries.

Now, let's look at the feasibility of retrofitting an existing "on-the-grid" home to an independently powered PV home. If we use the same Remote Home Kit 4 for the power system and assume that we will replace the battery bank twice in 30 years, the total cost of the system, exclusive of labor for installation and maintenance, is $17,695. Assuming an average of 5 hours per day of photovoltaic output, we will reap 1,679 kWh per year or 50,370 kWh over the 30-year period. This comes out at a net cost of $0.35 per kWh, seemingly well above the current national average of between $0.08 and $0.10 per kWh from the utility company. But wait! We all know utility rates increase. In Northern California, over the last four years, our rates have increased on average between 10% and 12% every year. At these conservative increases in utility company rates, it would only take 10 years before the grid power costs more than the PV-power in cents per kWh and only 15 years before the entire system begins to turn a monthly profit! This is only the half way point of a typical 30 year mortgage.

So the economics of solar alternative power make sense. Given an adequate wind supply or a prolific creek or spring, the numbers look even better than solar for wind generation or hydro-electric power production. As if the economics weren't enough to convince you, let's take a quick look at the environmental savings we can harness with solar-electric power. We did some calculations to determine the savings in electricity (kWh), barrels of oil, and pounds of carbon dioxide and sulfur dioxide. We calculated these savings figures for one 50-watt PV module and for the total amount of PV modules that you bought from us last year. You can see from the following table that the effect is significant.

Environmental Savings from Photovoltaic Modules

	Savings of One 50 Watt Solar Panel	Savings of Real Goods' Customers' Purchases in 1990-1991
Electricity saved per year	90 kWh	810,000 kWh
Electricity saved per life of solar panel	2,700 kWh	24.3 million kWh
Barrels of oil saved over life-time of solar panel	4.8 barrels	43,200 barrels
Pounds of Coal saved over life of solar panel	2,700 lbs	24.3 million lbs
Carbon Dioxide kept out of the air over life of solar panel	4,000 lbs	36 million lbs
Sulfur Dioxide kept out of the air over life of solar panel	23.3 lbs	210,000 lbs

Formulas upon which these calculations are based:
    1 PV module = 50 watts
    1 PV module will last 30 years
    11,000 BTUs to generate 1 kWh of electricity
    6.25 million BTUs = 1 barrel of oil
    Coal required to produce 1 kWh = 1 lb
    Carbon dioxide emissions = 1.5 lb/kWh

Think About It... The Real Cost of Oil

Oil and gas cost a buck a gallon, right?
Wrong.

The *price* of fossil fuel may be in the vicinity of a dollar, but the true *cost* is much more. An illustration may show the difference more graphically.

A widely reported recent study in a Snow Belt state pointed out that $10 million annually was spend on salt to clear the roads. However, the same study pointed out that the state spends $500 million annually to repair damage done to roads, cars, bridges, and the environment *by salt*. The *price* of the salt is $10 million, but the *cost* of using it is $10 plus $500 million, or $510 million.

An alternative chemical, having zero impact on roads, cars, and environment, is available. Its price is 10 times as much as salt ($100 million), but its total cost is the same as its price. Thus, using the alternative will generate $410 million of positive cash flow to the public coffers.

Simple decision, right? The proverbial "no-brainer."

(Unfortunately, the state decided to keep on using salt, using for justification the same logical gyrations that keep us thinking that oil is cheap. The imperfections of the political process are another subject, however.)

Think it through.

Oil costs the same as the spring water in my office bubbler. The spring water is bountiful, comes from the next town, requires no treatment or refining, and passes through a minimal distribution chain. And since the spring that produces the water is constantly being replenished, the cost of raw material is nothing.

Oil, by contrast, comes from many feet under the ground. It is a rare commodity, found in only a few locations. The supply is finite. Much of it must be purchased from foreign governments (who we know are becoming rich from the revenues), transported to refineries (themselves a great source of pollution), loaded onto supertankers, and shipped halfway around the globe. It then goes through several distribution tiers before winding up in your basement for the same price as that spring water from the next town. Two conclusions are possible:

Either the spring water business is the most profitable enterprise known to man and we are fools not to be in it, *or* the price of oil is significantly less than its cost.

What is the real cost of oil? The figure is so obscured that we can only guess at it. We know that the consumer in Europe who is buying oil that comes from beneath the same sands as does ours is paying $4.50–5.00 per gallon. Since the oil is the same, the only difference is how these governments and ours have chosen to depict the cost.

It is a simple matter to think of other factors that should be considered in the cost of oil. Two words, "Gulf War," could have substantial impact. What about the cost of protecting shipping lanes and cleaning up oil spills? What are the environmental costs of unbridled consumption encouraged by the artificially low price? You need more of an intellect than mine to properly quantify these into dollars, but it does not take a genius to recognize the cause and effect relationship between our current fiscal crisis (national deficit, collapse of the banking industry, and recession) and the artificially low price of oil.

The "Energy Crisis" of the 1970s, unpleasant as it was, shocked the nation to action. President Carter called the need for conservation "the moral equivalent of war."

And the programs worked, but we did not like the taste of the medicine, and promptly returned to the excessive "prosperity" of the 1980s, increasing our reliance on foreign oil with each successive year. The momentum on conservation was lost, solar tax credits were repealed, and the miles per gallon requirements for cars were eliminated. Many of us, but especially the Ivan Boeskys, Michael Milkens, and Donald Trumps of the world, thoroughly enjoyed the ride, but now most of us are now facing the harsh reality of our energy irresponsibility.

Don't look to the government to bail us out. George Bush has already demonstrated, with an energy policy that promotes the myth of cheap power, which side of the fence he is on. As for Congress, they are chameleons who will respond to the issues that the polls tell them will result in re-election.

That leaves it up to us. As a nation we have proved too individualistic to band together toward a common goal, unless a good war is involved. OK, let's go back to Jimmy Carter's concept and declare war, but a war where compact fluorescent lightbulbs replace bombs and electric vehicles substitute for tanks. It is a war that we can lose only if, as we did in the 1980s, we stop fighting.

Instant On!

I'm one of those New Age guys who has formed a little business that exists by networking electronically around the globe. My business creates no acid rain, no toxic waste, and no revenue (whoops, that's just in there for the sake of the IRS). My only luxury is a dispenser for spring water that comes from the very same town that I live in.

I do my bit to keep that big ozone hole in the sky from expanding. I live a life of parsimonious energy consumption. I don't drive a gas-guzzling car; I heat with wood; I even installed some of those $25 lightbulbs so that the earth would be a safer place for my children. I live lightly on the planet.

Imagine my horror and disgust to learn that, despite my pretentions, I am a complete power glutton, depleting the planet's finite resources even as I am writing about how to save them.

Remember my bubbler? Turns out the thing is plugged in, bleeding the resources from my local nuclear plant (which, as a right kinda guy, I oppose) so that I can have instant hot water for my herbal tea.

But the bubbler is just a tip of a fuel gluttony iceberg. The real sickening thud came the other night, as I was leaving my "energy efficient" office. I turned off the compact fluorescent overhead, and was greeted by the normal display of colored lights that tell me that all is well with my office.

Those lights, I suddenly realized, are sucking up electricity 24 hours a day. Despite good intentions, I am hemorrhaging power around the clock. This means that I am burning up fossil fuels (or creating nuclear waste) even as I lie in bed.

I counted the monitor lights that tell me that my appliances are at the ready, awaiting my DOS command. When I put my work space to sleep for the night, there are 16—count 'em one, two... sixteen lights.

Here's the breakdown: two on my laser printer, one on my color monitor, three on my computer (one for power, one for "turbo," whatever that means, and another for the hard drive).

Wait a minute, is this a joke? Tell me there's not a computer design nerd out there somewhere laughing up his sleeve. Tee-hee-hee. Let's put a "turbo" light on the box. People will actually think it means something. The next generation of boxes will probably have toggle switches reading "Fire retrorockets." Back to my monitor lights: I've got three, yeah three, monitor lights on my answering machine. It's a two-line phone, and one of the lights is on to tell me that the second line (which isn't even connected) is off. My fax machine has one light, to tell me that it is plugged in. My copying machine has two lights, plus a very colorful lime green, digital zero to tell me that nothing is being copied at the moment.

The remainder of my lights are for various surge protectors to tell me that they are ever-vigilant, ready to launch into action at a nanosecond's notice.

Of course it goes without saying that I have been advised by my computer gurus to leave my machines on at all times to protect them from the nasty surges of electricity unleashed every time you flip the on/off switch. Of course, if you leave the machines on, you are more vulnerable to the mega-surges that can come through the phone lines or power lines that can leave the inside of your machines in smoking ruins. (True story: I once returned a dysfunctional fax to be repaired. The company reported back that they could not honor the warranty because the machine had been hit by lightning. Inside my office?)

So, you need surge protectors to protect your equipment from surges because you leave the machines on all the time to protect them from surges. This circular logic reminds of an acquaintance who chain-smoked cigarettes in between cups of black coffee.

"Why do you drink so much coffee?" I asked.

"I have to keep drinking coffee because my throat hurts," he replied.

"Why does your throat hurt?"

"Because I smoke so many cigarettes."

"Why do you smoke so many cigarettes?"

"Because I'm jittery from all the coffee."

Oh. Back to my flagrant, irresponsible, planet-destroying energy consumption, as typified by the 16 monitor lights in my one little office.

I lead an enviable life. I can ride my bicycle to work. I can go cross-country skiing at lunch. I can send files winging to California while watching snowflakes drift down outside my window. But let's face it. In an instant-on society, we want to punch a button, and we don't want to wait. As my energy-oozing office proves, however, even the best intentioned can be energy hypocrites. We try to consider the impact of our actions on the next seven generations, but we can barely keep a nuclear family together, let alone think seven generations into the future.

Every convenience carries a price, even if it is hidden deep in the recesses of our power bills. And beyond the price is a true cost, to our health, to the environment, to the future. Eventually the bill has to be paid. Never was it a more appropriate time for the buyer to beware.

— Stephen Morris

System Design

Ninety percent of the work that we do for customers at Real Goods is explaining the feasibility of alternative energy systems. Our goal in this chapter is to take you through all the steps of planning and properly sizing your system. Please read it carefully, several times if need be. After you've completed the worksheets send them in to us or call one of our technicians to help you.

System Planning

It is important to understand what is to be accomplished with an energy system. There are inappropriate ways to make use of the technologies we have access to, and our greatest concern for our children's sake is to make use of all the world's resources in the most efficient way possible. This becomes obvious when the energy is developed by us personally, instead of produced miles away by some remote entity whose only contact with us is the monthly power bill they send.

Before we can design an energy system, we must dissect our energy use patterns. Energy usage can be broken down into three broad categories: thermal (heat) energy, electrical energy, and refrigeration. It is important that differences between these usages (commonly called **loads**) are understood. Our challenge is to use the most appropriate form of energy for the job, then to consume the lowest amount of that form of energy (maximize the appliance efficiency). This will reduce the size of our energy system and therefore reduce the capital expenditure for our system.

Thermal Loads

In a residence, large amounts of energy are devoted to heating air, water, food, and people. We can provide this heat in a number of different ways. Electricity generates heat very inefficiently. An electric power utility often takes fossil fuels and burns them to heat up water to make steam, which is then used to turn a generator.

The resulting electricity is then transmitted many miles to a home or business, where it is reconverted into heat. This conversion and transmission dissipates much of the energy in the fossil fuel. The system efficiency could have been much higher had we burned the fossil fuel directly at the home. Therefore, an important strategy for the independent energy user is to use fuels directly to produce heat. These fuels are typically propane, heating oil, wood, diesel, kerosene, natural gas, and agricultural wastes.

We hope that renewable fuels will soon dominate those consumed, and that we can add **hydrogen** and **biogas** (methane derived from agricultural resources) to the list. **Solar thermal** energy (as distinguished from **photovoltaic** or solar electric energy) can reduce our need for fuels to generate heat.

To summarize, if we need heat, we should use sources that provide heat directly, such as the sun, wood (biologically stored solar energy), or fossil fuels (ancient biologically stored solar energy) or even more intriguing fuels such as biogas or hydrogen, instead of electrical energy. In the case of the home independent of utility power, electrical resistance heat is cost-effective.

Electrical Loads

Electrical loads are ubiquitous in our culture, from radios and televisions to briefcase computers and microwave ovens. Electricity powers a mind-boggling variety of devices. We can access power from utility electrical grids, generators, solar systems, wind energy devices, and hydro-electric devices, to name a few. Which method is best for an application depends on many criteria, one of which is economic. The use of more efficient light bulbs, washing machines, or any other electrical load translates directly into energy savings. Choosing a more costly but efficient product over an inexpensive wasteful product will always be the better choice when all factors are considered.

The largest electrical load requirement in an independent energy home is often lighting. Major progress has been made in compact fluorescent lighting systems that are four to five times more efficient than the familiar incandescent bulbs, which have changed little since the time of Edison. By increasing the efficiency of lighting systems, we can substantially reduce our total electrical energy load without sacrificing our standard of living.

Refrigeration

Cooling our food is a modern-day "necessity" that has not been addressed by most appliance manufacturers in regard to energy efficiency. See our Refrigeration secton for a complete discussion of this important topic. Conventional air-conditioning of homes and businesses represents an enormous electrical load which cannot be satisfied by reasonably sized photovoltaic systems. However, air-conditioning is now possible with a new propane/natural gas cooling system from Real Goods. These units have a fan and pump system that can be powered by photovoltaics.

Minimizing Electrical Loads

Before investigating power generation strategies, we must define precisely how much energy is needed. If we fail to do so, the resultant energy system will be undersized or oversized, translating into potential failure to satisfy our needs, or a needless investment of resources.

The first step in minimizing electrical loads is to substitute all possible electrical loads with appliances that accomplish the same function without electricity. Every watt of electricity we avoid using is a watt that does not need to be generated, transmitted, regulated, stored or transformed. Amory B. Lovins of the Rocky Mountain Institute refers to these unused or eliminated loads as **negawatts**, and we can generate many simply by using more efficient appliances.

Our energy-efficient homes cannot afford the wasteful energy expense of electric ovens/ranges, water heaters, clothes dryers, and space heaters. Acceptable replacements must be found that employ natural gas, propane, kerosene, wood, diesel, solar thermal, hydrogen, or biogas. Microwave ovens represent a reasonably efficient method of cooking and help reduce the electrical load in the kitchen.

Send a copy of your energy needs to us if you would like our technical staff to help you design an energy system.

Calculating Electrical Loads

The electricity we can purchase from an electrical utility and the appliances that operate with this power are rarely given much thought. Every appliance needs particular electrical characteristics feeding it to operate properly, and to provide improperly defined electricity may ruin it. Most appliances require **alternating current** (AC), a form of electricity that flows and ebbs between two wires. The effect is "back and forth" and oscillates 60 times a second in power provided by utilities in the United States (50 cycles per second in many other countries). As the utility is supposed to be operating continuously, this energy is not stored, but is always available. Fossil-fuel generators are designed to simulate this AC power.

If we do not or choose not to operate all our electrical loads off of a utility or fossil-fueled generator, we must access electricity that has been stored in a **battery**. Batteries do not have the pulsing of AC power, but a steady push in one direction called **direct current(DC)**. Automotive lights and stereos represent a few examples of appliances that operate off of battery stored DC power. It is important to be aware that AC and DC appliances are not interchangeable, and will not work properly if connected to an inappropriate power source.

We *can* operate AC equipment with battery-supplied power if we use a device called an **inverter**, which transforms the power from DC to AC. There is a cost associated with this transformation, although the tremendous advances in inverter design have now made it possible to run all of the loads in the house as AC loads. This is preferable for many due to the breadth of products available as AC products, whereas DC products are fewer and often more expensive. This was unthinkable only a few years ago, when inverters were tremendously inefficient and expensive.

We must determine how much electrical power we anticipate using every day. This is a critical determination, and can only be calculated by examining one's lifestyle in detail. The following pages lay out the method for determining total energy consumption and exploring methods of generating power on site. Refer to the sample worksheet to get an idea of how to do this properly. In the appendix, you will find several worksheets that you can fill out and send in.

Definitions:

Volts = electric potential, similar to pressure in water systems

Amps = electrical current flow, similar to gallons/min. in water

Watts = rate of energy production or consumption = volts x amps, similar to speed (rate of movement; miles per hour)

Watt-hours = total power produced or consumed = watts x hours

Kilowatt = 1,000 watts; 1 kilowatt-hour (1 kWh) = 1,000 watt-hours

AC = Alternating current; electrical energy form as is available from utilities, generators, and inverters

DC = Direct current; electrical energy form as is stored in batteries

Typical Wattage Requirements for Common Appliances

(always use the manufacturer's specs if possible)

Description	Watt-Hour
Refrigeration:	
22 cu. ft. auto defrost	490
(approximate run time 14 hrs per day)	
12 cu. ft. Sun Frost refrigerator	58
(approximate run time 6-9 hrs per day)	
Standard freezer	750
(runs approximately 14 hrs per day)	
Sun Frost freezer	88
(runs approximately 6–9 hrs per day)	
Water Pumping:	
AC Jet Pump (1/4 hp), 165 gal per day, 20 ft. well depth	500
DC pump for house pressure system	60
(typical use is 1-2 hr per day)	
DC submersible pump	50
(typical use is 6 hr per day)	
Entertainment/Telephones:	
TV (25-inch color)	130
TV (19-inch color)	60
TV (12-inch black & white)	15
Satellite system, 12 ft dish with auto orientation/remote control	45
VCR	30
Laser disk	30
Stereo (Avg. volume)	15
CB	10
Cellular telephone	24
Radio telephone	10
Electric player piano	30
General Household:	
Typical fluorescent light (60W equivalent)	15
Incandescent light of 60W brightness	60
Intellevision	27
Electric clock(s)(3)	12
Clock radio(s)(2)	10
Iron (electric)	1500
Clothes washer	1450
Dryer (gas)	250
Central vacuum	750
Furnace fan	500
Alarm/security system	3
Air conditioner	1500/ton
Kitchen Appliances:	
Dishwasher	1500
Trash compactor	1500
Can opener (electric)	100
Microwave (.5 cu. ft.)	750
Microwave (.8 to 1.5 cu. ft.)	1400
Exhaust fans (3)	144
Coffee pot (electric)	1200
Food processor	400
Toaster	1200
Coffee grinder	100
Blender	350
Food dehydrator	600
Mixer	120
Range, small burner	1250
Range, large burner	2100
Office/Den:	
Computer/Monitor/Modem	80
Ink jet printer	35
Dot matrix printer	200
Laser jet printer	1500
Electric eraser	100
Phone dialer	4
Typewriter	200
Adding machine	8
Electric pencil sharpener	100

Some appliances and equipment can run directly from the power produced from solar modules. These "sun-synchronous" or "array-direct" devices are listed under "Solar Cooling" or "Water Pumping".

Hygiene:

Hair dryer	1500
Waterpik	90
Whirlpool bath	750
Hair curler	750

Miscellaneous:

Electric blanket	120
Garage door opener	300
AC table saw, 2.0 hp	2400
AC grinder, 1/2 hp	1080
AC lathe, 3/4 hp	1400
AC drill, 1/8 hp	300

See page 498 in the Appendix for a blank worksheet to fill out.

System Demand Planning Chart

Name: **Meg. A. Power**　　Site Location: **Santa Fe, NM**

Appliance*	Qty		Wattage (volt x amp) Mult by 1.1 for AC		Hrs per Day		Days per Week	÷7	=	Avg Watt-Hrs/Day
		x		x		x		÷7	=	
lights	3	x	18	x	4	x	7	÷7	=	216
TV	1	x	60	x	2	x	2	÷7	=	34.3
stereo	1	x	75	x	1	x	4	÷7	=	42.9
tooth cleaner	1	x	15	x	.25	x	7	÷7	=	3.8
		x		x		x		÷7	=	
		x		x		x		÷7	=	
		x		x		x		÷7	=	
		x		x		x		÷7	=	
		x		x		x		÷7	=	
		x		x		x		÷7	=	
✓ fan	1	x	24	x	8	x	5	÷7	=	137.1
✓ pump	1	x	80	x	1	x	7	÷7	=	80
		x		x		x		÷7	=	
		x		x		x		÷7	=	
		x		x		x		÷7	=	
		x		x		x		÷7	=	
		x		x		x		÷7	=	

*(✓) Check all DC appliances

Step 5: Maximum AC wattage at one time	200	Total Watt-Hour Per Day Load	514.1

Total Watt-Hr Per Day Load		Battery Inefficiency Factor		Total Corrected Watt-Hours Per Day
514.1	x	1.25	=	642.5

Generator Direct Loads

Appliance	Qty	Wattage (Volts x Amps)	Hours Per Day	Days Per Week
washer	1	550	1	5
well pump	1	750	2	1
table saw	1	400	.5	3
dryer	1	300	1	5

Worksheet Instructions

To make calculations as accurate as possible, we have designed an easy-to-use worksheet. You will find tear-out copies of this worksheet in the appendix. It can be filled out and mailed to our technical department for system design assistance. See the sample worksheet that we've filled out. The following steps will assist you in filling out your worksheet.

Step 1: List the type, number, wattage, and hours per day each load or appliance will operate. You can determine power consumption of appliances in a variety of ways. Some have the wattage directly listed (60 watt light bulb). Many will require you to calculate the wattage by examining the appliance, and finding a small plate that lists the amps and voltage at which the appliance operates. The wattage is determined by multiplying these together (0.5 A x 115 V = 57.5 Watts). Appliances that will be run off a generator should be listed in the section at the bottom of the worksheet.

Step 2: Multiply quantity, wattage, and hours/day for each load and carry over into watt-hours/day column.

Step 3: Add the watt-hours/day sums together to determine your theoretical total watt-hours/day required for your home electrical system.

Step 4: Multiply by 1.25 to account for battery charging inefficiencies, as the process of charging batteries is not 100% efficient. The result is the amount of power you will need to produce daily to satisfy your electrical loads.

Step 5: Calculate your maximum concurrent appliance load. Because you may wish to operate several appliances at the same time, you will need to calculate the maximum AC wattage required at any one time. Think about which appliances you will want to operate concurrently (washing machine, dryer, stereo, etc.). Add up their total wattage and enter it in the box at the bottom of chart. This total is necessary for proper inverter sizing.

Step 6: Often, a home has certain electrical loads (a washing machine or table saw, for example) that will tax the battery capacity excessively. These loads can be run directly off of a fossil-fueled generator (an almost essential component in an independent energy system), sparing the batteries. This strategy is often employed if you have a small system. If your generator is adequately sized, you can charge batteries while doing the laundry, killing two birds with one stone. List all the appliances and tools that you might run directly from your generator in the generator section at the bottom of the chart.

Congratulations! You have just completed one of the most important parts of the puzzle.

Determining Your Battery Size

Power demands can be satisfied in different ways. You can run a generator. The cost is initially low, although the consumption of fuel is high, and the noise and servicing can be annoying. Running a generator to power a small pump or lamp is extremely inefficient.

Normally, we store the energy in batteries, allowing power to be drawn independently of our battery recharging. You can use your generator to recharge these batteries, or find another method, such as photovoltaic modules, wind energy devices, or hydro-electric systems.

The amount of energy stored in batteries is referred to as *amp-hours* stored at a particular voltage. For instance, a 12-volt battery might be said to have 100 amp-hours of storage (this is *not* the same as *cold-cranking amps* — see our battery section for details): 100 amp-hours x 12 volts = 1,200 watt-hours of stored energy from a fully charged battery to a fully discharged battery. You will want to have several days of energy stored in your battery bank to provide a cushion or reserve from which you may draw if your daily consumption fluctuates (which it will).

To determine the battery capacity required, fill in the Battery Sizing Chart in the Appendix. (We've provided a sample at the end of this section). First, divide the total corrected watt-hours needed per day from your System Demand Worksheet by the voltage of the battery system, resulting in amp-hours consumed per

If you are totally baffled, please give our technical staff a call. It's important that you understand these energy fundamentals to provide a foundation on which your energy system will be built.

day. Multiply the amp-hours per day by the number of days capacity desired in the battery bank to arrive at the total amp-hours storage.

We suggest maintaining 3 to 5 days' worth of power (sometimes referred to as 3–5 days of *autonomy*). Humans have great ability to vary their energy consumption. They throw parties; they go on vacation; they build weekend projects; and they vacuum the house before company comes over. Also, the generator may break down, or the sun might not shine. A reserve is a wise investment. You can literally say that you are saving (energy) for a rainy day!

Deep-cycle lead-acid batteries are the most common battery employed in independent energy systems. These are related to automotive batteries but are much better suited for operation at a variety of states of charge. It is recommended that one not discharge these batteries by more than 50% of their capacity. Lead-acid batteries can be discharged as much as 80%, although their life will be shortened if this is done repeatedly.

To determine the optimal battery capacity, divide the amp-hour subtotal by the maximum depth of discharge. Because you should not use the battery's entire capacity regularly, we must over-size the battery. For most battery systems divide by 80% or .80. With nickel-cadmium batteries, we could omit this last step, as they are somewhat immune to damage from excessive discharge, but also subject to shorter life with increased depth of discharge.

We now know two critical factors: how much energy we use per day, and the capacity of our battery bank (as expressed in amp-hours and volts). Now we can decide on which means are best to fill those batteries with energy.

See page 499 in the Appendix for a blank worksheet to fill out.

Battery Sizing Worksheet		
Total Corrected Watt-Hours per Day (from Demand Worksheet)	643	Wh/Day
System Voltage (usually 12 or 24)	+ 12	Volts
Load Amp-Hours per Day	= 53.6	Ah/Day
Days of Storage	x 4	Days
Amp-Hours Battery Storage Capacity	= 214.4	Amp-Hours
Maximum Depth of Discharge	+ .8 (80%)	Percent
Total Battery Capacity	= 268	Ah at 12 Volts

Charging the Batteries

Fossil-Fueled Generators

A generator can be considered the training wheels for an independent energy system. The dependency will drop off as other charging sources, such as solar, wind, or hydro can be employed. It is often cost-effective to provide 90% of needed energy from renewable sources and the remaining 10% by supplemental means.

In some applications, generators provide the only practical way to charge batteries. Despite their well-known flaws (noise, fuel cost, etc.), they do have the flexibility of permitting battery charging (or directly operating electrical loads) day or night, rain or shine. To charge batteries with a generator, you will need a **battery charger**, which transforms the 120V alternating current that generators are designed to produce into low-voltage direct current appropriate to replenish batteries.

Large storage batteries should only be charged slowly over several hours, because the electrochemical transformation proceeds at a modest rate (new battery research focuses on ways to accelerate the recharge process). The rate becomes even slower as the battery bank approaches a full state of charge. As a result, generators make excellent chargers when batteries are depleted, but poor chargers for topping-off. Where generators are the only charging source, batteries typically are less than fully charged, resulting in shortened life. The last 10% of charge, or topping off, is an ideal use for photovoltaic modules, as they quietly pump energy into batteries for many hours.

Hydro-Electric Systems

It is the fortunate homesteader who can take advantage of this tremendous source of power. The primary advantage of hydro is the fact that it is a constant 24-hour-a-day charging source, which can maintain batteries at a high state of charge, maximizing their life. The constant charging also allows for a smaller battery, as the necessity to store for the rainy day will be diminished. If you have a stream or spring with at least 15 feet of vertical drop on your property, it is worth your time to investigate.

Wind Generators

Wind generation captivates intense interest, and can be a tremendous source of power. If you live in a high-wind area, it can make good sense to investigate wind generation. Important considerations are site evaluation, energy conservation, and proper choice of equipment. If you have an average monthly wind speed of 8–14 miles per hour, wind generation is worth considering. See the Wind Generator section for details.

Photovoltaics

Photovoltaics are the power source of choice, used by over 90% of our customers. Sunlight is energy that is permanent, free, nonpolluting, totally quiet, and universally available. The sun is a giant power plant that generates radiant energy at an enormous rate estimated to be a staggering 110 trillion kW. Even though less than one billionth of this energy is intercepted by the earth, every square meter of the earth's surface facing the sun is estimated to receive about 1,000 watts at mid-day.

Supplying Your Power Needs

Now that you have calculated your total corrected watt-hours per day, from the Systems Demand Planning Chart, and the required size of your battery bank, from the Battery Sizing Calculation Worksheet, it is time to figure out where all the watts will come from to replenish your battery bank.

Charging with a Generator

If we wish to supply all of our energy with a generator, we must determine a maximum charger size because too large a charger can damage batteries and too small will require excessive charge times.

To determine an optimum battery charger size, divide battery capacity by 10. This will give us the optimum charge rate and charger size. Purchase a battery charger close in amperage to this number. Refer to the Total Corrected Watt-hour per day from the System Demand Planning Chart. Divide this number by the system voltage to determine amp-hours per day of

electrical consumption. When this number is divided by the battery charger amperage then the result is the approximate number of hours required to replace the watt-hours per day .

Charging with a Hydro System

If you have a hydro-electric potential, please fill out the hydro analysis worksheet and have us calculate your potential. Once this is done, you will know the watt-hours per day to be expected from the hydro system. This can be directly subtracted from your Corrected Watt-hours per day total. Any remainder will need to be fulfilled by another charging source. If the watt-hours per day generated by the hydro is in excess of that required per day, then you will not generally need any other charging sources.

Charging with a Wind Generator

First, an average windspeed at the site must be determined. Compare windspeed with watt output, as expressed by the wind generator manufacturer's literature. Multiply by 24 hours. The resulting watt-hours per day can be subtracted from the corrected total watt-hours per day.

Charging with Photovoltaics

Local solar conditions vary greatly due to local weather, vegetation, buildings, seasons, and geographies. The sizing method that follows is very general, using only average solar gain as the basis for determing the number of modules. In some cases, the average is not the best for sizing a system (for instance, a cabin used only in the summer months).

To give you a general idea of how to size a photovoltaic system, review your "Total Corrected Watt-hours/day" from the sizing worksheet. The Solar Sizing map in the Appendix will give you the average number of hours of sun per day at your location in the US. For foreign locations, contact one of our technicians. We have this many hours during which we must replace this many watt-hours. We divide the total watt-hours required by the number of hours of sun per day to result in the minimum **array** watt output.

Oversize the array by 15 percent to account for module inefficiencies, temperature-induced

The Real Goods technical staff has accurate solar information for hundreds of sites. Additionally, our computer programs can help tailor systems to a great number of circumstances.

voltage drop, or dirty panels. Select a photovoltaic module, and divide the array wattage by the wattage of the moules: (array wattage x 1.15)/wattage per module = number of modules. (See the sample worksheet).

A PV Power System For Your Home

A properly planned photovoltaic system consists of six basic components: (1) a solar array (solar panels) based on the number of watt-hours used each day, with back-up or secondary generator (wind, hydro, etc.), (2) the choice of a roof, ground, pole, or passive tracker mounting, (3) a charge controller and safety disconnect system, (4) a metering panel, (5) battery storage, and (6) an inverter (if AC power is desired).

How to Determine Which PV Voltage Is Best

All PV manufacturers make modules in several voltages for different applications. This needn't be confusing. Three "standard" voltages have emerged over the years: **14.5V** (Siemens M-65, Kyocera K-45), **16V** (Hoxan 4810, Siemens M-75, Kyocera K-51), and **17.1V** (Siemens M-55, Solarex MSX-60) with slight variations in voltage for different manufacturers. In the electricity vs. water analogy, voltage represents pressure or potential energy. Think of the battery as a pressure tank and the PV panel as the voltage pump.

To fill the 12V battery, higher pressure (voltage) must be pumped into the tank (battery) to stabilize it at 12 volts. A typical 12V battery reaches its optimum charge around 14.5V. Thus if we use a 14.5V PV panel to charge the battery system there is virtually no room for voltage loss or voltage "drop." 14.5V PV modules are called "self-regulating" because theoretically they taper the charge down to a minimum when the battery reaches full charge.

The 14.5-volt potential is considered inadequate to overcharge a battery. These self-regulating modules (Siemens M-65, Kyocera K-45) have been heavily promoted over the years by the RV (recreational vehicle) industry as the panacea for full-time RVers. There is one serious flaw. RVs typically hover in warm climates like the Southwest desert where the high temperatures decrease the PV module's voltage output. This can easily cause the lower 14.5V

modules to begin regulating too early, seriously diminishing the output when the battery is in its final charging cycle.

Other potential voltage loss can occur through long wire runs with improperly sized wire. Even the slightest voltage drop can cause significant power loss when the "pump" is only 14.5V to begin with. The only time we recommend using a 14.5V module is where the user has no intention of ever increasing to more than two modules, where the wire is sized appropriately, and where temperatures do not exceed 100°F.

For 95% of the systems we design we recommend using a 16V solar panel like the Hoxan 4810, Siemens M-75, or Kyocera K-51. These modules must have a charge controller to prevent overcharging the battery beyond 14.5 volts. Operating without a controller, particularly when you're using all the power you're generating, will not harm the batteries, but if you go away for several weeks at a time, thus removing a balancing load, your batteries will get seriously overcharged. On a one or two-module system, a charge controller may not be absolutely necessary.

As a rule of thumb, you can leave a one-module system for a week without damaging a battery, provided the battery is rated at 100 amp-hours or more. We recommend placing a switch in the line to one module of a two-module system to disconnect one of them when leaving a system unattended for up to one week. If a system of any size will be left for more than a week, install a charge controller.

The 17.1V modules are appropriate for battery charging in high temperature situations where the temperature may substantially drop the voltage, and in array direct applications, particularly water pumping.

Dear Dr. Doug:
Can I Mix & Match Different Solar Panels?

Answer: It is fine to mix different solar panels together in the same array, but try to keep the voltages within 5–10% of each other. If you put a 17.5V panel in parallel with a 14.5V panel the overall voltage will decrease to 14.5V. You won't hurt anything — you'll only lose a little power.

Power Supply Worksheets

Battery Charger

Total Battery Capacity in Amp-Hours	268 Amp-Hours	
	÷ 10	
Maximum Battery Charger Amperage	= 26.8 Amps	
Select a charger with the closest maximum charge rate		See the Todd Battery Chargers on Page 203
MODEL	30 AMPS	@ 15.5 VOLTS

See page 500 in the Appendix for a blank worksheet to fill out.

Generator Supply

Total Corrected Watt Hours per Day (from Demand Worksheet)	642.5 Wh/Day	Charger Amps	30 Amps	
System Voltage	÷ 12 Volts	Volts	x 15.5 Volts	
Load Amp-Hours per Day	= 54 Ah/Day	Watts	= Watts	
Charger Amperage	÷ 30 Amps	Gen. Derate (prevent overload)	x 1.25	
Approximate Hours of Generator Run Time	= 1 2/3 Hours	Minimum Generator Size For Battery Charging Only	= 581 Watts	

Photovoltaic Supply

Total Corrected Watt-Hours per Day (from Demand Worksheet)	642.5 Wh/Day	
Hours of Sun per Day (Solar Installation)	÷ 6 Hours	See Sun Map in Appendix on Page 490
Minimum Array Wattage Subtotal	= 107 Watts	
Photovoltaic Derate Factor	x 1.15	
Array Wattage Total	= 123 Watts	
Watts per Module	÷ 48 Watts	See Module Specs Beginning on Page 149
Number of Modules Required	= 3	Round up (to even number if 24 volt system)

Wind Power Supply

Average Windspeed at Site (mph)	15 MPH	
Wattage at This Windspeed	85 Watts	Write or Call for Manufacturers Spec's
Hours Running	x 24	
Expected Watt-Hours per Day	= 2040 Wh/Day	
Total Corrected Watt-Hours per Day (from Demand Worksheet)	642.5 Wh/Day	
Watt-Hours per Day from Wind	−2040 Wh/Day	
Watt-Hours To Be Satisfied By Other Sources	= 1397 Watt-Hours	

Hydro Power Supply

Total Corrected Watt-Hours per Day (from Demand Worksheet)	642 Wh/Day	For More Detailed Information on Your Hydro-Electric Potential Order Our Hydro-Site Analysis
Watt-Hours Produced by Hydro	−3456 Watt-Hours	See Hydro Sizing Chart on Page 191 for Estimate
Watt-Hours To Be Satisfied By Other Sources	= 2814 Watt-Hours	

AC or DC: That is the Question

When planning an alternative power system, one question that continually comes up is whether a certain appliance should be powered by alternating current(AC) or direct current (DC). Every system is owner specific, so there is no single answer to this question. However, we can provide information on this subject that will enable a person to make the right choices.

Unfortunately, most AC appliances are produced with one thing in mind: low cost. Even with the overwhelming evidence that we are in the midst of an energy crisis, there is still little concern for making products that are energy-efficient. Many DC appliances can perform the same tasks as their AC counterparts while consuming only a fraction of the power, but the initial price is higher. Often, the overall system price is lower with the more expensive DC appliance, because fewer solar modules are needed.

Advances in inverter technology have increased the options for anyone living off the grid. It is important here to understand a few characteristics of inverters in general. An inverter's function is to efficiently convert DC power, provided by the system, into AC power for "normal" household appliances. For a complete discussion of inverters, see the Off-the-Grid living section.

Although today's inverters are more efficient than their ancestors, they still lose from 5% to 20% of the power in the process of converting it. As system components, they are extremely reliable today, but they do require periodic servicing and replacement. When this happens, AC appliances will be temporarily out of commission while DC appliances will be humming along.

Inverters are generally most efficient when they are operating at about one-tenth or more of their rated output. If the load is extremely small, they are less efficient and waste a significant amount of energy. A small load that may be on by itself for extended periods of time is called a *phantom load* and is extremely undesirable. In this case it is best to use a DC appliance.

Many times there will not be a choice. Simply stated, there is not nearly as large a selection of DC appliances as there is for AC appliances. The most common situations where these choices come up are lighting, refrigeration, entertainment, fans, and pumps. We will address these subjects separately for convenience and simplicity.

Lighting

Lighting is perhaps the most confusing area of all. The variety and quality of the products available is constantly changing.

Until recently, the three strikes against DC lighting have been the limited selection, comparatively poor quality, and higher prices. At least one of these, quality, has been overcome recently. With DC lighting, you will save the 5% to 20% of power consumed by the inverter.

There is another issue called *power factor*, which has been greatly overlooked in the past. In a DC circuit, power is the product of voltage (volts) and current (amps); in an AC circuit, because voltage and current may not coincide in timing or phase, power can be less than that product. Power factor converts the product of voltage and current to actual power. For those on grid power, it is not a concern; they only get charged for energy used before considering power factor. For those living with an alternative energy system, this means that they need to account for power factor and the true wattage of an appliance when calculating system loads and sizing the system. If you choose AC lights, it is best to pick the ones with the highest power factor (as close to 1 as possible). For example, if a 15 watt compact fluorescent has a power factor of 0.5, then it will actually consume 30 watts. This is still better than incandescent lighting by a long shot. This does not apply to DC ballasts, because current and voltage are always in phase.

Manufacturers are always improving power factor (and harmonic distortion). In a couple of years, this should be a very small or non-existent problem.

The power factor issue is really only a concern if lighting is a major portion of the load. If lighting is not a significant portion of the total load, it may be best to utilize the larger selection of AC lights that work fine.

One other advantage to DC lights is that they do not interfere as much with radio and TV reception since they do not require an inverter, which can cause interference. For a complete discussion on reception and interference, see the inverters chapter.

DC lights are more expensive. Until recently, the investment hasn't been worth it, because the ballasts were low quality. The new DC ballasts in production are extremely high quality. They start the light instantly and are not very sensitive to temperature.

The vast majority of remote energy systems today use AC compact fluorescent lights. Perhaps the best systematic approach is to use a blend of both types. Because of the inefficiency of large inverters when there is only one small load, we recommend using a DC light for reading at night when there are no other loads on. The Co-Pilot lights are superb for this purpose.

Refrigerators

The only electric refrigerator that should be considered for a remote energy system is a Sun Frost refrigerator. The only other option is a propane refrigerator. The advantages of the Sun Frost usually outweigh the advantages of using propane. Once you have decided to go with a Sun Frost, you must decide which type of current to run it on. Sunfrost models are either AC or DC.

According to the makers of Sun Frost, AC and DC models consume the same amount of energy. The only difference between the two, concerning energy usage, is the efficiency of the inverter. This can be significant, as a refrigerator can be on up to 12 hours a day, every day. Also, if your inverter should need servicing, a DC refrigerator will still be in operation.

Stereos

Most inverters today do not produce a true sine wave. While they will operate almost all appliances with no problem, there can be a small 60-cycle hum or buzz in the background due to this. Additionally, inverters can cause interference in reception, particularly in AM radio. Almost all high-quality stereos will produce clean and clear sound on an inverter.

To avoid AM interference and possible 60-cycle hum, we suggest using high-quality car stereo equipment. Since they operate at 12 volts, the sound will be extremely clean. Typically, they use one-tenth the power of an AC stereo. There are a wide variety of premium-quality products available, including power boosters for higher volume.

Televisions

The vast majority of TVs will operate properly on inverters, and there is certainly a better selection in AC. There are some cases where the inverter can cause interference, particularly on the lower channels. This is a result of the type of inverter being used, the sensitivity of the TV, and the quality of local reception. This problem does not occur at all with certain types of inverters. Before buying an inverter, ask about possible interference.

There are some TVs (including a few high-end Sony units) that simply do not like to be run on inverters. Symptoms include lines across the screen, or a smaller viewing area.

Even DC televisions can experience interference if your inverter is on for some other reason. Power consumption is not really an issue with TVs, as most DC televisions are small, use relatively little power and provide a good quality picture. Sometimes AC TVs are more efficient than the same size DC model even when inverter inefficiency is taken into account. Black and white units typically consume much less power for the same size screen.

Pumps and Fans

Although pumps and fans are two different appliances, the rationale for recommending DC is the same for both. AC pumps and fans are not made with energy-efficiency in mind; the designers are more concerned with keeping the cost down. Although most AC pumps and fans can be run on an inverter, the energy use is excessive. As with refrigerators, it is cheaper in the long-run to buy the DC efficient model.

Remember, the only type of lighting that is appropriate for an independent power system is fluorescent lighting.

The Synergistic Effect

It is important to understand that the system as a whole should be considered when choosing which appliances are AC or DC. Because inverters have a tendency to interfere with AM reception, keep in mind that even with a 12-volt car stereo, you won't be able to get AM radio if an inverter is running the lights.

Some of these choices can get a bit confusing. If you feel overwhelmed, feel free to ask a Real Goods technician. We can give you the facts, and leave the choices up to you.

The McNeir family on their solar-powered boat in Raleigh, North Carolina. George and Hilliary McNeir have created a home that floats quietly and gracefully along the eastern waterways. The electricity from the solar modules atop the roof of the cabin permits extended cruising without the need for refueling stops.

Photovoltaics

A photovoltaic device or silicon solar cell converts light into DC (direct current) electricity. Photovoltaic modules have proven to be reliable electrical power producers in thousands of terrestial and space applications. It does not use heat from the sun as does thermal solar hot water. In fact, the higher the ambient temperature, the less efficient a solar electric cell becomes. The most common commercially available solar cell is a small wafer or ribbon of semiconductor material, usually silicon. One side of the semiconductor material is electrically positive (+) and the other side is negative (–). When light strikes the positive side of the solar cell, the negative electrons are activated and produce electrical current.

Photovoltaic Technology

When a group of solar cells are connected or the semiconductor ribbon material is applied to a predetermined surface area, a solar module is created, and the cells are encapsulated under tempered glass or some other transparent material. Quantitative electrical output is determined by the number of cells or ribbon material connected together within the module and then further determined by the number of modules connected together. More than one module connected together is called a solar array.

There are two primary photovoltaic technologies in use today: crystalline and amorphous silicon. Crystalline silicon was developed for the space program in the early 1950s and has actually seen very little improvement in over 30 years. Silicon is grown in large blocks, and the thin wafers that become the solar cells are sliced off. The conversion efficiency of commercially marketed crystalline silicon is about 13%. The great advantage of crystalline silicon is that there has been virtually no evidence of voltage degradation over time since the 1950s! Most reputable manufacturers warrant their crystalline modules for a minimum of 10 years, yet expect a life in excess of 40 years! Crystalline modules can be found in two varieties: single crystal (like Hoxan and Siemens) and polycrystalline (like

Kyocera and Solarex). There is very little if any difference between them in performance.

Amorphous silicon appears to be the wave of the future. It is created by a diffusion process whereby the silicon material is vaporized and deposited onto a glass or stainless steel substrate. Because of this process, there are unlimited possibilities for solar generation using amorphous silicon. Skylights, bay windows, office windows on skyscrapers, roofing material, and sunroofs on vehicles are all potential electrical generators with this new technology. Amorphous silicon, unfortunately, at this point has a maximum of 6% conversion efficiency, requiring twice the surface area as crystalline silicon to produce the same amount of power. Also, thus far with research and development still in its infancy, amorphous silicon has degraded significantly over time. Arco Solar introduced its first amorphous G-4000 power module in late 1986 only to recall it a year later. Nevertheless, the future is very bright for amorphous PV technology. Because of the manufacturing process, amorphous is far cheaper to produce than crystalline technology, particularly when quantities increase. The flexibility of amorphous PV also brings significant advantages, allowing the modules to be custom fitted to countless applications. One further advantage of amorphous is its property of greater output in low light conditions.

Gallium arsenide is one more technology of photovoltaics that is currently under intense research. It has the advantage of conversion efficiencies in the 30% range, but thus far is too expensive to be practical.

The 10 or 12-year warranty that comes with standard modules is a reflection on the reliability and confidence the manufacturers have in these amazing devices.

Tilt Angle for PV Module Mounting

Your PVs will operate at optimum when placed perpendicular to the path of the sun. Always mount PVs facing due south (due north in the southern hemisphere). Do not assume that **magnetic south** is the same as true south. An inspection of the map of magnetic declination (deviance of true north from magnetic north) will show roughly how far out of whack one can be! Change the tilt angle at least two times every year if possible. At the summer solstice set them at the local latitude minus 15°, and at the winter solstice at latitude plus 15°.

Many new PV users have eliminated this step by sizing the PV array for the worst-case winter scenario. This means setting the array to 50-55° in a location at 40° north latitude and never changing it. More power will be delivered if changed four times a year. Modules placed within 5° of perpendicular to the sun operate within 0.9962%± maximum allowable output. Beyond a 10° error, dramatic losses occur.

Dear Dr. Doug:
Do I Need a Blocking Diode?

Answer: Customers get differing opinions on whether or not to use a blocking diode on solar systems. A blocking diode is an electronic device that allows current to flow in only one direction, like a one-way check valve in a water system. Without this device, a small amount of power will flow backwards at night from the batteries to the PVs where it will dissipate. In performing the diode's function, some power loss occurs during daytime charging from the slight forward resistance of the diode. Generally you will lose more power through the diode than you will lose in night-time losses for systems. Most photovoltaic charge controllers will already have some mechanism designed into them to prevent nighttime reverse current leakage. It should be emphasized that this does not represent a major loss in most systems, and the effect has been somewhat blown out of proportion.

For a diagram on the proper wiring of photovoltaic panels to your battery, see page 488 in the Appendix.

A List of PV Absolutes

Do not use concentrators or reflectors to enhance module output unless you are located in a very cold climate. Modules operate best from 63° to 120°F. Concentrators will decrease the efficiency of the module due to heating.

Do not place plexiglas, glass, etc. as a shield in front of the panels. This will filter the light, increasing the temperature, decreasing the output.

Do not mount the panels flat on the roof. They need air circulation under the panels.

Do not place any part of the module in the shade. No light, no electricity. This includes power and telephone lines.

Do determine where the best solar window is at your home site and put the modules there.

Do wash off the modules from time to time. *This is the only regular maintenance requirement.*

On the following pages you will find a selection of photovoltaic modules by different manufacturers, most of whom we have worked with for many years. These modules are highly reliable and warranted by their respective manufacturers for up to 12 years; however you can expect a 40-year life span or more with a possible power degradation of 10% over that time. There are no residential systems that have been in service for 40 years, so we really can't say for sure just how long a module will last, but it looks like the life expectancy is substantial.

Hoxan PV Modules

Hoxan operates the world's largest and most highly automated PV manufacturing plant. Hoxan PV modules are very close in looks and performance to Siemens modules. All Hoxan modules come with a full 10-year power output warranty. These have been our standard modules in our remote home packages throughout the 1990s and we're very pleased with their performance as well as the company's commitment to quality.

H-4810 Hoxan

The Hoxan H-4810 is nearly identical in output to the Siemens M-75. It uses 36 high-efficiency square single-crystal cells on a thick anodized aluminum frame with highly transparent tempered glass. The 4810 produces 48 watts of power at 16.2 volts to generate 3.0 amps of current. The 4810 measures 16.5" wide by 37.25" long by 1.25" thick.

HOXAN 4810		
rated watts	48.0	@ 25°C
rated power	volts: 16.2	amps: 3.0
open circuit volts	21.3	@ 25°C
short circuit amp	3.2	@ 25°C
size (LxWxD)	37.2 X 16.5 X 1.3	
construction	single crystal, tempered glass	
warranty	10 years	

11-401 Hoxan 4810 (Qty 1–3)   **$349**
          (Qty 4–19)   **$339**
          (Qty 20+)   **$329**

Siemens Photovoltaic Modules

Recycled Arco ASI 16-2000

We're conservatively rating these modules at 32 watts even though at 25°C the manufacturer rates them at **35 watts**! At 47°C (typical summertime temperature) they're rated at 31.5 watts. These modules are from a demonstration power plant and have seen 7 to 8 years of service. You can safely expect another twenty to thirty years of service. All modules are given the standard laboratory flash test to insure that substandard units are weeded out. There is one terminal per polarity at opposite ends of the module. Frames are standard anodized aluminum with tempered glass glazing. If you need heavy charging in hot weather our higher voltage recycled Quad-lams offer better performance. But if your loads are primarily residential which peak in the winter, then these tried and proven modules offer an unbeatable bargain. 5 year warranty to produce within 15% of 32 watts.

Mounting structures start on page 167 and Charge controllers start on page 207

ARCO SOLAR 16-2000		
rated watts	35	@ 25°C
rated power	volts: 15.5	amps: 2.26
open circuit volts	20.5	@ 25°C
short circuit amps	2.55	@ 25°C
size (LxWxD)	48 x 12 x 1.5	
construction	single crystal, tempered glass	
warranty	5 years	

11-125 Arco ASI 16-2000 (Qty 1-3)   **$169**
          (Qty 4-19)   **$159**
          (Qty 20+)   **$149**

Recycled Arco Quad-Lam Modules

We find our recycled Arco Quad-Lams to be the most powerful charging configuration of any used module system we've ever offered (or seen offered). This Quad (4-module) configuration makes for a much higher voltage charging system that really pumps out the power, especially in hot climates. We are giving the Quad-lams a conservative real world rating of 100 watts (this is 20 to 30 watts less than their "official" rating). Peak power is at 17.7 volts and 5.6 amps. Four single lams must be wired in series to charge a 12V system. Modules have individual anondized aluminum frames and weather-tight junction boxes, with secure electrical connectors. Instructions and diagrams are included for wiring. The modules measure 11-7/8 x 48¼ inches each, are six to seven years old, and were used on a power plant in Southern California. All modules are warranted to put out within 15% of 100 watts for a period of 5 years.

The modules have a brown tinge to them resulting from the discoloration of the EVA encapsulated layer. All modules using EVA (currently all major manufacturers) will be affected similarly, with higher module temperatures accelerating this process. Once the cells have browned there usually is no further discoloration or loss of power. Tests have shown that the worst case is a power reduction of 10% from a brand-new to a browned module. Therefore what you buy now as a recycled module will have minute, if any, future power loss. Beware of other, less expensive "Lam systems" on the market. Our Quad-lams are hand selected for the highest possible wattage and quality.

We occasionally come up with used or "recycled" modules that are very attractively priced and will last for many years. Call for information

Siemens M-75

The Siemens M-75 produces 48 watts at 15.9 volts. The M-75 is efficient, attractive, easy to install, and comes with a wired-in bypass diode in each junction cover. The M-75 consists of 33 cells in series. Each module comes with an easy-to-understand instruction manual and a 10-year warranty. With its high amperage (3 amps at load) it is an excellent module for stand-alone systems. UL Listed. American made. 10-year warranty.

RECYCLED QUAD-LAMS		
rated watts	100	@ 25°C
rated power	volts: 17.7	amps:5.6
open circuit volts	25.0	@ 25°C
short circuit amps	5.9	@ 25°C
size (LxWxD)	approx. 48 x 12 x 1.4	
construction	single crystal, tempered glass	
warranty	12% of 100 watts 5 years	

11-134 Recycled Arco Quad-Lams (100 watts)
Quantity 1–3	$449
Quantity 4–19	$429
Quantity 20+	$399

SIEMENS M-75		
rated watts	48	@ 25°C
rated power	volts: 15.9	amps: 3.02
open circuit volts	19.8	@ 25°C
short circuit amps	3.4	@ 25°C
size (LxWxD)	48 X 13 X 1.4	
construction	single crystal, tempered glass	
warranty	10 years	

11-101 Siemens M-75 (Qty 1–3) $449
(Qty 4–19)	$439
(Qty 20+)	$429

Siemens M-40

The Siemens M-40 is an underrated version of the M-75. It is exactly the same size as the M-75, but when Siemens tests their modules, those that come out less than 48 watts get classified as M-40s, guaranteed to produce at least 40 watts at 15.7 volts for a current of 2.55 amps. UL Listed. Ten year warranty. American made.

SIEMENS M-40		
rated watts	40 +	@ 25 C
rated power	volts: 15.7	amps: 2.55
open circuit volts	19.5	@ 25 C
short circuit amps	3.0	@ 25 C
size (LxWxD)	48 X 13 X 1.4	
construction	single crystal,tempered glass	
warranty	10 year	

11-102	Siemens M40 (1–3 qty)	$369
	(4–19 qty)	$359
	(20+ qty)	$339

Siemens M-50

The M-50 is an underrated version of the Siemens M-55, producing at least 48 watts at 17.3 volts for a current at load of 2.78 amps. It is identical in size and configuration to the M-55. UL listed. Ten year warranty. American made.

SIEMENS M-50		
rated watts	48	@ 25°C
rated power	volts: 17.3	amps: 2.78
open circuit volts	21.6	@ 25°C
short circuit amps	3.2	@ 25°C
size (LxWxD)	50.9 X 13 X 1.4	
construction	single crystal,tempered glass	
warranty	10 year	

11-106	Siemens M-50 (Qty 1–3)	$419
	(Qty 4–19)	$409
	(Qty 20+)	$389

Siemens M-55

The M-55 is Siemens' most powerful standard module, producing 3.05 amps with 53 watts at 17.4 volts, and consisting of 36 cells in series. It is ideal for water pumping applications where higher voltage is required, or for providing extra voltage for long wire runs. It is the best module to use in extremely hot climates as high-temperature voltage drop is kept tolerable. UL listed. Ten year warranty. American made.

SIEMENS M-55		
rated watts	53	@ 25°C
rated power	volts: 17.4	amps: 3.05
open circuit volts	21.7	@ 25°C
short circuit amps	3.4	@ 25°C
size (LxWxD)	50.9 X 13 X 1.4	
construction	single crystal, tempered glass	
warranty	10 years	

11-105	Siemens M-55 (Qty 1–3)	$469
	(Qty 4–19)	$459
	(Qty 20+)	$439

Mounting structures start on page 167 and Charge controllers start on page 207

Siemens M-65

The M-65 is designed primarily for RV, marine, and remote home usage; when only one module is employed. The M-65 consists of 30 cells wired in series. It is a "self-regulating" module that decreases its current output from 3 amps to less than 1/2 amp when the battery approaches full charge, eliminating the need for a charge controller, but often seriously limiting the module's output. Battery must be 83 amp-hours or larger or a charge control will be necessary. Not recommended for very hot climates where temperatures often exceed 100°F. The Siemens M-65 produces 43 watts at 14.6 volts for a current of 2.95 amps. UL listed. Ten year warranty. American made.

Most utilities would be delighted to buy surplus electricity generated by rooftop PV arrays during peak usage hours. They tell us it's the regulators that don't want it happening: they might not get their pound of flesh. Keep an eye on your local PUC.

SIEMENS (ARCO) M-65		
rated watts	43	@ 25°C
rated power	volts: 14.6	amps: 2.95
open circuit volts	18.0	@ 25°C
short circuit amps	3.3	@ 25°C
size (LxWxD)	42.6 x 13 x 1.4	
construction	Single crystal, tempered glass	
warranty	10 years	

11-103 Siemens M-65 (Qty 1–3) $399
(Qty 4–19) $389
(Qty 20+) $369

Siemens M-35

The M-35 is an underrated version of the Siemens M-65 and it produces a guaranteed 37 watts at 14.5 volts making 2.56 amps of current. It is identical in size and configuration of the M-65. UL listed. Ten year warranty. American made.

Siemens M-20

The M-20 is a compact, 20-watt self-regulating module ideal for RVs, boats, and remote homes where needs are minimal, use is intermittent, or space is limited. As the battery approaches full charge, the M-20's current output decreases from a rate of 1.37 amps to less than ¼ amp, eliminating the need for a charge controller. Siemens recommends at least 70 amp-hours of battery storage for each M-20 module. 14.6 volts, 1.37 amps. UL listed. Five year warranty. American made.

SIEMENS M-20		
rated watts	22	@ 25°C
rated power	volts: 14.6	amps: 1.5
open circuit volts	18.2	@ 25°C
short circuit amps	1.65	@ 25°C
size (LxWxD)	22.4 X 13 X 1.4	
construction	single crystal, tempered glass	
warranty	5 years	

11-107 Siemens M-20 $229

SIEMENS M-35		
rated watts	37	@ 25 C
rated power	volts: 14.5	amps: 2.56
open circuit volts	18.1	@ 25 C
short circuit amps	3.0	@ 25 C
size (LxWxD)	42.6 X 13 X 1.4	
construction	single crystal, tempered glass	
warranty	10 year	

11-104 Siemens M-35 (Qty 1–3) $349
(Qty 4–19) $339
(Qty 20+) $319

Solarex PV Modules

MSX60 Solarex Module

The MSX60 is Solarex's production solar panel. Solarex Corporation uses photovoltaic power to help build their modules in the U.S. The Solarex line offers exceptional performance in high temperature climates, as all the modules offer 16.8 volts or higher output. Solarex uses polycrystalline cells coated with a titanium dioxide antireflective material. The MSX60 is built with a strong, rugged frame of corrosion-resistant, bronze-anodized aluminum. It will produce 60 watts of power at 17.8 volts to produce 3.5 amps of current. Ten-year warranty. Dimensions: 43.65"L x 19.76"W x 2.13"D and weighs 16 pounds.

SOLAREX MSX-60		
rated watts	60	@ 25°C
rated power	volts: 17.1	amps: 3.5
open circuit volts	20.8	@ 25°C
short circuit amps	3.8	@ 25°C
size (LxWxD)	43.65 X 19.76 X 2.13	
construction	multicrystal, tempered glass	
warranty	10 year	

11-501 Solarex MSX60 (Qty 1–3)   **$439**
     (Qty 4–19)   **$429**
     (Qty 20+)   **$399**

MSX56 Solarex Module

The MSX56 will produce 56 watts of peak power at 17.8 volts to produce 3.37 amps of current. It comes with standard Solarex features including the ten-year warranty. It measures 43.65"L x 19.76"W x 2.13"D and weighs 16 pounds.

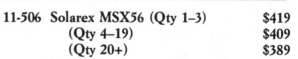

SOLAREX MSX-56		
rated watts	56	@ 25°C
rated power	volts: 16.8	amps: 3.35
open circuit volts	20.8	@ 25°C
short circuit amps	3.6	@ 25°C
size (LxWxD)	43.65 X 19.76 X 2.13	
construction	multicrystal, tempered glass	
warranty	10 year	

Mounting structures start on page 167 and Charge controllers start on page 207

11-506 Solarex MSX56 (Qty 1–3)   **$419**
     (Qty 4–19)   **$409**
     (Qty 20+)   **$389**

MSX64 Solarex Module

The MSX64 is Solarex's highest power photovoltaic panel. Solarex uses polycrystalline cells coated with a titanium dioxide anti-reflective material (which also turns them a beautiful deep blue). Has tempered glass cover and bronze anodized extruded aluminum frame. Spacious and convenient junction box takes a wide variety of wire terminations including bare wire. Produces 64 watts at 17.5V and 3.66 amps. Ten year warranty. 43.65"L x 19.76"W x 2.13"D.

11-505 Solarex MSX64 (Qty 1-3)   **$469**
     MSX64 (Qty 4-19)   **$459**
     MSX64 (Qty 20 +)   **$429**

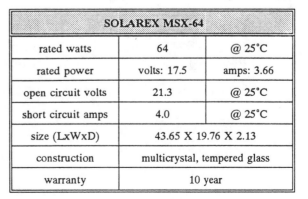

SOLAREX MSX-64		
rated watts	64	@ 25°C
rated power	volts: 17.5	amps: 3.66
open circuit volts	21.3	@ 25°C
short circuit amps	4.0	@ 25°C
size (LxWxD)	43.65 X 19.76 X 2.13	
construction	multicrystal, tempered glass	
warranty	10 year	

MSX40 Solarex Module

The MSX40 is similar in output to the old Arco M-73. It produces 40 watts at 17.2 volts to deliver 2.34 amps of current. It comes with the standard 10-year Solarex warranty. 30.07"L x 19.76"W x 1.97"D. 10 pounds.

SOLAREX MSX-40		
rated watts	40	@ 25°C
rated power	volts: 17.2	amps: 2.34
open circuit volts	21.1	@ 25°C
short circuit amps	2.53	@ 25°C
size (LxWxD)	30.07 X 19.76 X 1.97	
construction	multicrystal, tempered glass	
warranty	10 year	

11-503 Solarex MSX40 (Qty 1–3) $329
(Qty 4–19) $319
(Qty 20+) $299

MSX30 Solarex Module

The MSX30 will supply small to moderate energy demands. This panel is identical in output to the unbreakable SX-30, but with tempered glass glazing, bronze anodized frame, and standard junction box for wiring. It will produce 30 watts of peak power at 17.8 volts and 1.68 amps. Weight is 8.5 lbs.

SOLAREX MSX-30		
rated watts	30	@ 25°C
rated power	volts: 17.8	amps: 1.68
open circuit volts	21.3	@ 25°C
short circuit amps	1.82	@ 25°C
size (LxWxD)	23.3 x 19.76 x 2.13	
construction	multicrystal, tempered glass	
warranty	5 year	

11-507 Solarex MSX30 (Qty 1-3) $399
(Qty 4-19) $389
(Qty 20+) $369

SX30 Solarex Unbreakable Module

The "Lite" series of Solarex modules are made without glass and are very thin, lightweight, and highly portable. Typically they're used for boating, camping, RVs, scientific expeditions, railroad signaling, and mobile communications. This series of panels is great for recharging laptop computers and video cameras in remote locations. The SX30-Lite produces 30 watts of peak power at 17.8 volts for a current of 1.68 amps, weighing in at less than 5 pounds! One-year warranty. 24.25"L x 19.5"W x 0.38"D. 4.5 pounds.

SOLAREX SX-30		
rated watts	30	@ 25°C
rated power	volts: 17.8	amps: 1.68
open circuit volts	21.3	@ 25°C
short circuit amps	1.82	@ 25°C
size (LxWxD)	24.25 x 19.5 x .38	
construction	multicrystal, no glazing	
warranty	1 year	

11-511 SX30 Solarex $299

SX18 Solarex Unbreakable Module

The SX18-Lite produces 18.6 watts of power at 17.8 volts, for a current of 1.06 amps. 19.63"L x 17.63"W x 0.38"D. 3.5 pounds. One-year warranty.

SOLAREX SX-18		
rated watts	18.5	@ 25°C
rated power	volts: 17.5	amps: 1.06
open circuit volts	21.0	@ 25°C
short circuit amps	1.16	@ 25°C
size (LxWxD)	19.63 x 17.63 x .38	
construction	multicrystal, no glazing	
warranty	1 year	

11-512 SX18 Solarex   **$239**

SX10 Solarex Unbreakable Module

The SX10-Lite produces 10 watts of peak power at 17.5 volts for a current of 0.57 amps. 17.63"L x 10.63"W x 0.38"D. 2 pounds. One-year warranty.

SOLAREX SX-10		
rated watts	10	@ 25°C
rated power	volts: 17.5	amps: .57
open circuit volts	21.0	@ 25°C
short circuit amps	0.60	@ 25°C
size (LxWxD)	17.63 x 10.63 x .38	
construction	multicrystal, no glazing	
warranty	1 year	

11-513 SX10 Solarex   **$139**

Solarex SA5 Module

The Solarex SA5 is nearly identical to the old Arco G-100. It produces 5 watts at 17.5 volts to make 300mA of current. It is attractively priced and ideal for small applications and science projects. Size: 14.14"L x 13 -14"W x 1.08"D, weighs 2 pounds.

11-521 SA5 Solarex   **$89**

Epos Amorphous Module

The Epos Amorphous PV module is rated at 4 watts peak, producing 14.5 volts at a current of 330 mA. It is attractively priced and ideal for small applications and science projects. Perfect for fence chargers and the Mighty Mule gate opener. Dimensions: 12" x 12".

11-120 4 watt Epos   **$79**

See the article "Electronic Toys in the Bush" on pg. 160 for more how-to info on running your laptop or video camera while away from the grid.

To order any of these products
or for more technical information
call us toll-free at
1-800-762-7325

Kyocera
Photovoltaic Modules

Kyocera is a Japanese company with San Diego PV-powered manufacturing facilities for their PV modules. They make an excellent polycrystalline panel with a conversion efficiency over 13% (similar to Hoxan and Siemens) and an industry-leading 12-year warranty.

Kyocera K51

The K51 is Kyocera's standard "building block" module similar to the Hoxan 4810 and the Siemens M-75. It produces 51 watts at load using 16.9 volts for an amperage of 3.02 amps. Twelve year warranty.

Kyocera K45

The K45 is very similar to the Siemens M-65. It is a self-regulating module with an optimum voltage of 15.0 volts. Its output is 45.3 watts for a maximum current of 3.02 amps. It is ideal when only one or two modules are used, since a charge controller need not be employed. The same cautions apply as to the M-65s: don't use in extremely hot climates or when wire isn't adequately sized. Twelve year warranty.

KYOCERA K-45		
rated watts	45.3	@ 25°C
rated power	volts: 15.0	amps: 3.02
open circuit volts	18.9	@ 25°C
short circuit amps	3.25	@ 25°C
size (LxWxD)	34.6 X 17.5 X 1.4	
construction	multicrystal, tempered glass	
warranty	12 years	

11-202 Kyocera K45 (Qty 1–3)   **$409**
      (Qty 4–19)   **$399**
      (Qty 20+)   **$379**

KYOCERA K-51		
rated watts	51.0	@ 25°C
rated power	volts: 16.9	amps: 3.02
open circuit volts	21.2	@ 25°C
short circuit amps	3.25	@ 25°C
size (LxWxD)	38.8 X 17.5 X 1.4	
construction	multicrystal, tempered glass	
warranty	12 years	

11-201 Kyocera K51 (Qty 1–3)   **$419**
      (Qty 4–19)   **$409**
      (Qty 20+)   **$389**

Kyocera K63

The K63 is a very high voltage panel with an optimum voltage of 20.7 volts. Its output is 62.7 watts at 3.03 amps. If three K63s are placed in series, they can charge a 48V system. The extremely high voltage gives the K63 very limited application. Twelve year warranty.

KYOCERA K-63		
rated watts	62.7	@ 25°C
rated power	volts: 20.7	amps: 3.03
open circuit volts	26.0	@ 25°C
short circuit amps	3.25	@ 25°C
size (LxWxD)	47 X 17.5 X 1.4	
construction	multicrystal, tempered glass	
warranty	12 years	

11-215 Kyocera K63 (Qty 1–3)   **$549**
        (Qty 4–19)   **$539**
        (Qty 12+)   **$519**

Uni-Solar Flexible PV Modules

Since Sovonics stopped producing lightweight, durable, flexible solar panels, there has been no supply of modules that people can conveniently take with them backpacking, on boats, or even on airplane rides. We are happy to announce that the manufacturers are back in business with a new company name, Uni-Solar. Their amorphous silicon modules are flexible, so you can roll them up and transport them just about anywhere. They have grommeted holes in the corners for easy mounting wherever the sun is shining. Use them with one of our 12-volt nicad battery chargers (50-214) to charge batteries much faster than the solar chargers we sell. Or simply plug one into any 12-volt appliance with the appropriate adapter plug. In this case, you must make sure that the wattage of the panel is sufficient for the power usage of the appliance. They work splendidly for powering "boom boxes" on the beach. Take one on your camping trip, and you won't be left in the dark.

They are so rugged that you can place one on the deck of your boat and walk on it with no damage. They have even survived bullet holes without significantly affecting output. So throughout treks, voyages, or skirmishes, you'll never be without watts from the sun. Three sizes are available, according to how much power you need. All modules come with a 3-year warranty.

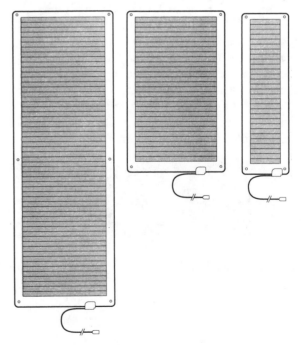

11-220 MBC-525 22W (51"x16")   **$395**
11-221 MBC-262 11W (28"x16")   **$245**
11-222 MBC-131 5.5W (28"x8")   **$145**

10-Watt Amorphous PVs

We made a large purchase of brand new 10-watt PV modules and are offering them at a very attractive price. These modules are amorphous and may degrade in output by as much as 20% over the lifetime of the panel. Originally rated at 11 watts; we are conservatively calling them 10-watt modules. They measure 12-1/8 by 36-1/8 inches and have a polycarbonate frame. Output is 14.5 volts at 759 mA.

11-555 Amorphous 10W   **$79**

Car Charger

Our car charger is a self-contained, self-regulated, portable battery charger. When exposed to the sun, it provides a continuous trickle charge to your battery in order to maintain optimal power and performance. Batteries typically leak or discharge power at a rate of up to 15–40% per month. Parking a car for a long period always carries the risk of finding a dead battery when you return. The charger is very conservatively rated at 1 watt (although we consider them to be 2 watts), and all you need to do to use it is place it on your dashboard and plug it into your cigarette lighter. An 8 foot cord is included. Size: 13½"L x 7"W x ½"D. No more dead batteries when you leave your car at the airport. It works great for tractors and farm vehicles too. We are offering a great deal because of a large volume purchase.

11-121 Solar Car Charger   **$39**

Solarex Battery Mate

The battery mate is designed to sit on the dashboard of your car and is ideal for keeping your battery charged if your car, tractor, or boat is parked for a long time. It puts out 1½ watts at 14½ volts for 100 milliamps. You'll never have to worry about dead batteries when you leave your car at the airport for two weeks.

11-122 Battery Mate   **$59**

Sovonics 1-watt Panel

We've acquired a number of these handy little thin-film, flexible panels that are great for the hobbyist. They can be used in a number of applications, such as building your own battery charger, powering a walkman or a small radio or a toy. They are waterproof and have solder tabs. Each panel produces approximately 1 watt and is rated at approximately 350 milliamps @ 3.3 volts. It measures 8-1/8 long by 6-3/8 wide.

11-131 Sovonics 1-watt   **$14**

Solargizer Battery Conditioner

This is the only solar-powered device that can actually recondition lead-acid storage batteries. During extended periods of nonuse, a battery's porous lead plates become clogged with a residue that resists removal by normal charging methods. The result is a battery that cannot supply the current necessary to start an engine. The Solargizer's unique electronic circuit stores current from the solar panel and discharges it at high amperage into the battery. This burst of energy clears the plates and keeps the battery ready for quicker starts. For use on any type of starting battery; cannot overcharge. Comes with heavy-duty, 3-foot cable, and copper-plated battery clips.

11-126 Solargizer   **$59**

Solar Pathfinder—Site Analyzer

One part of designing a system is picking the best place to mount the panels. Also, you may want to know if there are any good sites at all on your property. This simple-to-use tool eliminates all the guess work. Place it anywhere and it will instantly show you the sun's path across the sky for every day of the year. If there is a telephone pole or a tree blocking the way, the Pathfinder will tell you what time the sun will be blocked by it and for how long. Put it on your kitchen table and it will tell you when the sun will shine on a plant there all year round. We have investigated several of these instruments and found this one to be the simplest to operate and by far the most versatile. It even works great in cloudy conditions, and even in moonlight! When you have taken a reading, you can record it on paper and compare results from other sites. Includes data for your specific latitude; data for other latitudes is available. No solar technician or installer should be without this extremely valuable and finely made tool. Useful for photovoltaic, solar hot water, and passive solar applications. Please specify latitude at which Pathfinder will be used.

63-348 Solar Pathfinder   **$189**

Electronic Toys in the Bush

Or how to take your computer, video camera, etc. out into the field with you

Have you ever wondered how you're going to power your laptop computer or video camera while taking your dream vacation (or sabbatical) backpacking across the Sierras or bushwhacking through a tropical rain forest? Well, you could hire a burro and pack along 600 pounds of batteries, but that's not an elegant solution. The technical staff at Real Goods has been presented with this problem with increasing frequency, and we've developed a variety of elegant, light-weight solutions.

The first problem is the lack of any standard voltage for laptops or cameras. All these devices use some kind of rechargeable battery pack. These packs vary between 3 and 15 volts. If your electronic device manufacturer offers an optional 12 volt DC charger (or better yet a separate charging stand), get one! This will make recharging simple (this charger is designed to plug into a standard cigarette lighter socket in a vehicle). It's better if you can get a separate charging stand with a spare battery or two. You can then operate with one battery while a second one is recharging. Now instead of plugging into a vehicle, we're simply going to plug into a lightweight unbreakable solar module. The Solarex Lite series of modules is what we recommend. They're mounted on a tough fiberglass backing with no heavy, breakable glass glazing. The Lite series is available in 10, 18, and 30 watt sizes. (Item #'s 11-511, 11-512, & 11-513.) The SX-10 (10 watt) is usually sufficient to recharge a laptop battery with 6 to 8 hours of sunlight. Heavy users may want to step up to the SX-18 (18 watt). Video cameras are heavier power consumers, you'll probably need the 18 or 30 watt size for camera use. These modules come with about 5' of flexible wire attached, (you can splice on more if necessary), but keep the distance from module to appliance at 15' or less. Finally you'll need a female cigarette lighter plug to match the male plug on the charger. Our item #26-108 at $3 does the job neatly.

What if you can't get a 12 volt charging adaptor? What if your unit will only recharge from a 120 volt AC source? Not to fear! We can still get you power in the bush, but it's not going to be quite as simple and efficient. In this case you need to insert a small inverter between the solar module and your charger. The PowerStar 200 Pocket inverter (item # 27-104) works beautiful-

See page 235 for the Power Star 200 inverter.

See pg. 154, 155 for the Solarex "Lite" series of modules.

ly here. The PowerStar will accept the variable input from the module and turn out regulated, precise 117 volts AC. Power output will depend on power input. This may also be the simplest solution if you have equipment with differing voltage needs. Call our tech staff if you have further questions.

Pre-Packaged PV Power Systems

The beauty of solar electric power systems lies in their modularity. Solar panel installations are like building-block constructions that can always be added to and enhanced in the future. Unlike most consumer items that have obsolescence built in, you can use your solar panels virtually forever and add on as power needs and budget capabilities expand.

We'll be happy to sit down with you and plan out your system no matter how small or large, either in person or by phone. To make getting started easier, we've assembled some pre-packaged kits that are properly sized for various applications and that include the necessary hardware to bring you solar electricity. We've been selling these pre-engineered systems for four years now and have received rave reviews from purchasers.

Siemens Freedom Line Solar RV Kits

The following RV kits are packaged by Siemens Solar and are designed for modest battery charging applications. All Siemens RV kits contain 25 feet of wire, installation accessories (such as fuses and mounting hardware) and thorough instructions.

RV Kit 1

This kit will keep fresh starter batteries when vehicles are left unused for prolonged periods. The Siemens G50 solar module will trickle charge a starting battery for your RV, boat, or tractor so that when your ready to play, your rig will be as well.
Components: Siemens G-50 solar panel, 25 ft wire, installation accessories & instructions.
Power produced: 12 watt-hours per day
12-101 RV Kit 1   **$109**

RV Kit 2

With twice the output of the G50 module used in the RV1, the RV 2 will recharge depleted batteries when recharge time is not a factor. This will allow for occasional small appliance use (lights, radio).
Components: Siemens T-5 solar panel, 25 ft wire, installation accessories & instructions.
Power produced: 25 watt-hours per day
12-102   RV Kit 2 $169

RV Kit 3

The Siemens M20 module is the little brother to the full size modules used in large systems. A reasonable minimum system for the full time RV user, the 20 watt module can provide power for the most frugal.
Components: Siemens M-20 solar panel, 25 ft wire, installation accessories & instructions.
Power produced: 110 watt-hours per day
12-103   RV Kit 3 $449

RV Kit 4

The largest of the kits which makes use of *self regulating* modules, the RV 4 would recharge a fully depleted 100 amp-hour battery with a week of good sun. Usage: Battery charger plus power for small and large appliances.
Components: Siemens M-65 solar panel, 25 ft wire, mounting hardware, installation accessories and instructions.
Power produced: 175 watt-hours per day
12-104   RV Kit 4 $599

Siemens Freedom Line High Power Kits

Siemens now offers two new higher power kits which will allow better summer performance.

RV Kit 5

The Siemens RV kit 5 makes use of a full size and voltage M75 module controlled by a SCI Mark III charge controller with status meters. The flush mounted control will reliably protect batteries while informing one of battery state of charge and solar panel status. The controller can accept the input from as many as four full size solar modules.
Components: Siemens M75, Mark III controller, installation hardware, mounting brackets, wire, and instructions.
Power produced: 225 watt-hours per day
12-105   RV 5 Kit $799

RV Kit 7

The RV 7 is an expander kit for the RV 5 and will double the output of the RV 5 kit. The RV 7 should not be installed without a charge controller.
Components: M75 module, hardware, mounting brackets and module interconnect wire.
12-106   RV 7 Kit (RV 5 Expander) $689

**To order any of these products
or for more technical information
call us toll-free at
1-800-762-7325**

Remote Home Kits

The typical alternative energy home (there are now over 50,000) uses between a third and a tenth of the energy consumed by a conventional home.

We've spent considerable time and energy fine-tuning our Remote Home (RH) Kits over the last four years. We've produced a comprehensive and easy to understand owner's manual and installation guide that includes complete instructions and safety guidelines for all the Remote Home Kits. We've refined the kits themselves to a state we're very happy with, including our much larger and deluxe Remote Home Kit 4 for larger power consumers who are looking for state-of-the-art equipment. All power figures are calculated at 5 peak sun hours per day. It is important to remember that large electric resistance heating loads such as water and space heaters and standard refrigeration and air conditioning equipment represent excessive and inappropriate loads for these systems. If you'd like to receive a copy of the owner's manual and preview our systems, please send in $5 (see book section). *We reserve the right to change or substitute components as availability dictates.*

On Substitutions: We realize that everyone's needs are different and that our kits may not fit perfectly for all systems. With this in mind, we welcome substitution, wherever you think it's necessary. Please call us for a modified price quote and feel free to let us help engineer your system — we've got lots of experience! If you want to deduct an item from a Remote Home Kit, deduct its list price less 10% from the total price of the kit. This is because the package price is discounted.

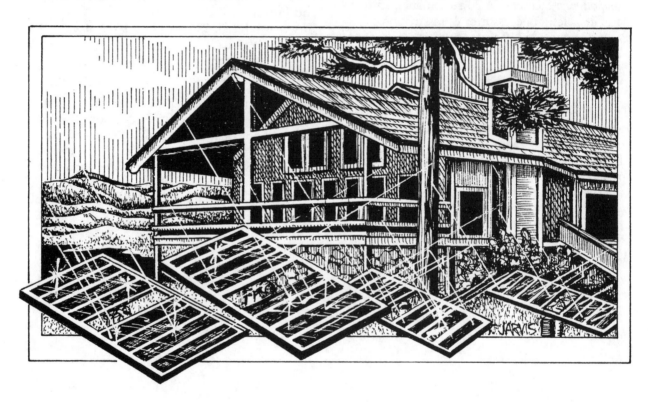

All our Remote Home Kits comply with current National Electrical Code requirements. See pg. 442 for further discussion.

Remote Home Kit 1 (RH-1)

Our basic power setup enhances your part-time summer or weekend cabin, larger RV or boat. It will operate the following DC equipment: sound system, TV, cassette player, lighting, and even some water pumping.
Components:

2 ea. 48 watt PV modules
1 ea. M8 Bobier charge controller
1 ea. Two-panel mounting structure
■ 2 ea. 6V x 220 amp-hour deep cycle battery
1 ea. #4 Battery interconnect
1 ea. 10-16V voltmeter
1 ea. 0-10A ammeter
1 ea. 30A, 2-pole disconnect (safety switch)
1 ea. Newark eight position fuse box
1 ea. 15 position buss bar
1 ea. Remote Home Kit Owner's Manual

Power produced: 480 watt-hours/day

12-201	Remote Home Kit 1 (RH-1)	$1095
12-211	RH-1 Kit without batteries	$945

Batteries are shipped freight collect

Note: the mounting structure provided with the RH-1 Kit is a roof or ground mount not intended for RVs. If you are planning RV use talk to our techs for the appropriate mounting structure.

Remote Home Kit 2 (RH-2)

This package is a real workhorse for small, high efficiency remote homes which makes use of extensive electrical conservation measures already in place or planned, such as propane refrigerator/freezer, cook stove and water heater, as well as high efficiency lighting. With solar power you can operate lights, 120V or 12V TV/VCR, satellite dish, computer system, limited water pumping, sewing machine, drill, small vacuum, small microwave, juicer, blender and any 120V appliance up to 800 watts for 30 minutes. Many of these kits have been purchased as the ultimate Emergency Power System.
Components:

4 ea. 48-watt PV modules
1 ea. latitude adjustable four-module mount
1 ea. Trace C-30A charge controller
1 ea. 30A 2-pole fused disconnect w/fuses
1 ea. UL-listed DC load center
2 ea. 15 amp circuit breakers
1 ea. Digital Volt/Amp Monitor
■ 4 ea. 6V, 220 amp-hour deep cycle batteries
4 ea. #4 battery interconnects
1 ea. Trace 812 12V to 120V power inverter
1 ea. Remote Home Kit Owner's Manual

Power Produced: 960 watthours per day

12-202	Remote Home Kit 2 (RH-2)	$2,995
12-212	RH-2 Kit with no Batteries	$2,675

Batteries are shipped freight collect

Need to pump water too? Give our tech staff a call to spec out a water pumping system to go with your Remote Home Kit.

Trace 2012 Upgrade — Many customers have found our RH-2 Kit to be properly sized for their needs but want to be able to run larger appliances than the Trace 812 inverter will allow. We offer the 2012 upgrade for these situations and we include two #4/0 five foot inverter cables and a 400 amp fused disconnect for the inverter. Remember that your system will almost always grow in the future, which makes this option especially attractive.

12-222	RH-2 Kit w/Trace 2012	$3,595
12-215	RH-2 kit w/ 2012 no batteries	$3,295

Batteries shipped freight collect

Trace 2012 w/ Standby Option — Many customers want to have the Trace inverters fitted with the battery charging option, which will allow them to charge batteries automatically when a backup generator is engaged. This is a factory installed option only. Trace engineering recommends a 6.5 kilowatt generator to take full advantage of their 110 amp charger. Ask our technical staff for detail.

12-236	RH-2 with Trace 2012SB	$3,795
12-217	RH-2 w/2012SB no batteries	$3,495

Remote Home Kit 3 (RH-3)

The two most popular accessories for RH Kits are Sun Frost refrigerators on page 277 and tankless water heaters on page 315.

This kit is designed to satisfy the needs of moderately sized high efficiency homes. It will deal admirably with all the usual daily demands for electricity. When our RH-3 is installed, your household will be able to enjoy the benefits of a full size vacuum cleaner, washing machine, and microwave oven. You can have a large color TV, and a computer for business and games. This system will also drive most power tools. All in all, this kit really demonstrates how satisfactory off-the-grid living can be. Call our experienced technicians toll-free for details.

Power produced: 1,920 watt-hours/day
Components:

8 ea. 48 watt PV modules
2 ea. four-module mounting structure
1 ea. Trace C-30A charge controller
1 ea. UL-listed DC load center
2 ea. 20 amp circuit breakers
1 ea. 2-pole fused disconnect w/30A fuses
1 ea. Digital Volt/Amp monitor
■8 ea. 6V, 220 Amp-hour deep cycle batteries
10 ea. #4 gauge battery interconnects
1 ea. Trace 2012 2,000-watt power inverter
1 pr. Trace 2012 battery/inverter cables
1 ea. 400A DC Fused disconnect
1 ea. Remote Home Kit Owner's Manual

12-203	Remote Home Kit 3 (RH-3)	$5,495
12-213	RH-3 with no Batteries	$4,875
12-223	RH-3 with 2012SB inverter	$5,695
12-233	RH-3 w/2012SB w/o batteries	$5,045

All Batteries are shipped freight collect

Remote Home Kit 4 (RH-4)

In the world of photovoltaic home systems, this is a giant. It will power a large energy efficient home with all the comfort and convenience of a city home hooked to the power grid. But there's one very big difference: no power company meter will sucking dollars out of your pocket, more and more, year after year. Use your vacuum cleaner, washing machine, microwave oven, entertainment equipment and computers with the confidence that you have a substantial energy system composed of the best components available. With 12 48 watt modules mounted on a Wattsun tracker, this system will out perform our 15 modules with a fixed orientation. The RH-4 is based on a massive 750 amp-hour, 24V chloride battery bank, and makes use of the extra-beefy Trace 2500 watt inverter. The precision module tracker has allowed us to gain an additional 1000 watt-hours /day for hundreds of dollars less!

Power produced: 4600 watt-hours/day
Components:

12 ea. 53 watt PV modules
1 ea. 12 module active tracker
1 ea. SCI Mark III Volt/Amp Meter
1 ea. Trace C30A charge controller
1 ea. Chloride 740 amp-hr @ 24V battery
1 ea. Trace 2524 inverter
1 ea. 400A DC fuse assembly
1 ea. UL-listed DC load center
2 ea. 20 amp circuit breakers
1 ea. 30A fused disconnect w/fuses
1 ea. Remote Home Kit Owner's Manual

12-204	Remote Home Kit 4 (RH-4)	$10,695
12-234	RH-4 with Trace 2524SB	$10,895
12-237	RH-4 with no batteries	$7,695
12-238	RH-4 w/2524SB w/o batteries	$7,895

Batteries are shipped freight collect

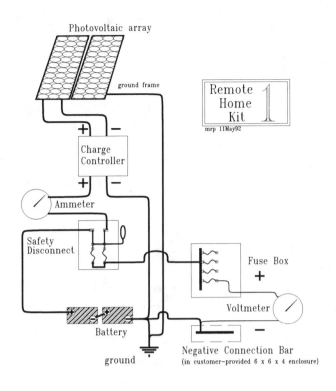

Photovoltaic array

ground frame

Remote
Home
Kit 1
mrp 11May92

Charge
Controller

+ −

+ −

Ammeter

Safety
Disconnect

Battery

ground

Fuse Box
+

Voltmeter
−

Negative Connection Bar
(in customer-provided 6 x 6 x 4 enclosure)

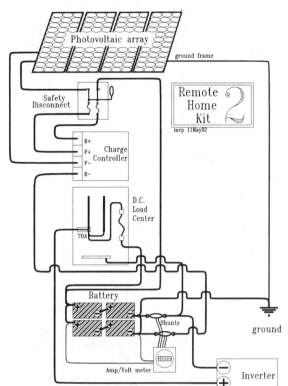

Photovoltaic array

ground frame

Remote
Home
Kit 2
mrp 11May92

Safety
Disconnect

Charge
Controller
B+
P+
P−
B−

D.C.
Load
Center

70A

Battery

Shunts

ground

Amp/Volt meter

Inverter
−
+

*See our Remote
Home Kit
Owner's
Manual on
page 408.*

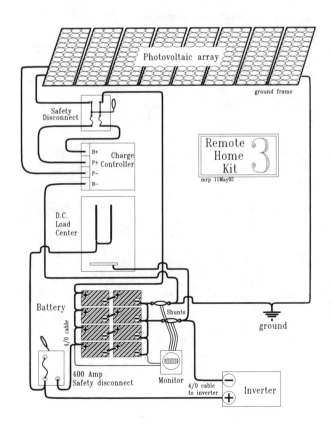

Photovoltaic array

ground frame

Safety
Disconnect

Remote
Home
Kit 3
mrp 11May92

Charge
Controller
B+
P+
P−
B−

D.C.
Load
Center

Battery

Shunts

4/0 cable

400 Amp
Safety disconnect

Monitor

ground

4/0 cable
to inverter

Inverter
−
+

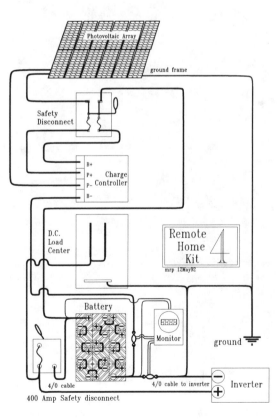

Photovoltaic Array

ground frame

Safety
Disconnect

Charge
Controller
B+
P+
P−
B−

D.C.
Load
Center

Remote
Home
Kit 4
mrp 12May92

Battery

Monitor

ground

4/0 cable

400 Amp Safety disconnect

4/0 cable to inverter

Inverter
−
+

"Sailors are in an ideal position to use alternative energy. Sailing itself is one of the oldest uses of renewable clean energy. Our vessel combines the best of modern technology, with a strong link to the ancient heritage of maritime wind power use," say Greg and Bonnie Chaney of their boat. The 52-foot Cutty Sark plies the Alaska seas, with a 12-volt wind generator providing the electrical power.

PV Mounting Structures

When planning a photovoltaic system, the mounting structure for the solar panels is a very important piece of equipment. The basic choices are:

- Home built
- Fixed mount
- Fixed mount with adjustable tilt angle
- Manual tracker
- Passive tracker
- Active tracker

Each system is site-specific and owner-specific, and will dictate a different mounting decision. The most common issues are discussed below:

To Track or Not To Track

Solar panels produce the most energy when situated perpendicular to the sun's rays. A tracker is a mounting structure that follows the sun and keeps the panels in the optimum position for maximum power output. Passive trackers use no electricity to do this, but follow the sun on one axis—east to west—by the effect of the temperature created by the sun's movement. Active trackers use a very small amount of energy from the panels, and track on two axes, east-west as well as elevation of the sun in the sky.

Tracking is only an option if there are no obstacles blocking the array early in the morning and late in the afternoon. These are the times that a tracker can add a fairly large amount of energy into the battery bank. With a fixed mount, we are only concerned with 3 to 4 hours before and after solar noon. Only if a solar site will receive sunshine beyond this time period should someone consider a tracker. Additionally, trackers are not as effective in the winter, because the sun's position doesn't cover as large an arc as it does during the summer. Solar siting instruments can show you exactly when the sun will be blocked by any obstacle for any time of the year. Trackers are not recommended for systems used primarily during the winter months, particularly in northern latitudes.

A manual tracker is only recommended for a system whose owner is regularly present and willing to move the panels three or four times a day. Even if the tracker can't be adjusted all the time, it can be left in a good fixed position.

Active trackers begin operating immediately at sunrise, and are not affected by cold temperatures or high winds. If your site experiences any of these conditions, the advantages of active over passive tracking are enhanced.

Fixed Mounts

If a site is not appropriate for a tracker, a fixed mount can be used to hold the panels just fine. Some mounts can be seasonally adjusted to account for the sun being higher in the sky in summer and lower in winter. These come highly recommended over the completely fixed mounts, which do not allow you to adjust tilt angle. A simple adjustment performed two to four times per year will noticeably increase your output. Fixed mounts can be used on the roof, on the ground (use concrete), on top of a pole, or on the side of a pole. All poles should be placed into the ground with concrete.

The next issue for those who have a fixed mount is how to position the panels. First, determine which direction is true south. Use the magnetic declination map in the appendix, along with a compass. If the chart tells you that your magnetic declination is X degrees east, then true south is X degrees in the eastern direction. If the declination is X degrees west, then true south is X degrees in the west direction. True south is the direction that your panels should face.

For a fixed mount that cannot be adjusted, tilt the panels at an angle equal to the latitude (in relation to a horizontal line). If the mount can be adjusted (which is recommended), add 15 degrees in the winter and subtract 15 degrees in the summer. Spring and fall are best at the angle of local latitude.

See the Solar Path Finder on pg. 159.

Building Your Own Mount

The mounting systems available today are relatively inexpensive for the job they perform and the length of time that they will last. However, for those do-it-yourselfers, we offer a few suggestions from Jon Vara:

One strong, versatile, and extremely simple way to build array mounts is with slotted steel angle, which consists of sections of 12 or 14 gauge steel angle, with regularly spaced slots and holes already in place. It's usually purchased in ten or twelve-foot lengths, which can be cut with a hack saw and bolted together, erector-set fashion, with nuts and bolts. To find a local supplier, look in the yellow pages under "Shelving", "Racks" or "Material Handling Equipment". You may be able to get used angle at a discount.

It is not recommended to build a mounting structure out of wood. Solar panels can last more than 50 years, and even treated wood will not last that long.

Dear Dr. Doug:
Should I use a tracker or a fixed mount?

Answer: First you need to thoroughly analyze your site and power needs. Trackers can squeeze about 40% more energy annually out of a solar array, but mainly during the summer and only if the site is right. Don't over-buy for your system just because you think tracking is sexy. There are, however, many cost-effective applications for tracking systems. To determine if your site is worthy, see above, "To Track or Not To Track."

I love the idea behind your company. Here's one of the many solar powered navigational markers on Rainy Lake, northern MN, bordering Canada. All summer the 12 watt panel I purchased from you kept up both deep cycle batteries on my sail boat. Thanks. This summer I'll be developing my land, going with all solar power. I'll be in touch about that. —J. Wydra, Keewatin, MN

Wattsun Tracker

Wattsun's new active tracker will provide up to 50% more power than can be generated from a fixed mount, and 25% more than freon type passive trackers. The Wattsun uses no freon but has two worm-gear actuators that give a pointing accuracy of ±0.1 degrees in either axis. East-west range is 120° and elevation range is 65°, meaning no seasonal adjustment is needed. It will withstand winds up to 120 mph. About 1/2 hour after sunset the tracker returns to the sunrise position with the power supplied by the built-in Nicad power pack. Power for the tracker is tapped from one of the modules, with a consumption of 1 watt (a 12-hour day will use 1 amp-hour for the day). The frame is constructed out of aircraft-quality anodized aluminum. Most models mount on top of a 5-inch steel pipe (pipe not included). Nearly all models can be shipped via UPS, for added savings. This is the best tracker on the market. Full 10-year limited warranty.

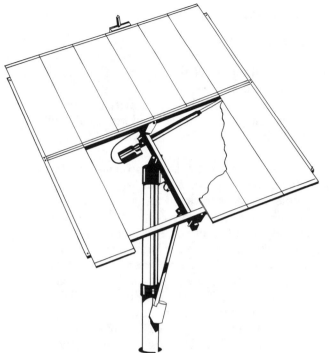

■13-236	4 panel (Siemens M75,M55)	$695
■13-237	6 panel (Siemens M75,M55)	$795
■13-220	8 panel (Siemens M75,M55)	$1,094
■13-221	12 panel (Siemens M75,M55)	$1,372
■13-222	16 panel (Siemens M75,M55)	$1,696
■13-223	20 panel (Siemens M75,M55)	$2,120
■13-238	4 panel (Kyocera K51,K63)	$695
■13-239	6 panel (Kyocera K51,K63)	$795
■13-224	8 panel (Kyocera K51,K63)	$1,135
■13-225	12 panel (Kyocera K51,K63)	$1,426
■13-226	16 panel (Kyocera K51,K63)	$1,848
■13-227	21 panel (Kyocera K51,K63)	$2,310
■13-240	4 panel (Solarex MSX53,56,60)	$695
■13-241	6 panel (Solarex MSX53,56,60)	$895
■13-228	8 panel (Solarex MSX53,56,60)	$1,135
■13-229	12 panel (Solarex MSX53,56,60)	$1,426
■13-230	16 panel (Solarex MSX53,56,60)	$1,848
■13-231	21 panel (Solarex MSX53,56,60)	$2,310
■13-242	4 panel (Hoxan H4810)	$695
■13-243	6 panel (Hoxan H4810)	$795
■13-232	8 panel (Hoxan H4810)	$1,135
■13-233	12 panel (Hoxan H4810)	$1,426
■13-234	16 panel (Hoxan H4810)	$1,848
■13-235	21 panel (Hoxan H4810)	$2,310

"The Wattsun is the most effective PV tracker I have ever seen. Its performance is reliable and precise. I have never been excited enough by a PV tracker to install one in our system. The mechanical vagaries seemed to decrease the inherent reliability of the PV system. Wattsun has changed my mind. They have made a PV tracker we can rely on." Richard Perez Home Power Oct./Nov. '91

To order any of these products or for more technical information call us toll-free at
1-800-762-7325

Zomeworks Track Racks

See chart on pg. 172 for mounting pole size and hole recommendations.

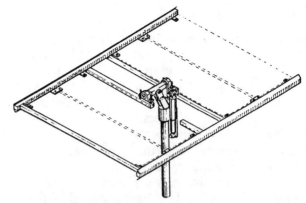

The Zomeworks passive solar tracker will increase the output of your PV panels by up to 50% at the peak of the summer. There are no drive motors, gears, or pistons to wear out. They are as dependable as gravity and the heat of the sun. Easy to install and seasonally adjustable, the Track Rack comes with a 10-year warranty.

The north-south axis is seasonally adjustable for top performance all year long. All refrigerant joints are silver soldered and there are no plastic drive components. The Zomeworks track rack is welded together and constructed of painted mild steel. Stainless steel racks are available for extra cost. The racks are designed to withstand 30 pounds per square foot wind loading and winds up the 125 mph. The specially-made shock absorbers dampen motion in high winds.

Tests conducted over a 12-month period by New Mexico State University at 34° latitude showed an improvement in electrical output by tracking around an adjustable north-south axis of 29% over a fixed latitude mount. The improvement ranged from 19% in November to 42% in June and July. This means that using an eight-module track rack is like having 1.6 extra modules in the winter and 3.2 in the summer.

■13-101	1-2 Panel (Siemens M55,65,75)	$360
■13-111	1-2 Panel (Hoxan H4810)	$375
■13-121	2 Panel (Kyocera K51/K63)	$385
■13-131	1-2 Panel (Solarex MSX60)	$385
■13-102	4 Panel (Siemens M55,M75)	$590
■13-112	4 Panel (Hoxan H4810)	$600
■13-122	4 Panel (Kyocera K51)	$600
■13-123	4 Panel (Kyocera K63)	$735
■13-132	4 Panel (Solarex MSX60)	$735
■13-103	6 Panel (Siemens M55,M75)	$705
■13-113	6 Panel (Hoxan H4810)	$715
■13-126	6 Panel (Kyocera K51)	$715
■13-127	6 Panel (Kyocera K63)	$865
■13-133	6 Panel (Solarex MSX60)	$850
■13-104	8 Panel (Siemens M55,65,75)	$845
■13-114	8 Panel (Hoxan H4810)	$870
■13-128	8 Panel (Kyocera K51)	$870
■13-124	8 Panel (Kyocera K63)	$1,075
■13-134	8 Panel (Solarex MSX60)	$1,040
■13-105	12 Panel (Siemens M55,65,75)	$1,040
■13-115	12 Panel (Hoxan H4810)	$1,085
■13-125	12 Panel (Kyocera K51)	$1,085
■13-129	12 Panel (Kyocera K63)	$1,315
■13-135	12 Panel (Solarex MSX60)	$1,300
■13-106	14 Panel (Siemens M55,65,75)	$1,235

(all Track Racks shipped freight collect from New Mexico)
Stainless steel track racks are available.
Call for quote
■**13-171 Marine Bearings** **$20**
Must be installed a time of purchase.

This graph and table show the results of a year-long test conducted by the New Mexico Solar Energy Institute at the Southwest Residential Experimental Station.

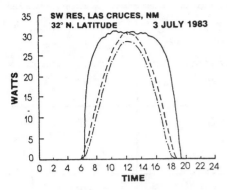

SW RES, LAS CRUCES, NM
32° N. LATITUDE   3 JULY 1983

——— Tracking, tilt = 5° South, Output = 338 watt-hours
– – – Fixed, tilt = 5° South, Output = 241 watt-hours
–·–· Fixed, tilt = 32° South, Output = 212 watt-hours

AVERAGE DAILY ENERGY OUTPUT (WATT-HOURS)

	TRACKING Adj. Elev* Angle	FIXED Adj. Elev* Angle	FIXED Fixed Elev Angle	TRACKING Fixed Elev Angle
JUL 1983	270	207	184	242
AUG 1983	263	204	195	252
SEP 1983	219	179	178	219
OCT 1983	209	172	172	204
NOV 1983	226	196	185	215
DEC 1983	183	160	149	168
JAN 1984	193	167	155	177
FEB 1984	255	207	197	244
MAR 1984	271	211	209	265
APR 1984	304	224	219	296
MAY 1984	286	217	200	267
JUN 1984	250	194	172	225
YEAR LONG DAILY AVERAGE	**244**	**195**	**185**	**231**

* 6 Adjustments in one-year test period

Rapid Response West Canister Shadow Plate

Zomeworks has developed this shadow plate, which speeds morning tracker wake-up. It's available for Track Racks in the field and as an option for new Track Racks. This wider shade is made from polished aluminum in order to reflect more sunlight onto the canister. We have found that this special shadow plate will cut wake-up time by approximately 40%. *Specify the mount and type of module you have!*

■**13-161 Shadow Plate 8/12/14 module $135**
■**13-162 Shadow Plate 4-6 module $105**
■**13-163 Shadow Plate for 2 module $75**
Shipped freight collect from New Mexico

Roof-Mounted Zomeworks Track Racks

The Zomeworks Track Racks for roof-mounted photovoltaic arrays are made with "driver" and "slave" axles which hold four modules each. Each driver axle can operate itself plus one slave axle. Pairs of driver and slave axles are linked together on a roof. A minimum of eight-feet is required between racks on the roof. Roof-mount trackers are made to mount at the angle of the existing roof slope. If a different elevation angle is desired, custom end-stands will be required.

Example: To track 4 modules order one driver axle; to track 8 modules, order one driver axle and one slave axle; to track 12 modules, order two driver axles and one slave axle.

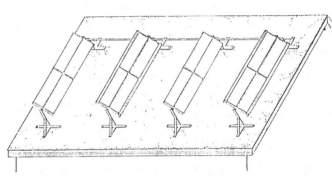

■**13-201 Driver Axle for Siemens $665**
■**13-202 Slave Axle for Siemens $275**
■**13-203 Driver Axle for Hoxan $705**
■**13-204 Slave Axle for Hoxan $290**
■**13-205 Driver Axle for Kyo K-51 $710**
■**13-206 Slave Axle for Kyo K-51 $290**
■**13-211 Driver Axle for Solarex $730**
■**13-212 Slave Axle for Solarex $320**

All shipped freight collect from New Mexico.

Both Wattsun and Zome-works use Schedule 40 steel pipe for mounting their trackers and pole top mounts. Since this pipe is commonly available at all plumbing outlets, there's no sense shipping it. 4 and 6 panel Wattsuns use 3" pipe. 8 panel Wattsuns may be either 3" or 5", you must specify. 10 panel and larger Wattsuns use 5" pipe.

Specifications For Pole Mounted Track Racks (TRPM) & Top of Pole Racks (FRPT)

TRPM =
Tracking Rack
Pole Mount
FRPT = Fixed
Rack Pole Top

	model	steel pole size (O.D.)	minimum pole height above ground	hole depth	hole diameter (fill w/concrete)
TRPM01,02/ AR, SX, HX, KY	FRPT01/AR	2" SCH 40 (2.38")	48"	30"	10"
	FRPT02/AR 01/SX 01/KY 01/HX	2" SCH 40 (2.38")	48"	30"	10"
	FRPT02/SX 02/KY 02/HX	2" SCH 40 (2.38")	48"	30"	10"
TRPM04/ AR, HX, KY51	FRPT04/AR 03/SX 03/K51 03/HX	2.5" SCH 40 (2.88")	48"	32"	12"
	FRPT04/K51 04/HX	2.5" SCH 40 (2.88")	48"	32"	12"
TRPM06/ AR, HX, KY51	FRPT04/SX 06/AR 04/K63	3" SCH 40 (3.5")	48"	35"	12"
TRPM04/ SX, KY63	FRPT06/K51 06/HX	3" SCH 40 (3.5")	52"	35"	12"
TRPM08/ AR, HX, KY51	FRPT08/AR 06/SX 06/K51	4" SCH 40 (4.5")	56"	44"	16"
TRPM06/ SX, KY63	FRPT08/K51 08/HX	4" SCH 40 (4.5"	58"	44"	16"
TRPM12/ AR, HX, KY51	FRPT12/AR 08/SX 08/KY63	5" SCH 40 (5.56")	66"	50"	18"
TRPM08/ SX,KY63	FRPT12/K51 12/HX	5" SCH 40 (5.56")	66"	50"	18"
TRPM14/AR TRPM12/SX	FRPT14/AR 12/SX	6" SCH 40 (6.63")	66"	52"	20"

AR = Arco (Siemens)
SX = Solarex
HX = Hoxan
KY = Kyocera (or "K")

Ground or Roof Mount
Fixed Zomeworks Racks

Zomeworks has added to its line of mounting structures these sturdy, economical racks made of heavy gauge mild steel, painted with a satin black urethane paint. Their telescoping struts offer quick and easy seasonal adjustment from 15 to 65 degrees. They're designed to withstand 30 pounds per square foot of wind loading and winds up to 125 mph. Stainless steel module mounting hardware is provided with each rack. The customer or installer provides the anchor bolts. They are designed for ground or roof mounting.

In the first series of fixed mounts the modules are mounted with their length horizontal and they are stacked vertically. Listed below are the number and type of modules each rack will hold. The K series refers to Kyocera K-51s or K-63s.

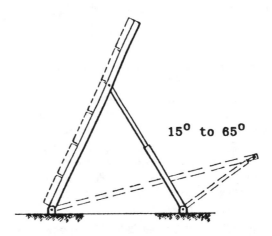

15° to 65°

The next series of mounts is the "low-profile" series. Module length is vertical and they are stacked side by side.

- 13-521 8 Siemens (Low Profile) $215
- 13-522 6 Hoxans or 6 K-51 (Low Profile) $215
- 13-523 6 Solarex (Low Profile) $230
- 13-524 9 Siemens (Low Profile) $225

The last series of mounts stack the modules vertically. They are stacked half on top and half on the bottom.

- 13-531 8 Siemens $265
- 13-532 6 Solarex or 6 K-63 $265
- 13-533 8 Hoxan or 8 K-51 $280
- 13-534 10 Siemens $290
- 13-535 12 Siemens $310
- 13-536 12 Hoxans or 12 K-51 $325
- 13-537 12 Solarex or 12 K-63 $340
- 13-538 14 Siemens $340

All shipped freight collect from New Mexico.

These are the mounts we use with our Remote Home Kits.

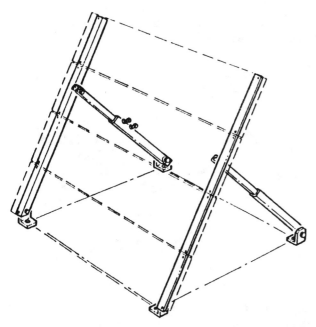

- 13-501 2 Siemens $82
- 13-502 2 Hoxans, 2 K51, or 2 K63 $98
- 13-503 2 Solarex $100
- 13-504 3 Siemens $98
- 13-505 3 Solarex $115
- 13-506 4 Siemens $110
- 13-507 4 Hoxans, 4 K51, or 4 K63 $120
- 13-508 4 Solarex $125
- 13-509 6 Siemens $128
- 13-510 5 Hoxan or 5 K-51 $133
- 13-511 6 Hoxan or 6 K-51 $200
- 13-512 5 Solarex or 5 K-63 $200
- 13-513 6 Solarex or 6 K-63 $225
- 13-703 1 set(4) Quad Lams $89

Top-of-Pole Fixed Zomeworks Racks

Zomeworks makes a standard pole-mounted fixed rack of painted hot rolled steel and heavy-duty pipe gimbals with horizontal and vertical axes of rotation which are locked in place by hex-head set-bolts. These racks are easier to install than lean-to style racks because careful alignment is not necessary. You can manually track the sun by rotating the rack on it's mounting pole. Stainless steel module mounting hardware is provided with each rack. Poles are not included. *Item numbers below show how many and which module the rack will hold. Please be sure to specify which type of panel you will be using.*

Top of pole mounts are excellent for shedding snow.

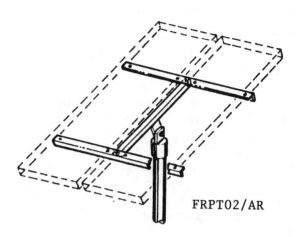

FRPT02/AR

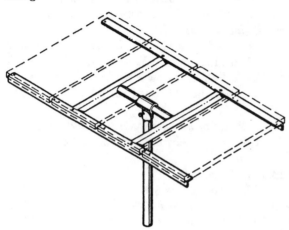

■13-400	1 Siemens	$60
■13-401	2 Siemens	$65
■13-406	2 Hoxan or K-51	$72
■13-407	2 MSX-60 or K-63	$82
■13-408	3 Siemens	$90
■13-402	4 Siemens	$165
■13-409	3 Hoxan or K-51	$155
■13-410	3 MSX-60 or K-63	$165
■13-411	4 Hoxan or K-51	$180
■13-412	4 MSX-60 or K-63	$195
■13-403	6 Siemens	$195
■13-413	6 Hoxan or K-51	$250
■13-414	6 MSX-60 or K-63	$275
■13-404	8 Siemens	$285
■13-415	8 Hoxan or K-51	$305
■13-416	8 MSX-60 or K-63	$340
■13-405	12 Siemens	$440
■13-417	12 Hoxan or K-51	$470
■13-418	12 MSX-60	$500
■13-419	14 Siemens	$490

All shipped freight collect from New Mexico

Side-of-Pole Zomeworks Racks

These racks are adjustable from 0° to 90°. Two hose clamps securely hold the rack to the side of a 1-1/2 to 3-inch OD pipe. The item codes below with the prices specify which modules each rack will hold. Be sure to specify which panel you will be using.

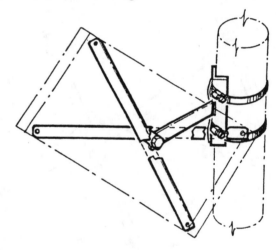

■13-301	1 Siemens M-20	$55
■13-311	1 Siemens M65-M75-M55	$65
■13-315	1 Hoxan or K-51	$70
■13-316	1 MSX-60 or K-63	$75
■13-321	2 Siemens M65-M75-M55	$120
■13-322	2 Hoxan or K-51	$145
■13-323	2 MSX-60 or K-63	$150
■13-331	3 Siemens M65-M75-M55	$160
■13-341	4 Siemens M65-M75-M55	$190

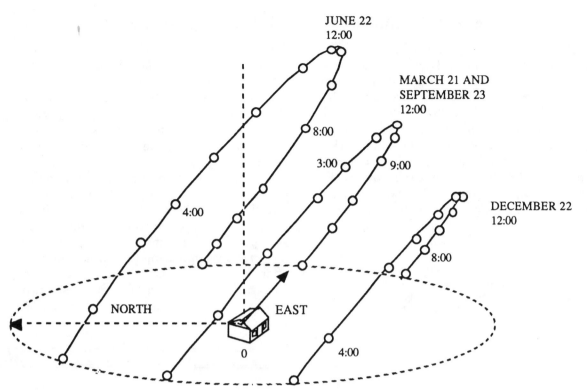

JUNE 22
12:00

MARCH 21 AND
SEPTEMBER 23
12:00

8:00

3:00   9:00

DECEMBER 22
12:00

4:00

8:00

NORTH   EAST

0   4:00

Seasonal Sun Trajectories at 40°N Latitude.

Solarpivot PV Pole Mount

Solarpivot is a simple, sturdy pole mount for
your PV panel(s), which allows you to manually
track the sun. Whether you wish to adjust your
panels throughout the day, or merely seasonally,
this mount is effective and inexpensive. It is
constructed completely out of structural aluminum up to 3/8-inch thick with all stainless steel
bolts that securely mount to any post (steel or
wood). It is engineered to withstand 70-mph
winds. It pivots without touching pins, bolts, or
wingnuts, moving easily to aim it anywhere you
want. The unit is shipped fully assembled, except for the aluminum arms that the panels bolt
to. Mounted by a frequently used path or door,
it takes only seconds to track the sun for maximum energy production. No tools are needed.
Solarpivots are currently in use from Hawaii to
New York and from Texas to Alaska. The Solarpivot currently fits 1 to 4 Siemens M-series
modules, 1 to 3 Solarex, 1 to 3 Hoxan modules
or 1 to 3 Kyocera K-series modules. **You must
specify which modules when ordering.**

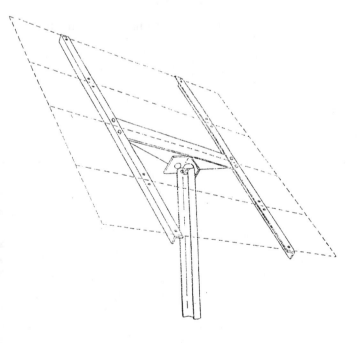

■13-905 **Solarpivot Mount** $139

Aluminum Universal Solar Mounts

For a simple installation, it's hard to beat the economy of our aluminum PV mounting structures. Each mount will hold up to four Siemens (Arco) M-Series modules, three Kyocera modules, 3 Solarex modules (MSX-53, 56, 60), or 3 Hoxan modules. Aluminum angle rails adjust to three seasonal tilt angles. It mounts on a deck, wall, or roof. All stainless steel bolts and hardware is included to attach your solar panels. The installer provides his own anchor bolts. Instructions are included. Specify module type.

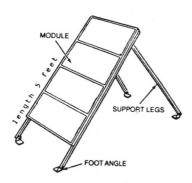

13-702 Solar mount for 4 Modules $109

RV Mounting Brackets

Here is a very simple mounting bracket for the top of a recreational vehicle. Two Z-brackets are made of 14 gauge anodized aluminum and will hold one Siemens module (or any other module if you redrill the holes).

13-901 RV mounting brackets (2) $17

Courtesy The Solar Electric Independent Home Book

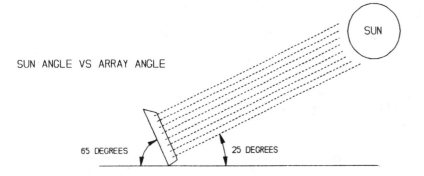

RV Mounting Structures

When you're on the road, it is impossible to know which direction you will be facing when you stop. Until now, the only way to mount your solar panels has been to attach them flat on the RV's roof. This results in getting significantly less power than the panels put out when perpendicular to the sun's rays. Now, thanks to Zomeworks, there is an affordable and simple way to mount your solar panels to the top of your RV, and you can adjust the angle and direction easily. Always drive with them in the flat position, but when you stop, you can raise your panels and adjust them laterally. If you're not interested in making the adjustment, we also offer fixed flush mounts. All units are constructed of aluminum and available for one or two panel systems. If you have more than two panels, use more than one mount.

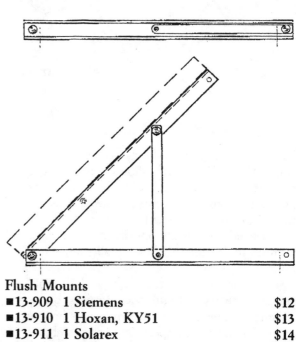

Flush Mounts

■13-909 1 Siemens		$12
■13-910 1 Hoxan, KY51		$13
■13-911 1 Solarex		$14
■13-912 2 Siemens		$20
■13-913 2 Hoxan		$23
■13-914 2 Solarex		$25

Tilting Mounts

■13-915 1 Siemens		$40
■13-916 1 Hoxan, KY51		$45
■13-917 1 Solarex		$49
■13-918 2 Siemens		$58
■13-919 2 Hoxan, KY51		$65
■13-920 2 Solarex		$72

Gasoline Generators

For many years the traditional gasoline or diesel powered generator was the *only* economical "alternative" for generating electricity. Fuel was so cheap it offset the maintenance costs. However, since 1973 fuel costs have increased while inflation has sent generator prices and maintenance costs skyrocketing. Presently, the cost of electricity from a high-quality, long-life generator, including fuel, maintenance, and amortization over 15 years, is close to $0.50 per kilowatt. This is almost double the cost of photovoltaic power.

The traditional generator still plays an important part in the production of affordable alternative energy, but in a more energy-sensible way. Now, most remote home generators are operated only occasionally to run high power consumption tools and battery chargers.

The key to this change has been the development of efficient inverters, and high-powered battery charging devices that plug into traditional AC generators. When the generator operates, the charger can deliver up to several hundred amps of DC power into a battery system.

With the proper size generator you can run the battery charger once a week, do other chores, such as water pumping, vacuuming, or clothes washing and charge your batteries at a very rapid rate simultaneously. Such periodic use of a generator prolongs the life of the machine and drastically reduces fuel and maintenance costs, not to mention avoiding the hideous noise pollution of the internal combustion engine! It provides abundant AC electricity for major appliances and tools without interfering with the full enjoyment of a quiet low-voltage system.

Most significantly, the use of a "photo-gen" (photovoltaic and generator) hybrid system can reduce the cost of a photovoltaic system by as much as two-thirds.

Many people begin remote home life with a traditional generator, using it to power construction on their home and for water pumping, the lifeblood of any homestead. Photovoltaic, wind, or hydro-electric power is added later as budgets permit, until the generator is relegated to a complete backup status.

Before choosing your generator, consider the following:

- Local regulations
- Kilowatt capacity
- Fuel type
- Optional features

Since generators produce AC electricity, they are governed by local electrical codes. The first step is to learn about those regulations, then use a qualified licensed electrician if necessary to do the installation.

The kilowatt and surge capability of the machine is another important consideration. Generally speaking, the average homestead will need in the range of 5,000 to 7,000 watts capacity to do the kind of simultaneous tasks typical of a remote home. Most generators are rated by surge power: continuous output is usually somewhat less.

The rated output of a generator assumes operation at sea level. For each 1,000 feet of altitude, derate the output by 3-1/2%. (A 3,000-watt generator operating at 2,000 ft altitude will deliver only 2,790 watts.) For LPG (propane) fuel, derate the output another 10%.

Generators that burn diesel fuel are more economical than those that burn gasoline. Maintenance is lower because diesel machines are more efficient and avoid conventional igni-tion systems. However, the initial cost is higher, and diesels tend to be noisier and more polluting.

Conversion to LPG (propane) or natural gas can be economical. Clean burning, propane-fired generators nearly double average engine life, and reduce maintenance and repair costs. Since many homes use propane for cooking and refrigeration, it can be convenient to use the same fuel.

Some of the built-in features and options available include:

- **Remote, electric, or automatic load demand starting.** This is necessary if you choose a photo-gen system, and is convenient in any installation.

- **Continuous duty rating.** This feature ensures efficient operation at the fully loaded rate on a constant basis.

- **Full wattage output.** Beware of generators

that distribute 120VAC on two circuits. This means you get only half the rated capacity on one. (Example: inexpensive generators rated at 2,000 watts and equipped with two outlets are really capable of delivering only 1,000 watts per outlet, making it impossible to run a 2,000-watt tool.)

• **High motor starting.** Special windings deliver extra surge power for induction and capacitor motors.

• **Inherent voltage regulation.** The voltage output adjusts automatically to match the required load.

• **Auto idle control.** Fuel consumption is reduced by the generator dropping to idle when no load is sensed. One caution here: a low-wattage motor alone (such as used in a can opener) may not cause the generator to react sufficiently. To correct, simply add a 25-watt bulb to the load. (Better yet, what are you doing with an electric can opener in an alternative energy system? Use a manual can opener!)

Buying a generator can be a lot like buying an automobile. The body is there, but many desirable extras drive up the cost. The basic price of a generator rarely includes the charging battery (for electric start models) or gauges to help monitor the operation. All this should be considered.

Real Goods and Generator Sales

Generators have traditionally been sold for extremely low profit margins by small dealers who are capable of servicing them. Over the years we at Real Goods have endeavored to provide Onan, Kohler, Makita, Honda, Winco, and many other generators to our customers. The problem has been the great difficulty in remaining cost competitive. For this reason we have decided to provide you with the name of a very reputable and very reasonably priced local dealer in Ukiah. It isn't easy for us to refuse your money, so please be conscientious enough to tell this dealer that you have been **Referred by Real Goods.** This way we will get a small sales commission for our referral. Check first locally to see if you can get a good price and equally or more important *good service.* If

neither of these options works for you we'll be happy to sell you a generator as part of an overall renewable energy system. But please don't price shop Real Goods with quotes for individual generators ordered alone (except Honda). Here is the name of the local dealer that we recommend:

Martin's Electric (Rich Martin)
920 Waugh Lane, Ukiah, CA 95482
707/463-1800
Lines: Onan, Kohler, Wisconsin Robin, Multi quip, Generac, Homelite, & Winco
Be sure to tell him Real Goods referred you!

Dear Dr. Doug:
What are the best brands of generators to look for? Is Onan really the **only** generator worth considering?
Answer: Onan is certainly one of the best, but if you need a generator less than 100 hours per year, it may not be necessary to invest in such a high-quality unit. Honda makes a number of high-quality, lower priced generators. Anything powered with a Wisconsin-Robin or Kohler engine is a good choice also.

Dear Dr. Doug:
How come generators cost so much? I can pay anywhere from $400 to $4,000 for a 5,000 watt generator.
Answer: Beware the cheapo generator! You get what you pay for. The cheapo $400 generators use inexpensive, undersized "throwaway" engines. Life expectancy is very short, usually under 500 hours, starting is hard, fuel consumption is high, and the noise level is deafening. Repair parts are unavailable (that's why they're called throwaway in the industry). Quality products such as the Onan offer a 20-year-plus life expectancy of efficient, quiet, and reliable service. It's a case of Mercedes-Benz versus Yugo.

Bicycle Powered Generator

This 100-watt bicycle powered generator/battery charger appears to be the only rugged and dependable small generator manufactured that can be run by a bicycle. It comes with a stand built to accept any three or ten-speed bicycle. Also included is the generator, an ammeter, and all the wires needed to hook up to your battery system. The bicycle generator will produce 8 amps at 12VDC when pedaling at full strength. A human adult can only provide about 100 watts (1/7 of a horsepower) for any period of time. This isn't much, but combined with the 12V batteries, it is plenty for dependable clean lights, pumps, fans, or entertainment centers.

*We have several customers who put their kids (or themselves) on a strict discipline—No TV unless you generate the power yourselves on the bicycle generator. It Works!*Will not work with exercise "bikes". Only for full-size bicycles. No knobby "off-road" tires. (Makes generator bounce too much).

63-308 Bicycle Generator   **$295**

Thermoelectric Generators

Global Thermoelectric Inc. is a manufacturer and supplier of highly reliable, low-maintenance power systems. Global's thermoelectric generators can provide power for applications requiring from 15 watts to several thousand watts.

These generators are operating worldwide for environmental data gathering equipment, remote telecommunications, radios and radio repeaters, navigational aids, and impressed current cathodic protection.

Thermoelectrics convert heat directly into electricity without moving parts. Global manufactures portable and stationary thermoelectric generators that operate on gaseous fuels such as propane or natural gas or on liquid fuels such as diesel or kerosene.

Global's generators have a 20-year design life and each generator requires less than 1 hour of maintenance per year on gaseous fueled generators. The generators are designed to operate unsheltered in ambient temperatures ranging from –65°C to +65°C.

Each unit is available from 15 watts to 550 watts, from 2.5 volts DC to 48 volts DC. Call for specs and pricing.

Output/Watts	Propane (gal/day)		Price
108W @ 12V	3	■	$ 5,500
220W @ 12V	3	■	$ 8,050
480W @ 12V	20	■	$16,500
550W @ 24V	20	■	$16,500
480W @ 48V	20	■	$16,500

**To order any of these products
or for more technical information
call us toll-free at
1-800-762-7325**

ALTERNATIVE ENERGY SOURCEBOOK   179

Honda Generators

Refer to "Large Battery Chargers" on page 203 and "Inverters" on page 231 to help determine your generator wattage needs.

If you're interested in a Honda generator, we have worked out an arrangement with a local and very reputable Honda dealer to provide you with the best priced Honda generators shipped freight collect anywhere in the continental United States. All Honda generators are prepped by factory-trained technicians and are fully tested before shipment. For optimum performance and safety, please read the owners manual before operating your Honda Power Equipment. Connection of generator to house power requires a transfer device to avoid possible injury to power company personnel.

In the price list below the number indicates the output capacity in watts.

Model	Price
■EG650A	$ 439
■EM650A	$ 519
■EX650A	$ 579
■EX1000A	$ 669
■EG1400XA	$ 695
■EM1600XA	$ 799
■EG2200XA	$ 819
■EM2200XA	$ 949
■EX3300SAAI	$1689
■EG3500XA	$1100
■EM3500XA	$1360
■EM3500SXA	$1469
■EX4500SA	$2149
■EG5000XA	$1539
■EM5000SXA	$1860
■EX5500KIA	$2795
■ES6500KIA	$2549
■EB12DAGAI	$7529

Wind Generators

The following article was submitted by Mike Bergey of Bergey Windpower. It offers a very thorough introduction to wind energy.

A Little History

The wind has been an important source of energy in the U.S. for a long time. The mechanical windmill was one of the two "high-technology" inventions (the other was barbed wire) of the late 1800's that allowed us to develop much of our western frontier. Over 8 million mechanical windmills have been installed in the U.S. since the 1860's and some of these units have been in operation for more than a hundred years. Back in the 1920s and 1930s, before the REA began subsidizing rural electric coops and electric lines, farm families throughout the Midwest used 200-3,000 watt wind generators to power lights, radios, and kitchen appliances. The modest wind industry that had built up by the 1930's was literally driven out of business by government policies favoring the construction of utility lines and fossil fuel power plants (sound familiar?).

In the late 1970s and early 1980s intense interest was once again focused on wind energy as a possible solution to the energy crisis. As homeowners and farmers looked to various electricity producing renewable energy alternatives, small wind turbines emerged as the most cost effective technology capable of reducing their utility bills. Tax credits and favorable federal regulations (PURPA) made it possible for over 4,500 small, 1-25 kW, utility-intertied wind systems to be installed at individual homes between 1976 and 1985. Another 1,000 systems were installed in various remote applications during the same period. Small wind turbines were installed in all fifty States. None of the small wind turbine companies, however, were owned by oil companies or other industrial giants, so when the federal tax credits expired in late 1985, and oil prices dropped to $10 a barrel two months later, most of the small wind turbine industry once again disappeared. The companies that survived this "market adjustment" and are producing small wind turbines today are those whose machines were the most reliable and whose reputations were the best.

Windfarms

Starting in the early 1980's larger wind turbines were developed for "windfarms" that were being constructed in windy passes in California. In a windfarm a number of large wind turbines, typically rated between 100-400 kW each, are installed on the same piece of property. The output of these units is combined and sold under contract to the utility company. The windfarms are owned by private companies, not by the utilities. Although there were some problems with poorly designed wind turbines and overzealous salesman at first, windfarms have emerged as the most cost effective way to produce lots of electrical power from solar energy. There are now over 16,000 large wind turbines operating in the California windfarms and they produce enough electricity to supply a city the size of San Francisco. Large wind turbine prices are coming down steadily and even conservative utility industry planners project massive growth in windfarm development in the coming decade, most of it occurring outside California. One recent study actually called North Dakota the "Saudi Arabia of wind energy." With the federal governments "hands-off" energy policy, however, a key question is whether the thousands of large wind turbines that will be installed in the years ahead will be built in the U.S. or imported.

The Cost Factor

Photovoltaic technology is attractive in many ways, but cost is not one of them. Small wind turbines can be an attractive alternative to those people needing more than 100-200 watts of power for their home, business, or remote facility. Unlike PV's, which stay at basically the same cost per watt independent of array size, wind turbines get less expensive with increasing system size. At the 50 watt size level, for example, a small wind turbine would cost about

$9.00/watt compared to $6.60/watt for a PV module. This is why, all things being equal, PV is less expensive for very small loads. As the system size gets larger, however, this "rule-of-thumb" reverses itself. At 250 watts the wind turbine costs are down to $3.50/watt, while the PV costs are still at $6.60/watt. For a 1,500 watt wind system the cost is down to $1.93/watt and at 10,000 watts the cost of a wind generator (excluding electronics) is down to $1.15/watt. The cost of regulators and controls is essentially the same for PV and wind. Somewhat surprisingly, the cost of towers for the wind turbines is about the same as the cost of equivalent PV racks and trackers.

For homeowners connected to the utility grid, small wind turbines are usually the best "next step" after all the conservation and efficiency improvements have been made. A typical home consumes between 800-2,000 kWh of electricity per month and a 5-10 kW wind turbine or PV system is about the right size to meet this demand. At this size wind turbines are much less expensive.

Wind Energy

Wind energy is a form of solar energy produced by uneven heating of the Earth's surface. Wind resources are best along coastlines, on hills, and in the northern states, but usable wind resources can be found in most areas. As a power source wind energy is less predictable than solar energy, but it is also typically available for more hours in a given day. Wind resources are influenced by terrain and other factors that make it much more site specific than solar energy. In hilly terrain, for example, you and your neighbor may have the exact same solar energy. You could also have a much better wind resource than your neighbor because your property is on top of the hill or it has a better exposure to the prevailing wind direction. Conversely, if your property is in a gully or on the leeward side of a hill, your wind resource could be substantially lower. In this regard, wind energy must be considered more carefully than solar energy.

Wind energy follows seasonal patterns that provide the best performance in the winter months and the lowest performance in the summer months. This is just the opposite of solar energy. For this reason wind and solar systems work well together in what are called "hybrid systems". These hybrid systems provide a more consistent year-round output than either wind-only or PV-only systems.

Wind Turbines

Most wind turbines are horizontal-axis propeller type systems. Vertical-axis systems, such as the egg-beater like Darrieus and S-rotor type Savonius type systems, have proven to be more expensive. A horizontal-axis wind turbine consists of a rotor, a generator, a mainframe, and, usually, a tail. The rotor captures the kinetic energy of the wind and converts it into rotary motion to drive the generator. The rotor usually consists of two or three blades. A three blade unit can be a little more efficient and will run smoother than a two blade rotor, but they also cost more. The blades are usually made from either wood or fiberglass because these materials have the needed combination of strength and flexibility (and they don't interfere with television signals!).

The generator is usually specifically designed for the wind turbine. Permanent magnet alternators are popular because they eliminate the need for field windings. A low speed direct drive generator is an important feature because systems that use gearboxes or belts have generally not been reliable. The mainframe is the structural backbone of the wind turbine and it includes the "slip-rings" that connect the rotating (as it points itself into changing wind directions) wind turbine and the fixed tower wiring. The tail aligns the rotor into the wind and can be part of the overspeed protection.

A wind turbine is a deceptively difficult product to develop and many of the early units were not very reliable. A PV module is inherently reliable because it has no moving parts and, in general, one PV module is as good as the next. A wind turbine, on the other hand, must have moving parts and the reliability of a specific machine is determined by the level of the skill used in its engineering and design. In other words, there can be a big difference in reliability, ruggedness, and life expectancy from one brand to the next.

Towers

A wind turbine must have a clear shot at the wind to perform efficiently. Turbulence, which both reduces performance and "works" the turbine harder than smooth air, is highest close to the ground and diminishes with height. Also, wind speed increases with height above the ground. As a general rule of thumb, you should install the wind turbine on a tower such that it is at least 30 ft above any obstacles within 300 ft. Smaller turbines typically go on shorter towers than larger turbines. A 250 watt turbine is often, for example, installed on a 30-50 ft tower, while a 10kW turbine will usually need a tower of 80-100 ft.

The least expensive tower type is the guyed-lattice tower, such as those commonly used for ham radio antennas. Smaller guyed towers are sometimes constructed with tubular sections or pipe. Self-supporting towers, either lattice or tubular in construction, take up less room and are more attractive but they are also more expensive. Telephone poles can be used for smaller wind turbines. Towers, particularly guyed towers, can be hinged at their base and suitably equipped to allow them to be tilted up or down using a winch or vehicle. This allows all work to be done at ground level. Some towers and turbines can be easily erected by the purchaser, while others are best left to trained professionals. Anti-fall devices, consisting of a wire with a latching runner, are available and are highly recommended for any tower that will be climbed. Aluminum towers should be avoided because they are prone to developing cracks.

Remote Systems Equipment

The balance-of-systems equipment used with a small wind turbine in a remote application is essentially the same as used with a PV system. Most wind turbines designed for battery charging come with a regulator to prevent overcharge. The regulator is specifically designed to work with that particular turbine. PV regulators are generally not suitable for use with a small wind turbine. The output from the regulator is typically tied into a DC source center, which also serves as the connection point for other DC sources, loads and the batteries. For a hybrid system the PV and wind systems are connected to the DC source center through separate regulators, but no special controls are generally required. For small wind turbines a general rule-of-thumb is that the AH capacity of the battery bank should be at least 6 times the maximum charging current.

Being Your Own Utility Company

The federal PURPA regulations passed in 1978 allow you to interconnect a suitable renewable energy powered generator to your house or business to reduce your consumption of utility supplied electricity. This same law requires utilities to purchase any excess electricity production at a price ("avoided cost") usually below the retail cost of electricity. In about a dozen states with "net energy billing options" small systems are actually allowed to run the meter backwards, so they get the full rate for excess production.

These systems do not use batteries. The output of the wind turbine is made compatible with utility power using either a special kind of inverter or an induction generator. The output is then connected to the household breaker panel on a dedicated breaker, just like a large appliance. When the wind turbine is not operating, or it is not putting out as much electricity as the house needs, the additional electricity is supplied by the utility. Likewise, if the turbine puts out more than the house needs the excess is instantaneously sold to the utility. In effect, the utility acts as a very big battery bank.

Hundreds of homeowners around the country who installed 4-12 kW wind turbines during the go-go tax credit days in the early 1980's now have everything paid for and enjoy monthly electrical bills of $8-30, while their neighbors have bills in the range of $100-200 per month. The problem, of course, is that these credits are long gone and without them most homeowners will find the cost of a suitable wind generator prohibitively expensive. A 10 kW turbine (the most common size for homes), for example, will typically cost $19,000-24,000 installed. For those paying 10 cents/kilowatt-hour or more for

Real Goods turns principle into practice: a useful selection of the things you are looking for, with essential background information, at fair prices, guaranteed, anywhere you want them delivered.
- Amory Lovins, in the Foreword

electricity in an area with an average wind speed of 10 mph or more, and with an acre or more of property, a residential wind turbine may be worth considering. There is a certain thrill that comes from seeing your utility meter turn backwards.

Performance

The rated power for a wind turbine is not a good basis for comparing one product to the next. This is because manufacturers are free to pick the wind speed at which they rate their turbines. If the rated wind speeds are not the same then comparing the two products is very difficult. Fortunately, the American Wind Energy Association has adopted a standard method of rating energy production performance. Manufacturers who follow the AWEA standard will give information on the Annual Energy Output (AEO) at various annual average wind speeds. These AEO figures are like the EPA Estimated Gas Mileage for your car, they allow you to compare products fairly, but they don't tell you just what your actual performance will be ("Your Performance May Vary").

Wind resource maps for the U.S. have been compiled by the Department of Energy. These maps show the resource by "Power Classes" that mean the average wind speed will probably be within a certain brand. The higher the Power Class the better the resource. We say probably because of the terrain effects mentioned earlier. On open terrain the DOE maps are quite good, but in hilly or mountainous terrain they must be used with great caution. The wind resource is defined for a standard wind sensor height of 33 ft (10 m), so you must correct the average wind speed for wind tower heights above this height before using AEO information supplied by the manufacturer. Wind turbine performance is also usually derated for altitude, just like an airplane, and for turbulence.

As a rule of thumb, wind energy should be considered if your average wind speed is above 8 mph for a remote application and 10 mph for a utility-intertied application.

Keeping Current

The best way to keep current with the progress of wind energy development, both small and large scale, in the U.S. is to join the American Wind Energy Association (777 North Capitol St., NE, Suite 805, Washington, DC 20002; (202) 408-8988). A $35/year individual membership brings a newsletter and an opportunity to help push legislation to promote the increased use of wind energy. - **Mike Bergey.**

Site Evaluation

First, inquire about local building codes. Many communities require permits for structures higher than 20 or 25 ft. Wind plants make noise, so consider your neighbors, too. A little discussion beforehand may eliminate future problems.

An average monthly wind speed of 8 to 14 miles per hour (mph) will be needed. Small wind plants like the Wind Baron Neo require a minimum of 11 mph average monthly wind speed.

Most wind plants are mounted on top of sturdy towers 30 to 85 feet high to take maximum advantage of prevailing winds. Roof mounting is not recommended. A wind generator transmits a great deal of vibration and noise throughout a building, and one can even lose a roof in high winds.

An anemometer measures wind speed. As the anemometer turns, it measures the distance a column of air moves over a site in a given period of time and registers the count on a digital meter. Daily readings are advisable and should be totaled each month for at least three months prior to installation.

If you have trouble assembling weather information, surface weather observations from local airports, the Coast Guard, and forestry people can be obtained on microfiche for a moderate charge from the U.S. Department of Commerce, Environmental Data and Information Service, National Climatic Center, Asheville, NC 28801. A microfiche film report is much cheaper than its printed counterpart. Microfiche can be viewed on a special machine that many libraries, Ace Hardware, and auto parts stores have.

Before selecting the equipment, here are some additional site guidelines:

1. The wind plant should be located within 100 feet of the house if possible because of potential voltage drop in the transmission lines of DC power generators. Longer distances will mean larger wiring and higher installation costs.

2. Nothing should interfere with maximum winds. Extend the tower so that the wind plant reaches 20 feet above any obstruction for at least 500 feet all around. This will reduce or eliminate turbulence from trees and roof tops, which can negatively affect the output while increasing stress on the machine. Remember, trees grow taller.

3. Towers must be strong and must be properly guyed with cable and grounded against lightning strikes. A wind generator's rated output is usually for sea level air density. Higher altitudes have lower air density and require higher wind speeds to achieve a given output. Temperatures also affect output. This Density Ratio Altitude (DRA) chart will approximate the real output of our wind generators according to your site's altitude. We are using 60°F to provide a typical norm.

Altitude	DRA (60°F)
Sea level to 2,500 ft	1.000
5,000 ft	0.912
7,500 ft	0.756
10,000 ft	0.687

From the chart, take the DRA figure nearest your own site's altitude and multiply it by the rated output of a wind generator. As an example, a Wind Baron rated at 400 watts at a given wind speed at sea level would deliver 302 watts at 7,500 feet altitude (400 W x 0.756 DRA = 302.4 W).

Although a wind-plant's tower might appear to be a good grounding system, it is not. The concrete base and anchor points are poor conductors, so the tower needs to be grounded with an 8-foot copper grounding rod and connecting strap.

Batteries for Wind Generators

The battery system size should be designed to allow you to operate under average loads for at least 3 days with no wind power. If wind is to be your only source of power, be careful not to plan too large a battery capacity, that is, one that could not be recharged from the wind within 4 or 5 days. Otherwise your battery system will be in danger of never attaining a fully charged state. Use true deep-cycle, high-antimony, lead-acid batteries for a stand-alone wind system. See the Battery chapter for more details about battery types and selection.

Wind Baron NEO
500 to 600 Watt Wind Generators

The Wind Baron NEO represents the next generation technology beyond Windseeker. The NEO voltage output has increased substantially but the price has stayed the same. This reliable and affordable small wind turbine is specifically designed for use in independent homes, telecommunications and water pumping systems. The NEO begins generating power in lighter winds (7½ mph cut in speed) and continues efficient power production to its maximum of 575 watts at 30 mph (12 volt system). If your annual average wind speed is at least 8 mph the NEO will work for you.

The Wind Baron NEO is extremely light (20 lbs), made of corrosion resistant cast aluminum and stainless steel. The NEO uses patented windings, state-of-the-art (Neodymium Iron Boron) permanent magnet brushless design. The NEO has a unique fail-safe governor design that prevents damage in extreme wind conditions (up to 120 mph) yet allows for continued turbine operation at up to 80% of full output during furling. A "smart" regulator is built into each unit, which compensates for wiring line loss and provides a conditioning tapered charge to batteries. The NEO is available in 12, 24, 36, 48, 90, 120, and 180 voltage configurations. Voltage is pre-set at 14.8 for 12 volt, and 29.5 for 24 volt. Voltage may be adjusted at the factory to accommodate various battery types and special systems. Mounting is designed to fit onto standard 2 inch Schedule 40 steel pipe. Tower heights can go up to 63 feet with simple guyed Schedule 40 pipe. The NEO comes with a comprehensive owners manual that gives complete step by step instructions on installing, siting, lightning protection, tower construction, trouble shooting, maintenance and using your NEO in hybrid systems.

The Wind Baron NEO is available with a Marine Upgrade option for corrosive environments. This option includes a sealed face plate, yaw shaft, stator, bearing, and miscellaneous hardware. If you are installing within ½ mile of a coast or on a boat this option is required.

The Wind Baron NEO is available with an Industrial Upgrade option for mountain top applications (usually telecommunications) where wind speeds are greater than 120 mph, or annual average wind speeds are in excess of 20 mph, or locations that experience severe icing. To order this option please consult Real Goods with site specifications. This option includes polyurethane leading edge impregnated into the blade during manufacturing, double tail, sealed face plate, yaw shaft, stator bearings and miscellaneous hardware.

The Wind Baron NEO is a rugged yet simple small wind turbine, offering a wide range of applications at an affordable price.

16-101 **Wind Baron NEO 12V/500W**	**$875**
16-102 **Wind Baron NEO 24-180V/600W**	**$925**
16-103 **NEO Marine Upgrade (12V)**	**$945**

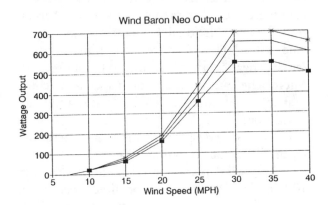

Bergey Wind Generators

For those with the need or the ability to generate more wind power than the Wind Baron can offer, we're happy to offer the Bergey Windpower line with two basic models at 1,500 and 10,000 watts. Bergey is one of the few American windplant manufacturers that has survived the post-1986 tax credit crash and is still thriving and growing. Both of these units are wonderful in their simplicity: no brakes, pitch changing mechanism, gearbox, drive shaft, or brushes. There is an automatic furling design that forces the generator and blades sideways at wind speeds over maximum production, but still maintains maximum rotor speed, so the generator keeps producing. Start-up speed is 7 Mph, cut-in wind speed is 7½ mph, maximum wind speed is 120 mph. The 1,500 watt unit weighs 275 lbs.

At the SERI (Solar Energy Research Institute) test site in Rocky Flats, Colorado, a Bergey 10 kW unit survived wind gusts of 120 mph—or possibly more as SERI's anemometer was carried away at that point. In an independent test of four wind power manufacturers by Wisconsin Power & Light over a 3-year span, Bergey scored a top 99.4% availability. Down-time was only due to a direct lightning strike. Recommended maintenance is this: —once a year, walk out to the tower and look up. If the blades are turning, everything is OK! Tower not included in price.

■16-201 **Bergey 1500W Turbine (24V)** $3,545
■16-202 **Bergey 1500W Turbine (12V)** $3,775
Shipped freight collect from Oklahoma
■16-205 **Control Unit (12/48/120V)** $1,095
■16-206 **Charge Control Unit (24V)** $975
Call for pricing and additional information on these and larger units – 60' to 120' towers also available.

We strongly recommend you monitor your potential wind site for at least 1 year before investing in a large wind system. See our accumulating anemometers on the next two pages.

8000 Series Wind Hawk Anemometer

The 8000 Series Wind Hawk is ideal for wind site surveys for home, farm, and small business wind power applications. Wind turbine owners find it invaluable for monitoring turbine generation, power production, and wind variations. The 8000 continuously computes and displays eight functions: present windspeed, peak gusts, average windspeed, power density, elapsed time, hour of peak gust, hours above cut in and wind status. Comes complete with sensor, 60 feet of cable, stub mast, AC adapter, instructions.

63-355 **The 8000 Series Wind Hawk** $395

The 2100 Totalizer Anemometer

The 2100 Totalizer is a moderately-priced instrument that accurately determines your average windspeed. The Totalizer is an odometer that counts the amount of wind passing through the anemometer to accurately indicate average windspeed in miles per hour; simply divide miles of wind by elapsed time. A 9-volt alkaline battery (included) provides 1-year of operation. The readout can be mounted in any protected environment. Complete kit includes a Maximum #40 Anemometer, 60-feet of sensor cable, battery, stub mast for mounting, and instructions.

63-354 The 2100 Totalizer   **$219**

Hand-Held Windspeed Meter

This is an inexpensive and very accurate windspeed indicator. It features two ranges, 2–10 mph and 4–66 mph with a included chart; so conversions to knots can be made. Speed is indicated by a "floating" ball viewed through a clear tube. Many of our customers have been curious about the wind potential of various locations and elevations of their property, but are reluctant to spend big money for the bulk of the anemometers on the market, just to find out. Here is an economical solution. Try taping the meter to the side of a long pipe, (a piece of tape over the finger hole for high range reading) and have a friend hold it up while you stand back and read with binoculars. This will be very helpful in determining the wind speed of higher elevations above ground. Complete with protective carrying case and cleaning kit.

The Sou'wester Anemometer

The Sou'wester is a low-cost instrument perfect for those who want basic information about their current windspeed. Easy to install, and powered by a self-generating anemometer, the Sou'wester displays current windspeed on an easy-to-read dial. Comes complete with a wind sensor, 60' of cable, stub mast and instructions.

63-353 The Sou'wester   **$139**

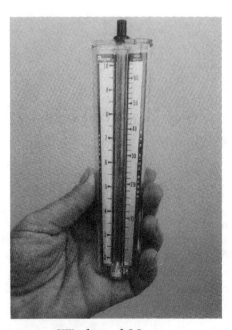

63-205 Windspeed Meter   **$19**

Hydro-Electric Systems

Water wheels working in locations between Maine and Georgia numbered in the thousands during the nineteenth century. Used mostly in mills, their construction required both money and hard work, even without being able to generate electricity.

Among the most important technological developments of the last century was the water wheel designed by Lester Pelton in 1880. The Pelton wheel uses cuplike buckets at regular intervals. A jet nozzle protruding from the end of a pipe shoots water into the buckets with enough force to turn the wheel. The wheel then turns a generating device. Pelton wheels are designed to work best with lower volume and higher pressure (or "head").

Hydro-electric technology can take advantage of falling water to generate electricity economically.

One such low-voltage machine is the Burkhardt or Harris turbine, which operates with a Pelton wheel. These turbines are low-cost, reliable, and use only modest amounts of water. If you have the water source, nothing pays back as quickly as small-scale hydro.

A hydro-electric system interfaces well with photovoltaics, providing much needed power in the winter months, when water is usually abundant and sunshine scarce.

Many user-installed Burkhardt and Harris Peltons in the mountainous portions of the United States and Canada have been in operation for years with a high degree of customer satisfaction. These hydro turbines are relatively inexpensive because they require more head and don't have to handle large volumes of water to produce a given amount of power. They do not work on low-head, high-volume installations. For these applications, see Dr. Doug's comments.

Overview

Designed for light to medium household power requirements, hydro-electric systems are appropriate for small streams that produce from 4 to 150 gallons of water per minute (gpm) and where the head (vertical fall) is a *minimum* of 20 feet (at 100 gpm). With less head, more flow is required to produce a given wattage. These turbines work best with high-head, low-volume situations.

The standard system is designed for charging DC batteries from a modified automotive alternator. 12, 24, 36, or 48 volt DC battery configurations may be charged from a Burkhardt or Harris system. The Burkhardt and Harris turbines are not designed to produce the 120/240-volt 60-cycle AC used by the typical home. One reason is that AC electricity cannot be stored. Another is the high cost of stabilizing the alternating current at precisely 60 Hz in the face of continuously changing loads. The DC power output of these turbines can range from 1,000 to 30,000 watthours per day.

Site Evaluation

The operation of a water-driven generator requires water pressure, which builds under the force of gravity as water drops through a pipeline. You need to know both the head and flow to determine if there is enough force available in your stream to produce usable electricity.

Measuring the head of your water source means calculating the vertical distance water falls from a dam or other source to the spot where you will install the hydro-electric system. It's good to install the turbine as close to the batteries as possible. A distance greater than 500 feet usually translates into large-gauge, costly wire to transport the electrical current with minimum power loss. (For every 100 feet of distance, you have 200 feet of wire.)

Begin your measurements at the projected turbine site. Work back toward the water source using this primitive but effective process. You will need a carpenter's level and a long board or straight stick. Take the board or stick, stand it upright and mark it at eye level. Then measure the length of the stick from your mark to the ground. Let's say it is 5 feet. Now stand on your turbine site with the measuring stick. Using the carpenter's level look uphill across the mark, on a horizontal line, toward your water source. Pick a landmark on the same level as the mark. Walk over to it, place the stick, and repeat the process. When you have done it 15 times, using the 5 ft. mark, you will have measured a water head of 75 feet. If you have a friend with a stick marked at the same height, you can speed things up by sighting on each other's landmarks, exchanging places as you go up the hill.

(Of course if you or a friend have access to aircraft equipment, borrowing or renting a sensitive altimeter will really speed things up.)

Once you have checked for sufficient head, you need to check the water's flow, which should be the minimum necessary to produce usable power (see System Potential chart). To measure the flow, block the stream, then take a length of 1-1/2 inch pipe and stick it into the water right at the point where you plan to have the dam. If either the head or volume falls short of the minimum requirement, you can balance one against the other to a degree. Greater head permits less volume, and greater volume reduces the head requirement.

The batteries should be located at the house for two good reasons: (1) to keep them warm and retain maximum storage capacity, and (2) to minimize voltage drop from the batteries to the house. It's most efficient to transmit the highest voltage possible. The charging voltage being higher than normal battery voltage, less voltage loss will occur during transmission through a given wire size. Such loss should not exceed 4% for best results.

An exception to this reasoning would be if you have an all AC house. In this case, it might be wise to place both the batteries and your inverter at the hydro site and to transmit 120V-AC to the house.

What Your Hydro System Will Cost

Total hydro system costs are determined by four factors:

1. Size of turbine: Turbine costs vary from $875 to $1,250 depending on the number of nozzles you need. Higher output systems also require greater battery capacity.

2. Length of pipe: This varies from under $25 for small, short, steep pipes in mountainous country to over $1,000 for large, long pipes on flat land.

3. Distance from turbine to point of use: Wire cost is dependent on system output, transmission distance, and voltage. It can vary from under $10 to over $1,000.

4. Power conditioning equipment: All but the smallest systems need voltage regulation to protect batteries from overcharging. Costs range from $300 to $500. Inverters are often needed to power AC loads in a house for a cost of $500 to $1,200.

Here is a hydro cost breakdown for a typical mountain cabin:

Site Conditions:

Head	100 ft.
Flow	15 gpm
Pipe length	300 ft.
Pipe size	2" PVC
Distance to battery	30 ft.
Output	100 watts

Costs:

Pipe	$100
Turbine	$875
Regulator	$340
Batteries	$180
Wire, etc.	$ 50

Total Costs	**$1,545**

Note that this system will produce 100 watts times 24 hours or 2,400 watt-hours per day, or 200 amp-hours per day at 12 volt. To generate this much power using PV would require 11 48-watt PV panels generating for 5 hours per day at a cost in excess of $3,800 for the solar panels alone.

2525, Butte, MT 59702, (800)428-2525 or Canyon Industries, P.O. Box 574, Deming, WA 98244, (206)592-5552.
(Tell them that Real Goods sent you.)

Dear Dr. Doug:
I've got 3,000 feet of frontage on the Mississippi River. There isn't much drop, but boy do I have the volume! Is there anything I can do to make useful power out of this?

Answer: A classic example, I'm glad you asked this question! The small Pelton wheel turbines that we sell work best with low volume and high head. Pelton wheels can't handle large quantities of water. In your situation you've got high volume and low head, a situation requiring machinery that can handle huge quantities of water. Water is very heavy and as the machinery grows, so does the cost. For many sites, like the Mississippi above, the cost of the equipment may outweigh any possible benefits.

For larger systems or low-head systems (both AC and DC) you can contact: National Appropriate Technology Assistance Service, P.O. Box

Dear Dr. Doug:
I've got this great idea. I've calculated that if I take the electrical output from one of your hydro-electric generators and use it in a high-efficiency water pump to push water back up the hill to run through the generator again I can get about 1,000 watts of free energy for every 5,000 gallons I cycle. What do you think?

Answer: Congratulations! You've just invented a perpetual motion machine with an energy bonus thrown it. While we appreciate the ingenuity, there's a fatal flaw in all plans of this type: efficiency. The very best, most carefully load-matched positive displacement water pumps approach 50% efficiency. Same goes for the hydro-generator, if we can extract 50% of the theoretical energy available we're doing good. All of which means that in this less-than-perfect world, your plan won't work. (But if you get a working model be sure to let us know!)

Piece by piece, with that uniquely American blend of idealism and intense pragmatism, [the Real Goods staff] quietly built, and continue to build a grass-roots energy revolution.
- Amory Lovins, in the Foreword

Sizing Your System's Potential

Short of having our computer size your system's potential, the chart below will give you a good idea of how much output in watts you may expect for different gpm (gallons per minute), for varying feet of head. Multiply the number of watts by 24 watt-hours to get the total number of watts generated in one day. Divide this by DC system voltage to get your amp-hour per day output.

Feet of Head

G.P.M.	25	50	75	100	200	300	600
3	-	-	-	-	40	70	150
6	-	-	10	20	100	150	300
10	-	15	45	75	180	275	550
15	-	50	85	120	260	400	800
20	25	75	125	190	375	550	1100
30	50	125	200	285	580	800	1500
50	115	230	350	500	800	1200	-
100	200	425	625	850	1500	-	-
200	275	515	850	1300	-	-	-

For Standard Delco, expect 20% to 30% lower output.

Hydro Site Evaluation

In order to accurately size your hydro-electric system we need specific site information. We take this information and feed it into our computer with our hydro sizing program. Send in the following information with an SASE and $10 and we'll size your potential site and recommend a system for you. Allow two weeks processing time.

1. Your head, or drop in elevation from source to turbine site.
2. Flow in gallons per minute to be used. (120 gpm is maximum usable)
3. Length, size, and condition of pipe to be used, (only if using already existing pipe).
4. Distance from turbine to point of power use.

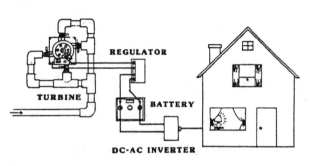

17-001 Hydro Site Evaluation $10

Burkhardt and Harris Turbines

When calculated on a cost-per-watt basis, the average hydro-electric generator costs as little as one-tenth as much as a solar (photovoltaic) system of equivalent power and can even be cheaper than grid power. Solar only generates power when the sun is shining; hydro generates power 24 hours a day.

The generating component of the Burkhardt and Harris turbines is an automotive alternator (Delco or Autolite, depending on system requirements) equipped with custom-wound coils appropriate for each installation. The rugged turbine wheel is a one-piece Harris casting made of tough silicon bronze. There are hundreds of these wheels in service, with no failures to date.

The aluminum wheel housing serves as a mounting for the alternator and up to four nozzle holders. It also acts as a spray shield, redirecting the expelled water into the collection box.

Burkhardt and Harris turbines are available in several different nozzle configurations to maximize the output of the unit. The particular number of nozzles that you need is a function of

We refer repeatedly to Burkhardt and Harris as if the companies were interchangeable. In reality Don Harris is Harris Hydroelectric and John Takes is Burkhardt Turbines. Both companies use virtually the same components, have the same prices, and the same warranty. Both are well proven reliable systems.

See the Enermaxer charge controller on pg. 213. This is the only controller to use with hydro systems (your excess energy can be used to heat your water!).

the available flow (gpm) and the existing pipe diameter. Here is a chart with some general rules for sizing the number of nozzles on the system you will need, but bear in mind that we need to size your system exactly.

gpm	Number of Nozzles
5 to 30	1
30 to 60	2
60 to 120	4

Here are the output limits for the Standard and optional High Output alternators. Please note that if your site will exceed the wattage limits for the Standard alternator, you will need to purchase the High Output option. The High Output option will not increase wattage output for a smaller site.

Alternator output limits		
	12 volt	24 volt
Standard:	400 watts	700 watts
High Output:	750 watts	1500 watts

■17-101	1 nozzle Turbine	$875
■17-102	2 nozzle Turbine	$995
■17-103	4 nozzle Turbine	$1,250
■17-131	High output alternator	add $275
■17-132	24V operation	add $79
■17-133	Low head (less than 60')	add $79
■17-134	Extra nozzles	$6

Pelton Wheels

For the small hydro-electric do-it-yourselfer, we offer reliable and economical Pelton wheels. Harris silicon bronze Pelton wheels resist abrasion and corrosion far longer than polyurethane or cast aluminum wheels. A nozzle jet of up to 1/2 inch can be used. Designed with threads for Delco or Motorola alternators.

■17-202 Silicon Bronze Pelton Wheel $345

Batteries

Batteries are often the most misunderstood component of an alternative energy system. Batteries are the energy storage component that accumulate small quantities of energy as it is slowly generated by various generating devices such as PV modules, wind, or hydro plants. This stored energy buffer runs the household at night or during extended periods when there is no energy input such as a long string of cloudy days in a PV system. Batteries can be discharged rapidly to run much larger loads than the energy source can run alone, such as pumps or motors. If caution is not used, however, impromptu welding and even fires can occur.

Compared to the electronic marvels in the typical alternative energy package, the battery seems primitive in design. There is much research, and some promising new technologies are on the horizon. But for now we must coexist with traditional battery technology—a technology that is nearly 100 years old, but is tried and true, and requires surprisingly little maintenance.

One aspect of the battery's capacity for storing energy is rated in amp-hours: one amp delivered, for one hour equals one amp-hour. To know how much total energy is delivered it is necessary to know at what voltage the amp-hour(s) are delivered. Battery capacity is listed as so many amp-hours times such and such a voltage.

Engine-starting batteries are rated for how many amps they can deliver at a cold temperature, or Cold Cranking Amps. This rating is not relevant for storage batteries. Beware of any battery that claims to be a deep-cycle storage battery and has a CCA rating.

Battery manufacturers typically rate their storage batteries at a 20-hour rate. For example; our golf cart batteries are rated at 220 amphours. It will produce 11 amps for 20 hours. The 20-hour rate is the standard we use for all the batteries in the *Sourcebook*. This rating is designed only as a means to compare different batteries to the same standard, and is not to be taken as a performance guarantee.

Batteries are electrochemical devices, and are sensitive to temperature, charge-discharge cycles, and age. The performance you will get from your batteries will vary with location, climate, and usage patterns. A battery rated at 200 amp-hours has approximately twice the capacity of one rated at 100 amphours.

Lead-acid is also in the laboratory—this old dog can still learn some new tricks.

Most electrical appliances are rated with *wattage* which is energy consumption per unit of time. Again, one watt delivered for one hour equals one watt-hour of energy. Wattage is the product of current times voltage. This means that one amp used at 120 volts is the same wattage as 10 amps used at 12 volts. It is independent of voltage. A watt at 120 volts is the same amount of energy as a watt at 12 volts. To convert a battery amp-hour capacity to watt-hours, simply multiply the amp-hours times the voltage and the product is watt-hours. Or, to figure out how much battery capacity it will take to run an appliance for a given time, take the appliance wattage times the number of hours it will run to get the total watt-hours, and then divide by the battery voltage to get the amp-hours. For example, running a 100-watt light bulb for 1 hour uses 100 watt-hours. Running off a 12-volt battery will consume 8.33 amp-hours (100 watt-hours/12 volts).

Batteries are not perfect containers for storing the energy of our power system. For every 1.0 amp-hour you remove from the batteries, it is necessary to pump about 1.25 amp-hours back in. This figure varies with temperature, battery type, and age, but is a good rule of thumb by which to calculate approximate battery efficiency.

There are two different kinds of electrochemical batteries in common use for alternative energy systems: Lead-acid, the most common, and nickel-cadmium (nicads).

Lead-acid Batteries

The lead-acid battery cell uses two lead plates of slightly different composition suspended in a dilute sulfuric acid solution called the electrolyte. Lead and sulfuric acid can be dangerous and need to be treated with great respect. Be sure to turn in old batteries for recycling at many auto parts stores or recycling centers. A majority of the lead in new car batteries comes from recycled batteries. Lead-acid batteries produce approximately 2 volts per cell. Individual cells are series-connected to produce the desired voltage. For instance, a 12-volt battery has six individual cells that are connected together in the battery case to produce 12 volts.

Lead-acid batteries produce hydrogen gas during charging, which poses a fire or explosion risk if allowed to accumulate. The hydrogen must be vented to the outside if the batteries are kept indoors. Lead-acid batteries will sustain considerable damage if they are allowed to freeze. A fully charged battery can survive temperatures as low as -30° F, but as the state of charge falls sulfur begins to chemically bond with the lead. The electrolyte becomes closer and closer to plain water as the state of charge falls. At 50% charge level a battery will freeze at approximately 15°F. This is the lowest level you should ever let your batteries reach. If freezing is a possibility, the batteries can be kept indoors (remember to vent!) if the house is occupied full time. If it's an occasional use cabin, the batteries may be buried in the ground within an insulated box.

Lead-acid batteries age in service. Once a bank of batteries has been in service for 6 months to 1 year, it generally is not a good idea to add more batteries in series to the bank. Any new batteries will perform at the level of the worst battery in the bank. All lead-acid batteries in a bank should be of the same capacity, age, and manufacturer.

See page 487 for a drawing of a recommended battery enclosure.

The state of charge of lead-acid batteries can be monitored either with a hydrometer by sucking up a sample of electrolyte, or very simply with a voltmeter. Using a hydrometer is the most accurate way to monitor the battery condition, but it's messy and potentially hard on your clothes. (Watch out for drips!) The battery voltage is a measure of the charge level of the battery. The voltmeter is only accurate when the battery is in a state of rest. Voltage is quite elastic and will stretch upward when a charge is applied, or stretch downward when a load is applied. The battery needs to have been sitting at rest for at least one hour to obtain a reading that is indicative of the charged state. Digital meters are highly recommended for their high degree of accuracy and their ability to read fine differences.

Batteries are built and rated for the type of "cycle" service they are likely to encounter. A cycle is defined as the time elapsing between the point when a charged battery is discharged to a specific level to the point when it is recharged back to its starting level. Cycles can be "shallow," reaching 10% to 15% of the battery's total capacity, or "deep," reaching 50% to 80% of total capacity. Batteries designed for shallow-cycle service will tolerate few, if any, deep cycles without sustaining internal damage. Batteries used in remote power systems must be capable of repeated deep cycles without ill effects.

Several types of lead-acid batteries are presented here in order of worst to best.

Car Batteries

The most common type of lead-acid battery is the automotive battery, often called the starting battery. This type of lead-acid battery has many thin lead plates and is designed to deliver hundreds of amps for a few seconds to start a car. Starting batteries are only designed to cycle about 10% of their total capacity and to recharge quickly from the alternator after cycling. They are *not* designed for the deep cycle service demanded by remote home power systems, and will fail quickly when used in this inappropriate fashion.

"RV" or "Marine" Deep-Cycle Batteries

This generic category includes all of the 12 volt batteries that Sears, Montgomery Ward, K-Mart, etc. sell as "deep cycle," "RV," or "marine" batteries. They are always 12 volt, and between 80 and 120 amphour capacity. These batteries are a compromise between starting batteries and true deep-cycle batteries. Many of them are actually put into starting battery service. They will give far better deep-cycle service than starting batteries, and may be the ideal choice for a beginning system that you plan to expand later. Life expectancy for these batteries is typically 2 to 3 years.

Gel Cell Sealed Batteries

Gel cell sealed batteries have the acid either gelled or put into a spongelike filler. They have the advantage of being completely sealed. They can operate in any position, even sideways or upside down, and will not leak acid or gas. Because the electrolyte moves more slowly, these batteries cannot tolerate high rates of charging or discharging. Their sealed construction, which makes them ideal for some limited applications, makes it impossible to check individual cell conditions with a hydrometer. We recommend this type of battery only in situations where hydrogen gassing during charging cannot be tolerated or in conditions where the battery needs to fit into unique, tight spaces. Boats and UPS computer power supplies are the most common uses. Life expectancy is 2 to 3 years for most gel cell batteries.

"Telephone Company" or Lead-Calcium Batteries

During the past 10 years telephone companies have been upgrading much of their switching equipment from the old style 48-volt relay type to newer solid-state equipment. When a telephone station is changed over, the monster battery bank that ran the old equipment is discarded or recycled. Many of these *shallow-cycle* lead-calcium batteries are finding new homes in remote power systems. The typical life expectancy for these batteries is 15 to 20 years, although there are some on the market that claim 50 years or more. These batteries will work fine in remote power systems, IF you treat them carefully. These are shallow-cycle batteries that rarely experienced more than a 15% cycle in telephone service. If you are careful never to discharge them deeply, these batteries can give years of excellent service. As usual there is a trade-off with phone company batteries. While they can sometimes be found free for the hauling, their sheer weight and size makes them difficult to contend with. Some of these batteries weigh close to 400 lbs. per 2-volt cell! Because cycle capacity is limited to 15 or 20% you have to buy, move, and install 4 or 5 times more battery mass than is required for deep-cycle batteries.

True Deep-Cycle Batteries

True deep- cycle batteries are specifically designed for storage and deep-cycle service. They tend to have larger and thicker plates. They are available in 6-volt and 2-volt configurations for ease of movement. Once in place the multiple batteries are series and parallel wired for your basic system voltage. These batteries are built to survive hundreds of 80% cycles, though for best life expectancy we recommend 50% as the normal maximum discharge. Save the bottom 30% for emergencies. The three commonly available batteries within this group are the "golf cart" types with a 3 to 5-year life expectancy, the L-16 series with a 7 to 10-year life expectancy and industrial Chloride batteries with a 15 to 20-year life expectancy.

Nicad Batteries

The nickel-cadmium battery reaction is very different from the lead-acid battery reaction. These cells use a base, potassium hydroxide, for electrolyte instead of an acid. The electrolyte does not enter into the chemical changes with the plate materials; it only acts as a transfer medium for the electrons. So the specific gravity of the electrolyte does not change with the state of charge. The positive pole is composed of nickel compounds, the negative pole is composed of the metallic element cadmium. Cadmium is a highly toxic element and should be regarded with respect and caution.

Nicads are relatively new in the alternative energy business. Accordingly, some strong opinions have been formed with some "experts" loving them and some hating them.

There is considerable controversy concerning the life expectancy of the common pocket plate nicads. Pocket plate construction is the nicad battery type most likely to be found in storage battery service. Life expectancy can vary from a few hundred cycles to tens of thousands of

See pg. 200 for Gel Cell batteries. See pg. 199 for Electric Golf carts and L-16 batteries. See pg. 201 for Industrial Chloride batteries.

cycles depending on how deeply the cells are cycled. Contrary to popular belief, these cells do not like to be deeply cycled. Major manufacturers rate pocket plate cells at approximately 500 cycles with 80% depth of discharge (DOD) cycling. This is not much better than a Trojan L-16 lead-acid battery, which costs a fraction of nicads.

There *is* a major difference in the way nicads age and die as compared to lead-acid batteries. Lead-acid batteries gradually decrease in capacity over time until they reach around 80% of new capacity. From this point battery capacity starts dropping very rapidly. Within a couple months of this point, the lead-acid battery will be unable to hold any charge. Battery capacity acts like it fell off a cliff. Nicads die more gracefully. There is no sudden cliff they drop off; capacity just gradually diminishes. The rate at which it diminishes is relatively fast with DOD cycles in excess of 50%, and is almost negligible with DOD cycles of less than 20%. Nicad battery manufacturers consider a battery to be "used up" at 80% of original capacity. Many alternative energy users probably have nicad battery banks that are far below the 80% mark. But because these batteries still perform at a satisfactory level, do hold a charge, and otherwise perform well, why complain? Unless you're tracking capacity with an expensive cumulative amp-hour meter the slow loss of capacity will probably never be noticed.

See pg. 202 for FNC batteries (sit down first!)

At this point we should mention the new fiber-nickel-cadmium battery. This battery shares the same chemistry as conventional nicads, but the construction of the plates and cells are a major innovation. Rather than plates of active material the FNC battery uses plastic fibers, like a Brillo pad, which hold the active materials. This increases the surface area available for interaction and eliminates the need for graphite fillers, which can react and create potassium carbonate poisons in the battery. The result is outstanding battery life, in excess of 15,000 cycles at 25% DOD, and high current flows to meet inverter surge demands. The FNC batteries are a considerable step beyond the typical pocket plate nicad batteries.

Nicads have some great advantages:

• They can be extremely long lived if cycled at about 30% capacity or less. Deeper cycling will cut sharply into life expectancy.

• Nicads have flat discharge voltage curves. The voltage on nicads remains constant until the last 10% of capacity. This characteristic is easier on inverters and other voltage-sensitive appliances.

• Batteries of different sizes, capacities, and ages can be added together, something that is absolutely forbidden with lead-acid batteries.

• Nicads do not lose capacity at low temperatures as quickly as lead-acid batteries do. They can tolerate freezing with no adverse effects, although they won't work while frozen.

Nicads also have some big disadvantages:

- The single biggest disadvantage is **cost.** Nicads are very expensive initially. New nicads may be 3 to 10 times as costly as comparable lead-acid batteries. We sometimes offer used nicads, which are somewhat less expensive.

- Nicad batteries display the same voltage from 10% to 90% of charge capacity. This means that a voltmeter cannot be used to gauge state of charge. It is necessary to use a sophisticated and expensive amp or watt-hour accumulator to accurately judge how much energy you have in storage.

- Nicads are incompatible with lead-acid batteries. You can't mix the two battery types in a single system.

- Most inverters and charge controllers now on the market won't handle the higher voltages that nicads prefer when charging.

- Even though the batteries may live for long periods, they need periodic electrolyte changes. Potassium Hydroxide is difficult to handle and highly toxic. Great caution must be used while juggling heavy batteries to pour out the old electrolyte. This is probably not a good home project.

Helpful Hints for Lead-Acid Batteries

1. Batteries will self-discharge and can sulfate when left for long periods unattended and where no charge controller or trickle charge device is employed. Disconnect them in this case. However, self-discharge is not prevented by disconnecting batteries. Clean, dry battery tops is a must as a preventative measure.

2. Keep batteries warm. At 0° F, 50% of their rated capacity is lost.

3. All batteries must be vented to prevent potential explosion of hydrogen and other gases.

4. Do not, under any circumstances, locate any electrical equipment in a battery compartment.

5. Wear old clothes around electrolyte solution as they will soon be full of holes.

6. Be careful of metal or tools falling between battery terminals. The resulting spark can cause a battery to be destroyed or damaged.

7. Baking soda neutralizes battery acid, so keep a few boxes on hand.

8. Check the water level of your battery once a month. Excessive water loss indicates gassing and the need for a charge controller or voltage regulator. Use distilled water only.

9. Protect battery posts from corrosion. Use a professional spray or heavy-duty grease or vaseline. If charging has ceased or been reduced and the generator checks out, chances are corrosion has built up where the cable terminal contacts the battery post, preventing the current from entering the battery. This can occur even though the terminal and battery post look clean.

10. Never smoke or carry a lighted match near a battery, especially when charging.

11. Take frequent voltage readings. The battery manufacturer will provide instructions about the voltage of a cell when their battery type is fully charged. The voltmeter is the best way to accurately monitor a battery.

12. Batteries gain a memory about how they are used. Large deviations from regular use after the memory has been established can adversely affect performance. That incurred memory can be erased by discharging the battery system 95%. Recharge it to 140% of its capacity at a slow rate, then rapidly discharge the battery again completely and recharge to normal capacity. A generator/battery charger combination should be used for this procedure.

13. Use fewer, larger cells in series rather than lots of small cells paralleled. For example use two 6V batteries in series rather than two 12V batteries in parallel. It is safer in the case of a shorted cell, you have half the chance of random cell failure, half the maintenance effort, and in general, thicker and more rugged plates — certainly more rugged than popular marine or RV deep cycle batteries.

See page 481 for further articles on lead-acid batteries.

Although Real Goods dedicates much time and energy educating consumers about batteries, due to weight and shipping considerations it may not be appropriate for us to sell you one. Nonetheless, regardless of your purchase decision, we want you to select the right type and size of battery. Surprisingly, with the high volume discounts we receive from our freight company, we are often cost-competitive with locally purchased batteries. This is especially true for our large chloride storage batteries. We recommend that you check locally for smaller batteries before purchasing from us, where the freight can be considerable.

Dear Dr. Doug:

I'm just getting started on my power system. Should I go big on the battery bank assuming I'll grow into it?

Answer: The answer is yes and no. Yes, you should start with a somewhat larger battery bank than you absolutely need. Over time most folks find more and more things to use power for once it's available. But if this is your first venture into remote power systems and battery banks then we usually recommend that you start with some "trainer batteries." So no, don't invest too heavily in batteries your first couple of years. The golf-cart type deep-cycle batteries make excellent trainers. They are modestly priced, will accept moderate abuse without harm, and are commonly available. You're bound to make some mistakes and do some horrible things with your first set of batteries. You might as well make mistakes with inexpensive batteries. In three to five years, when the golf cart type trainers wear out, you'll be much more knowledgeable about what you need and what quality you're willing to pay for.

Dear Dr. Doug:

The battery bank I started with two years ago just doesn't have enough capacity for us anymore. Is it OK to add some more batteries to the bank?

Answer: Lead-acid batteries age in service. Generally it isn't recommended to add new cells in series after a battery bank is 6 months to 1 year old. The new batteries will be dragged down to the performance level of the worst cell in the bank. Different batteries have different life expectancies, so we really need to consider how long the bank should last. For instance, it might be acceptable to add more cells to a large industrial chloride set at 2 years of age since this set is only at 10% of life expectancy. Whereas a Sears "deep cycle" at 2 years of age is at 100% of life expectancy. Don't add to your battery bank once it's past 10% of life expectancy. Better to just run them out or trade with a neighbor. All batteries in a bank should be the same age, capacity, and manufacturer.

Nicads are a different story. It's OK to add more capacity at any time (you can afford it), and it doesn't have to be the same capacity cells.

Dear Dr. Doug:

I keep hearing rumors about some great new battery technology that will make lead-acid batteries obsolete in the near future. Is there any truth to this, and should I wait to invest in batteries?

Answer: Ah, pie in the sky someday... The truth is there are at least a dozen battery technologies in the laboratory now. Some of them look very promising, but none of them are going to give lead-acid a run for your money within 5 or 10 years. Lead-acid is also in the laboratory—this old dog can still learn some new tricks. Lead-acid technology is going to be around, and is going to continue to give the best performance per dollar for a long time to come.

"Golf Cart" Type Deep Cycle Batteries

These are the batteries we spec out for our Remote Home Kits 1, 2, & 3. They are an excellent "trainer" battery for folks new to battery storage systems. First time users are bound to make a few mistakes as they learn the capabilities of their systems. These are true deep cycle batteries and will tolerate some 80% cycles without suffering unduly. Even with all the millions of dollars spent on battery research, conventional wisdom still indicates that these good old lead-acid "golf cart" batteries are the most cost-effective solution to battery storage. Because of the relatively low cost these batteries are hard to beat for small to medium sized systems. Typical life expectancy is approximately 3 to 5 years. Rated at 6 volts, 220 Amp-Hours capacity. Size is 10.25"L x 7.25"W x 10.25"H. Each Battery weights 63 pounds.

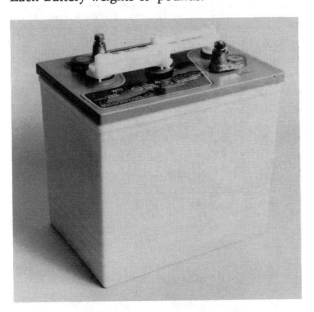

15-101 6 Volt/220 Amp-Hour Battery   **$89**
Shipped Freight Collect from N. California

L-16 Series Deep Cycle Batteries

For larger systems or folks who are upgrading from the "golf cart" batteries, these L-16's have long been the workhorse of the alternative energy industry. L-16's are rated at 350 Amp-Hours at 6 volts capacity, and typically last from 7 to 10 years. The larger cell sizes offer increased ampacity and lower maintenance due to larger water reserves. Size is 11-3/4"L x 7"W x 16"H. They weigh 128 pounds each.
15-102 L-16 Battery   **$229**
Shipped Freight Collect from N. California

"RV or Marine" Deep Cycle Batteries

These batteries are ideal for small or just starting out systems. They represent a big step up from ordinary starting batteries (which should never be used in storage systems). These are 12 volt batteries with high antimony plates for good deep cycle service. Normal life expectancy is 18 to 24 months. The 24-SRM is rated at 85 amp-hours and measures 10.25"L x 6.7"W x 8.5"H. The slightly larger 27-SRM is rated at 105 amp-hours and measures 13"L x 7"W x 9"H.
15-103   Interstate 24-SRM   **$85**
15-104   Interstate 27-SRM   **$95**
Shipped Freight Collect from N. California

Take good care of your batteries! See the Control and Mounting section starting on pg. 207.

See our battery care and interconnect equipment on pg. 204-205.

To order any of these products
or for more technical information
call us toll-free at
1-800-762-7325

Prevailer Gel Cell Batteries

Prevailer gel cells, made by Sonnenschein, are truly 100% maintenance-free. Prevailer batteries can be charged and discharged more frequently than any other battery because the lead-acid electrolyte is totally gelled. Gel cells are leak-proof, sealed for life, and can operate in any position (even upside down) with an extremely low self-discharge rate. These batteries can be taken down to a 100% depth of discharge and can sit totally discharged for 30 days with absolutely no damage. They are one of the only batteries rated for use in extreme cold environments. To summarize some of the advantages of gel cells (to justify their high cost!) over standard lead-acid batteries:

- They don't require acid checks or watering
- They can withstand shock and vibration better
- They can be stored up to 2 years without re charging
- They tolerate extreme cold and extreme heat
- There is virtually no gassing or emission of corrosive acid fumes

For best life expectancy these batteries must be charged at lower voltages than standard lead-acid batteries. Peak voltages of 13.7 to 14.1

Gel Cells are not a good choice for PV based charging systems which need to cram the maximum charge rate in the available sun time.

Nickel Metal Hydride Battery Pack

Our 3.4 amp-hour Nickel Metal Hydride battery pack represents a breakthrough in portable power. It can be clipped to a belt or placed in a leather/canvas portable case and used to power any number of 12-volt appliances or small AC loads through a small inverter. Portable cellular phones, radios, and hobby toys are examples of applications, as well as diagnostic equipment requiring a reliable portable power source. Nickel Metal Hydride is nontoxic and has no "memory effect," a characteristic of conventional nickel-cadmium rechargeable batteries to lose capacity if they are repeatedly recharged before fully discharging. It's almost three times the capacity of similar sized nicad batteries. It measures 1½" x 3¼" x 6½", weighs 2½ lbs, and has a female lighter plug attached. As with nickel-cadmium cells, this battery uses nickel for the positive electrode, but it uses metal hydrides as the negative electrode. The disposal of these new rechargeables is much safer because they contain no cadmium. Though with thousands of cycles before wearing out, disposal time is years in the future.

Want to be a walking Christmas tree? Plug the NMH battery pack into our 12 volt Christmas lights and you're good for 3 hours of party time.

maximum are recommended. All Prevailer batteries come with a 3-year warranty. Send SASE for more information.

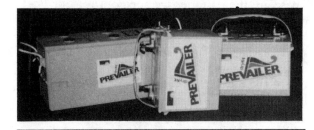

Prevailer Gel Cell Batteries				
Maximum Dimensions				
	Length (in.)	Width (in.)	Height (in.)	Weight (lbs.)
32 Amphour	7.75	5.2	7.25	24
46 Amphour	9.4	5.5	9.25	39
66 Amphour	10.9	6.75	9.9	54
85 Amphour	12.75	6.75	9.9	64
165 Amphour	20.75	8.5	10.0	135
210 Amphour	20.75	11.0	10.0	168

■15-204 Gel Cell 32 Amp-Hour(12V) **$89**
■15-205 Gel Cell 46 Amp-Hour(12V) **$135**
■15-206 Gel Cell 66 Amp-Hour(12V) **$175**
■15-203 Gel Cell 85 Amp-Hour(12V) **$195**
■15-207 Gel Cell 165 Amp-Hour(12V) **$395**
■15-208 Gel Cell 210 Amp-Hour(12V) **$495**
Shipped freight collect from Florida
Allow 4 to 6 weeks for delivery

15-551 Nickel Metal Hydride Battery **$225**
15-552 120V Adapter **$19**

Our Best Industrial Quality Chloride Batteries

Pacific Chloride industrial batteries are among the best batteries that we sell. They come with a five-year warranty, and with proper maintenance will easily last 15 years or more.

Each cell is individually packaged in a steel case with two lifting handles coated with acid resistant paint. Moving these cells is a relatively safe and easy two person task, even over rough terrain.

The advantage of using Chloride 2-volt cells for alternative energy storage is, increased performance and battery reliability. This is achieved by reducing:
- the number of cells needed to supply the required amp hour capacity,
- the number of inter-cell connections subject to corrosion, and
- the number of cells that need to be watered.

Pacific Chloride cells are specifically constructed for high performance. They are designed to be cycled to 80% depth of discharge for 1,500 cycles giving you more usable capacity than is available in other lead acid batteries. The leakproof cell covers are thermally bonded to the cell jar, eliminating the expense, labor and mess of re-sealing. The cell terminals have leak proof seals and the positive terminals have floating bushings designed to permit normal positive plate growth without damaging the integrity of the cell jar. The plate separators and grid structure have been designed to reduce internal resistance, increase kilowatt-hour output and increase realiability. All inter-cell connectors with stainless steel hardware are included. The inter-cell connectors are made of premium quality, very flexible copper welding cable with plated lugs, all designed for easy installation and reduced corrosion.

The Real Goods Pacific Chloride Alternative Energy System Batteries are priced as 6 cell (12 volts) batteries. For 24 volt systems you will need to purchase two 6 cell systems and connect all of the cells in series. These batteries are totally recyclable. Call for Spec Sheet and Brochure.

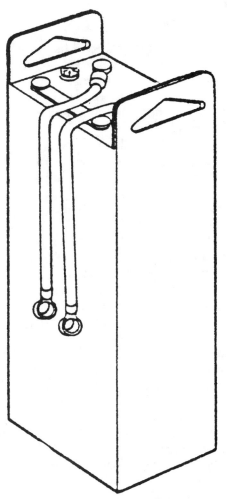

■15-420 420 Amp-hr Chloride Battery $1,425
■15-421 525 Amp-hr Chloride Battery $1,595
■15-422 635 Amp-hr Chloride Battery $1,745
■15-423 740 Amp-hr Chloride Battery $1,895
■15-424 845 Amp-hr Chloride Battery $2,025
■15-425 950 Amp-hr Chloride Battery $2,195
■15-426 1055 Amp-hr Chloride Battery $2,395
■15-427 1160 Amp-hr Chloride Battery $2,595
■15-428 1270 Amp-hr Chloride Battery $2,795
■15-429 1375 Amp-hr Chloride Battery $2,895
■15-430 1480 Amp-hr Chloride Battery $3,055
■15-431 1585 Amp-hr Chloride Battery $3,275
■15-432 1690 Amp-hr Chloride Battery $3,555

Shipped freight collect from Northern California
Allow 4-6 weeks for delivery
Chloride batteries are made of lead and lead is a commodity subject to price fluctuations. Check before purchasing batteries!

Hoppecke Fiber Nickel-Cadmium Batteries

It is widely recognized that lead-acid batteries are the weakest link in stand-alone PV power systems. Recently, PV system designers have turned to nickel-cadmium batteries as an alternative. Many systems now are using recycled "pocket plate" nicad cells.

However, pocket plate cells can have a problem with potassium carbonate formation from the graphite filler they need for high current performance. Because of this, pocket plate nicad cells can require electrolyte replacement as frequently as every 5 years.

Fiber nickel-cadmium (FNC) batteries are made with nickel-plated plastic fibers (like a Brillo pad) that hold the active materials. This gives high surface area and excellent conductivity with no graphite filler. The only carbonate formation is from atmospheric CO_2, and that is negligible.

These batteries were developed in the 1970s in the Volkswagen/Mercedes Research Laboratories in Germany for electric vehicle use.

FNC cells have the following advantages:

• No carbonate problem.

• Outstanding life—15,000 cycles (30 years) in typical use (see graph).

• Low maintenance. Optional *Aquagen* recombiners reduce water loss by 99%. Up to 5 years between waterings.

• Superior high current performance—batteries can be sized to the amp-hours needed, not the inverter current draw.

• High efficiency—Ampere hour efficiencies of 98% are achieved if most operation is kept is below the electrolysis range (1.47 V/cell). Recommended voltage regulation is 1.48–1.55 volts per cell.

• Wide temperature range(–40°to +130°F; temperatures below –18°F require special electrolyte). Batteries can be installed in unheated spaces, they have 95% capacity at 32°F. Unharmed by freezing.

• FNC cells can be fully discharged and left at any state of charge without damage. A low voltage cutout (0.8 V/cell min) is recommended, however, in normal use to prevent battery bank imbalance.

• These are *new*, not recycled units.

Despite their initial cost, FNC batteries are very cost-effective. As battery banks can be sized for just the storage required, rather than for peak current or conservative cycle depth, sizes are often smaller than lead-acid equivalents. The extremely long life and sustained performance (as opposed to the inevitable degradation of a lead-acid bank) easily justify the initial cost in many applications.

Hoppecke FNC batteries are manufactured in the USA and Germany and must be shipped freight **prepaid** from Butler, New Jersey. Please call for a freight quote. Allow 4-6 weeks for delivery.

		Amphours	Weight	Price
■15-440	FNC Battery	21	35	$469
■15-441	FNC Battery	42	57	$839
■15-442	FNC Battery	63	61	$999
■15-443	FNC Battery	84	89	$1,399
■15-444	FNC Battery	105	94	$1,569
■15-445	FNC Battery	126	115	$1,949
■15-446	FNC Battery	147	123	$2,049
■15-447	FNC Battery	167	161	$2,459
■15-448	FNC Battery	167	150	$2,249
■15-449	FNC Battery	188	152	$2,499
■15-450	FNC Battery	209	171	$2,799
■15-451	FNC Battery	209	186	$2,559
■15-452	FNC Battery	241	227	$2,959
■15-453	FNC Battery	251	202	$3,399
■15-454	FNC Battery	293	240	$3,699
■15-455	FNC Battery	283	235	$3,499
■15-456	FNC Battery	314	240	$3,949
■15-457	FNC Battery	335	264	$4,299
■15-458	FNC Battery	366	251	$4,449
■15-459	FNC Battery	419	337	$4,799
■15-460	FNC Battery	450	348	$5,199
■15-461	FNC Battery	492	363	$5,999
■15-462	FNC Battery	544	376	$6,899
■15-463	FNC Battery	589	376	$7,499

(Pricing is for 10 cell/12V Batteries— double the price for 24V systems)

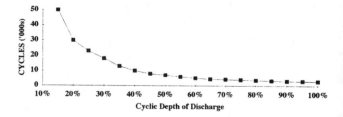

Todd Battery Charger

The new battery charger manufactured by Todd Engineering (and also marketed and re-labeled by Heliotrope General) is the best battery valued charger we've found to date for a stand alone charger. Our most popular model is a 75 amp high-output 15.5 volt, regulated voltage charger. It puts out 70-80% of rated output at just under 14V. It begins to taper only after 14V is achieved, allowing for battery charging in hours instead of days, **greatly reducing generator time and fuel costs.** Another big plus for this solid-state charger is its amazing efficiency, *using a scant 1050 watts for the 75 amp model!* The Todd delivers close to 90% efficiency, compared to the best transformer based chargers which can't top 60%. This means more amps into your battery and less watts out of your generator. The 75 amp model can deliver full output, on generators as small as 2000 Watts. With larger generators, this charger leaves you with enough reserve power to do other chores such as laundry, pumping, and vacuuming. It's a very versatile charger allowing series connection for 24V charging or unlimited paralleling to achieve as many amps as desired (making it far more reasonable than IBE). It is compact (15.5" x 7.25" x 4") and light (9#), can be wall mounted for added convenience and is warranted for one year. The low voltage model (14.0V) is designed for charging using standard AC power and can be left hooked up indefinitely for unattended operation. The medium voltage model (15.5V) is designed for faster and more efficient charging using generator power. The high voltage model (16.5V) is designed for charging Nicad batteries only. All three models come with an automatic temperature sensitive 2-speed fan. The charger is also available in a 45 amp model (using less than 1000 watts) and a 30 amp model (using less than 600 watts.)

15-631	75 amp Charger (14V)	$329
15-629	75 anp Charger (15.5V)	$329
15-633	75 amp Charger (16.5V)	$329
15-637	45 amp Charger (14V)	$249
15-628	45 amp Charger (15.5V)	$249
15-636	45 amp Charger (16.5V)	$249
15-639	30 amp Charger (14V)	$169
15-627	30 amp Charger (15.5V)	$169
15-638	30 amp Charger (16.5V)	$169

Battery Care Accessories

Charg-Chek Battery Tester

The Charg-Chek is a very simple hydrometer that accurately tests the state of charge within the temperature extremes of –40° to 130°F. It is easy to use, easy to read, and features an unbreakable rubber tip, leakproof container, and pocket protector. You read the state of charge by the number of balls floating—one ball floating is 25% charged, four balls floating indicates a full charge.

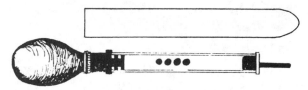

15-701 Charg-Chek Mini Hydrometer $5

874-L Battery Hydrometer

This full-size specific gravity tester is accurate and easy to use in all temperatures. Specific gravity levels are printed on the tough, see-through plastic body. It has a one-piece rubber bulb with a neoprene tip.

15-702 Full Sized Hydrometer $6

Battery Service Kit

This kit has everything you need to service a battery. It contains an open-end wrench to remove terminal bolts and nuts, a wire brush, a battery post and terminal cleaner, a safety grip battery lifter, a terminal clamp lifter (a must if you've ever tried to remove a corroded battery terminal), an angle nose plier, a terminal clamp spreader, and a box wrench.

15-703 Battery Service Kit $49

Post & Clamp Cleaner

The most basic battery tool for cleaning your posts and clamps.

15-704 Post & Clamp Cleaner $4.50

Battery Maintenance Kit

This kit contains one 1¼ ounce aerosol can of NCP-2 battery corrosion preventative, one 2½ ounce aerosol can of Noco battery bath cleaner, and one ST-11 pouch of two terminal protectors. The protectors are impregnated with a corrosion preventative and slip over the battery post before the terminal is attached.

15-705 Battery Maintenance Kit $11

Battery Treatment Kit

This kit includes a 1 ounce tube of corrosion preventative, an applicator brush, and two protection rings. It's a good idea to put a treatment kit on every new battery to prevent corrosion. Designed to fit round style posts only.

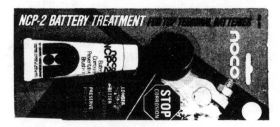

15-706 Battery Treatment Kit $4

Battery Terminals

Pro-Start Battery Terminals

Pro-Start solid brass noncorrosive battery terminals will last a lifetime, making a positive sealed connection between the battery and electrical system. The positive connection and higher conductivity allows more energy to be solidly stored in the battery. Solid brass terminals have 3.5 times higher conductivity than lead connections. This elimination of corrosive oxides allows faster power flow back into the battery from any charging source. Pro-Start terminals can extend battery life potential by 50%, increase available power by 23%, and forever end cable replacement and costly maintenance. Each terminal comes with a grease fitting which is filled with grease upon installation. When purchasing your battery system, we highly recommend Pro-Start. Designed to fit round automotive style battery posts only (cannot be used on Chlorides, Gel Cells, or nicads).

15-781 Pro-Start Terminals (set of 2) $19

Wing Nut Battery Terminals

This is the terminal to use to convert conventional round post-type batteries to a wing nut connection adapter with a 5/16 inch stud.

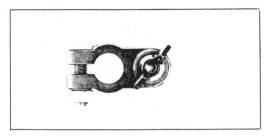

15-751 Wing Nut Battery Terminal $2

Insulated Battery Cables

For interconnecting batteries in series or parallel you need these insulated battery cables. Cables are 4 gauge stranded copper. Our most popular seller and what is used most frequently in putting 6V batteries in series are the 16 inch cables.

15-776 4 Gauge Insulated Cable, 16" $3
15-777 4 Gauge Insulated Cable, 24" $3.50

To order any of these products
or for more technical information
call us toll-free at
1-800-762-7325

Notes and Workspace

Controllers, Metering, & System Safety

We must manipulate energy to prevent it from running amok. Left to its own devices, energy can behave as the Sorcerer's Apprentice, mindlessly performing certain chores to the detriment of others more essential. This section deals with "balance of system" components. These are electronic brains and sensors that take the pulse of our systems and let us become masters of the available power.

Some photovoltaic systems are models of simplicity; if the sun shines, the fan ventilates the attic. Controls, monitors, and batteries may be superfluous in such simple applications. Most systems, however, require more attention.

It is *strongly* recommended that anyone who will be dealing with electrical systems purchase at least an inexpensive, hand-held digital multimeter. These are indispensable for tracking down wiring problems or even saving your life by letting you know that a wire is "hot." More expensive models will allow complex measurements, but the real benefit is that this device will provide a window into your system. Digital multimeters can be purchased at reasonable cost at electronic equipment outlets such as Radio Shack.

Controllers

The correct controller can maximize the effectiveness of an energy system. Charge controllers (also called "voltage regulators") prevent batteries from being overcharged and damaged. Pump controllers maximize pump efficiency, and can turn pumps on or off according to need or power availability. Some controllers can do a primary task and then switch to a secondary task when the first has been satisfied.

When charging batteries, a voltage in excess of that of the battery is required. One might imagine filling a barrel of beer. One must hold the source at least above the level in the barrel to siphon liquid in, or above the lip of the

barrel to pour. Once the barrel is full, to continue adding beer will just make a mess.

With this in mind, think of controllers as the brains that keep batteries from overcharging and meters as gauges of how full our barrels are, how much and how fast we are pouring, and how much we have consumed.

In choosing a controller, make sure voltage specifications are compatible with the charge source and the battery (or load) to be controlled. The controller must also have the ability to handle the amperage to be passed through it. In many installations, the addition of PV modules can be anticipated, and a controller correctly sized to handle all the expected amperage will save money in the long run.

Controller manufacturers offer a wide variety of options, such as temperature compensation (maximum battery voltage will differ at different battery electrolyte temperatures), user adjustment, load controlling, and more.

The choice of controller is most often determined by the method of charging. It should not be assumed that a controller appropriate for a photovoltaic array will work equally as well with a wind or hydro generator.

About Diodes

A fascinating piece of photovoltaic folklore can be observed in the number of people obsessed with the murky subject of diodes and PV modules. A diode is an electrical component that functions as a one-way valve for electricity; the current may flow past the diode in one direction, yet be prevented from flowing in the other direction. Diodes are a standard tool in the electrical engineer's toolbox, and come in a great many forms for different applications. In fact, a single photovoltaic cell (a module is composed of cells electrically connected together) is very similar to a diode.

It is a widespread myth that huge amounts of energy will flow out of a battery to a photovoltaic array at night unless a diode is installed. This "reverse-current leakage" has an origin in fact. However, the amount of energy lost is usually insignificant, unless the array is very large and the nights very long. The standard use of diodes is unnecessary and can substantially reduce system performance. Furthermore, nearly all photovoltaic charge controllers are designed to provide protection for this by incorporating the correct diode or more commonly a relay.

Monitoring

Not all independent energy systems require monitoring devices. Without minimal information, however, the homeowner has limited ability to troubleshoot problems or control energy usage.

The primary monitor on battery storage systems is a **voltmeter**. This relatively inexpensive piece of equipment reveals the momentary voltage of the whole electrical system; the battery, the charge sources, and the electrical loads taken together. The voltmeter reading, however, can be misinterpreted unless one understands the principles at work.

Voltage can be compared to the pressure in a balloon. As we push air into the balloon, the pressure rises and the potential energy is increased. As we let air out of the balloon (use energy), the pressure will drop. There can be a difference in the system pressure (or system voltage) when a system in flux is compared to the same system storing the same energy under static conditions. The voltmeter does not tell us if the system is static (nothing coming in or going out) or in flux (having a flow in or out). The voltage reading from a voltmeter will have greater value if we know the degree of flux the system is experiencing. We monitor this flux with an **ammeter**.

Ammeters measure amperes, a quantitative flow of electricity. In all power systems we measure the flow of electricity from the charging source (photovoltaic array, hydro or wind generator, battery charger) into the batteries. A second ammeter is used to measure amps flowing out of the battery to the various electrical loads (pumps, lights, stereos, etc.). These two meters in combination with the voltmeter provide the most essential information to the

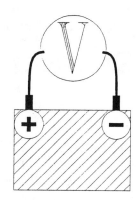

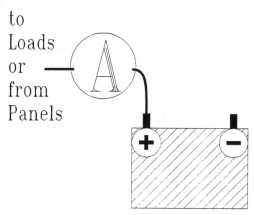

to Loads or from Panels

health of a system.

Sophisticated devices, called accumulating amp-hour meters, exist which take all the information from these sources and process it to a bottom line of how much energy is left in the battery bank (akin to a "gas gauge"). Other pieces of more esoteric information can be generated, depending on the interest of the user and the degree of involvement desired.

Electrical Safety

We have found that fuses represent only one component of a safe system. A properly designed system will perform within reasonable expectations, and have mechanisms to protect the system, the user, and the environment from malfunction. Additionally, local authorities and insurance companies have a stake in safe operation. As the independent energy industry matures, meeting the priorities of all concerned parties will become increasingly important.

Real Goods provides the safest systems possible in our prepackaged systems. While a prepackaged approach may remove some of the educational stimulation of planning and designing your own power system, this is an area where our extensive experience in component-matching is invaluable. Safety is too important an area in which to cut corners or make a simple mistake.

The greatest dangers in any electrical installation are **electrocution** and **short circuits**. Electrocution is the seizure of the heart resulting from the flow of electricity through it. Short circuits are the uncontrolled flow of energy which may cause fires that threaten person and property.

One can minimize the possibility of electrocution just as one can the chance of being hit by lightning. The key is to understand the location of potential (voltage) so that one does not become a conductor for the electricity. The hand-held voltmeter provides simple, yet critical information to the system user.

Having a properly grounded system and using components specifically designed for electrical applications can greatly contribute to safety. Wiring and components must be easily disconnected from the battery bank, where all the energy (and therefore, danger) is stored.

The charge source is another potentially dangerous voltage source, and should have a convenient means of disconnection from the rest of the system.

Short circuits are prevented by **over-current limiting devices**. The **fuse** is a sacrificial component consisting of a wire or wires that will predictably disintegrate (blow) when a known current passes through. The fuse is sized slightly (25%) above the acceptable number of amps anticipated under normal conditions. Fuses come in many varieties designed for specific applications. It should not be presumed that fuses of the same amperage rating are interchangeable if voltage ratings or codings differ.

A more elegant solution than a fuse is the **circuit breaker**. These can be reset instead of needing replacement. Circuit breakers are widely used in AC (alternating current, as is provided by utilities) systems, but are only recently available for DC (battery) systems. Make sure that any device you choose is certified for the task by a proper testing laboratory, such as Underwriters Laboratories.

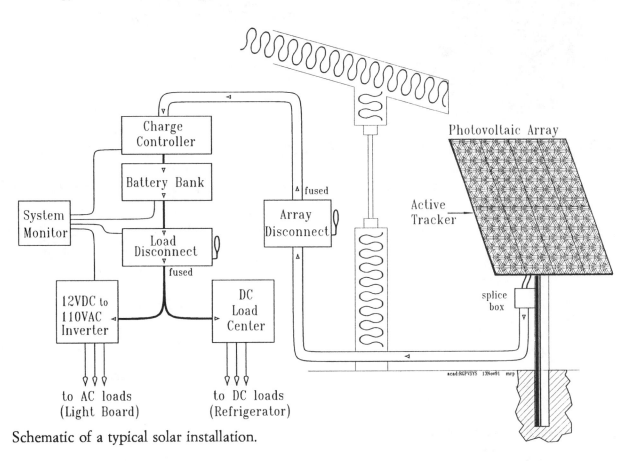

Schematic of a typical solar installation.

Charge Controllers

Trace C-30A Charge Controller

The Trace C30A Charge Controller has quickly become our best selling basic charge controller for systems with up to 30 amps of charging current and it can be wired in parallel for arrays over 30 amps. It is the charge controller we've chosen for both reliability and economy in our Remote Home Kits #2, #3, and #4. It can be operated from either 12 or 24-volt battery systems. It comes factory preset for high voltage disconnect and low voltage return, but it's also user adjustable (requires a digital voltmeter). The C-30A features a night-time disconnect eliminating the need for a diode and an "equalizing charge" switch that allows for occasional heavy charging of the battery bank. One-year full warranty.

25-103 C30A Charge Controller $95

The providers of the famous UL listing are the best known, but not the only lab that certifies electrical components. The Real Goods technical staff is well versed in the proper mating of controllers with chargers; if unsure about an application, call us!

Trace C-30 Load Controller

This quality Trace product is very similar to the best selling C-30A, but without the automatic night time disconnect or equalizing features. The C-30 is a battery voltage controlled relay. User-adjustable high and low voltage switch points allow optimum settings for differing systems. Range is approximately 10 to 16 volts (or double for 24 volt). Like the "A" unit, lightning and input protection is provided by a 56 V transorb (a lightening arrester) and a 30-amp slow-blow fuse. Internal electronics are reverse polarity protected and switchable for either 12 or 24 volt operation. Large terminals accept from #14 thru #4 AWG wire. The C-30 can be configured for either charge control (low voltage connect/high voltage disconnect), or load control (low voltage disconnect/high voltage connect). This unit can be used to protect batteries from being overly discharged, to trigger low voltage alarm systems, and to turn pumping systems on and off—The list is only limited by your imagination. One-year full warranty.

25-100 Trace C-30 Load Controller $95

To order any of these products
or for more technical information
call us toll-free at
1-800-762-7325

M8 / M16 Sun Selector Charge Controller

If you don't need to regulate a full 30 amps of solar power, then the M8 or M16 is the way to go. The M8 and M16 are full-feature sophisticated charge controllers for a very reasonable price. Made by Sun Selector Electronics, who also make the ingenious "Linear Current Boosters," these fine charge controllers eliminate the internal or external need for blocking diodes. They are specifically designed to eliminate battery gassing and maintenance. The M8 is rated at 8 amps maximum current, but realistically will handle up to 10 amps, so it is the ideal controller for up to three 48-watt PV modules. Both the M8 and the M16 measure only 2" x 2" x 1¼" and are simple to install with just four wires to connect. Both units are totally encapsulated for outdoor or marine installations. The M16 will handle conservatively up to 16 amps of charging current. Both units feature four LED indicators, which inform the user of a wide variety of system conditions. Sun Selector makes excellent charge controllers for a very reasonable price. Five-year full warranty.

25-101	M8 8-Amp 12V Controller	$65
25-102	M16 16-Amp 12V Controller	$79
25-128	M8 8-Amp 24V Controller	$65
25-104	M16 16-Amp 24V Controller	$79

SCI Mark III 30-Amp Charge Controller

The SCI Mark III/30 provides a two-step charging and is available in 12V or 24V models. The SCI Mark III/30 will charge with 100% of available charging power up to a factory preset voltage of 14.8 V; here it switches to float charge mode charging at 3 amps and tapering the batteries to 14.1 volts. The 100% charging rate is resumed any time the battery voltage drops below 12.7 volts. A low voltage warning light indicates when the battery voltage falls below 11.5 volts. The Mark III/30 comes mounted on a 5¼" X 8½" flush mount faceplate. The optional knockout box will fully enclose the unit. The Mark III/30 has reverse-current protection eliminating the need for a diode. Five-year full warranty.

25-113	SCI III 30 amp 12V Controller	$119
25-118	SCI III 30 amp 24V Controller	$119
25-114	Single SCI Knock-out Box	$25
25-115	Double SCI Knock-out Box	$35

SCI Mark III 15-Amp Metered Charge Controller

For smaller systems this attractive full-featured charge controller contains a blocking diode, circuit breaker, LED charging light, and built-in volt and ammeters. It is available in 12 or 24 volts and will regulate up to 15 amps of incoming power. It uses only 28 mA of current. The Mark III/15 contains everything you need in a small charge controller.

25-116	SCI III 15-Amp 12V Controller	$149
25-117	SCI III 15-Amp 24V Controller	$149
25-114	Single SCI Knockout Box	$25

GenMate Auto-Start Controller

Our GenMate is a super-versatile computerized generator controller that can be customized to automatically start and stop nearly any electric start generator. Flipping tiny switches programs the GenMate—a one-time operation. This controller can be told to automatically start the generator when the batteries get below a certain voltage or (for generator powered pumping) when the water tank is too low, and then turn off when the task is complete. You can also select a prestart warning beeper and the number of start cycle retries to attempt before sounding a start failure alarm. To use this controller, be sure that your generator is   equipped with a low oil cutoff switch and automatic choke. 12 or 24 volt operation. Full five-year warranty.

25-170	GenMate Auto-Start Controller	$439

Heliotrope CC-60 Charge Controller

The CC-60 is a 45 amp controller that can easily be upgraded to 60 amps. It is a non-relay pulse width modulation controller that is selectable for either 12 or 24 volt operation. An LCD digital meter is incorporated into the front cover and is used via a rotary switch to allow readings of "Array Voltage", "Battery Voltage", & "Charging Current". Battery charge level is adjustable in 16 steps from 13.5V to 16.5V in .2 volt increments (x2 for 24 V). Nicad battery users take note! Has indicator lights for charging, charged, low battery voltage warning (adjustable from 10.5 to 11 volts), & overtemp shutdown. Accepts wire sizes up to 250MCM. 60 Amp Shottky blocking diode is built-in. Size is 11" x 9" x 3.75". Pro-rated 10 year warranty.

25-105	CC-60 Heliotrope Controller	$295
25-106	Fan-60 (to expand to 60 amp)	$45

Enermaxer Universal Voltage Controller

The Enermaxer Universal Battery Voltage Controller is an excellent state-of-the-art voltage controller and has been extremely reliable. *It is the only voltage controller that will work with many different charging sources at once (wind, hydro, PV).* It is a parallel shunt regulator that uses the excess power of the charging source to perform useful work such as heating water, space heating, powering fans, pumps, and lights. It is made of all solid-state components so that there are no relay contacts to wear out. Amperage is reduced gradually to provide a complete finishing charge and then accurately maintain a user-selected float voltage. The 12 and 24-volt configurations will handle 50 amps, and the 32 and 36-volt configurations will handle up to 45 amps. Two-year warranty. When ordering the Enermaxer, keep in mind that you must have an external load at least equal to your charging capacity. This is usually performed with either water-heating elements or air-heating elements that draw 15 amps each (24V elements are also 15 amp). *Always specify voltage when ordering.*

Many of our customers order the water heating elements separately. They are ¾ inch male pipe thread and measure 5½ inches from threads to tip, stainless steel housing. Higher amperage elements available by special order.

If using photovoltaic panels with this controller, you need the Schottky diode kit listed below installed between the panels and batteries.

25-107	Enermaxer Controller—12/24V	$249
25-108	Water Heating Element—12V	$89
25-130	Water Heating Element—24V	$89
25-109	Air Heating Element	$25
25-704-KT	60-Amp Diode w/Heat Sink	$25

PPC PV Power Controller

The PPC, made by Specialty Concepts (SCI), is a professional-grade charge controller system, with a full array of bells and whistles that brings systems together. It is available for 12, 24, 36, and 48 volt systems with models for 30 or 50 amps of charge current. It consists of a series-relay battery charge regulator with low voltage load disconnect, a load circuit breaker, system status lights, and full metering. The lights indicate "charging" and "load disconnect" conditions, and meters monitor battery voltage and charging current, providing system status information at a glance. The PPC is housed in a waterproof NEMA enclosure and has a terminal block with adaptors for up to 6 gauge wire. Five-year full warranty.

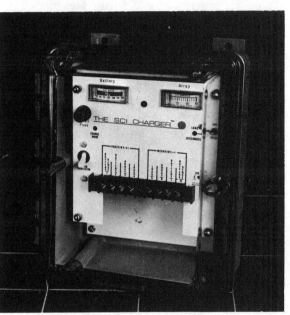

25-141	PPC Controller—12V	$395
25-142	PPC Controller—24V	$395
25-143	PPC Controller—36V	$595
25-144	PPC Controller—48V	$595
25-145	PPC 50-Amp option—12V & 24V	$125
25-146	PPC 50-Amp option—36V & 48V	$185
25-147	PPC Temperature Comp.	$25

Read our Safe Installation Practices Guide located at the end of the Sourcebook for many helpful tips.

Linear Current Booster

The Linear Current Booster (LCB) is an electronic interface between solar modules and load devices such as a water pumps or ventilator fans. The device provides surge current (amperage) for the purpose of starting motor loads operated directly off of PV modules. This offers substantially improved performance for array direct pumping systems by allowing the pump to start earlier in the day and run longer through cloudy weather. We consider a current booster to be standard equipment on most pumping systems.

The LCB must be set to the proper solar module operating voltage. This can be accomplished by ordering the unit preset from the factory, or by setting the unit in the field when the unit is ordered as a "T" (tunable) model. The basic operating range is from 9 to 38 VDC with a maximum open circuit voltage of 50 VDC. All LCB 20's are tuneable (no option).

An option for the LCB is the RC (remote control) feature. An extra wire is brought out of the unit which, when shorted to PV(–), will turn the LCB and the motor load off. The remote control wire can be actuated by a float switch or remote on-off switch. The **water level sensor** (WLS-1) is an inexpensive sensor that works with the RC option to keep a water tank full. The sensor is waterproof and requires no power supply. When the water touches the probes, the LCB shuts off the pump! It does not offer differential level adjustment. All sensors can be virtually any distance from the LCB and only requires a pair of inexpensive telephone wires between them.

All Sun Selector LCBs can be configured both in parallel and series to achieve any needed level of voltage or current. The first number in the model indicates maximum input amperage. Five-year full warranty on all LCB' options.

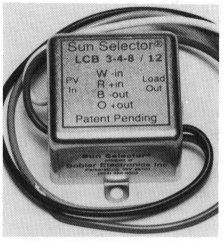

25-133	LCB 3M (12V)	$69
25-133-RC	LCB 3M (12V) RC	$79
25-139	LCB 3M (24V)	$69
25-139-RC	LCB 3M (24V) RC	$79
25-133-T	LCB 3M Tunable	$79
25-133-T-RC	LCB 3M Tunable RC	$95
25-134	LCB 7M (12V)	$99
25-134-RC	LCB 7M (12V) RC	$119
25-138	LCB 7M (24V)	$109
25-138-RC	LCB 7M (24V) RC	$119
25-134-T	LCB 7M Tunable	$119
25-134-T-RC	LCB 7M Tunable RC	$129
25-126	WLS-1 (option)	$29

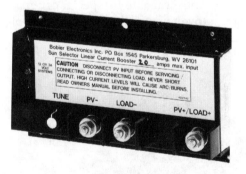

25-135	LCB 20	$375

Metering

Emico Analog Meters

We have found the ammeters and voltmeters manufactured by Emico to be the best all-around meters that work in the 5% accuracy range. All ammeters operate without the need for external shunts. The meters measure 2¼ x 2¼ inches. The expanded scale voltmeters are very easy to read and extremely accurate. Rather than reading the low end of the scale (0–9 volts) which never registers in a 12V system or 0–18V in a 24V system, the expanded scale voltmeters (10-16V and 20-32V) show you only what you need to see and give a more accurate reading. All voltmeters should be installed with an inline fuse.

25-301	10–16V Voltmeter	$19
25-302	22–32V Voltmeter	$19
25-311	0–10A Ammeter	$19
25-312	0–20A Ammeter	$19
25-313	0–30A Ammeter	$19
25-315	0–60A Ammeter	$19
25-408	Inline fuse holder	$3

Equus Computer Voltmeter

This Equus voltmeter will give you your battery voltage with a digital LCD display to one decimal place at a glance. It comes with a programmable bar graph that shows the state of charge and an internal light that allows easy reading day or night. The meter mounts in a 2-1/16 inch hole. It runs on only 12 milliamps. 12 volt only.

| 25-341 | Equus Digital Voltmeter | $49 |

SCI Mark III Battery Monitor

Available in either 12 or 24 volt. A rotary selector switch on this combination volt/amp meter lets you change the large LED display between: incoming amperage up to 100 amps, battery voltage, and outgoing amperage up to 500 amps (with the addition of an optional shunt). A high voltage alarm light indicates when battery voltage exceeds 15.5 volts, and a low voltage light turns on at 11.5 volts. Both are user adjustable. The Mark III Monitor comes mounted on a 5¼ x 8½ inches flush-mount faceplate. The optional knock-out box will fully enclose the unit. This battery monitor can be coupled with the SCI III charge controller in a double box for a very attractive and compact battery control center. (When using the 500-amp shunt, the 500-amp display reads 1/10 of actual value, showing 25.0 per 250 amps.) 100 amp shunt is supplied with the unit. Five-year full warranty.

This SCI monitor is standard equipment with all of our Remote Home Kits.

25-361	DM III—24V Battery Monitor	$169
25-362	DM III—12V Battery Monitor	$169
25-351	100-Amp Shunt	$39
25-364	500-Amp Shunt	$79
25-114	Single SCI Knockout Box	$25
25-115	Double SCI Knockout Box	$35

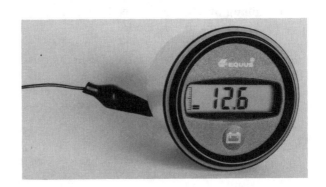

Cruising Equipment AmpHour Meters

These digital LCD backlit meters made by Cruising Equipment allow you to monitor your system at a glance. The meter displays amphours consumed during discharge cycles with a minus sign. During recharge cycles it counts back to zero and on into a plus scale if charging continues into an overcharged state. When dis-charging begins from overcharge, it automatically resets to zero. The meter automatically compensates for battery charging efficiency, approximately 1.2 amphours must be replaced for every 1.0 amphour removed. This charging efficiency factor can be adjusted in the field to compensate for aging batteries, etc. This sort of monitoring is a necessity with nicad battery banks.

The AmpHour *plus* meter tracks cumulative amp-hours like the standard meter above and also reads instantaneous battery voltage or instantaneous amperage flow. This combines all necessary monitoring functions into a single easy-to-use and easy-to-read unit. There is a "setup" procedure on the AmpHour+ and AmpHour+2 models that allows the meter to be customized to your particular battery capacity and charging system. It can be wired with tiny 18-gauge wire, so your system monitor can be mounted in the living room or kitchen where it's easy to keep an eye on. The + and +2 models automatically adjust the charging efficiency factor. At the end of every charge cycle, the factor is averaged with past charge efficiency history. There is a feature that will display the percentage of battery capacity remaining.

The +2 model has all the capabilities of the + model above, and it will monitor two battery banks of different capacities. Each bank may be setup individually.

The +2 model also serves as a Battery/Source monitor. With this option the battery #2 switch position monitors the input from an energy source and accumulates up to 9,999 amphours before rolling over to zero. It will not monitor two battery banks in this mode.

All of these meters measure 4½ x 3 x 1¼ inches, and are available in either 12 or 24 volt DC. All units come with a 500-amp shunt, and the +2 model comes with two shunts. Warranty for all models is 18 months.

The Reagan Administration's rollback of efficiency standards for light vehicles immediately doubled oil imports from the Persian Gulf, effectively wasting exactly as much oil as the govern-ment hoped could be extracted each year from beneath the Arctic National Wildlife Refuge.
- Amory Lovins, in the Foreword

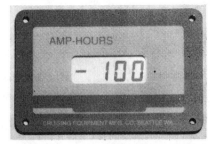

25-350	Amp-Hour Meter (12V)	$195
25-347	Amp-Hour Meter (24V)	$195
25-348	Amp-Hour Meter+ (12V)	$325
25-358	Amp-Hour Meter+ (24V)	$325
25-359	Amp-Hour Meter+2 (12V)	$399
25-349	Amp-Hour Meter+2 (24V)	$399
25-353	800 Amp Shunt	$89

Energy Monitor

The Energy Monitor measures volts and amps, and tracks amp-hours into and out of the battery through the use of a powerful microcomputer using an accurate crystal timebase. Because the Energy Monitor is built with precision circuits, no calibration is required, yet the instrument is the most accurate and stable amp-hour meter available. The recharge factor, which varies from battery to battery, is even adjusted automatically by the microcomputer. The Energy Monitor operates from either 12 or 24 volts, and measures 400 amps full scale with a resolution of 0.1 amp. The unit comes standard with equal distribution backlighting. Drawing less than 0.015 amp, it consumes a mere 0.36 amp-hour per day. Full one-year warranty.

25-365	Energy Monitor	$299

Power Meter 30

The Power Meter 30 is the first meter that looks like it belongs in your battery-powered home, RV, or boat, mounting just like your thermostat. It's designed to record and give you up-to-the-minute status on your entire electrical system. From across the room the Power Meter's LEDs will tell you at a glance if your battery voltage is adequate, if the inverter or generator are on or off, and if your solar array is producing power (in three different colors).

For a more in-depth accounting, press one or both of the Power Meter's two keys and learn what the solar array current is (up to 250 amps with optional shunts) and what the cumulative amp-hours or watt-hours are. It also will tell you how long the generator has been running since you last turned it on, or the cumulative run-time, a great feature for maintenance schedules and oil changes.

Available on the system battery channel is a *battery voltage* reading and a *battery voltage high* (from last reset), *low,* and *average* reading. On the inverter and generator channels, both voltage and frequency are available. Inverter voltage is RMS calibrated and generator voltage is peak.

Standard equipment is a backup battery that is used by the Power Meter should the system battery be dead or disconnected. The backup battery channel will display the current voltage of the internal 9V alkaline battery, or the voltage of a second external battery should the user prefer to connect another battery to the Power Meter. It works on 12 or 24 volt systems.

To top off the list of features, the Power Meter has built-in alarms that can be set to go off under many different circumstances. Current draw is a mere 20 mA. Shunts are used on the solar input channel and are NOT included. Solar array current shunts must be ordered separately. Full one-year warranty.

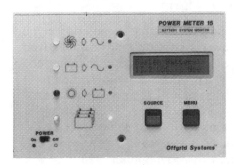

25-356	**Power Meter 30**	**$295**
25-357	**250-Amp Shunt**	**$49**
25-354	**30-Amp Shunt**	**$ 49**

SPM 2000 Amp-Hour Accumulator

The SPM 2000 is ideal for monitoring 12 or 24-volt systems. Its two channels allow you to simultaneously measure and record power produced and energy used. The digital LCD display accurately tells you battery voltage, amps, watts, amp-hours, and watt-hours for two separate inputs or outputs. You can monitor two different inputs like your PV array on 1 and your wind generator on 2 or you can monitor two different loads, like your lights on 1 and your Sun Frost refrigerator on 2. This is a great educational device for us on our showroom floor! It handles up to 199.9 amps (two shunts included), and 10–35 volts. It accumulates up to 999,999 amp-hours. The SPM 2000 was designed to be virtually foolproof with regard to installation. It is accurate to 1% on amperage. Do not expose to rain or moisture. Two-year full warranty.

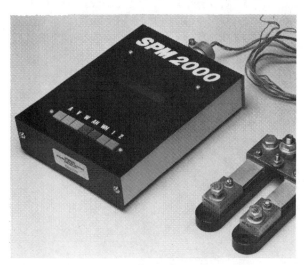

25-355 SPM 2000 Amp-Hour Accumulator $495

Low Battery Alarm

Over-discharging your batteries can cripple your 12V home power system or your boat. Inadvertent deep cycling of batteries drastically reduces their life. This alarm signals audibly when battery voltage falls below 11.5 volts, so you can turn off lights and appliances before damage is done. The alarm is preset to come on at 11.5V, but is adjustable. The alarm's dimensions are 2 X 1 X 4 inches. Voltage: nominal 12V.

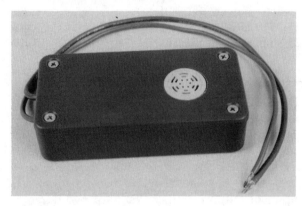

25-342 Low Battery Alarm $39

Voltage Alarm

This alarm automatically detects hazardous high and low voltage fluctuations in your 12V system or on your boat, warning you of problems with your batteries, alternator, or charging mechanism. The alarm has adjustable high and low settings, with both a visual and an audible alarm. At the factory-set low point of approximately 11.7V, a red LED and buzzer are activated. At the high setting of 14.6V, the yellow LED lights up. At 15.1V, the buzzer sounds. Voltage: nominal 12V.

25-343 Voltage Alarm $75

Timers, Photoswitches, Load Diverters & Rheostats

Spring Wound Timer

Our timers are the ultimate in energy conservation, they use absolutely no electricity to operate. Turning the knob to the desired timing interval winds the timer. Timing duration can be from 1 to 12 hours with a hold feature that allows for continuous operation. This is the perfect solution for automatic shutoff of fans, lights, pumps, stereos, VCRs, & Saturday morning cartoons! It's a single pole timer, good for 10 amps at any voltage, and it mounts in a standard single gang switch-box. A brushed aluminum faceplate is included. One-year warranty.

25-400 SpringWound Timer $29

24-Hour 12V Programmable Timer

This timer will operate any voltage appliance up to 20 amps, including lights, pumps, fans, and appliances. It will also start and stop remote generators. It comes complete with a 1.2V nicad battery giving it a 200-hour power reserve in case your battery is dead. It allows for up to three separate on-off operations per day. The timer is run by a 12V quartz movement and best of all it draws only 0.01 amp.

25-401 Programmable 12V Timer $109

Outdoor Lighting Timer

This solid-state timer is designed to sense darkness and then automatically turn on the lights or other loads. The loads are then allowed to run for an adjustable period of time (10 min. to 15 hrs). The 12-volt timer can operate up to 8 amps, enough for several low-wattage walkway lights or two 50-watt porch lights. It can prevent deep battery discharging, and because it senses darkness your outdoor lighting will automatically change with the seasons. Operates on only 12 mA. Five-year full warranty.

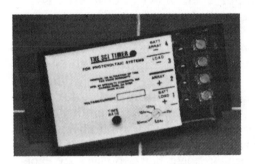

25-402 Outdoor 12V Lighting Timer $89

Day/Night Switch

Another way to operate outdoor lighting is by sensing battery voltage and turning off the lights when the batteries discharge to a specific voltage. This does just that. This 12V 10-amp switch turns on the lights at dark and shuts them off at a user-adjustable (11.5V to 13.5V) voltage level. It uses only 12 mA to operate. Five-year full warranty.

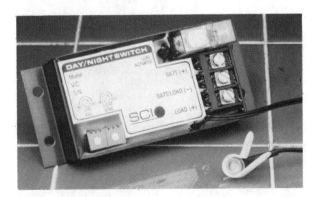

25-403 Day/Night Switch $89

Conserve Switches

Conserve switches are variable-speed control switches (rheostats) for 12V applications such as lights, fans, and motors, using less than ½ watt to control them. They are available in a 4-amp and 8-amp configuration. These work great on incandescent and halogen lamps and also on our 12V fans. The 4-amp switch comes either with or without the forward-reverse switch. The 8-amp switch does not have the reverse option. The forward-reverse feature is useful for summer-winter reversal of air direction in fans.

24-101 4-Amp Switch $27
24-102 4-Amp Switch (w/forward-reverse) $29
24-103 8-Amp Switch $39

To order any of these products
or for more technical information
call us toll-free at
1-800-762-7325

Safety Disconnects

Every professional PV system should employ a fused disconnect for safety. The function of this component is to disconnect all power generating sources and all loads from the battery so that the system can be safely maintained and disconnected in emergency situations with the flick of a switch. We carry several different safety disconnects for different sized applications. Fuses should be sized according to the wire size you are using.

We recommend using a two-pole disconnect for the charging side of your batteries. One pole is used for the line between the panels and charge controller, and the other between the controller and the batteries. make sure this fused disconnect can handle the maximum amperage of your array output.

Another fuse or fused switch can be used to handle the discharge of your battery.

The 60-amp three-pole fused disconnect should be employed when the Enermaxer voltage regulator is used, which requires the third pole for disconnection.

24-201   30-Amp 2-Pole Safety Disconnect $59
24-202   60-Amp 2-Pole Safety Disconnect $105
24-203   60-Amp 3-Pole Safety Disconnect $139
Fuses sold separately (see below)

Fused Disconnect

At last, a fused disconnect guaranteed to put a smile on any inspectors face. This space saving, 400 amp disconnect can handle a 2,000-amp surge for three minutes, to protect your batteries and inverter. Features include a sealed, oil-filled 2,000-amp switch that eliminates arcing. UL-listed fuse is rated to 125VDC, 20,000 AIC, and wire lugs that can take up to 300 MCM wire.

This is the Fused Disconnect your full size inverter has always dreamed of. Standard equipment on our RH3 and RH4 kits.

Load Centers & Transfer Switches

Automatic Transfer Switch

Transfer switches are designed as safety devices to prevent two different sources of voltage from traveling down the same line to the same appliances. This transfer switch made by Todd Engineering will safely connect an inverter and an AC generator to the same AC house wiring. If the generator is not running, the inverter is connected to the house wiring. When the generator is started, the house wiring is automatically disconnected from the inverter and connected to the generator. A time delay feature allows the generator to start under a no-load condition. It comes in both a 110V model and a 240V model. Each will handle up to a maximum of 30 amps. These switches are great for applications where utility power may only be available a few hours per day, or if frequent power outages are experienced.

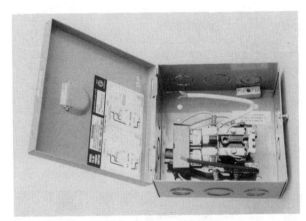

23-121   Transfer Switch—30A, 110V $85
23-122   Transfer Switch—30A, 240V $125

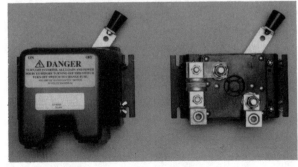

24-206   400A Fused Disconnect $139

Code Approved DC Load & Distribution Centers

Being on the leading edge of the PV industry, we at Real Goods think it's time to move toward safer, code-approved equipment. We pledge to continue to work with responsible manufacturers to comply with the new electrical code requirements for DC systems. Although code-approved equipment can often be more costly, we feel when weighed together with safety factors it is a small price to pay for insurance.

We have worked with a manufacturer to engineer and build this new low-voltage DC distribution center designed to meet NEC code for solar photovoltaic systems. It meets all safety standards and utilizes only UL listed components. *All current-handling devices have been UL approved for 12V through 48 volt DC applications.* Enclosures are also available by special order that will handle up to 16 branch circuits with a maximum of 70 amps on an individual branch and up to 200 amps total for all branches. The enclosure and wiring lugs have been designed to handle the larger wires used in low-voltage DC applications. Extra wire bending room and multiple knockouts up to 2½ inches are standard features. All distribution centers are shipped with a 200A main fuse installed, breakers must be ordered separately. Load center does not have the 200A main fuse.

23-123	**8-Circuit Distribution Center**	**$250**
23-126	**8-Circuit Load Center**	**$129**
23-131	**10-Amp Circuit Breaker**	**$11**
23-132	**15-Amp Circuit Breaker**	**$11**
23-133	**20-Amp Circuit Breaker**	**$11**
23-134	**30-Amp Circuit Breaker**	**$11**
23-136	**40-Amp Circuit Breaker**	**$11**
23-135	**50-Amp Circuit Breaker**	**$11**
23-137	**70-Amp Circuit Breaker**	**$29**
23-139	**200-Amp Main Fuse**	**$45**

See the SWRES Electrical Code in the Appendix

APT Power Center IV

The APT Power Center IV makes installing a solar electric system fast, easy, economical, and up to code. It greatly reduces the number of components by including all the required DC safety and control equipment in a single enclosure. Simply wall mount the power center, connect the solar array, batteries, inverter, and grounding rod, and the system is operational and installed to the National Electrical Code Guidelines. Some of the prewired components installed in the Power Center include: main fused disconnect, charge controller, load center with breakers, metering and shunts, lightning protection, and fused charger inputs. All of the user connections are labeled and provided with oversized lugs and terminal blocks to make the installation quick and trouble-free, even for a novice installer.

The Power Center IV saves money by reducing the installation time and eliminating costly wiring mistakes. Every power center is prewired and fully tested for maximum reliability and correct operation. The compact design and heavy-duty industrial components increase performance and efficiency by reducing power loss in wiring. Virtually any system design can be configured in the compact enclosure with the options available. Please call for design assistance; prices start at $1,200.

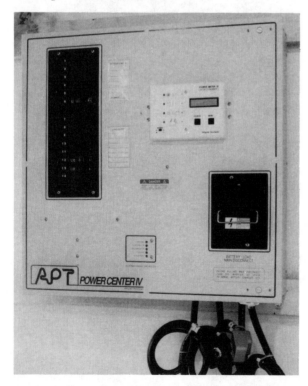

25-171 APT Power Center   Call for Pricing

12 Volt/30 Amp Relay

This general purpose DPDT power relay has a 12 volt/169mA pull in coil. It can be used to remotely switch loads larger than your 12 volt timer, load disconnect, float switch, etc. can handle alone (up to 30 amps per pole). It can also be used to switch AC circuits from DC powered timers or sensors. Because this is a highly adaptable Double Pole, Double Throw relay it is possible to do multiple switching tasks simultaneously.

25-112 12V/30A Relay $25

Fuses & Fuseholders

Fuses and circuit breakers are circuit protectors. Their only function is to break an electrical circuit if the current (amps) flowing in that circuit exceeds the rating of the device. Any size fuse may be used safely as long as its rating is lower than the maximum ampacity of the smallest wire in the circuit.

Most fuses and breakers will pass three times their rated current for a few seconds, but they will open the circuit immediately in the event of a short circuit, which draws hundreds or even thousands of amps. This is necessary so that the fuse will melt or the breaker will open before the wire catches fire in the event of a short circuit. Fuses or breakers should be installed as close to the batteries as possible, and absolutely before wires pass through a flammable wall. The chart below gives the maximum ampacity for various sizes of copper wire.

Some types of wire can handle slightly more current, and all types of wire can handle much more current for very short amounts of time. For example, a 2,000-watt inverter can surge to 6,000 watts for a few seconds when starting large motors. During these few seconds, a 12V inverter is drawing 600 or more amps from the battery, but 4/0 wire and a 250-amp fuse will work fine.

Wire gauge	Maximum ampacity
14	15
12	20
10	30
8	45
6	65
4	85
2	115
0	150
2/0	175
4/0	250

If you choose not to use our highly recommended Safety Disconnects listed above, you must at a bare minimum fuse your system both between your solar array and your battery, and between your battery system and the distribution panel or load. This can be done cheaply and easily with our FRN-R Fuse Blocks, available for 30A fuses or for 60A fuses. Don't forget to order fuses for your Disconnects!

24-401	30A Fuseholder	$5
24-402	60A Fuseholder	$9
24-501	30A Fuse	$4
24-502	45A Fuse	$7
24-503	60A Fuse	$7

Newmark Fuse Box

This is our bare-bones basic fuse box. It consists of an eight-position fuse block riveted onto a hinged metal box with three knockouts in place for wiring. It's made to accept AGC-type fuses up to 20 amps. It has a screw terminal for positive input from the battery and one screw terminal for each of the eight fused outputs. This fuse box only provides space for the positive wires, see our 15-Position Bus Bar below for the negative wires.

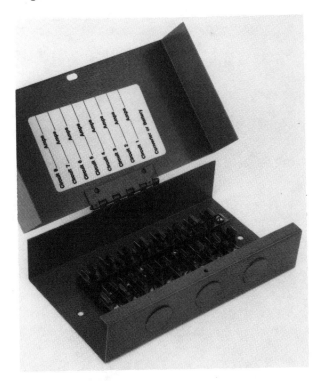

24-411 8-Position Fuse Box (Newmark) $15

Class T Fuse Block with Cover

For those who want to protect their mid-sized inverter, but can't afford the switch, we also offer this 110-amp fuse with block and cover. This is a slow-blow fuse that will blow immediately if your battery is short circuited, but will allow several minutes of up to 150 amp use. It is easily installed on a wall or shelf with the provided screws. Dimensions are 2"W X 2½"H X 5½"L, including the cover. Rated for DC voltage up to 125 volts, it accepts up to a 2/0 gauge cable.

24-507	Fuse Block w/ Fuse & Cover	$49
24-508	110 Amp Class T Fuse	$18

Here's the fuse protection your mid-sized 400 to 800 watt inverter has dreamed of.

Fuses for the Nemark are sold on page 224

Overcurrent Device Comparison

	Device	Protects low AIC components from high short circuit currents	Arcing is suppressed to reduce explosion hazard with batteries	UL rated for DC applications	Provides the maximum AIC rating available and required by the NEC	Can protect DC load centers from high short circuit conditions
FUSES	CLASS T	YES	YES	YES	YES	YES
FUSES	ANN	NO	NO	NO	NO	NO
FUSES	NON	NO	NO	NO	NO	NO
BREAKERS	Heinemann GJ	NO	NO	YES	NO	NO
BREAKERS	Square-D QO	NO	NO	YES	NO	NO
BREAKERS	Heinemann AM	NO	NO	YES	NO	NO

AGC Glass Fuses

These glass fuses go into the Newmark fuse box. The most common size for systems is the 20A fuses. They are available in the following amperages: 2A, 3A, 5A, 10A, 15A, 20A, 25A and 30A. They come in a box of five, and the fuses measure 1¼"L x ¼" in diameter and are good up to 32V maximum.

24-521 AGC Glass Fuses—2 amp, 5/box $2
24-522 AGC Glass Fuses—3 amp, 5/box $2
24-523 AGC Glass Fuses—5 amp, 5/box $2
24-524 AGC Glass Fuses—10 amp, 5/box $2
24-525 AGC Glass Fuses—15 amp, 5/box $2
24-526 AGC Glass Fuses—20 amp, 5/box $2
24-527 AGC Glass Fuses—25 amp, 5/box $2
24-528 AGC Glass Fuses—30 amp, 5/box $2

Inline Fuse Holder

There is no excuse for not fusing your equipment with a fuseholder that installs this easily. This bayonet-type fuse holder has 15 inches of #14 wire in a loop that can be cut anywhere along the length for versatility. This is a "universal" type with three different springs for any length AGC style fuse. Always keep a couple of these on hand. Recommended for fusing your analog voltmeter.

25-408 Inline Fuse Holder $3

15-Position Bus Bar

You can use this 15-position bus bar as a common negative for any fuse block.

24-431 15-Position Bus Bar $9

To order any of these products
or for more technical information
call us toll-free at
1-800-762-7325

Adapters

Wall Plate Receptacle

Our most popular basic 12V receptacle. It is made of a break-resistant plastic housing and all-brass metal parts and fits a standard junction box up to 10 amps. **Specify brown or ivory.**

26-201-B Wall Plate Receptacle (Brown) $4
26-201-I Wall Plate Receptacle (Ivory) $4

Replacement Plug

The SP-36 is the simplest 12V plug on the market. The plug is supplied with two solderless terminals for easy attaching of terminals to wire. The terminals are inserted into the rear of the plug and lock into place.

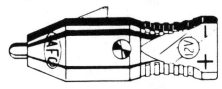

26-101 Replacement Plug (SP-36) $1

Heavy-Duty Replacement Plug

The SP-20 is designed for heavier duty applications than the SP-36. It is supplied complete with three styles of end caps to accommodate coil cords, SJ, SJO, and round wire. The plug is unbreakable and can accommodate currents of 8 amps under continuous duty.

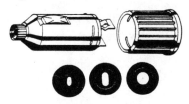

26-102 Heavy-Duty Plug (SP20) $2

Fused Replacement Plug

The 80KF is a fused adapter plug kit with a unique polarity reversing feature. It comes complete with a 2-amp fuse and four different sizes of snap-in strain relief wire connectors to accommodate gauges from 24 to 16. It is rated at 8 amps continuous duty and can be fused to 15 amps to protect any electronic device.

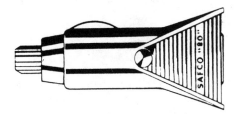

26-103 Fused Replacement Plug (80KF) $3

Adapter (#70)

This simple durable plastic adapter plug produces an elegant conversion from any standard 110V fixture. Simply plug the 110V plug into the base on the #70 adapter (no cutting!), insert a 12V bulb, and you're in 12-volt heaven for cheap!

26-104 Adapter (#70) $2.75

Triple Outlet Box

The triple outlet box permits the use of three plugs at one time. The plug is rated at 10 amps, comes with heavy-duty 16 gauge wire, and a self-adhesive pad in back.

26-105 Triple Outlet Box $9

Double Outlet

The double power outlet permits the use of two 12V products at once out of one outlet. The two receptacles are connected to the adapter plug with short lengths of 16 gauge wire. Rated at 10 amps.

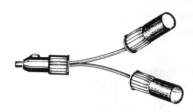

26-107 Double Outlet $6

Extension Cord Receptacle

This receptacle, made of all brass parts in a break-resistant plastic housing, mates with all 12V cigarette lighter type plugs. It comes with a 6 inch leadwire.

26-108 Extension Cord Receptacle $3

Non-Fused Cordset

This replacement cord set is handy where the existing cord set has been damaged or worn. It consists of a male cigarette lighter plug on one end and is attached to 8 feet of coiled 20 gauge, polarity coded wire, the free end of which is pre-stripped.

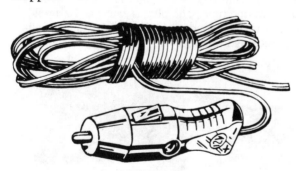

26-109 Non-Fused Cordset $2.75

Fused Cordset

The fused cordset is identical to the 26-109 listed above except that it comes with a fused male plug on one end.

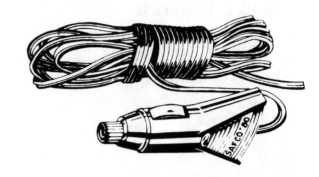

26-110 Fused Cordset $4

Extension Cords with Battery Clips

These unique cords have heavy clips for attachment directly to a battery. They increase the use and value of 12-volt products like lights, TV sets, radios or appliances. *The maximum amperage that can be put through the 1 foot cord is 10 amps and the maximum that can be put through the 10 foot cord is 4 amps.*

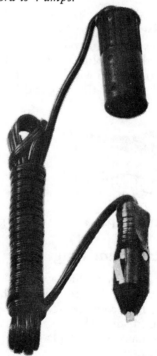

26-111 Extension w/Battery Clips—1 ft. $6
26-112 Extension w/Battery Clips—10 ft. $7.50

Extension Cords

These low-voltage extension cords come with a cigarette lighter plug on one end and a 12V male plug on the other. They are available with either a 10 foot cord or a 25 foot cord. *The maximum current recommended is 4 amps for the 10 foot cord and 2 amps for the 25 foot cord.*

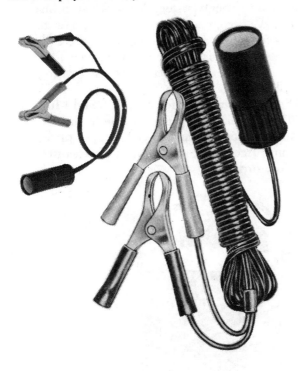

| 26-113 | Extension Cord—10 ft. | $6 |
| 26-114 | Extension Cord—25 ft. | $7 |

UL Listed Switches and Receptacles

In keeping with our trend of offering code-approved equipment for all of our systems, we now have a line of UL listed DC rated 15A switches and 20A receptacles. All are heavyduty commercial grade. Although expensive, these rugged components may be the only way to keep your local building inspector happy (and maybe even give you better sleep at night). They are light years ahead of conventional RV grade plugs and switches. All are ivory colored and mount in standard electrical boxes.

26-310	Single pole 15A Switch	$6
26-311	3-Way 15A Switch	$8
26-312	Single 20A Receptacle	$7
26-313	Duplex 20A Receptacle	$15
26-314	Switch Plate	$1
26-315	Single Receptacle Plate	$1
26-316	Duplex Receptacle Plate	$1
26-317	Heavy Duty 15A Plug	$15

DC Power Converter

This converter allows you to operate a wide variety of DC powered products, including a portable stereo, cassette player, video cassette recorder and other applications from a standard automobile dash receptacle. One convenient switch enables the output voltage to be set to 6, 7.5, 9, or 12 volts DC from a 12-volt DC input, negative ground. Output polarity may be reversed and the LED indicator shows when the adapter is in use. The complete unit, which is UL listed, includes a three amp fused universal plug, six-foot cordset and an assortment of four popular polarity-reversible coaxial power plubs. The fuse is replaceable.

| 26-121 | Converter 60F-3 | $15 |

Suntronics 12-Volt Brass Outlets

Suntronics makes a very attractive line of solid brass outlets and switches designed for the 12V market. Constructed of .036 solid brass, they are rated at 20 amps for the outlets and 15 amps for the switches.

26-221	One Plug	$7
26-222	Two Plugs	$10
26-223	One Switch	$7
26-224	Two Switches	$11
26-225	Four Switches	$15
26-226	1 Switch w/1 Plug	$11
26-227	1 Switch w/1 Fuse	$9
26-228	1 Switch, 1 Fuse, 1 Plug	$12
26-229	1 Plug w/1Fuse	$10
26-230	2 Switches w/2 Fuses	$15
26-231	2 Plugs w/2 Fuses	$16

Wire

Voltage drop must always be considered when sizing wire for any installation. Please refer to the voltage drop charts on page 489.

Type USE Direct Burial Cable

Type "USE" (underground service entrance) cable is moistureproof and sunlight resistant. It is recognized as underground feeder cable for direct earth burial in branch circuits and is the only wire you can use exposed for module interconnects. It is approved by the National Electrical Code and UL. It is resistant to acids, chemicals, lubricants, and ground water. Our USE cable is a stranded single conductor with a sunlight-resistant jacket. It is much more durable than standard romex.

		Price/Foot
26-521	USE #10 Wire	$.40
26-522	USE # 8 Wire	$.60

Copper Lugs

We carry very heavy duty copper lugs for connecting to the end of your large wire from #4 gauge to 4/0 gauge (#0000). The hole in the end of the lug is 3/8 inch diameter. *Wire must be soldered to copper lugs.*

26-601	Copper Lug—#4	$2
26-602	Copper Lug—#2	$4
26-603	Copper Lug—#1	$4
26-604	Copper Lug—#1/0	$4
26-605	Copper Lug—#2/0	$5
26-606	Copper Lug—#4/0	$6

Nylon Coated Single Conductor Wire

This is your basic single conductor wire for connecting the discrete components of your alternative energy system. **We stock it in black only.** We recommend you use red tape on the ends for positive and white tape on the ends for negative. Wire is stranded copper, type THHN and should never be used in exposed wiring applications (conduit only).
The minimum order for #16, #14, and #10 is a 500 foot roll. Wire gauge size of #8 and larger can be ordered in any length. Copper prices change frequently; please call to check prices before ordering.

		Price/ ft
26-534	#8 THHN Primary Wire	$0.55
26-535	#6 THHN Primary Wire	$0.65
26-536	#4 THHN Primary Wire	$0.90
26-537	#2 THHN Primary Wire	$1.35
26-538	#0 THHN Primary Wire	$2.20
26-539	#00 THHN Primary Wire	$2.75

#16, #14, #12 and #10, must be ordered in 500 ft rolls

26-531	#16 TFFN Primary Wire (500')	$50
26-532	#14 THHN Primary Wire (500')	$60
26-530	#12 THHN Primary Wire (500')	$70
26-533	#10 THHN Primary Wire (500')	$110

Split Bolt Kerneys

Split bolt kerneys are used to connect very large wires together or large wires to smaller wires. You must always wrap the kerney with black electrical tape to prevent corrosion and the potential for short-circuiting.

26-631	Split Bolt Kerney—#6	$5
26-632	Split Bolt Kerney—#4	$6
26-633	Split Bolt Kerney—#2	$7
26-634	Split Bolt Kerney—#1/0	$9
26-635	Split Bolt Kerney—#2/0	$14

Solderless Lugs

These solderless lugs are ideal for connecting large wire to small connections or to batteries.

26-622	Solderless Lug (#8,6,4,2)	$2
26-623	Solderless Lug (#2 thru 4/0)	$7

Crimp-on Terminals

While of course it's always better to solder your connections or to use split bolt connectors and electrical tape, many of you are still going to fall back upon the old RV and vehicle type crimp-on connectors. They come in various sizes to accommodate different gauge wires.

26-640	Ring Terminal—#8, Stud 10	$1
26-642	Ring Terminal—#8, Stud 1/4"	$1
26-645	Ring Terminal—#10/12, Stud 10	$.50
26-646	Ring Terminal—#10/12, Stud 1/4"	$.50

Crimping Tool

This crimping tool cuts wire and strips insulation. It also crimps wire terminals.

26-501	Crimping Tool	$7

To order any of these products
or for more technical information
call us toll-free at
1-800-762-7325

Notes and Workspace

Inverters

Most of the world uses an Alternating Current (AC) electrical system. The electrical current flow changes direction (alternates) 50 or 60 times per second. This type of electricity is transmitted easily with no loses over large distances, but needs to be transformed for storage. Storage is a necessity for remote power systems. Batteries convert Direct Current (DC), electrical energy into chemical energy for storage.

Most alternative energy systems collect and store energy as DC. When you have a battery full of DC energy and a house full of AC appliances, how do you bridge the gap? Inverters are the answer. These incredible devices convert low-voltage DC to high-voltage AC. Just 10 years ago, high efficiency, long-lived devices like these were only a dream. Thanks to rapid advances in solid state technology these dreams are now realized.

To understand how inverters work it is first necessary to explain AC power.

Note the **Inverter waveform** illustration that shows what AC power looks like displayed on an oscilloscope. Conventional utility power is delivered as a *sine wave*. Notice the smooth curving waveform. This is the kind of waveform that all AC appliances are designed to accept.

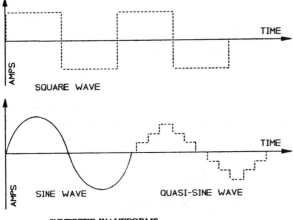

INVERTER WAVEFORMS

Now look at the *square wave*. As you can see, this is a rough approximation of a sine wave. It is AC power because it does alternate, but the waveform is only crudely similar to a true sine wave. Square wave is the type of AC power that

the first generation of electronic inverters produced 15 to 20 years ago. Some inexpensive inverters that produce square wave AC power are still sold. They work fine for running motors but are insufficient for most solid state equipment. They are also very inefficient, drawing 20 to 50 watts just to be turned on (even with no loads). When running a load they typically operate at less than 80% efficiency, wasting a great deal of energy.

The third wave form is *quasi-sine wave* or modified sine wave. This waveform has multiple steps and closely approximates a true sine wave. Modified sine wave is the form of power that modern inverters produce and that is acceptable to most AC appliances. Modern inverters are from 80% to 98% efficient and draw minuscule amounts of power while in standby mode.

Non-sine wave forms read differently with inexpensive voltmeters. Normal utility power is labeled as "115 volts." The peak is usually 150 to 180 volts. A complex formula called root mean square, or RMS is used to compute the average voltage. When a standard voltmeter is set to read AC voltage, it is assumed that the voltage is sine wave. When the meter is used to read inverter (modified sine wave) output, it will read very low, typically 80 to 100 volts. This is normal.

An expensive RMS corrected meter is needed to accurately read any waveform other than standard sine wave. All the modern inverters hold voltage to a maximum plus or minus 2% tolerance. Utilities usually allow 10%, sometimes as much as 15% variation.

Modern modified sine wave inverters come in a great variety of wattage capacities. Wattage ratings can be deceiving as most inverters can sustain temporary loads of several times their normal capacity. Some manufacturers slightly overrate their products by giving wattage outputs that can really only be supported for a limited time. Thus, if your total of wattage requirements will be 600 watts continuously, don't buy an inverter rated for exactly 600 watts unless you enjoy sitting in the dark.

Most modern inverters reach their peak efficiency from approximately 20% to 50% of full rated load. They will perform adequately at higher or lower percentages of full load, but total efficiency isn't quite as good. The efficiency graphs for PowerStar and Trace illustrate this.

Modified sine wave inverters come in three basic sizes; small inverters up to about 300 watts, medium-sized units of 300 to 1,500 watts, and large units of 1,500 watts and up. The larger sizes buy you more than extra wattage capacity. The small inverters have the least sophisticated waveform and options. As the inverter size goes up the quality of electrical output generally improves as well. The largest inverters tend to have many options and accessories that can make life easier, or increase the usefulness of the inverter.

Modern inverters aren't without warts and blemishes. Here are the drawbacks that most retailers would rather not tell you untill after the sale.

Power Consumption

Inverters shouldn't be used to run small wattage loads like clocks or telephone answering machines. Although a large inverter can idle at 0.3 watt, that same inverter can draw 10 to 15 watts to simply "wake up" and run a 2-watt load. Inverters run most efficiently with loads of about 20 watts or more. "Switchers" like Statpower and PowerStar are the exception, they are always on.

Radio Frequency Interference, or RFI

All inverters broadcast radio "noise," especially on the AM band. Don't expect to listen to AM radio with an inverter operating. Ham and amateur radio operators take note. RFI is sometimes a problem with small inverters and certain TVs as well. Keeping the inverter at least 15 feet away from the TV helps, so does a correct choice of TV. RFI is rarely a problem with FM radio (but read about audio interference below). Twisting the inverter input cables and keeping them as short as possible will limit or sometimes cure RFI problems.

Recently, we have discovered some audio equipment designed with SCRs in their circuits. Generally, SCRs will not survive inverter wave forms if not a true sine wave. If in doubt, call your manufacturer's technical support.

Audio Interference

Modified sine wave is close to, but not exactly the same as true sine wave. Some audio amplifiers don't like the little stair-steps in the waveform and you'll get some 60-cycle buzz through the speakers. It doesn't hurt the equipment, but it is annoying. Some equipment gets the buzz, some doesn't depending on the quality of the power supply inside the stereo. Even within a single brand name, some units are fine and others will buzz. We have found that price is not an indicator as to the quality of the power supply (manufacturer claims aside). Older equipment often has fewer difficulties than newer equipment. If you like to have the radio on for background noise buy one of the excellent DC-powered stereos; they consume 1/10th the power of a big inverter driven AC stereo. There is some excellent car stereo equipment on the market now if you want higher performance.

Waveform Problems

Sophisticated equipment sometimes uses SCRs (Silicon Controlled Rectifiers), or Triacs, as filters on the incoming power to protect from spikes and surges. Computer laser printers and fax machines are the most notable examples. SCRs will not accept modified sine wave; it looks like "dirty" power to them. If exposed to modified sine wave the SCR device may be destroyed, resulting in serious damage. Do not expose this equipment to inverter power until you've received positive assurance from the manufacturer that the equipment will function properly. More than one expensive laser printer has been lost by uninformed customers.

The Many Benefits of Using Inverters

Wiring: With an inverter as the primary power conditioner, the house only needs to be wired **once**. That wiring can be conventional 12-2 and 14-2 romex with conventional inexpensive outlets and switches. Enjoy the advantages of mass production! Anyone who's wrestled with the 10 gauge wire needed for DC, and paid $6 to $9 for a cigarette lighter plug outlet will see the immediate benefits.

Lighting: This usually accounts for 50 to 75% of the total load in an alternative energy home, so you want to get the most from your precious watts by using fluorescent lights. A 12-volt DC compact fluorescent light currently costs $45 to $65, and selection is limited. The same item in 120-volt AC costs $17 to $30 and comes in a wide variety of sizes, configurations, and aesthetic packaging. If you're buying five lights or more, going AC will save you enough money to pay for a small inverter.

Entertainment Equipment: Televisions, VCRs, satellite systems, stereos, CD players, computers, and most printers all run happily on inverters with the limitations previously mentioned. Be aware that entertainment equipment often has "phantom loads," small constant drains that keep your inverter from dropping into its power-saving "standby" mode. Phantom loads consume precious watts and give nothing in return. Appliances like instant-on TVs and VCRs with clock timers are good examples. Install these appliances in a switched outlet so they can be positively shut off when not in use.

Kitchen and Laundry Appliances: Microwaves, blenders, mixers, toasters, juicers, food processors, garbage disposals, trash compactors, most washing machines, all gas dryers, and yes, even hair dryers can be run with inverters, provided the inverter and battery bank have sufficient capacity. Electric ranges or other 220 volt appliances are **not** possible. (The reason the appliance is 220 volt is because it's an energy hog in the first place). While it's technically possible to run conventional AC refrigerators on inverters, their exorbitant power consumption makes it impractical.

Power Tools: Large stationary equipment like 10 or 12 inch table saws or band saws are best run with a generator. The 2,000 watt and larger inverters will easily run any hand-held power tools; however, the energy consumption of these loads tends to run down battery banks quickly. If tool use is intermittent, then use of an inverter is preferable to leaving the generator running constantly.

Water Pumps and Large Motors: Water pumping is an application that's best accomplished with DC power. Submersible water pumps are the **most** difficult AC motors to start. While the most popular Trace 2012 will positively start a 1/3-hp pump, and will probably start a 1/2 hp pump, this may not be the wisest strategy. AC submersible pumps aren't designed with energy-efficiency as a primary consideration. An AC pump will consume 3 to 4 times as much power as a DC powered pump to produce the same amount of water. This holds true for both submersible and surface pumps. While it's possible to run your AC water pump off an inverter, we only recommend it as an interim measure until you can afford something better. Large AC motors, particularly inductive or split-phase types, will draw up to 6 times their rated running wattage while starting. This means a typical 1/2 hp motor may require up to 6,700 watts while starting. Only the biggest inverter and battery bank will handle loads like this.

Clocks and Answering Machines: Up until very recently low wattage, but constantly "on" loads weren't efficient to run on AC. The answering machine might only need a couple watts, but to keep the inverter turned on costs 10 to 15 watts. The new PowerStar series of upgradable inverter, has changed all that. This inverter runs just as efficiently at 10 watts as at 1,000 watts. If you've got power to spare, you can run your old clock radio relatively efficiently now.

Air Conditioners and Electric Heaters: Typically these appliances aren't practical to run on most alternative energy systems. Technically, the inverter could be sized to handle the load, but the exorbitant amount of energy that these appliances draw generally limits their use to folks with really big hydro systems. Now, with the Real Goods gas-fired chiller, air conditioning can be an off-grid reality.

Times have changed. Solid state electronics are part of our lives, and reliable, efficient inverters are part of that. Choosing 120-volt AC allows the remote homeowner to enjoy the cost benefits and variety of mass production. It allows you to wire your house with conventional equipment and methods (a **real** blessing if and when the building inspector comes around). This is a case of using technology to make your life better, simpler, and cheaper.

Inverter Helpful Hints:

1) Keep the inverter and the battery close together, but **not** in the same compartment. Ten feet is the maximum allowable distance, 5 feet is better. High-voltage, low-current AC transmits easily, low-voltage, high-current DC transmits poorly. Use big wire to connect the inverter and battery. Always twist the DC cables together. Trace makes the best hook-up cables for the larger inverters.

2) Plan for future growth. Most of the small to medium sized inverters can't be upgraded or stacked. On the other hand, large inverters don't run small loads as efficiently. Sometimes it's nice to start with a smaller model and keep it to run smaller loads, while your newer, bigger model takes care of the larger appliances.

Dear Dr. Doug:
Does my inverter really need a fuse to protect it? The factory manual doesn't say much about fuses. Why does your fused disconnect cost so much? Can't I use a regular fuse?
Answer: The 1990 National Electrical Code requires that overcurrent protection and a disconnecting means be provided for all photovoltaic, power conditioning (inverter), and storage battery circuits. This is only common sense. Why gamble with the possibility of burning down your house? Manufacturers tend to be vague on this point because until recently there was no fuse/switch combination that didn't introduce unacceptable amounts of voltage drop during surge loads. Even the smallest amount of voltage drop drastically affects the inverter's ability to meet a surge. Our fused disconnect is expensive because the switch contact area is massive. Not even the largest commercial AC switches have this much surface contact!

See the Fused Disconnect for full size inverters on page 220

You can't use regular AC fuses for two reasons: surge loading and arcing. Large inverters like the Trace 2012 are capable of safely handling up to 800 amps and commonly surge to 350 or 400 amps when starting large motors. Any fuse used under these conditions must be a slow blow type.

DC current flow, because it's traveling in one direction only, is very difficult to stop in a short-circuit situation. AC current, because it stops and reverses direction 60 times a second, tends to be self snuffing. If an AC-rated fuse or circuit breaker is used in a DC circuit, it may be **unable to stop** a short circuit! The fuse will blow, or the breaker will trip, but the electricity simply jumps the gap and keeps right on flowing! Don't get cheap with safety equipment. Large DC-rated fuses, like the one in our fused disconnect, contain a special filler material to positively snuff the current flow when the fuse opens.

Dear Dr. Doug:
My TV reception is noticeably worse when I'm using the inverter. It's better when the generator is running, but that's wasteful and noisy. Got any ideas?
Answer: This sounds weird, but if you twist your inverter **input** cables together, it will probably improve the TV reception. This phenomenon is called wave cancellaton.

Dear Dr. Doug:
Can I sell my excess power back to the utility?
Answer: This is a perennial favorite. Everybody has a fantasy of getting money back from the power company. Technically, yes you can. If you are a reliable alternative energy power producer the utility must, by federal law, buy your energy. However...(you all knew this was coming), your power must conform to their standards and have certain safeguards and protective devices. On a practical level, you have to be a fairly large scale producer to make the venture worthwhile. It is not feasible with any of the current generation of inverters we sell in the catalog. Perhaps in a few years it will be common for homeowners to sell into the grid during the day, and draw that power back at night, like a perfect battery.

PowerStar Upgradable Inverters (400W-700W-1,300W)

PowerStar recently introduced the very first low-wattage upgradable inverters. The three models, which all use the same case, are rated in terms of their continuous true RMS power capability. The *continuous* power rating on the smallest inverter is 400 watts; the first upgrade takes you up to 700 watts continuous; and the final upgrade takes you up to 1,300 watts continuous. The actual upgrade can be performed by the factory in a few days. The price for the upgrade is simply the difference between the two units. *Therefore, there is no cost penalty for buying a smaller unit and later upgrading it, making these inverters extremely versatile!*

The small 400-watt unit is suitable for computer systems, power tools, and small appliances. Its 3,000-watt surge capacity can start and run ¼-horsepower motors.

The first upgrade—the 700-watt unit will run a 500 watt microwave, a vacuum cleaner, a hair dryer (on medium), or a small coffee pot or hotplate. It is preferred for power tools or appliances that run heavily loaded for long periods of time. Like the smaller unit, the surge capacity is 3,000 watts.

The second upgrade—the 1,300-watt unit will continuously operate a full-size microwave, any 1,300-watt appliance, or a circular saw.

The following specifications apply to all three units:
Input Voltage: 10.5 to 16.5 volts DC
Output Voltage: 120 volts AC true RMS ± 5%
Output Frequency: 60 Hz, modified sinewave
Idle Current: 90 mA DC
Low Battery Warning: Audible alarm below 10.9V input
Over Voltage Shutoff: 16.8 volts ± 2%
Indicator: A green micropower lamp shows AC output present
Over Temp. Protection: Proportional power reduction
Overload Protection: Dual mode limiter enables operation of appliance rated higher in power than inverter. A moderate overload lowers the output voltage. A severe one causes shut down. Reset by cycling the power switch.
Remote Control: A built-in connector socket has provisions for a remote switch as well as a hard-wired AC output.
Efficiency: Over 90% at half rated power
Size: 3.15" x 3.3" x 11"
Weight: Less than 5 lbs. Warranty: Two years.

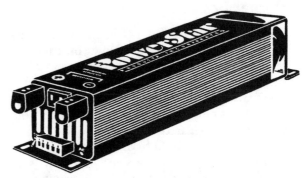

Note the input voltage on the Power Star upgradable series, these inverters are designed for nicad or lead-acid batteries.

27-105	PowerStar 400-W (UPG400)	$499
27-106	PowerStar 700-W (UPG700)	$599
27-107	PowerStar 1,300-W (UPG1300)	$899

PowerStar 200 Pocket Inverter

This compact inverter quickly became our best seller. It's the best mini-inverter in the industry. It is ideal for powering most 19 inch color TVs, personal computers, drills, VCRs, guitar amplifiers, video games, stereos, and lots of other small appliances *directly from your car or 12V system.* It's great for travelers to carry in their cars to power these standard appliances. It comes with a cigarette lighter plug that plugs into any cigarette lighter and will deliver 140 watts of 120V-AC power continuously from your 12V battery. It will provide 400 peak watts and 200 watts for over 2 minutes. Amp draw at idle is 0.25 amp (3 watts). PowerStar has recently made an improvement to the 200-watt inverter. In case of overload, the unit now safely delivers as much power as it can, into that overload. One year parts and labor warranty.

The Powerstar works great for watching videos on our boat. It's cheaper than a DC TV too! Susan Freeman, Venice, CA.

27-104	PowerStar 200 Inverter	$149

Statpower ProWatt 125
Pocket Power Inverter

The Statpower ProWatt 125 is one of the smallest inverters on the market. Now you can run most 13-inch color TVs and even some 19-inch color TVs off of this 90% efficient modified sine wave inverter that fits in your coat pocket and weighs only 16 ounces. Simply plug it into your vehicle's cigarette lighter or your 12V home outlet and you'll have instant AC power up to 125 watts continuous and 400 watts surge. It will operate standard and fluorescent lights, sound and video equipment, data and communications equipment, small appliances, and other electronics items.

The no-load current draw is only .06 amp. It comes with a low battery alarm set at 10.7 volts to avoid surprise shutdowns, and an auto shutdown will activate at 10.0 volts to prevent harm to the battery. One-year replacement warranty.

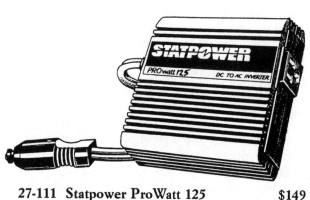

27-111 Statpower ProWatt 125   **$149**

Statpower ProWatt 250
300-Watt Inverter

This Statpower inverter fills the gap left since *Heart Interface* discontinued its 300-watt inverter. It will surge to 500 watts, produce 300 watts for 10 minutes, 250 watts for 30 minutes and 200 watts continuously. It has a low battery cut-out with audible alarm at 10V. It is voltage regulated and frequency controlled and features all of the protection features of its big brother, the ProWatt 1500. It comes with an 18 inch cigarette lighter cord and measures only 6" x 4½" x 1½"— it fits in the palm of your hand! The Prowatt 250's precisely regulated output allows you to run your sensitive electronic equipment. With an efficiency of 90%, almost all of your battery's current is converted into usable AC power. One-year replacement warranty and six additional months limited warranty. No-load power draw: 0.15 amp or less.

Statpower ProWatt 800

Statpower introduced this upgrade to the popular ProWatt 600 in March 1992. The ProWatt 800 is only a fraction of the size and weight of competitive size inverters. This 90% efficient inverter produces temporary power levels much higher than its 800 watt continuous rating. The ProWatt 800 will surge to start ½ hp motors, and will produce 1,000 watts of power for 10 minutes. LED bar graphs, unique to Statpower inverters, provide an excellent visual representation of the power being drawn and the state of the battery charge. These displays eliminate the guesswork that owners of other inverters have to endure when using high-powered equipment. Dimensions are 3" x 9" x 10". This unit weighs just over five pounds. No-load current draw is less than 300 mA, AC hardwire capable, and an 18 month limited warranty.

27-112 Statpower ProWatt 800   **$495**

27-109 Statpower ProWatt 250   **$229**

Statpower ProWatt 1500 Inverter

Statpower introduced this new and remarkable inverter in early 1992. Here are the incredible specifications: it is rated at 1,500 watts continuous output, and is capable of delivering 1,800 watts for 30 minutes or 2,000 watts for 10 minutes. Unlike any other comparable inverter on the market, it will sustain an astounding ¾ HP motor surge! Input voltage is 10–15 VDC, low battery alarm at 10.7 V, low voltage shutdown at 10 V. It features protection circuitry for over temperature, overload, short circuit, and reverse polarity. Some of the other features include AC hard wire capability, remote on/off capability, and remote display capability. Dimensions: 3"H x 9"W x 16"L. Weight is a scant 10 pounds! It has a no-load current draw of less than 0.6 Amp. All this at a price that beats all competition. 18 month limited warranty.

27-110 ProWatt 1500 $799

Exeltech Sine Wave Inverter

A reasonably priced **true sine-wave** inverter has come to the market. This new product from Exeltech is tailor-made for sensitive electronic equipment that has previously suffered from the modified sine-wave inverters. Finally, one can listen to audio gear without the annoyance of a background buzz.

The Exeltech models SI-250 and SI-500 are designed to handle continuous loads of 260 and 510 watts, respectively. The sophisticated circuitry has output-short, output-overload, and input-reverse polarity protection.

The 250-watt unit operates with an efficiency of 78-82% and the 500-watt operates with an efficiency of 73-77%. Typical no-load draw is 1 amp for the SI-250 and 2 amps for the SI-500. These are not for whole house applications, where more efficiency and better pricing can be had with modified sine wave inverters. Rather, these will be used in those particular applications where the shortcomings of most other inverters are all too obvious. Full one year warranty.

Applications for the Exeltech are numerous: color monitors, stereos, electronic instruments, satellite systems, VCRs, TVs, tape players and test equipment.

27-120 Exeltech SI-250(12V) $395
27-121 Exeltech SI-250(24V) $395
27-122 Exeltech SI-500(12V) $595
27-123 Exeltech SI-500(24V) $595

Most inverters will create AM radio interference. There are steps that can be taken to minimize this. Please refer to the text at the beginning of the inverter section for this information. Further, our tests have shown that most inverters may exhibit some visual "noise" with particular television sets. Unfortunately due to the vast number of televisions on the market, the only way to figure out which ones your inverter will work with, you must test it individually with your own television.

**To order any of these products
or for more technical information
call us toll-free at
1-800-762-7325**

Trace Inverters

Trace inverters have turned the inverter industry upside down in the past seven years. They have been consistently one of the best selling items in our catalog. Customer satisfaction with Trace is close to 100%, and the folks at Trace are a delight to deal with, always willing to lend a patient ear to a confused customer. Just a short few years ago Heart Interface was the big name in innovative inverter technology, but it seems now that Heart has all but bowed out of the remote home market, choosing instead to concentrate on marine applications. Trace has proved itself again and again to be the most efficient, durable, and reasonable (in dollars per watt) inverter on the market.

Trace inverters are designed with a wide range of user-installable options, allowing them to grow in sophistication and power as your needs change. The "Standby" option provides a powerful, internal, programmable, automatic battery charger that works well with standby generators allowing backup power for your solar array. If your power requirements increase over time, the "Stacking Interface" option allows an additional unit to be operated in parallel, doubling your output power capability.

Perhaps the greatest feature of Trace inverters is their microscopic "No-Load Power Drain." The 2012, Trace's 2,000-watt inverter, has a no-load current of only 0.03 amp or 1/3 of a watt! This means you can leave the inverter turned on for an entire day and only use 8 watt-hours of power or 2/3 amp-hour! (It's even less for the 812 inverter—0.02 amps.)

Trace's protection circuits are automatically resetting. If an overload occurs, the inverter shuts down until the condition is corrected and then resumes operation. It does not need to be manually reset. Trace inverters employ a unique "impulse phase correction" circuit allowing the inverter to closely duplicate the characteristics of standard 120V grid power. With this design approach, the limitations of the modified sine wave format are largely overcome. Protection circuitry carefully monitors the following conditions: over current, over temperature, short circuit, reverse output voltage application (output connected to grid), induced electrostatic charge (lightning), high battery, and low battery.

Trace uses a very large number of output transistors (44 FETs) and specially wound transformers, which results in exceptional high power

Here is a detailed photograph of the six Trace 2248 inverters interfaced together in series/parallel configuration for the North Carolina Biodome project that Real Goods designed in 1987.

performance. The 2012 will surge to well over 6,000 watts of power.

Here is a partial list of appliances typically used with the Trace inverter:

Stereos	Sewing machines
Floodlights	Electric typewriters
Ceiling fans	Night lights
Vacuum cleaners	Shavers
Compact disc players	Hair dryers
Trash compactors	Blenders
Satellite dishes	Ice makers
Color TV's	Toasters
VCR's	Microwave ovens
Coffee makers	Electric blankets
Fluorescent lights	Refrigerators
Washing machines	½-hp deep well pumps
Electronic musical instruments	
Computers and peripherals	
All hand held power tools	

Trace 812 Inverter

The 812 is a welcome addition to the Trace family. It will produce a 2,400 watt surge, 800 watts for 30 minutes, 650 watts for 60 minutes, or 575 watts continuously. It uses only 0.21 watt (0.017 amp) on standby. It will power TVs, VCRs, computers, and test equipment, and it will also power most vacuum cleaners, Champion juicers, and most microwave ovens. For an extra $100 you can order an optional 25-amp battery charger in the "Standby Option." This internal 25-amp battery charger comes with a 30-amp transfer switch that switches back and forth between battery power and grid/generator power. The battery charger requires 2,000 watts of generator power at a minimum to operate it. The current is not adjustable but maximum battery voltage is. Also available is the RC/3 Remote Control option allowing you to turn the unit on and off from a remote location (although with the extremely low no-load power drain you can let the unit stay on all day for around 0.4 amps-hours). The Trace 812 comes with our very highest recommendation. Two-year guarantee. Weight: 14 pounds.

We now offer a 100-amp fuse with holder for installing between the battery and the Trace 812. This will help you comply with state and local electrical codes.

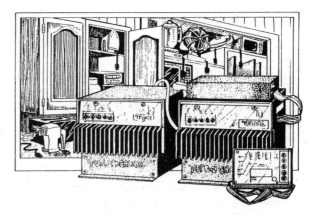

Trace 724 Inverter

The Trace 724 was created as a direct response to the demands of our customers who prefer to base their systems upon a 24V battery bank. Power ratings show a 425-watt continuous operation with a very strong 2,100-watt surge available, for starting inductive loads. The 724 will produce 1,000 watts for 7 minutes, 600 watts for 55 minutes, 500 watts for 90 minutes and 425 watts continuous. Idle current remains at an incredibly low 0.017 amp. A 12-amp battery charger is available as is a remote control (RC/3) option that will allow remote monitoring via an LED and will also allow the user to turn the 724 on and off. Warranty is 2 years and the weight is 14 pounds.

27-211	812 Trace Inverter	$550
27-212	812SB Inverter w/battery charger	$650
27-214	Remote control for 812	$50
27-215	100-amp fuse w/holder for 812	$35

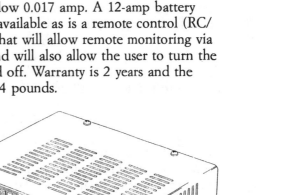

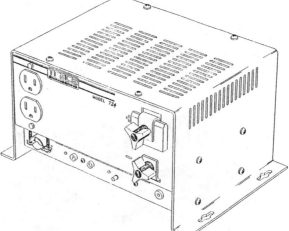

27-221	724 Trace Inverter	$625
27-222	724 Inverter w/battery charger	$725
27-214	Remote Control for 724	$50

Trace 2012 Inverter

The Trace 2012 is by far our best selling inverter. It will produce 2,000 watts of power at 120 VAC from a 12VDC battery source for up to ½ hour with a 6,000+ watt surge for starting induction and other high surge motors. Continuous power is 1400 watts without turbo fan or 2000 watts with the fan. The no-load power drain is the best in the industry—a scant 0.36 watt (.03 amp). The unit measures 6.9"H x 11.4"W x 12.4"D. We strongly recommend you go over the options available very carefully to tailor the Trace 2012 inverter to your needs. The 2012 is also available with 230VAC output at 50 or 60 Hz for export use (order the **2012/E**). As with all Trace products, a 2-year warranty is included.

Standby versions of 2000 series inverters are now equipped with an extremely sophisticated multi-stage 120 amp battery charger. Adjustable controls allow tailoring the charge cycle to the needs of your battery bank. Temperature compensation and an equalization cycle complete the package which is an incredible bargain at $220. Trace series 2000 inverters now have an improved hardwire system utilizes an externally accessible 30-amp terminal block and wire clamps for strain relief that accommodate wire sizes up to #10 gauge. All units with the Standby option will also include an externally accessible 30-amp circuit breaker on the AC input. An AC outlet will no longer come as standard equipment, but is available as an option.

The Trace 2000 series inverters can be installed as a "stacked pair" using the optional SI/B stacking interface.

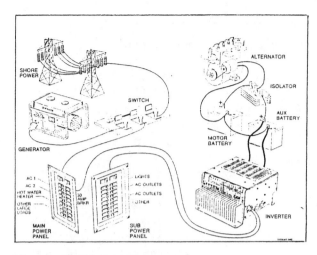

| 27-201 | 2012 Trace Inverter | $1,090 |
| 27-202 | 2012SB w/battery charger | $1,310 |

Trace 2524 Inverter

The Trace 2524 will provide 2,500 watts of power at 120VAC for ½ hour and 1500 watts continuous from a 24V battery source. The no-load current is 0.018 amp. The unit will provide 2 to 3 minutes of surge power to 6,000+ watts. It's available with 230VAC output at 50 or 60 Hz for export use (order the **2524/E**).

| 27-203 | 2524 Trace Inverter | $1,350 |
| 27-204 | 2524 w/battery charger | $1,570 |

Trace 2248 Inverter

The Trace 2248 will provide 2,200 watts of output power for 30 minutes and 1,400 watts continuous at 120VAC from a 48VDC input. It has a surge capacity of 6,200 watts for a 2 to 3 minutes. The no-load current is only 0.012 amp.

| 27-205 | 2248 Trace Inverter | $1,500 |
| 27-206 | 2248 Inverter w/battery charger | $1,720 |

Trace 2232 Inverter

The Trace 2132 will provide 2,100 watts of output power for 30 minutes and 1,400 wats continuous at 120VAC from a 32VDC input. It has a surge capacity of 6,200 watts and a no-load current of 0.016 amp.

| 27-207 | 2132 Trace Inverter | $1,400 |

Trace 2236 Inverter

The Trace 2236 will provide 2,200 watts of output power for 30 minutes and 1,400 watts continuous at 120VAC from a 36VDC input. It has a surge capacity of 6,000 watts and a no-load current of 0.014 amp.

| 27-209 | 2236 Trace Inverter | $1,450 |
| 27-210 | 2236SB w/battery charger | $1,670 |

Options for Trace Inverters

Standby Battery Charger

Standby versions of the Trace 2000 Series Inverter now include a sophisticated multi-stage 120 amp battery charger with features seldom found even the most expensive stand alone chargers.

Three stage charging — an innovative, multi-stage controller design regulates the charger's output to reflect the battery's ability to accept a charge. This controlled progression from Bulk charge, to Absorption, to the Float stage assures a rapid and complete charge sequence while maximizing battery life.

Front Panel Controls — Easily accessed controls allow the user to fine tune charger parameters to specific battery cofigurations and environments. Individual controls are provided for bulk charge voltage, float voltage, bulk charge rate, return from bulk amps timer and temperature compensation. All control settings are factory adjusted to insure excellent, out-of-the-box performance.

Equalization Cycle — A time limited equalization cycle facilitates the periodic preventive maintenance necessary to keep batteries at their peak performance level.

Time delay transfer circuitry — A 1 hp, 30 amp transfer relay detects the presence of grid or generator power and automatically changes the inverter to and from the battery charger mode. A 30-amp automatic transfer switch is standard with the battery charger that switches the unit from battery power to grid/generator power. Many of our customers purchase the battery charger option and use it in conjunction with their generators. Typically a generator is used for very large loads (water pumping, washing machine, etc). If you leave the generator on for a couple hours per day for that load, you can charge your batteries at the same time supplementing your solar panels.

The standby battery charger puts out the following maximum charge rates for the various sizes of Trace inverters:

Inverter	Maximum charge rate
2012	120 amps
2524	60 amps
2132	45 amps
2236	40 amps
2248	30 amps

One important note on the Standby Option: *You need 6,000 watts of power (a 6-kW generator) to utilize the full power of the battery charger.* If you have that much input power you'll get 120 amps (with the 2012) of charging capacity when your batteries are fully drained. However, don't fret if you only have a 4-kW generator (as most of us do) because the Trace battery charger can be front panel adjusted to limit charging amperage.

If you choose to add the standby option to an inverter at a later date the total price is $275, and the inverter must be returned to the factory.

Digital Voltmeter (DVM)

The Digital Voltmeter (DVM) is a must option for purchasers of the Standby charger. The meter monitors four conditions: battery voltage to tenths of a volt, the charge rate of the battery charger, the frequency of the generator in hertz (Hz), and the peak voltage of the charging source. Without this option on your battery charger, you're stuck with a lot of guesswork! The DVM can be easily installed in 5 minutes by the customer. The DVM is available as an option to all 2000 Series Trace Inverters.

27-301-12 DVM (Digital voltmeter)-12V $130
27-301-24 DVM (Digital voltmeter)-24V $130

Turbocharger (ACTC)

This is a thermally activated fan cooling option that increases the continuous power of all the Trace inverter models. It senses the heat-sink temperatures and automatically operates the fan. The continuous power rating is increased by 400 to 600 watts. It is user installable on all 2000 series units.

27-302 Turbocharger (ACTC) $120

Low Battery Cut-Out (LBCO)

The LBCO protects against over-discharging the batteries. The circuitry evaluates battery voltage and current draw to shut down the inverter in a low battery condition. User programmable and user installed.

27-303-12 Low battery cut-out—12V $65
27-303-24 Low battery cut-out—24V $65

The LBCO won't give a warning before shutting power off. For an early warning system see the low battery alarms on page 218.

Stacking Interface Module (SI/B)

The stacking interface module allows two inverters to be paralleled for double output power at 120 VAC. If both units have the battery charger option, the charging capability is also doubled. This is the option that Trace pioneered to give its units maximum flexibility. Both units must be identical models. If one has battery charger, than the other one must also. User installed.

27-308 Stacking Interface SI/B   **$275**

Remote Control

The remote control option is for use in installations where the inverter is not easily accessible. The old option RC1 has been discontinued in favor of the new and improved RC-2000, which comes complete with a multi function digital voltmeter (see DVM description). It provides a duplicate set of control switches and indicator lamps and is only available in models with the standby battery charger option. Option RC2 is a more basic remote with on-off control and an LED that indicates on-off and search mode states.

27-306 RC2 Remote Control   **$75**
27-309 RC-2000 Remote w/DVM for 2012 $250
27-310 RC-2000 Remote w/DVM for 2524 $250
27-316 RC-2000 Remote w/DVM for 2132 $250
27-317 RC-2000 Remote w/DVM for 2236 $250
27-318 RC-2000 Remote w/DVM for 2248 $250

See page 220 for complete copy on the fused disconnect.

120V-to-240V Transformer (T-220)

The T-220 is a 3,000-watt 1-hour transformer that may be configured by the user to function as a step-up or step-down autoformer, an isolation transformer or a generator balancing transformer. Many alternative energy homesteads have everything on 12V or 120V with the exception of the oddball 240V submersible pump. By installing the T-220 between the submersible pump and the pressure switch, the transformer will only come on when the pump is activated. 5,000 watt maximum surge. Circuit breaker protected.

27-307 T-220 Step-Up Transformer   **$295**

Battery Cables for Trace Hook-Ups

As we mentioned earlier, the wire sizing between the battery and the inverter is of critical importance and many people tend to undersize. There is a tremendous current draw in this area, and it's always better to err on the side of overkill. For this reason, Trace has decided to make up very heavy duty 4,000-strand, 4/0 (that's four-ought!) welding cable. They're available in either a 5-foot or 10-foot pair of terminated and color coded battery cables. Highly recommended to complete your Trace installation with security. Our 400 amp fused disconnect switch complies with code requirements and can handle a 800 amp surge without voltage drop. Rated for 2,000 amps capacity.

27-311 2 each 5-foot 4/0 cables   **$79**
27-312 2 each 10-foot 4/0 cables   **$125**
24-203 Fused Disconnect (400A)   **$329**

Heart Inverters

For several years Trace has dominated the inverter market for remote home applications. Heart chose instead to concentrate on the marine and RV markets. Last year, with its introduction of the Energy Management System (EMS), Heart made some strong design changes and engineering breakthroughs in an attempt to recapture some of its lost market share to Trace. We enjoy this friendly competition because it keeps both manufacturers on their toes and hastens the development of new technologies to the end-user! (Note that our inverter test was done prior to the introduction of the new Heart EMS line.)

The Heart Interface EMS line of inverters/battery chargers make use of state-of-the-art microprocessor-based control logic, developed by Heart Interface. This new design offers several departures from traditional inverter performance. These new inverters produce a multi-stepped output waveform, which more closely approximates a true sine wave. This unique waveform will provide more consistent peak-to-peak AC voltage values, and, among other characteristics, provides exceptional motor-starting power for high-torque appliances such as washing machines and refrigerators.

These new EMS systems from Heart Interface are UL listed. They are designed to conform to National Electrical Code standards and feature an easy-to-access hard-wiring compartment with input and output strain reliefs. Heavy-duty battery-cable lugs and an AC input/output terminal strip designed for 30 amp 10 AWG wire are provided. The Heart warranty covers parts, labor, and return shipping for 30 months. A cascade option is available that allows connecting multiple units together for increased inverter output and charging capacity. The EMS Spectator works with either of the two EMS models.

EMS 2800 Heart Inverter

The new EMS 2800 will produce 2,000 watts of silent AC power continuously. It comes standard with a 65-amp 12-volt battery charger and the same relay switching as the EMS 1800. Like the EMS 1800, the 2800 is UL listed and provided with AC reverse polarity, thermal, overload, and circuit breaker protection. These two new models also share the same waveform, load detection, and battery charging circuitry, with the EMS 2800 having nearly twice the motor-starting power. 30-month warranty.

27-403 EMS 2800 Heart Inverter $2395

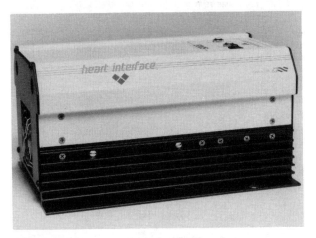

EMS 1800 Heart Inverter

The new EMS 1800 produces 1,500 watts of silent AC power continuously. It comes standard with a 65-amp 12-volt battery charger and features sealed, nitrogen-filled 30-amp relays for AC power source switching. It will operate sensitive electronics such as computers, TVs, and VCRs. The load detection circuit reduces power consumption in a no-load condition and responds almost instantly when an appliance is turned on. The unique waveform provides extremely high motor-starting torque. The EMS battery charger uses two-step regulation to initially provide constant current for a rapid recharge and then a constant, lower voltage float charge for safe long-term battery maintenance. The charging circuitry will compensate for low line voltage, maximizing the charging current when using small generators. Use of the EMS Spectator allows adjustment of charging parameters and digital monitoring of the system. 30-month warranty.

27-401 EMS 1800 Heart Inverter $1595

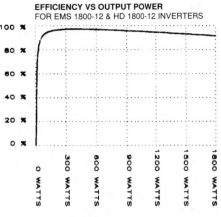

EFFICIENCY VS OUTPUT POWER
FOR EMS 1800-12 & HD 1800-12 INVERTERS

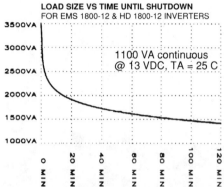

LOAD SIZE VS TIME UNTIL SHUTDOWN
FOR EMS 1800-12 & HD 1800-12 INVERTERS

1100 VA continuous @ 13 VDC, TA = 25 C

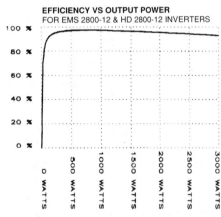

EFFICIENCY VS OUTPUT POWER
FOR EMS 2800-12 & HD 2800-12 INVERTERS

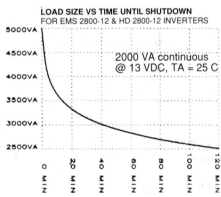

LOAD SIZE VS TIME UNTIL SHUTDOWN
FOR EMS 2800-12 & HD 2800-12 INVERTERS

2000 VA continuous @ 13 VDC, TA = 25 C

Heart Freedom 10

This is the latest entry from Heart and represents a major value in the inverter market. The Freedom 10 is an inverter/charger with a built-in 45 amp automatic three step battery charger that is programmable. The battery charger has a very impressive efficiency of about 70%, another inverter/charger breakthrough. Features include a .12 amp no-load power draw with instant on, 1000 watts continuous rating, surge to 3000 watts, exceptionally clean AC output (some of the best TV reception we have seen on inverters), automatic transfer switch with "power sharing" this feature allows the inverter, when in battery charging/transfer mode, to cut back on the power to the charger when a large AC load is present, normally this situation would cause the supply circuit breaker to trip from the overload. The optional remote "Power Window" displays the following read (both in and out), low battery, over-load, AC input, and charger construction. Dimensions: 10.4"D x 15"W x 7.2"H, weight is 43 pounds. Warranty is 30 months. The Freedom 10 gets a big "thumbs up" from Real Goods.

27-405 Freedom 10   **$895**

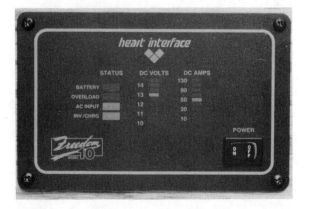

27-406 Remote Power Window   **$149**

Heart HF 12-600 Inverter

Heart Interface has discontinued its 300-watt unit (HF 12-300X) and replaced it with this 600-watt unit. It will run 600 watts for 25 minutes and surge to 1,200 watts. It uses only 0.7 watt (0.06 amp) on standby (idle loss) and reaches 90% efficiency at only 50 watts. It is an ideal unit for computers, printers, TVs, VCRs, and test equipment. Modified sine wave output. Guaranteed 1 year.

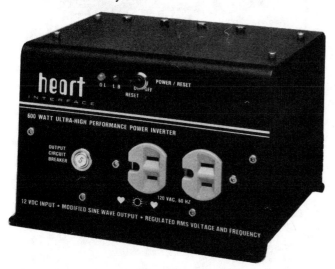

27-404 Heart HF 12-600X 600-Watt $675

Vanner Voltmaster Battery Equalizer

The Vanner Voltmaster allows you to upgrade to a larger 24-volt inverter system, but still use all your 12V appliances. The Voltmaster was designed to eliminate the overcharging of one battery in a split 24/12-volt system. All batteries will charge and discharge equally even when there is an unequal load. The device electronically monitors voltages of both batteries and transfers current whenever one battery discharges at a rate different than the other. By maintaining equalization down to 0.01 volt, the Voltmaster will extend battery life by preventing both over and undercharging.

All units are guaranteed for 1 year.

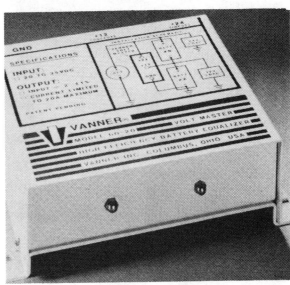

INSTALLATION SCHEMATIC

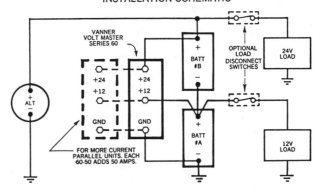

27-801	#60-10A– 10 amp draw	$340
27-802	#60-20A– 20 amp draw	$365
27-803	#60-50A– 50 amp draw	$495

Tripp-Lite Inverters

Tripp-Lite is our economy line of light-duty inverters. They produce 120VAC square wave current. Some models have frequency control and some do not. It would be hard to find a better inverter-cost-per-watt buy; however, this inverter will not operate induction or capacitor-start motors or any other reactive-load equipment. We have found that inexpensive square wave inverters will damage electronically ballasted compact fluorescent lamps. (Operation of lighting system is a poor idea with these inverters due to their low efficiency anyway.) The surge load capability is limited to 10% above each unit's rated power output.

Tripp-Lite inverters will operate AC equipment from 12, 24, or 32-volt battery systems: TVs, stereos, small hand tools, appliances, lights, minicomputers, and terminals. (Check wattage requirements on your appliances to determine the correct model for the application.)

Regulated output maintains voltage and frequency within a specified range across a wide variation of battery voltages. Output is compatible with most AC appliances and electronics. The input is protected against DC input line voltage spikes and transients. A sensor automatically turns off the inverter when battery voltage is low on most models. Reverse polarity protec

tion (most models) guards the inverters against improper connection to batteries.

Tripp-Lites are encased in rustproof anodized aluminum chassis with heat sink fins for convection cooling. Integral shock mounts reduce component vibration. All units come with a 1 year limited warranty. (For lower wattage models, see Powerstar and Statpower.)

27-603	PV400	$179
27-604	PV550	$245
27-605	PV550/Battery Charger	$325
27-606	PV250/Frequency Controlled	$245
27-612	PV600/Frequency Controlled	$425

To order any of these products
or for more technical information
call us toll-free at
1-800-762-7325

Off the Grid Living

In 1991, we orchestrated the first ever National Off-the-Grid Day. We designed the event as a tribute to the 50,000 American homes built without the noose of utility-supplied "grid" power. We wanted to drive home the message that it's time to change our nation's energy policy. Solar energy is no longer an environmentalist's pipe dream but a practical, economic technology. (In fact, one in every 2,000 American homes is currently independently powered.)

We weren't surprised to find the national media confused. The term "off-the-grid" is still foreign to most Americans. What did surprise us was the genuine curiosity and enthusiasm Off-the-Grid Day generated. Somehow we tapped into a dormant pioneer spirit with a lot of folks. For a few weeks the front office at Real Goods was like CNN Headquarters with requests for radio interviews rolling in from nearly every state, and reporters with notepads and cameras clicking away in our showroom.

The reporters wanted to understand this new off-the-grid mentality, while the people who read their stories or listened to the interviews just wanted to believe that it was possible. It was a fun and ego-flattering time that hopefully drew productive attention to the cause of Energy Independence.

The off-the-grid experience is at the very core of Real Goods. When we opened our first retail store, we catered to urban refugees who were conquering new frontiers in Mendocino County.

These people were heroes, albeit on a small and very private stage, hardly the stuff for the front page of tabloids.

Now, we are seeing a new category of eco-celebrity emerge, and the off-the-grid dweller is right at the eye of the storm.

What an ironic (not to mention amusing) twist of fate this is. Perhaps Elvis will be discovered living off-the-grid in Montana. Already the man who built a suburban off-the-grid home in Los Angeles has been featured in *People* Magazine. Hollywood stars Dennis Weaver and Ed Begley, Jr. have become as prominent for their energy-efficient lifestyles as their acting credentials. The irony of ironies will be when the act of escaping civilization becomes the ticket to fame and stardom.

Putting fantasies aside, Real Goods has always specialized in innovative, high-quality tools for independent living. What we are seeing at the moment, however, is that the market is moving toward us. Our customer survey in 1991 revealed that while about one-quarter of our customers live off-the-grid, a full 40% are planning to! The tools and appliances in this section are crafted from basic, time-tested technologies. All are geared specifically to the needs of the remote home, but they are not limited to energy-related products. Real Goods provides goods for all types of independent lifestyles, including the decadent! How you live your life is your choice; we just want to provide you with the means to do it.

Decadence is hardly the goal of the average Real Goods customer, but the days of huddling around a spare light bulb dangling from a wire are over. This is, after all, almost the twenty-first century. Why would humans have invented Macintosh computers, the Grateful Dead, Cuisinarts, and Super Mario Brothers if not to enhance life? These pleasures are not to be denied to people merely because they choose to live a life of energy sanity.

Way back in the early 1980s, just as The Sharper Image catalogs began to clutter our mailboxes, all that could be found for off-the-grid life was junky, plastic garbage designed for the RV or camping industry. Recent years have seen dramatic improvements in the goods available; moreover, this is a rare area where Americans clearly lead the technology brigade. (If recent inquiries from Japan are an indication, however, this competitive edge may not last long.)

Real Goods will do its part to hound manufacturers and to source new supplies in a continuing effort to improve the quality of off-the-grid life. This is a realm where our children can look forward to a better life than we enjoyed, due in part to our efforts. Maybe, for some in America, the dream is still alive.   – J.S.

POWER DOWN

Ms. Behnke's Cup of Coffee

Real Goods is not the type of company to shy away from controversy. This is, remember, the company that published a catalog with a C-5 transport disgorging troops onto the sand on the day the Gulf War began. It is curious, however, that one of the company's longest and most emotional controversies has revolved around a simple cup of coffee.

The readers and subscribers of Real Goods are by nature a feisty collection of fierce individualists. While the depth of their commitment can never be questioned, it often runs in equal and opposite directions. Witness the Ossengal Affair.

Ossengal is a spot remover. Works like a champ, too. It has been carried in the Real Goods catalog for several years, always producing steady, yet unspectacular sales. Completely organic, it is made from the bile of oxen. Ah, there's the rub.

Many Real Goods readers are ardent supporters of animal rights, and are firmly committed to the belief that products like Ossengal promote cruel and unnecessary slaughter of living beings. They let us know their feelings in highly charged attacks on Ossengal. This, in turn, resulted in equally articulate defenses of the product from the segments of the readership with different beliefs.

It was the Behnke case, however, that really opened our eyes to how the ramifications of a product can extend far beyond its original parameters.

Ms. Behnke is a Real Goods customer who wrote to express interest in participating in the first Off-the-Grid Day. She had one small problem, however, that she asked Real Goods to assist her with. How could she have a piping hot cup of coffee first thing in the morning on Off-the-Grid Day? We turned the challenge over to our readers, and they produced a dazzling array of responses:

"Quit drinking coffee!" wrote Ken Foster of Austin, TX. "Not only is it unhealthy for the body, but depending on where it is grown and how it is harvested, it can be an ecological nightmare. Go for a walk, a run, a bike ride, or do some yoga. These are much healthier alternatives, and they don't damage the environment."

Georg Feuerstein of Lower Lake, CA, suggested a similar stance: "Learn to live without this toxic substance." Then he admitted to enjoying a hot herbal brew himself. Also on the abstinence bandwagon were Alex Lange of Escondido, CA ("If you are so addicted to a substance that you wrack your brain trying to think of a way to get it, it can't be good for you.") and Dave Cutler of Montpelier, VT ("Coffee is bad for you and the third world peasants who have to pick the beans").

Chris Will of Portland, OR, was more sympathetic. "I would suggest the purchase of a quality stainless steel vacuum bottle. In the evening, preheat the bottle with boiling water when energy is abundant from your off-the-grid system, brew your coffee and place in the bottle. I've had coffee stay **hot** and palatable in one of these bottles for 24 hours and I'm a real coffee snob. Make sure there is very little airspace in the bottle—the coffee will stay hotter. Hot coffee with no waiting!"

A combination vacuum bottle/Coleman stove solution was offered by P.W. Back, writing from Waipio, a "magical valley on the Big Island of Hawaii." What we liked even more, however, was the accompanying description of the setting: "Sometimes I take the little stove and coffee fixin's down to a beautiful grassy knoll under the coconut palms and watch the Pacific Ocean wake up the black sand beach. You can't have a bad day after a start like this." (Except that, according to readers Foster and Feuerstein you've just consumed a cupful of toxic waste.)

The thermos solution was also favored by Amber Higgins (Flippin, AR), Sue Brown (Culver, NM) and Donald Vallere of Reading, PA, who included documentation of the fuel gluttony of an electric coffee maker. Bill Warren Mueller (Middlebourne, WV) opted for the thermos solution, as did "Easterner" Anthony (Millwood, NY), who wrote: "Try this. Buy a hot water bottle. In summer, fill it with cold water and take it to bed. Be cool. Very invigorating. By the A.M. it should be up to 98.6 degrees. Awesome!"

Paul Solyn from Athens, OH suggested several products to help Ms. Behnke—the Optimus 81 Trapper alcohol stove and the ZZ Zip Ztove which burns natural fuels like twigs, pine cones, and even animal droppings. Both stoves can boil a quart of water easily in under 10 minutes.

Also from the Heartland, Douglas Campbell seconded the idea of an alcohol stove using homegrown alcohol. His letter percolated with more ideas: "How about a methane burning stove? If commercial, drilled-out-of-the-ground methane is politically incorrect, how about landfill gas, compost gas, or sewage digester gas? Better yet, how about a nuclear coffee pot? Wait, don't hang up!"

He proceeded to describe just such a beast, adding, "Just think, no CO2, no electricity, no fuel, and it lasts a million years. P.S. I am not making any of this up." Less elaborate, fuel-efficient stove solutions were offered by Ann Shirako (Upper Lake, CA), Anne Clarke (Los Angeles), and Claire Van Wyngarden (Kansas) who enjoys her cup of hot brew after milking the goats in the morning.

Maybe, for some in America, the dream is still alive.

An even more grandiose set of solutions was faxed by James Fischer. He offers a menu of options. The "elegant, passive solar solution" is to build or buy a parabolic collector. The "brute force, 100% reliable solution" involves taking two tankless propane-fired water heaters and installing one downstream of the other.

Next comes the "low-tech minimal resources solution" (a can of Sterno), followed by the "unexpected side-effect solution," in which a container of water is suspended 1/4 inch above a DC-powered halogen bulb for 15 minutes.

Perhaps the tempest of "L'Affaire Behnke" was best put into perspective by Mr. Fischer's "last-ditch attempt solution": "Take a 1/2" or thicker metal bar, and toss it across the terminals of your main battery bank (disconnect your solar panel first!) Heat your water over the resulting fire. Once your coffee is brewed, leave the house, since it may be burning down around you."

Why such animated response? Perhaps Anthony Miksak of Caspar, CA summarized it best: "Somehow I think there are bigger problems in the world, but I CARE about Ms. Behnke and her cup of coffee because I like my steaming hot cup in the morning, too."

— Stephen Morris

Notes and Workspace

DC Lights

When designing an independent power system, it is essential to use the most efficient appliances possible. Generally, fluorescent lighting is the best and most efficient way to light your house.

Real Goods has, in the past, recommended against using DC lights in remote home applications except in cases where there is no inverter in the system. The greater selection, lower prices, and higher quality of AC products has made it hard to justify DC lights. However, recent issues have brought the question of using AC or DC lights back.

One of the greatest reasons not to use DC lights has been the poor quality of the ballasts for DC compact fluorescent bulbs. In our search to overcome this problem, we have uncovered a new source for premium-quality ballasts. There is also a variety of tube-type fluorescents, which are great for kitchen counters and larger areas. We also offer incandescent and quartz-halogen DC lights, but suggest caution in using these for alternative energy systems, where watts are precious.

PL Lamps

PL bulbs are more sensitive to voltage than other lightbulbs, and will not fire or will fire slowly at voltages below 12.0 volts. It also helps to ground them by touching the bulb, making them fire more rapidly (known as "stroking the lamp"). Some of our customers have suggested placing a 1/2" piece of aluminum foil around the two tubes of the lamp for quicker start.

Steve Willey of Backwoods Solar has come up with a novel solution to the slow PL firing problem. (However keep in mind that it will void your warranty.) Steve claims the solution will result in allowing every PL bulb to start every time without failure with any ballast! Here is how it's done with the two major PL bulb manufacturers, Phillips and Osram:

Osram: Cut off the "skin" from the bottom of the PL bulb's base, using a band saw, hacksaw, or grinding wheel. This is a very thin layer at the bottom — be very careful not to catch any wiring

when cutting. You will expose a glass cylinder (starting tube) and a grayish plastic cylinder (the capacitor). Cut one of the wires going to the capacitor and bend it out of the way so it doesn't touch other wires. Use glue to seal if you want.

Phillips: Cut as with the Osram, but in the Phillips PL bulbs, the capacitor is encased in a blue plastic unit. Snip one wire (any of the three going to the capacitor) and bend it back as with the Osram so that it doesn't touch any other wires. Use glue to seal if you want.

Prewired Screw-In Compact Fluorescent 12V PL Lamp Kits

These modular compact fluorescent lamp assemblies screw into standard light bulb sockets. They incorporate a 12V solid-state ballast, a standard edison base, a compact fluorescent lamp socket, and a long-lived 10,000 hour life PL lamp, which can be replaced when it burns out. *Remember, these lamps are for 12V use only!* **For all PBS series lights, the center tip pin must be positive.**
ATTN: All PL lamps require ballasts to operate

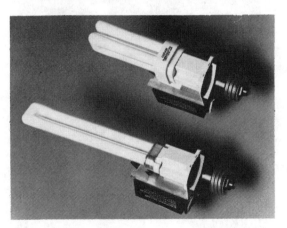

31-661	PBS-PL5	25W equiv.	5.25"x2"	$59
31-662	PBS-PL7	40W equiv.	6.5"x2"	$59
31-663	PBS-PL9	50W equiv.	7.75"x2"	$59
31-664	PBS-PL13	60W equiv.	8.25"x2"	$65
31-665	PBS-Quad9	50W equiv.	5.5"x2"	$65
31-666	PBS-Quad13	60W equiv.	5.75"x2"	$69

12 Volt Ballast/PL Lamp Kits

Use this kit to install energy-efficient **DC powered** compact fluorescent lighting into fixtures that cannot accommodate the bulkier size of the (above) PBS series or where the bare lamp may be visible and a neater look than that of the PBS series is desired. The RK-PL kits come with the appropriate Iota hard-wire ballast, compact fluorescent lamp, and PL lamp/edison screw-in base adapter. The ballast measures 3.2"L x 1.75"W x 1.37"D and may be installed in the base of adequately sized floor or table light fixtures, or in a junction box mounted in the wall or ceiling, a fixture, or on a fixture power cord. Please note: switching of power to the lamp must be done on the positive input lead to the ballast, so that ballast is not energized when the light is off. The height of the PL lamp/edison adapter base is the same as for the above comparable PBS series units, and the width is 1¼ inches.

For 12VDC use only

Lamp harp is too short to fit one of these PL lamp kits? See page 42 for longer lampshade harps.

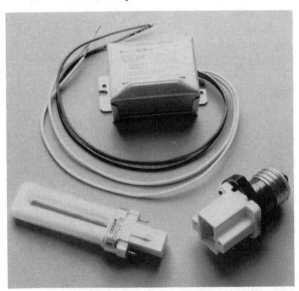

Do not put switches between the bulb and the ballast. The proper place for a switch is before the ballast only!

31-671	RK-PL5	$43
31-672	RK-PL7	$43
31-673	RK-PL9	$43
31-674	RK-PL13	$43
31-675	RK-Quad9	$49
31-676	RK-Quad13	$49

Iota 12V Ballasts

These high-quality ballasts replace the Sunalex ballasts that we have offered in the past and are a major step up in reliability. Power factor is not an issue with DC lights, so the net power used is equal to the amp draw times the voltage. Minimum recommended starting temperature is 50° F. According to the manufacturer, the recommended operating voltage for 12V models is 10.5V to 14V, although we have seen them work at higher voltages. Approximate lifetime is 10,000 hours, and all units come with a 1-year warranty.

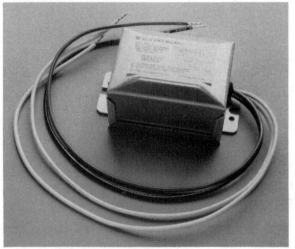

31-205	12V DC Ballast/PL5	$34
31-206	12V DC Ballast/PL7, 9, Quad 9	$34
31-207	12V DC Ballast/PL13, Quad 13	$34
31-208	24V DC Ballast/PL5	$55
31-209	24V DC Ballast/PL7, 9, Quad 9	$55
31-210	24V DC Ballast/PL13, Quad 13	$55
31-211	12V DC Ballast/F40	$65

Osram 12V Ballast/Bulbs

This is the state-of-the-art DC ballast. Although they are somewhat limited in application, they are the best DC ballasts we have found. The main limitation is the fact that they only can operate 4-pin, 9 watt compact fluorescent bulbs. These electronic ballasts are extremely efficient, drawing only 1 watt, will start instantly down to 0° F, and eliminate the need for copper grounding strips. Our preliminary testing has amazed us! Since they must be used with the 4-pin bulbs, there is no way to adapt them to a screw in base to insert them into a common "edison" plug. For convenience, we are selling these as complete kits, with a ballast, 4-pin PL hardwire socket and a bulb.

31-650 Osram Ballast $49

PL Twin-Tube Fluorescent Bulbs

The PL bulbs come in 5, 7, 9, and 13-watt versions. They can be purchased in three different configurations. First and least expensive is the bulb with a two-pin base as a replacement bulb for the RK or PBS series to fit into an edison adapter. Secondly, purchase the PL bulb with an edison base (either attached for a shorter overall length or with a separate edison base adaptor which makes the bulb longer overall but replaceable), and third purchase the bulb with an edison base AND with a frosted globe on the outside, which many people consider more attractive, although it cuts the light output down marginally. The quad tube bulbs are the answer to the common complaint that the bulbs are too long. They come with and without frosted globes. The Quad-9 consists of two 5-watt PL bulbs and the Quad -13 consists of two 7-watt PL bulbs. You can also order the edison base adapter separately that accepts a pin-based PL bulb on one end and screws into a standard edison base fixture. Also available is a hardwire PL socket for those who choose not to employ the edison base. **All PL bulbs must use a ballast!**

These pin base bulbs are also the replacement lamp for most AC compact fluorescents with replaceable lamps.

31-101	PL-5 w/pin base	$7
31-102	PL-7 w/pin base	$7
31-103	PL-9 w/pin base	$7
31-104	PL-13 w/pin base	$7
31-121	Quad 9 w/pin	$11
31-122	Quad 13 w/pin	$12
31-151	PL-5 w/separate edison base	$15
31-152	PL-7 w/separate edison base	$15
31-153	PL-9 w/separate edison base	$15
31-154	PL-13 w/separate edison base	$15
31-161	Quad 9 w/separate edison base	$18
31-162	Quad 13 w/separate ed. base	$18
31-156	PL-5 w/attached edison base	$13
31-157	PL-7 w/attached edison base	$13
31-158	PL-9 w/attached edison base	$13
31-159	PL-13 w/attached edison base	$13
31-165	Quad 9 w/attached ed. base	$23
31-166	Quad 13 w/attached ed. base	$23
31-131	Quad 9 w/ed. base and globe	$29
31-132	Quad 13 ed. base and globe	$29
31-402	Edison to PL adapter	$5
31-403	Hardwire PL Socket	$3

Sundancer

This handsome fixture is designed for indoor or outdoor areas where a clean, low-profile, surface mounted light is required. It may be mounted on the wall, ceiling, or underside of cabinets or shelves. The fixture incorporates a specular aluminum reflector, a 12-volt solid-state ballast, and a rugged, clear prismatic, polycarbonate lens, which provides a sparkling appearance and maximizes light distribution. Corrosion resistant materials are used throughout — outdoor rated. A gasketed lens keeps dirt and insects out. An elegant wood trim (T), which frames the lens, is available. Compact fluorescent lamp included. **12V DC only!** Dimensions are 4-5/6" wide, 2-3/8" deep, the PL7 (40W incan. equiv.) is 7" long and the PL13 (75W incan. equiv.) is 11" long.

31-331	ELDX-7 PL7	$79
31-332	ELDX-7T PL7	$95
31-333	ELDX-13 PL13	$79
31-334	ELDX-13T PL13	$95

("T" means with wood trim)

Quartz Halogen Lighting

Quartz halogen 12-volt lights burn hotter and brighter than incandescent lights. They can be up to 50% more efficient and lead a longer life. While not as efficient as PL or other fluorescent lighting, they lend themselves to situations where direct, bright lighting is called for, like desks, sewing tables, workshops, and spot lighting. For more information on Quartz halogen lighting see the AC Lighting chapter.

All the bedside reading lights in my house are copilots attached to the emergency power system. When the grid fails we feel smug. The well-focussed task lighting is perfect for reading and doesn't disturb anyone.

Quartz Halogens Inside Frosted Globes

A small cottage industry in the Northeast manufactures these incredibly ingenious 12-volt light bulbs. On the outside they appear identical to an incandescent, but they have the increased efficiency and longevity of quartz halogens. Multiply the watts on halogens by 1.5 and you'll get an idea of their equivalent light output compared to an incandescent lamp. **12V DC only!**

33-102	601F10 10W 0.8amp	$15
33-103	601F20 20W 1.7amp	$15
33-104	601F35 35W 2.9amp	$15
33-105	601F50 50W 4.2amp	$15

MR-16 Halogen Flood Lamp

The MR-16 is a very bright flood lamp with a faceted reflector and a quartz halogen bulb in the center and comes covered with a glass lens. It is ideal for situations where a bright light is desired but not overly directed to a pinpoint. The bulb is a 20-Watt halogen which draws 1.7 amps but is far brighter than you'd expect. Fits into an Edison base. **12V DC only!**
33-108   MR-16 12V Halogen bulb $25

Halogen Co-Pilot Tasklights

These rugged, compact, flexible 12V tasklights can be positioned precisely to deliver light where you need it, without getting in your way. Great for use by your bedside, kitchen sink, sewing machine, desk, or in your car, boat, or RV. Its 5-watt halogen lamps deliver a bright white light and operate approximately 30% more efficiently than standard incandescents. On-off switch and reflector is built into the cap of the lamp housing. Gooseneck stem is offered in three lengths for the hardwire mounted model and one length for the cigarette lighter plug-in model M. Matte black finish. The L-20 is our best seller with the K-30 running a close second. If you like to read in bed at night and your lighting system usess primarily 120V lights, your can avoid a phantom-load on your inverter by using this light. The 5 watt halogen bulb draws 0.42 amps. No minimum starting temperature. **12VDC only.**

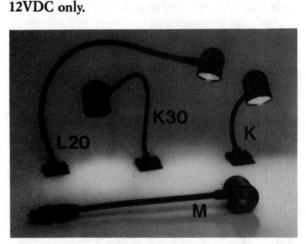

33-301	L20 20" neck	$26
33-302	K30 12" neck	$26
33-303	K 4" neck	$24
33-304	M 6" neck	$24
33-109	5W Replacement Halogen bulb	$9

Littlite High Intensity Lamp

The Littlite is a great gooseneck lamp for your desk, stereo, headboard, or worktable. It's available in a 12 inch or an 18 inch length. The "A" series comes with base and dimmer, 6-foot cord, gooseneck, hood, and bulb. Two pieces of snap mount are included for permanent mounting. Options include a weighted base (WB) for a movable light source, plastic snap mount with adhesive pads (SM), and an adjustable mounting clip (CL) that adjusts from 1/16 to 3/4 inch wide. Also available is a replacement halogen bulb (RHB). The WXF power transformer is available for 120V users to convert the Littlite from 12V to standard house current. Now available with a cigarette lighter plug. **12V DC only.**

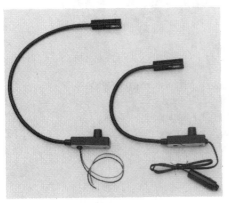

33-306	12" Littlite (L-3-12A)	$45
33-307	18" Littlite (L-3-18A)	$45
33-308	12" Littlite w/12V plug	$49
33-309	18" Littlite w/12V plug	$49
33-401	Weighted Base (WB)	$11
33-402	Snap Mount (SM)	$3
33-403	Mounting Clip (CL)	$12
33-109	Replacement Bulb (RHB)	$9
33-501	120V Transformer (WXF)	$14

Tail Light Bulb Adapters

This simple adapter has a standard medium edison base and accepts a standard automotive-type bulb. It's a very easy way to convert pole lamps to 12V. ½ inch long. *Bulb pictured but not included. Get taillight bulbs at your local auto supply store.*

33-404 Tail Light Bulb Adapter $7.50

Bullet Lights

Though not as efficient for general illumination, bullet lights are excellent for areas where a focused light is desired (over your bed or in a workshop or kitchen). The high-intensity bulb makes it a perfect reading light, yet it can be directed to give area lighting as well. A positive lock swivel assures illumination where desired. It comes in four different configurations: a single or double tube-shaped bullet light, or a single or double cone-shaped bullet light. The lights come in satin brass on a woodgrain base and have a positive lock swivel. Single lamps draw 1.4 amps at 12V, and the doubles draw 2.8 amps using 18-watt bayonet bulbs. **DC only.**

38-401 Bullet light — Double tube-shaped $18
38-402 Bullet light — Single tube-shaped $12

38-403 Bullet light — Single cone-shaped $12
38-404 Bullet light — Double cone-shaped $18

"These are incredible little lamps. My wife used to always get upset when I read in bed because the big overhead lamp kept her awake. Five months ago I installed the Littlite with the easily movable snap mount which attaches to just about anything. You can put the light up at night and take it down in the morning. With the halogen bulb, it casts a very directed light which enables me to read clearly at night while my wife sleeps on..."
J.S., Ukiah, CA

12V Night Light

This super efficient night light is ideal for the children's room or for shining directly down on the toilet seat at night. The Red MT 5000 light source is the most efficient light source on the market today. It won't destroy your night vision because red light doesn't wash out night vision like white light does. It also works well to light door locks or to mark trees, posts, or other areas on your driveway that you don't want to back into at night. The 12-volt Night Light draws only 0.2 amp, so it can be left on continuously with little effect on your power system. If operated for 24 hours a day it should last for 10 years.

37-315 12-Volt Night Light   **$35**

Conserve Switches

Conserve switches are variable speed control switches (rheostats) that are extremely useful for dimming 12V incandescent lights and some halogens *(don't attempt to use on ballasted fluorescent lamps!),* using less than 1/2 watt of power to control them. They are available in a 4-amp and 8-amp configuration.

Incandescent Lights

Clearly the undisputed energy hog of the lighting industry, incandescent lighting nevertheless captures a necessary niche in the low-voltage lighting market. They are very cheap, wonderfully easy to install, and if you have power to spare (hydro users take note) are the right tool for the job. Most people use them in the beginning of their off-line transformation or in rarely used spots like closets, bathrooms, and alcoves. **12 VDC only.**

Case Price (25) Mix or match $35

38-101	15W	1.2 amp	$2
38-102	25W	2.1 amp	$2
38-103	50W	4.2 amp	$2
38-104	75W	6.2 amp	$2
38-105	100W	8.3 amp	$2

24-101	4-amp Conserve Switch	$27
24-102	4-amp Conserve Switch	$31
	(with forward and reverse feature)	
24-103	8-amp Conserve Switch	$39

Standard Fluorescent Lamps

While not always as efficient as the new compact fluorescent lamps featured above, standard fluorescent lamps are nonetheless far more efficient than incandescents and halogens. We are featuring the Thin-Lites here, which are very reasonably priced and well built.

Thin-Lites

Thin-Lites are made by REC Specialties, Inc. Their 12-volt DC fluorescents are built to last, and are all U.L. listed. Easy to install, they have one-piece metal construction, nonyellowing acrylic lenses, and computer-grade rocker switches. A baked white enamel finish, along with attractive woodgrain trim completes the long-lasting fixtures, warranted for 2 years. All use easy-to-find, standard fluorescent tubes, powered by REC's highly efficient inverter ballast. For those that didn't know, bulbs used in DC fixtures are exactly the same as those used in AC fixtures and can be purchased at any local hardware store—only the ballast is different.

22-watt 12VDC circline. 9-1/2" diameter x 1-1/2" deep. 1.9 amps. Uses one FC8T9/CW fluorescent tube. Light output: 1,100 lumens. Lamp included.
32-109 #109C 22-Watt Thinlite   **$44**

30-watt 12VDC light. 18" x 5-1/2" x 1-3/8" deep. 1.9 amps. Uses two F15T8/CW fluorescent tubes. Light output: 1,760 lumens. Lamp included.
32-116 #116 30-Watt Thinlite   **$45**

32-watt 12VDC circline, 13-1/4" diameter x 2-1/4" deep. 2.6 amps. Uses one FC12T9/CW fluorescent tube. Light output: 1,900 lumens. Lamp included.
·32-110 #110 32-watt Thinlite   **$63**

15-watt 12VDC light. 18"x4"x1-3/8" deep. 1.26 amps. Uses one F15T8/CW fluorescent tube. Light output: 800 lumens. Lamp included.
32-115 #115 15-watt Thinlite   **$37**

Commercial and Industrial Lights: 150 Series

Our Thin-Lite 20-watt surface mount fluorescent is designed for practical lighting in commercial, industrial, and remote area site applications. It features anodized aluminum housings in a 2 foot length. It is shipped without the fluorescent tubes. This fixture utilizes standard AC 20-watt fluorescent tubes readily available worldwide. It is designed for remote switching.

15-watt 12VDC light. 18-1/8"x2-1/4"x2-7/16". 1.3 amps. Uses one F15T8/CW fluorescent tube. Light output: 870 lumens. Lamp included.
32-193 #193 15-Watt Thinlite $39

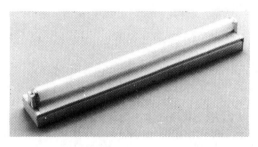

20-watt 12VDC light. Housing dimensions: 24"x3-3/8"x1-5/8". 1.6 amps. Uses one F20T12-/CW fluorescent tube. Light output: 1,250 lumens. Lamp not included.
32-151 #151 20-Watt Thinlite $39

Hi-Tech Styles

Thin-Lite Hi-tech styles were developed for both efficient and attractive lighting where maximum light is required. Anodized aluminum housings and clear acrylic diffuser lenses provide high light output on three sides. They are designed for commercial and industrial vehicles, and for use in remote area housing, schools, and medical facilities in conjunction with alternative sources of energy. Fixtures have almond end caps.

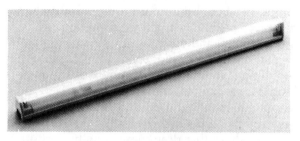

30-watt 12VDC light. 36-1/8"x2-1/4"x2-7/16". 2.1 amps. Uses one F30T8/CW fluorescent tube. Light output: 2,200 lumens. Lamp included.
32-197 #197 30-Watt Thinlite $55

Surface-Mount Lights

The ST 130 series are our most popular Thin-Lites. They are economically priced, practical lights that feature pre-painted aluminum housings and acrylic diffuser lenses. They are available as large as 5 feet long with two standard 40-watt AC fluorescent tubes to meet maximum lighting requirements. Where practicality is the principal consideration, the ST 130 series suit the requirement perfectly.

8-watt 12VDC light. 12-3/8"x2-1/4"x2-7/16". 0.9 amps. Uses one F8T5/CW fluorescent tube. Light output: 400 lumens. Lamp included.
32-191 #191 8-Watt Thinlite $37

30-watt 12VDC light. 18"x5-3/8"x1-3/4". 2.1 amps. Uses two F15T8/CW fluorescent tubes. Light output: 1,760 lumens. Lamp included.
32-134 #134 30-Watt Thinlite $42

40-watt 12VDC light. 24"x5-3/8"x1-3/4". 2.5 amps. Uses two F20T12/CW fluorescent tubes. Light output: 2,500 lumens. Lamp included.
32-138 #138 40-Watt Thinlite **$55.**

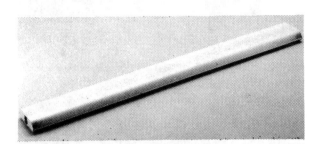

40-watt 12VDC light. 48"x5-3/8"x1-3/4". 2.9 amps. Uses one F40T12/CW fluorescent tube. Light output: 3,150 lumens. Lamp included.
32-139 #139 40-Watt Thinlite **$59**

Thin-Lite Ballasts

Thin-Lite ballasts convert a wide range of standard 120VAC fluorescent fixtures to 12V operation. One ballast is required for each fluorescent tube adaptation in a fixture unless otherwise noted. As a general rule, to find the amp draw of a particular light or inverter ballast, divide the watts by volts (W/V = A.). As an example, the 12VDC circline table lamp adapter Model 107 is rated at 22 watts: 22 watts divided by 12 volts = 1.8 amps. *Note: Thin-Lite ballasts are available by special order only; allow 2–4 weeks for delivery.*

Thin-Lite ballasts are also available in 24, 36, and 48 volt input. Call or write our technical department for details.

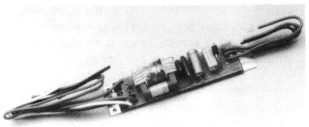

32-201	4 watt, single lamp	$29
32-202	4 watt, dual lamp	$29
32-203	22 watt, single lamp	$29
32-209	32 watt, single lamp	$29
32-210	8 watt, single lamp	$29
32-211	8 watt, dual lamp	$29
32-212	14 watt, single lamp	$29
32-213	14 watt, dual lamp	$29
32-214	15 watt, single lamp	$29
32-216	15 watt, dual lamp	$29
32-225	13 watt, single lamp	$29
32-226	13 watt, dual lamp	$29
32-247	30 watt, single lamp	$29
32-251	20 watt, single lamp	$29
32-252	20 watt, dual lamp	$29
32-253	40 watt, single lamp	$29

Note: Lamps and sockets not included.

To order any of these products
or for more technical information
call us toll-free at
1-800-762-7325

Elegant Brass Kerosene Lamps

We've recently located an importer of very high quality brass kerosene lamps from France. Our importer has been in the lighting field for 20 years, and every lamp is time tested. Spare parts and service are readily available on all lamps.

The Technology

The 1800s in Europe saw a period of revolution in indoor living after dark. The state of the art peaked with the invention of the Kosmos burner system and its derivative the Matador (flame spreader) burner, featuring symmetrical central drafting and an area-maximizing circular burning surface. Each of our lamps is equipped with these burners. With only the occasional replacement of wick or chimney, these lamps will provide decades of daily illumination. All use kerosene or lamp oil.

Lamp Patronne

Fitted with a Kosmos burner and chimney, our tallest lamp offers the optional elegance of a ball shade. The overall height is 19".

■35-325 Lamp Patronne complete $47
■35-323 Replacement Patronne Burner $17

Lamp Patronne with Etched Ball

This is the same lamp as above but with the beautiful 6" etched glass ball that fits over the glass chimney.

■35-320 Lamp Patronne w/Ball $75
■35-324 Spare Etched Ball $29

Lamp Concierge

With a Kosmos burner system and ball shade, this lamp has a pleasing profile and a very low center of gravity. The carrying handle is slotted for wall mounting. The overall height is 15".

■35-345 Lamp Concierge $49
■35-343 Replacement Burner $18

Lamp Concierge w/Etched Ball

This is the Lamp Concierge lamp listed above including the beautiful 6" diameter etched glass ball.

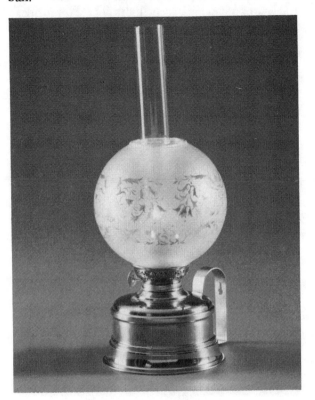

■35-340 Concierge w/Etched Ball $79
■35-344 Spare Etched Ball $29

Lamp Maitresse

Equipped with an elegant "bombe" chimney, this little lamp maintains an unobtrusive presence until it is needed; then it provides a light source exceeding conventional hardware store kerosene lamps. The overall height is 12".

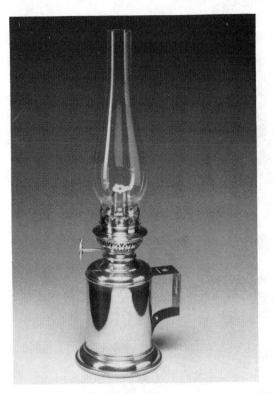

■35-330 Lamp Maitresse $45
■35-333 Replacement Burner $17

Spare Parts Packages

To assure uninterrupted service from your oil lamp, we recommend that you include a spare parts package with your purchase. Each package has two wicks and two chimneys.

■35-328 Patronne Parts $14
■35-346 Concierge Parts $14
■35-334 Maitresse Parts $14
■35-327 Vintners Parts $9

Vintners Lamp

The Vintners Lamp is a small kerosene lamp perfect to light your way when no other lighting is possible. One six ounce filling will last 20 hours. This is a functional and attractive addition to any vacation or solar home or for Off The Grid Day!

■**35-326 Vintners Lamp**   **$39**

12-Volt Christmas Tree Lights

The 12-volt Christmas lights were a great hit on my motorcycle last year. (With thanks to California climate.) -Dr. Doug

It just doesn't seem like Christmas without lights on the tree and these 12-volt lights will dazzle any old fir bush. These lights are actually great all year 'round for decorating porches, decks, and accenting homes and showrooms. The light strand is 20 feet long and consists of 35 mini, colored, non-blinking lights (amp draw=1.2A) with a 12-volt cigarette lighter socket on the end. Needless to say, you'll crank many heads on the freeway Christmas Eve when you plug a set of these into your car's cigarette lighter. (Note: you can't connect them in series like you can with many 120V lights).

Humphrey Propane Lights

Propane lighting is very bright and a good alternative if you don't have an electrical system. One propane lamp puts out the equivalent of 100 watts of incandescent light. The Humphrey 9T contains a burner nose, a tie-on mantle, and a #4 Pyrex globe. The color is "pebble gray." The mantles seem to last around 3 months, and replacements are cheap. *Note: Propane mantles emit low-level ionizing radiation due to their thorium content.*

35-101 Humphrey Propane Lamp (9T)   **$49**
35-102 Mantle for Propane Lamp (each)   **$1**

37-301 12V Christmas Tree Lights   **$12**

12V Appliances

Remington 12V Shaver

The Remington has two ultrathin micro screens for an extremely close shave, and it has a 12V lighter-type plug. It comes with a travel pouch, protective headcover, and a cleaning brush.

63-338 12V Remington Shaver $39

12V Hairdryer/Defroster

Our 12V hairdryer is probably most commonly used as a 12V defroster. It beats using your credit card for scraping icy windshields, and is found in many of our customers' glove boxes. It comes with a 54-inch cord and measures 4.5" X 4.5". It draws 12 amps from your 12V battery.

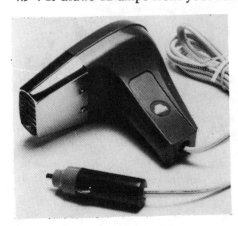

63-323 12V Hairdryer/Defroster $17

12V Bedwarmers

When it gets *really cold*, there is no other bed warming appliance that can put so much heat in your bed so fast as our bedwarmers. These units are far more efficient than electric blankets because they put all the heat *in* the bed rather than on top of the bed where it dissipates rapidly. The radiant heat keeps the bedding dry, and warms the mattress and spaces beside you. It gives you a soothing feeling of relaxation, like sitting in warm bath water. There are many sizes to choose from for different widths of beds. All units are 5 feet long. The rated amp draws are only valid while the heater is cycling on. Otherwise they draw nothing! 12V only.

63-328	B24 Bunk 24" wide 5.0 amps	$55
63-329	B48 Double 46" wide 6.7 amps	$65
63-330	B60 Queen 60" wide 6.7 amps	$89

12V Soldering Iron

This soldering iron comes complete with a 12V plug. Draw is 1.9 amps. A good basic tool for your 12V toolbox!

63-118 12V Soldering iron $18

"These bed warmers are everything they're cracked up to be! It makes getting into bed on a frosty night just like getting into a hot bath!" J.S., Ukiah, CA.

Important Note: If you are concerned about electromagnetic radiation (EMF) you should be aware that the 12V blankets are completely radiation free. For more thorough information on EMF see our book section on Cross Currents and The Body Electric.

12-Volt Waring Blender

This is by far the strongest, most durable, and most attractive 12V blender ever made. It was developed by Waring's commercial division. It draws up to 11.5 amps to blend a very hefty load. The base is made of chrome. The 45-ounce shatter-resistant plastic carafe has a snug-fitting vinyl lid with a removable center insert (so you can add ingredients while blending) which doubles as a 2 ounce measure. Stainless steel blades are removable for easy cleaning. A 15-foot cord with attached cigarette lighter plug are included.

Dear Real Goods, I am very glad that you are now carrying the 12VDC washing machine in your catalog. This machine has been keeping our clothes clean for over 5 years. It is not only efficient powerwise (5.5 amps), it is also economical, especially compared to laundromat costs. Its compact size is deceptive. I have washed area rugs, king size sheets, and a week's worth of bath towels in it. Another bonus is that I can use the discharge water for my trees and garden (biodegradable soap used), which is so important during this drought. This machine is the one that allowed us to decide to go entirely 12VDC in our home. K. Sanderson, Paicines, CA

63-314 12V Waring blender   **$139**

12-Volt Mini-Wash

The 12V Mini-Wash is the ideal appliance for the small homestead. It has a very efficient counter-rotating clothes drum housed inside a tough stainproof, dentproof, and rustproof polypropylene tank and lid assembly. The drum switches rotation direction approximately every 30 seconds and is gentle enough for delicate fabrics. The rotate and tumble action works well on even the dirtiest laundry without heavy power use. The washer draws a maximum of 5½ amps with full load. There is a 2–10 minute spring-wound automatic timer to time wash loads. The Mini-Wash will take up to 4.5 pounds of dry laundry in a single load and use a maximum of 3 gallons of water. This is a very simple washer; filling and emptying are done manually with the attached flexible hose. There is no spin cycle. Depending on dirt load, several loads may be run without changing the wash water. The safety lid automatically stops the machine when lifted. A standard 12V cigarette lighter plug is included. The entire washer is very compact, measuring just 20-inches high by 17-inches wide by 18-inches deep and weighs 17 pounds.

63-319 12V Miniwash   **$195**

Guzzle-Buster
Washing Machine Efficiency Kits

Here is a product that corporate America could take a lesson from. These Guzzle-Buster Kits are manufactured in a *totally solar powered factory*! This kit will cut your full size washing machine electrical consumption by an amazing 2/3 without affecting performance! If that sounds too good to be true, consider this from the Solar Technology Institute: "We measured the power consumption of the unmodified washer on four different inverters. Then we converted the washer...and ran it again on the same four inverters. The Maytag washer used one-third as much power after conversion." The STI students said "the installation instructions were the best they have ever used."

An ordinary washer is often the most stressful load your inverter is asked to handle. With the Guzzle-Buster installed your inverter will run cool and you'll be able to run other appliances simultaneously. We have found that mid-sized 600 to 800 watt inverters with surge ability will easily run a converted washer. The Kit is preassembled, leaving only two main parts to install in the washer. These parts consist of a custom-made DC motor that is unavailable anywhere else, and a control box that contains relays, controllers, and rectifier. They come with clear, concise, easy-to-follow instructions. The control box is outdoor-rated, and all components are robust and high-quality. Circuits are fuse protected and include a *gentle-start* feature that helps prolong the life of the washer transmission, motor, and relay switches. Grounding provisions are included to comply with code requirements. Kits come with a diagnostic switch and test-light to quickly determine if a washer malfunction is caused by the kit.

The 12-volt DC kit comes complete with a 50-amp wall receptacle, plug, and stout power cord. The 12-volt kit requires a 100 watt (or larger) inverter to run the washer timer. The inverter is not included. All kits are warranted for 2 years.

Guzzle-Buster Kits are available for the following full-size, top loading automatic washers. Kits are model specific. If your washer isn't listed the kit won't work! Pre-1985 Kenmore or Whirlpool washers can be identified by a 15-inch x 15-inch access panel on the back. Newer washers have no back side access.

63-171 12-Volt DC Kit   **$495**
(Will convert *any* Maytag or pre-1985 Kenmore /Whirlpool washer)
63-172 120-Volt Kit   **$475**
(For pre-1985 Kenmore/Whirlpool washers)
63-173 120-Volt Kit   **$495**
(All Maytags)

Gas Ranges

Our gas range comes with a universal orifice so it will accept either propane or natural gas, and we've culled out only the model without pilotless ignition so you don't need 120V AC to light them. We're selling them at a very attractive price. We checked with our local gas company and set our price 15%-20% below theirs.

30" Standard Range

The Crosley is a standard 30" range that will operate on either propane or natural gas. It comes with a standing pilot and is not "pilotless ignition." It measures 40"H x 30"W x 25-5/8"D. It can be ordered in either almond or white. Features include three walled contruction, removable oven door, deep porcelainized broiler pan, low heat oven control and variable broiler control. This unit normally retails for $495. We may substitute another brand of range as new ones become available.

■61-102   30" Standard Range   **$395**
Shipped freight collect
Specify White or Almond

To order any of these products
or for more technical information
call us toll-free at
1-800-762-7325

Cooling & Air Conditioning

CFC-Free Air Conditioning

Have you been doing without air conditioning because of CFC's and the ozone issue? The planet thanks you for that, but you no longer have to cook in the summer. The Real Goods gas-fired air conditioner is made for us by a reputable manufacturer. Absorption technology has been around for many decades and is well proven and reliable. There are no costly compressors present. Rather, a chrome-lined generator operates the unit; there are just three moving parts in the sealed system! Real Goods' chiller units are tough performers with cooling capacities from 3 to 50 tons. Energy-efficient heat exchangers and an automatic ignition feature promote energy savings even during the sweltering summer months. And vertical air discharge vents protect plants and shrubs from damage.

We also offer the Real Goods "AY" (All Year) chiller/heater that circulates hot water in the cold winter months and chilled water during the hot muggy summer months. Our off-the-grid customers will find that this is the only air conditioner that they can feasibly power. Features include simple installation, one gas hook-up (natural gas or propane), one single phase AC connection, one water supply and one water return.

You may use multiple fan coil units for precise control zone cooling. We are working with the manufacturer to supply the units with DC motors and controls. This will take a little while so keep an eye on our catalogs. For now the

three ton unit will work quite nicely on an inverter of 1500 watts or greater. Since this unit is so simple and trouble-free, it is backed by a 5-year guarantee, and for a little extra money it can be extended to 10 years. If the unit malfunctions it is replaced with a new one – not just repaired!

■62-310	Chiller – 3 ton	$4,395
■62-311	Chiller – 4 ton	$5,395
■62-312	Chiller – 6 ton	$5,995
■62-313	Chiller/Heater – 3 ton	$5,995
■62-314	Chiller/Heater – 4 ton	$6,995
■62-315	Chiller/Heater – 6 ton	$7,995

Shipped Freight Collect from Sacramento.

TYPICAL PIPING (GAS AND WATER)
SINGLE UNIT SINGLE ZONE APPLICATION

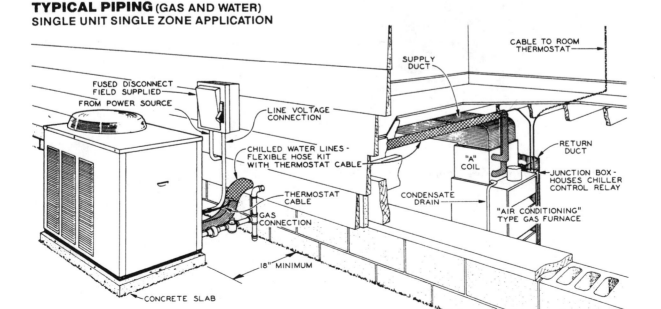

Solar Evaporative Air Conditioner

The Recair 18A2 turns hot dry summer air into a refreshing indoor climate, while filtering out dust and pollutants. It can cool up to 400 square feet, using only one 48-watt solar panel and 2–4 gallons of water per day to cool and clean the air. Unlike cellulose cooler filters, it will not promote bacterial growth or foul odors. It will move 750 cubic feet per minute (cfm) drawing 3 amps at low speed and 4.6 amps on high speed at 12VDC from one 48-watt solar panel. It is installed with simple hand tools and connected to a garden hose or the separately available pressure tank.

Recair is the most efficient cooler of its type on the market today. Conventional air conditioners recirculate the same stale air within your house. Recair uses outside air and filters out any impurities such as pollen or dust, while adding moisture to the air giving you a more healthful, cool indoor air environment.

Installed with a solar panel, Recair will cool when it is needed most. As the morning sun strikes the solar panel, Recair goes to work cooling your house. As the dry air is forced through the wet filter by the fan, the water is evaporated, which cools the outgoing air. The drier the air, the greater the temperature drop. The only drawback is that it will not work efficiently where relative humidity exceeds 40%.
Dimensions: 17½"W X 22¼"H X 20"D
Note: PV module not included.

...a band of pioneers in Northern California sustained their vision of a way to give everyone fair access to a diverse tool kit for energy self-sufficiency. Through their dedication, thousands of people have had the privilege of discovering that the energy problem, far from being too complex and technical for ordinary people to understand, is perhaps on the contrary too simple and political for some technical experts to understand.
- Amory Lovins, in the Foreword

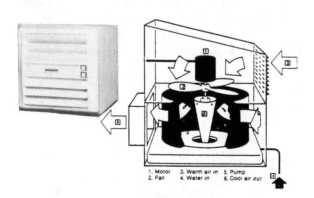

1. Motor 3. Warm air in 5. Pump
2. Fan 4. Water in 6. Cool air out

■64-201 Solar Evaporative Cooler $495

RV Evaporative Air Cooler

Hot, dirty air becomes clean, cool air as highly effective spin-spray action washes out dust, pollen, and impurities. The unit fits all standard 14" x 14" RV roof vents. Its streamlined styling and lower-profile contour offer reduced wind resistance. For longer life, the motor is cooled by dry air. Operation is exceptionally quiet. It produces 450 cfm, drawing only 2.1 amps on low speed and 750 cfm on high speed drawing 4.6 amps. The nonorganic industrial foam filter provides superior filtration and cooling, without bacterial growth and foul odors. The discharge grille, only 1½ inches deep, has individually adjustable louvers. The built-in reservoir fills automatically from the RV system or with a garden hose; can be hand-filled from inside the vehicle. The cooler weighs 16 pounds; dimensions are 35" x 22" x 10½". There's a 1 year warranty on all parts; 3 years on the water pump. Not recommended where average relative humidity exceeds 40%.

■64-204 RV Evaporative Air Cooler $429

Sunvent

The Sunvent is a compact solar-powered ventilating fan for any application where humid or stale air collects. It is great for boats, motorhomes, and travel trailers. It's simple to install, requires no electric hookup, is solar powered, and is completely weatherproof. It extracts 680 cubic feet of air per hour under normal working conditions. The Sunvent's simplicity is its greatest virtue, as it works whenever the sun is shining, pulling out hot air the fastest when the sun is at its highest. This is the perfect sun-direct application! It requires a 6 inch round mounting hole.

| 64-242 Sunvent Fan | $49 |

12V - 16 Inch Intake/Exhaust Fan

For moving large volumes of air, this 16 inch 12V fan can't be beat. It works on either 12 or 24-volt systems or in panel-direct applications. It has a 16-inch three-wing aluminum blade, and a 20-inch square steel venturi. An 18-inch hole is required for mounting. It uses a reversible permanent magnet 30-volt DC motor. At 12 volts, it moves 650 cfm and draws a meager 0.7 amp (8.4 watts). At 24 volts, it moves 1,300 cfm and draws 1.0 amp (24 watts). Ideal for greenhouse ventilation.

12V - 12 Inch Intake/Exhaust Fan

This is a smaller version of the 16-inch fan, it comes with the same reversible PM motor, has a 12-inch three-wing aluminum blade, and a 16-inch square steel venturi. It requires a 14-inch hole for mounting. At 12 volts, it moves 550 cfm and draws a meager 0.5 amp (6 watts). At 24 volts, it moves 800 cfm and draws 0.7 amp (16.8 watts).

12V Fans

These 12VDC axial fans are ideal for moving woodstove heat throughout the house. The brushless motor design minimizes electromagnetic interference and radio frequency (rf) interference. The fans have PBT plastic housings with permanently lubricated ball bearings. All motors are polarity protected. The voltage range for the nominal 12VDC fan is 6 to 16 VDC. The two smaller fans are 1 inch in depth and the larger fan is 1½ inches deep. Size shown is square.

64-211	4C909	15cfm	0.24amps	2-3/8"	$42
64-212	4C911	32cfm	0.25amps	3-1/8"	$42
64-213	4C918	105cfm	0.55amps	4-11/16"	$45

| 64-222 | 12" Intake/Exhaust Fan | $129 |
| 64-221 | 16" Intake/Exhaust Fan | $135 |

12V Ceiling Fan

We have researched the low voltage ceiling fan market over the years and carried several different brands. This ceiling fan is still the best. It can be operated on 12 volts only and draws ½ to 3/4 of an amp. It has three 44-inch wooden blades and a 10 x 8½' base which fastens to the ceiling with regular butterfly fasteners or may be screwed in to wood beams. Available in brown plastic or brass base. Pendants are not available.

64-231	Plastic Base 12V Ceiling Fan	$189
64-232	Brass Base 12V Ceiling Fan	$225
24-102	Variable Speed Control Switch	$28

Thermofor Solar Ventilator Opener

The Thermofor is a compact, solar-powered device which regulates window, skylight, or greenhouse ventilation according to the temperature using no electricity and requiring no wiring. You can set the temperature at which the window starts to open between 55° and 85°F. It will open any hinged window up to fifteen pounds a full 15 inches and can be fitted in multiples on long or heavy vents. These units are ideal for greenhouses, animal houses, solar collectors, cold frames, and skylights.

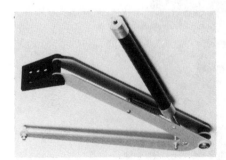

64-302 **Thermofor Solar Vent Opener** $59

12-Volt, Two-Speed Table Fan

This 8 inch two-speed oscillating fan is quiet and unobtrusive. It runs at 2,500 rpm and draws 1¼ amps at low setting, and 2 amps on high setting at 12V. It can be table or wall mounted. 12-volt only.

64-241 **Table Fan (12V)** $35

Entertainment & Electronics

Electronics items are difficult to sell in an annual catalog like our AE Sourcebook because models are changing fast as the technology continues to improve. We urge you to check with us on availability before ordering electronics products. Further, warranty problems are very difficult on electronic products. There is currently a trend in the industry to have manufacturers and their service centers deal exclusively with product warranty problems; eliminating the dealer from this loop. We have been forced to pay for certain warranty repairs and have had to reconsider our policy. The specific warranties on many products that you buy from us will be honored only by the manufacturer or their designated repair stations. For this reason **DO NOT SEND ANY ELECTRONIC ITEM DIRECTLY BACK TO US!** *We will simply return it to you and have you deal directly with the manufacturer wasting your postage and ours!* We are sorry for having to impose this policy. We carry electronics products as a convenience to our customers who are unable to procure them locally. We hope you understand.

Cordless AC/DC Freedom Phone

Our 12V/120V rechargeable Southwestern Bell Freedom Phone is the finest cordless telephone we've ever used! The leading consumer magazine rates this as the best cordless telephone on the market, bar none. It has two separate dialing pads, one on the remote handset and the other on the base unit itself. It has a hands-free speakerphone, volume control, and an 18-number memory. It comes with a built-in intercom and digital security coding. The cordless range is 1,000 feet. It's the only cordless we've ever used that doesn't have lots of static. A 12-volt power cord along with a 120V cord is supplied with the unit for recharging the handset on either 12V or 120V. *We're offering our AC/DC model for an attractive price.*

We have one of these units in our warehouse and it's the only cordless we've ever owned with no static, and never a customer complaint about it sounding funny. All around a great telephone!
J.S.

68-102 Cordless Freedom Phone   **$185**

Sony AC/DC Discman

Our Sony D180K portable Discman can be used in your car or your 12V or 120V home. It is a compact disc player with digital filtering, three-way repeat, shuffle play, and music search. It comes standard with a 12V plug and a cassette adaptor so that you can play CDs through your car stereo. It also comes with an audio system connection cable and AC adaptor. It will also run on four AA batteries. Warranty is 90 days for labor and one year for parts.

68-361 Sony AC/DC Discman   **$379**

RCA 13" AC/DC Color
TV with Wireless RemoteThis is the top-of-the-line RCA 13 inch ColorTrak TV with 24-button digital remote control. It comes standard AC or DC with a 12-volt power cord included. Gray texture finish. Size: 13-7/8"H x 14-1/2"W x 14-7/8"D. Amp draw at 12VDC is 5.5 amps. One-year warranty.

68-307 RCA 13" Color TV   **$449**

Entertainment Options

We stock varying types of 12-volt video cassette players, an AC/DC combination TV/VCP, and a basic 10 inch color TV. We don't list a specific model or brands, because availability changes so rapidly. We guarantee that we will be offering the best value that we can find on the market. **Call us to check on specific models, pricing, and specs.**

68-306	**10" Color 12V TV**	**$319**
68-309	**12" B&W 12V TV**	**$109**
68-308	**9" Color 12V TV/VCP**	**$575**
68-326	**12V VCP**	**$275**

To order any of these products
or for more technical information
call us toll-free at
1-800-762-7325

Refrigeration

The refrigerator is the most wasteful electric appliance found in a conventional household. The good news is that it is also the one that can generate the greatest energy savings, for the consumer individually and society as a whole. Unfortunately, to date there is only one small company that has taken the initiative to build refrigerators the right way.

There are two reasons a person might be interested in an energy-efficient refrigerator. First, for a home using a remote energy system, every watt of electricity produced must be savored and used as efficiently as possible. Secondly, "on-the-gridders" can save money by minimizing the consumption of costly and damaging planetary resources.

It helps to understand the workings of a refrigerator. A compressor (which produces heat as a by-product), pushes a refrigerant (freon) through condenser coils to transfer heat from inside to outside of the refrigerator. Oddly enough, conventional refrigerators locate the compressor underneath the unit, so that the waste heat rises into the food compartment. Thus, the system is constantly fighting itself and using more energy than is needed to perform the task of chilling the food. The origins of this design are legacies of the attitude that any design malfunction can be overcome by the addition of more power.

We offer several refrigerators with the compressor on top. The waste heat keeps our bodies warm rather than the food inside. Most of the models also have much more and better insulation than standard units.

With remote energy systems, it is possible but economically impractical to run a conventional refrigerator, because it would take so many more solar panels (the most expensive component of the system) to power a standard model. The investment in a super-efficient unit makes the total system cost less.

A second option for remote homes is a propane refrigerator. These can be a little less expensive than the super-efficient electric models, but have several disadvantages. One is size. The largest propane refrigerator available that we feel comfortable recommending is 8 cubic feet, half the size of a conventional refrigerator. Another is fuel. The expense and hassle of transporting propane is sometimes prohibitive. The cost of operation for a super-efficient model running on your alternative energy system is nothing!

For conservation-minded consumers, the Sun Frost refrigerator uses one-third to one-tenth the power of conventional models.

Instead of putting hot food in the refrigerator, let it cool first.

Dear Dr. Doug:

I am planning an alternative energy system. Should I buy an AC Sun Frost or a DC Sun Frost?

Answer: The Real Goods technical staff is in general agreement that there are several advantages to using a DC Sun Frost. First, the inverters available today are infinitely more reliable than the inverters of the past and we have very few problems with them. Still, they are an appliance, and as with all appliances, they stand some small chance of having to be repaired at some time. Wouldn't it be nice to still have an operating refrigerator while you are waiting for your inverter to be serviced?

Next, there is the issue of power consumption. Any time you run power through an inverter, you lose 5% to 20% due to the efficiency of the inverter. So, barring unusual circumstances (which we all know occur) we recommend using a DC model.

Dear Dr. Doug:

I plan on having an alternative energy system in the next 10 years, and I need to buy a refrigerator now. I understand that the Sun Frost is the way to go, but should I buy an AC model now and use an inverter later, or buy a DC model now and use a transformer until later?

Answer: This question comes up not only with refrigerators, but several other types of appliances. The key to the answer is that inverters are generally much more efficient than transformers. You would nullify the super-efficiency of the Sun Frost by running a DC model on a transformer. However, you could use one of the efficient Todd 14.0-volt battery chargers with a battery. Because of the above stated advantages of using DC models on independent energy systems, this would be our first suggestion. The next best set-up would be an AC Sun Frost.

Dear Dr. Doug:

Where can I get a larger propane refrigerator?

Answer: Unfortunately, the Servel (8 cu. ft) is the largest reliable propane model available. We have seen larger units, but they are of questionable quality.

Dear Dr. Doug:

Why are Sun Frosts so expensive?

Answer: Sun Frost refrigerators are expensive for a few reasons. First of all, they are custom made and of extremely high quality. The insulation they use is expensive, and there is a lot of it. Also, the cooling system is all copper on the larger units. Probably the biggest reason that they cost so much is that they make relatively few of them compared to other manufacturers. Any economist can tell you that the more you produce, the less each one will cost. So keep up the demand and maybe the price will come down!

Dear Dr. Doug:

Why should I spend so much money on a Sun Frost refrigerator for my solar system? Can't I just use the refrigerator I already have, or a less expensive one?

Answer: You would have to buy several more solar panels to run any other refrigerator. In the end, the solar panels cost more than the Sun Frost. Your only other choice is propane refrigerator, and the largest one is only 8 cubic feet.

Dear Dr. Doug:

Should I use a propane or efficient electric refrigerator for my remote home?

Answer: The choice is yours. In most cases, it is ultimately cheaper and easier to use the super-efficient electric model. Once you make your investment, there are no more added costs or hassles. Plus there is a larger selection of electric models.

Propane or electric: Which refrigerator should I choose for my independent energy home?

Our customers who pursue the off-the-grid utility-free path are often confronted with hard decisions that most utility power users have no need to consider. The means to keep food cold is one of these. Instead of strolling down to Sears and selecting the Kenmore with the color, features, or price you like, the off-the-grid homeowner must carefully weigh the energy requirements of the appliance.

Let us inspect the economic impact of three separate decisions made by the homeowner regarding refrigeration at his/her off-the-grid home.

Decision #1: You decide to buy an efficient Kenmore 16-cubic-foot refrigerator from Sears. Savings are considerable on the cost of the appliance, but the cost of the energy system to support it will be large:

Cost of appliance	$585
Kilowatt-hours/year	760
Cost of PV energy system	$5,500
5 year replacement batteries	$300
Up front cost	$6,085
Ten year cost	$6,385
cost per year	$640

Decision #2: You decide to buy two Servel propane refrigerators, each with a volume of 8 cubic feet.

Cost of appliances(2)	$2,790
Energy (propane) used gal/year	100
Price of propane/gallon	$1.50
Cost of propane over 10 years	$1,500
Up-front cost	$2,790
Ten-year cost	$4,290
Cost per year	$430

Decision #3: You decide to purchase a Sun Frost super-efficient refrigerator with a volume of 16 cubic feet (model RF-16).

Cost of appliance	$2,595
Kilowatt-hours/year	216
Cost of PV energy system	$1,225
5-year battery replacement	$175
Up-front cost	$3,820
Ten-year cost	$3,995
Cost per year	$400

(Note: this analysis assumes that all refrigerators are opened and closed the same amount. Also, the analysis does not factor in the interest costs of money.)

If you have extra room in freezer, fill it with plastic milk jugs filled with water. They will act as reservoirs and result in decreased energy use.

Sun Frost Refrigerator/Freezers

If every American home (of which there are 100 million) had a Sun Frost, our country would save 90 billion kWh every year. A large nuclear power plant generates approximately 5 billion kWh per year. This means that if every American home had a Sun Frost, we could immediately shut down 18 large nuclear power plants (or save $9 billion a year)!

The Sun Frost is a design breakthrough: for the first time it is practical to power a refrigerator on 12-volt DC. Refrigeration is typically the largest consumer of electrical energy in an energy-efficient home. Conventional AC refrigerators can draw 3,000 watt-hours (3 kWh) per day but the Sun Frost uses only 540 watt-hours (½ kWh) per day, making it by far the world's most efficient electric refrigerator. The larger units are cooled by two highly efficient top-mounted hermetically sealed compressors. The compressors and condenser are top mounted so they cool efficiently without having the waste heat re-enter the cabinet.

The walls of the refrigerator contain 3 inches of insulation and the freezer section up to 4½ inches of polyurethane foam. Frost buildup in the freezer is very slow because there is no air circulating between the refrigerator and freezer. The Sun Frost is quiet without the loud hum of conventional AC refrigerators. People have bought them before for this reason alone. Their silence is due to the design of the compact, efficient compressors.

The most popular unit, the single compressor RF-12, 12-cubic-foot refrigerator/freezer, draws 336 Watt-hours per day with a 70°F ambient temperature. Tests have shown the unit to use even less, which means that the entire refrigeration system can typically be powered from only two 48-watt PV modules! Although the initial price seems high, the Sun Frost is actually the cheapest source for electric refrigeration when amortized over a 10 year period.

All Sun Frosts are guaranteed for 2 years. Sun Frosts are available from 4 to 19 cubic feet, refrigerator only and freezer only models are also available.

There are several reasons why Sun Frost refrigerators are a worthwhile investment. For those on independent energy systems, it is clearly the best choice. If you are on the grid and only concerned about economics, the payback will vary greatly depending on how much you pay for electricity and to what you are comparing it. When weighed against the typical refrigerator already existing in a typical household, the payback time is extremely short, because efficiency has only begun to be addressed with conventional refrigerators recently. Compared to even the most efficient new models available today, the Sun Frost will in all cases pay for itself and save some money over its lifetime.

One great advantage is that all working parts

Ecologue, a consumer's guide to environmentally safe products, lists Sun Frost refrigerators as one of its ten best products. Sun Frost refrigerators have been shipped to over 50 countries.

can be easily accessed, serviced, and obtained for replacement if necessary. This extends the lifetime of the unit beyond others where the working parts are hidden underneath and difficult to fix.

Have you ever noticed that when you put something like broccoli, carrots, or romaine lettuce in your refrigerator, it becomes limp in just a few days (if that long)? The only way to prevent this is to use airtight containers for every article of food, which keeps oxygen from the food and decreases shelf life. The reason for this is that conventional refrigerators take moisture out of your food (and the air in the refrigerator) and convert it to frost in the freezer. Sun Frost has conquered this problem and maintains high humidity inside, with no air transfer between compartments. The results are much less frost and wilting, you don't have to use airtight containers, and your food will store longer.

The Sun Frost is not completely frost-free, although it creates *much* less frost than conventional models. The amount is somewhat similar to that of a partially automatic defrosting refrigerator, and the unique defrost system requires only a few minutes of labor.

In the list below, R stands for refrigerator, F for freezer. The RF19 has freezer and refrigerator compartments of equal size. The watt-hours per day energy consumption figures assume a 70°F ambient temperature, increase by about 50% for a 90°F ambient temperature. All models are 34½ inches wide and 27½ inches deep, height given by model.

All models are available in 12, 24, 120, and 220 volts. AC models have AC brushless compressors. DC models have DC brushless compressors. The entire cooling system contains only one moving part. The wattage is virtually the same for the various voltages.

Sunfrosts come standard in white but there are a wide array of colors available, almost any color of Formica or Navamar. For any color other than white there is a $100 charge. Sunfrost is also available with a wood finish of either birch or oak for an extra $150. Call for more details.

All Sun Frosts are made to order—allow approximately 6 weeks for delivery. Be sure to specify if you want the hinge on the right or the left and the voltage!

■62-134	RF-16	540 W-hr/day	62.5"	$2,595
■62-134AC	RF-16 - 120VAC		62.5"	$2,395
■62-135	RF-19	744 W-hr/day	64.0"	$2,795
■62-135AC	RF-19 - 120VAC		64.0"	$2,595
■62-125	R-19	360 W-hr/day	64.0"	$2,450
■62-125AC	R-19 – 120VAC		64.0"	$2,345
■62-115	F-19	1200 W-hr/day	27.5"	$2,795
■62-115AC	F–19 – 120VAC		64.0"	$2,645
■62-133	RF-12	336 W-hr/day	49.5"	$1,995
■62-122	R-10	180 W-hr/day	43.5"	$1,645
■62-112	F-10	660 W-hr/day	43.5"	$1,745
■62-131	RF-4	156 W-hr/day	31.5"	$1,395
■62-121	R-4	108 W-hr/day	31.5"	$1,395
■62-111	F-4	336 W-hr/day	31.5"	$1,395

05-215 Crating Charge (all models)   add $60
Shipped Freight Collect from No. California

Consider this: A Sun Frost RF-16 uses approximately 180 kWh of electricity per year compared to approximately 1080 kWh per year for a standard "energy saving" refrigerator. **This means the Sun Frost saves 900 kWh per year.**

Servel Propane Refrigerators

Servel has been a household word in gas refrigeration for over 50 years, when it was first marketed by the Swedish company Electrolux in 1925. In 1956, the rights to Servel were acquired by Whirlpool, which was unsuccessful with the unit, and it disappeared for 30 years. Now Servel is back with the state-of-the-art household propane refrigerator.

We've sold well over 800 of the new Servels since their reintroduction in 1989 and have one hooked up in our showroom. All continue to perform flawlessly and appear to be extremely well built. They draw less than 1½ gallons of propane per week! The body is all-white and the door is hinged on the right. The unit comes with four refrigerator and two freezer rust-proof racks. The spacious interior has two vegetable bins, egg and dairy racks, frozen juice rack, ice cube trays, and an ice bucket. An optional 4-year warranty can be purchased for $59.95 from the manufacturer. The total volume is 7.7 cu. ft. (6 for the refrigerator, 1.7 for the freezer). Total shelf space is 9.9 square feet. The overall dimensions are 57¾"H x 24¾"W x 24¾" deep. Net weight is 181 lb. It is operable on either propane or 120VAC (draw: 275 watts continous).

■62-300 Servel Propane Refrigerator $1,395
05-212 Add'l freight west of the Rockies $50
Shipped freight collect from California or Ohio

Dometic RC-65 Propane Deep Freezer

Dometic also makes kerosene powered refrigerators. Call for details.

The Dometic RC-65 is a very dependable deep freezer that gives efficient, trouble-free performance. Absolutely no electricity is required for this propane-powered unit, which consumes approximately 2-3 gallons of propane a week. It comes with thermostatic control, a convenient defrosting drain system, and a childproof lock. It also has a built-in spirit level to make sure the freezer is properly leveled, a thermoelectric safety valve for gas supply, a built-in easy to use flame-igniter, and a stainless steel wear strip for scratch-free loading and unloading. The RC-65 may also be converted, in less than 10 seconds, to a beverage cooler. Capacity is 5.5 cubic feet (155 liters). It measures 40.9"H x 37.4"W x 30.1"D.

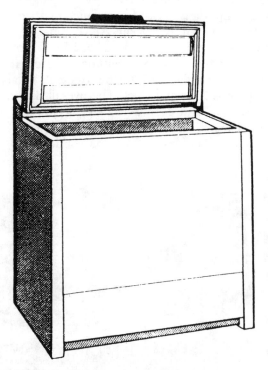

■**62-512 RC-65 Propane Freezer $1,895**
Shipped Freight Collect from Calif. or Ohio

Norcold Refrigerators

Norcold manufactures efficient refrigerators that run on 12VDC and AC/DC. The core of their 12V refrigerators is the 40-watt compressor unit manufactured by Sawafuji for Norcold for the last 25 years. While not as efficient as Sun Frost 12V refrigerators, they are more versatile and less expensive.

DE-251 Norcold

The 2.0-cubic-foot DE-251 is the best selling refrigerator in the entire Norcold line. It has one hermetically sealed 40-watt compressor. It is designed to be built in so the sides and top have been left unfinished, allowing you to slip it under a counter, into a closet, or into a wall. It draws 3.75 amps for 12 hours per day (540 watt-hours per day) and measures 20-1/4"H x 17-3/4"W x 20-3/8"D excluding door and flange, and weighs 56 lb. It is available as an AC/DC unit as the DE-251 or can be purchased as 12V only as the DC-254 (for cheaper!). One-year warranty.

■**62-431 Norcold DE-251 (AC/DC) $545**
■**62-432 Norcold DC-254 (12VDC only) $475**
Shipped Freight Collect from Ohio

**To order any of these products
or for more technical information
call us toll-free at
1-800-762-7325**

DE-704 Norcold

The DE-704 is an AC/DC 3-cubic-foot refrigerator that draws 3.75 amps approximately 12 hours per day (540 watt-hours per day). It has a cross-top freezing compartment. The cabinet is unfinished for built-ins. The unit measures 32-7/8"H x 22"W x 21-7/8"D excluding door and flange and weighs 85 lb. One-year warranty.

■62-420 Norcold DE-704   **$775**
Shipped Freight Collect from Ohio

DE-560 Norcold

The DE-560 is an AC/DC 6.2-cubic-foot, double door refrigerator/freezer. It has a 60-watt compressor and draws 7 amps for 12 hours per day (1008 watt-hours per day). The refrigerator has three removable shelves, twin crisper, and adjustable in-door storage. The big 40-lb capacity freezer also has in-door storage and two ice trays. The unit measures 52-1/2"H x 23-1/18"W x 23-1/4"D and weighs 132 lb. The cabinet is unfinished for built-ins. One-year warranty.

■62-410 Norcold DE-560   **$1,045**
Shipped Freight Collect from Ohio

MRFT Series Norcold AC/DC Refrigerator or Freezer Combinations

The MRFT series can be switched back and forth between refrigerator and freezer mode. They are chest-type units in charcoal gray color. The MRFT-630 is a 1.06-cubic-foot unit measuring 14-3/4"H x 25"W x 14"D, with a 50-lb capacity, and will hold two cases of beverages. The MRFT-640 is a 1.5-cubic-foot unit measuring 18-7/8"H x 24-13/16"W x 14-1/8"D, with a 75-lb capacity, and will hold three cases of beverages. Both the 630 and the 640 use a single 40-watt compressor and draw 540 watt-hours per day. The MRFT-660 is a 2.15-cubic-foot unit measuring 17-3/8"H x 31-1/8"W x 19-1/4"D, with a 100-lb storage capacity. It has a larger 60-watt compressor and draws 720 watt-hours per day. All units can be switched back and forth from refrigerator to freezer mode with the switch of a dial. One-year warranty.

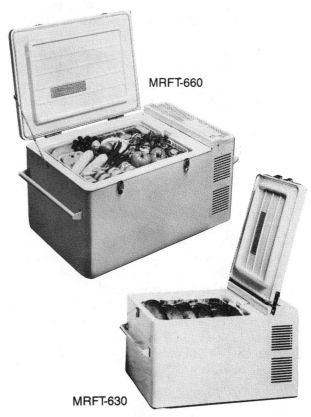

MRFT-660

MRFT-630

■62-441 Norcold MRFT-630   **$665**
■62-442 Norcold MRFT-640   **$745**
■62-443 Norcold MRFT-660   **$945**
Shipped Freight Collect from Ohio

Koolatron 12V Refrigerators

The Koolatrons are designed to plug into a vehicle on the way to the beach. For your home, you can run a Sun Frost RF 12 for the same amount of power a Koolatron requires.

Koolatron (of Canada) is the world's largest manufacturer of 12-volt coolers. The secret of their cooler is a miniature thermoelectric module that effectively replaces bulky piping coils, compressors, and loud motors used in conventional refrigeration units. For cooling, voltage passing through the metal and silicon module draws heat from the cooler's interior and forces it to flow to the exterior, where it is fanned through the outer grill. The Koolatron has a relatively low amp draw for a refrigerator, averaging 2.5 amps continous(maximum draw is 4.0 amps).

The unit will maintain approximately 40–50° F below outside temperature. It has an interior temperature indicator as well as a thermostat and can be used on 12VDC or adapted to 120-VAC (*by ordering the optional adaptor*). The unit is constructed of high impact plastic and super urethane foam insulation, and comes with a 10-foot detachable 12V cord. The Traveller II is 9"W x 12"H x 13"L, 0.25 cu. ft., holding 9 qts. and is warranted for one year. The Scotty II is 16"W x 15½"H x 18½"L, 0.9 cu. ft. holding 32 qts. and has a five year warranty on the cooling module, and the Caddy II is 16"W x 16"H x 21"L, 1.2 cu. ft., holding 36 qts. and has a 10-year warranty on its cooling modules.

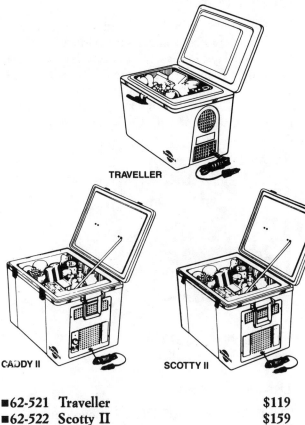

TRAVELLER

CADDY II   **SCOTTY II**

- ■62-521 Traveller $119
- ■62-522 Scotty II $159
- ■62-524 Caddy II $299
- ■62-525 AC Adaptor (for all models) $59

Mariner 12V Ice Cube Maker

Mariner 12V ice cube makers assure ice at all times. They are fully automatic with no special startup required. The model SFI-55 makes up to 13 lb per day (500 ice cubes) and stores them. The unit measures 24½"H x 14"W x 16½"D. It will run on either AC or 12VDC where is uses 6 amps at 12VDC and 2 amps at 120VAC to harvest the ice and only 1½ amps to maintain the ice cubes. It will harvest around one pound of ice per hour.

■62-531 SFI-55 (AC/DC) ice maker $1,195
Shipped Freight Collect from California

Water Pumping

If this were an ideal world we would all live in homesteads with a sweet flowing spring 50 feet to 100 feet higher up the hill. We'd pipe the water down to the house, Mother Nature providing water pressure free of charge. In this less-than-perfect world, however, we have to face the realities of pumping water.

Water is heavy stuff, 8.34 pounds per gallon. To lift and move this mass can absorb great amounts of energy. An essential formula is the relationship between elevation and water pressure: 1 psi (pound per square inch) equals 2.31 feet of lift. For instance, if you have a water tank 23 feet above your house it provides approximately 10 psi. "Normal" household water pressure is around 20 to 40 psi (although many folks get by with less, or more). A water tank 50 feet above your house provides 21.6 psi. (50 feet divided by 2.31 equals 21.6.)

Water is heavy stuff, 8.34 pounds per gallon.

Batteries are to be avoided in pumping systems whenever possible. When energy is run into and out of a battery, 25% is lost. It's better to take energy directly from a power source (PV modules, wind, hydro generator) to move water than to waste this efficiency with batteries. At the end of the day, you'll end up with 25% more water in the tank. You should install the largest storage tank that is practical. If your system has limited tank capacity, a battery system may be advisable.

In planning a photovoltaic water pumping system, the most efficient method is to connect the modules directly to the water pump. When the sun shines, water is delivered to a large storage tank or cistern. Household water pressure can then be supplied by gravity if the storage tank is 50 feet or higher than the house. Or if gravity pressure is inadequate, a separate battery operated pressure booster pump can increase the pressure. In some cases, due to climate, topography, or lot size, storage tanks are not practical. In these cases it is worth sacrificing efficiency to use the household battery bank to pump water into large pressure tanks, allowing pressurized water to be available 24 hours a day.

There are two basic pump classifications: centrifugal, and positive displacement. Within each classification are a variety of designs.

Centrifugal Pumps

Centrifugal pumps produce high gallon volumes with smooth, even flow, but at a low efficiency. These are the most common pumps in the AC world, because they are inexpensive and very low maintenance. A fan-shaped impeller spins the fluid out of the pump using centrifugal force. (Because they rely on centrifugal force to move the water, there is a great deal of "slippage" internally, resulting in low efficiency.)

Common centrifugal pumps are AC submersible pumps, pressure booster pumps, jet pumps, and swimming pool circulation pumps. We carry centrifugal pumps for hot water (solar) circulation systems. The gasoline-driven Tanaka Pressure pumps are also centrifugal.

Positive Displacement Pumps

Because energy is precious, and we want to wring the maximum output from each watt, most of the pumps we sell are positive displacement types. These pumps move fluid by forming a tight mechanical seal around a segment of fluid, and then expelling the fluid under pressure by mechanical means such as piston stroke, gear, or rotary vane. These pumps are highly efficient, but produce lower flow rates, and may introduce buzz or vibration into the plumbing system because they release the water in rapid discrete segments.

Common positive displacement pumps are diaphragm pumps, rotary vane pumps, piston pumps, and jack pumps.

The Slowpumps and the Flowlight Booster are rotary vane type pumps. Output is fairly smooth and quiet. They are capable of very high lifts, and in the case of the Booster pump,

For each psi of outlet pressure in a gravity system, 2.3 feet of head is necessary.

Flow: The measure of a pump output in gallons per hour (gph), or in gallons per minute (gpm).

Foot Valve: A type of check valve installed at the pump intake, usually with a strainer. Prevents loss of prime when the pump is higher than water source.

Prime: A charge of water required to start pumping action. Centrifugal pumps will not self-prime. Positive displacement pumps usually self-prime with a free discharge.

Head: A
measure of
pressure, the
vertical
difference
between the top
and bottom of
a water column
expressed in
feet. Pump
output may be
stated as head
or lift. Divide
head in feet by
2.31 to get
pounds per
square inch
(psi). For
example: 100
ft. of head
equals 43.29
psi. Conversely,
if you know the
psi a pump will
deliver,
multiply by
2.31 to get the
head.

Suction Lift:
When the
water source is
lower than the
pump, suction
lift is the
difference.
Theoretical
limit is 33 feet;
practical limit
is 10 to 15
feet, sometimes
less depending
on pump type.
Pumping action
creates a partial
vacuum and
atmospheric
pressure forces
water into the
pump. Suction
lift capability
of a pump
decreases
approximately
1 foot for every
1,000 feet
above sea level.

Static
Discharge
Head: The
vertical distance
from pump to
point of
discharge.

quite high flow rates. Their major drawbacks are that they are intolerant of dirt and abrasives and cannot be run dry. Service can only be done at the factory. High-quality filtration is a must!

---

*Energy is precious.
We want to wring the maximum
output from each watt.*

The Real Pump, the Solaram, the Solar Submersible series, and the gasoline driven Hypro series are all diaphragm type pumps. Diaphragm pumps work by driving an elastic diaphragm up and down with a piston. Water is forced in and out by a pair of check valves. One valve only lets water in, the other only lets water out. The efficiency of these pumps is very high; they can be run dry without damage and are tolerant of dirt and debris, though sometimes the check valves will hang up on a piece of dirt. Most of these pumps are easy to service in the field. The output tends to be fairly "buzzy".

Jack pumps are in a class by themselves. This is the type of pump that has seen use in windmills and hand pumps for generations. Most of the working components are surface mounted for ease of maintenance. The pump works by stroking a piston up and down inside a cylinder in the well. There are check valves top and bottom to allow water in and out. These systems are expensive, but life expectancies are extremely long and maintenance is minimal. Jack pumps can lift from as deep as 1,000 feet or, in lower lift situations, can provide as much as 10 gallons per minute flow. Jack pump systems must be sized and priced individually, as they are custom engineered for each installation.

Surface vs. Submersible Pumps

Pumps can be further classified as surface or submersible types. Surface pumps generally need to be mounted in a dry, secure location, no higher than 10 vertical feet above the water. (In some cases we can stretch this vertical rise, or suction lift, to 15 feet with very careful

attention to tight inlet plumbing and a high-quality foot valve, but problems with lost prime and poor pump performance may result.) Acceptable suction lift loses approximately 1 foot for each 1,000 feet of altitude.

Examples of surface pumps we carry are The Real Pump, Slowpumps, the Solar Force, the Solaram, and Flowlight Booster.

Submersible pumps are used when the pump cannot be situated within 10 feet of the water level, or when the pump has to be protected from freezing.

Examples of submersibles we carry are the Solar Submersible pump, and technically the Solarjack jack pump and the Bowjon Wind-Powered pump. These last two pumps have most of the working equipment at the surface for ease of maintenance, with only the actual pump submerged.

Sizing Your Surface Pump
(for solar direct applications)

• **Determining Needs** First, you need to know the total **vertical** lift necessary and the total gallons per day (gpd) you need. Water needs will vary from day to day and seasonally. It's best to size your pump generously unless you have a back-up source of power or water. Gallons per day is divided by hours of available peak sunlight at your pump site.(*For example, in the Central-Western USA, April thru September, assume an average of 7 peak sun hours per day. If using a tracking mount, assume 10 peak sun hours.*) This will give you the flow rate in gallons per hour you'll need for your pump. Armed with these two figures, your *vertical lift* and *gallons per hour* required, you are ready to wade into selecting a particular pump.

• **Pump Selection** Most of our pumps have performance charts which give you details of gallons per hour (sometimes gallons per minute, multiply by 60 for gallons per hour) at specific lifts. Find a pump model that fulfills your needs (and hopefully your budget).

• **Solar Panel Sizing** The pump performance chart will also give wattage required by the pump at the lift and flow rate you've selected. For PV direct systems (which are the most efficient) the total wattage of the PV array should exceed the wattage specified in the chart by at **least** 20% and up to 30%. This is necessary to derate the modules for the "real world". Modules are rated under perfect laboratory conditions that are rarely duplicated under real working conditions. A Linear Current Booster will prevent your pump from stalling under marginal light conditions.

• **Pump Voltage** Many pumps are available in a variety of voltages. If your pump is going to be run off your household battery bank then your options are limited. Generally speaking, the higher the pump voltage the less power is lost in transmission. However to run 24 volt pumps you need to buy PV modules in pairs, 48 volts requires sets of 4 modules (wired in series). This may not work out well with the required wattage to run the pump. Slowpumps are available in 90-120 volt AC or DC models by special order for very long wire runs.

Dear Dr. Doug:
I've got a well with an AC powered 1/2 hp submersible pump that I've been running off my generator. Can I get a large inverter and run this pump off my battery bank?

Answer: What you propose is possible; however, it'll cost approximately three times as much energy consumption as a Solar Submersible would for the same amount of water. Energy-efficiency is not a primary design goal for AC pump engineers. I only endorse a system like this as a temporary set-up. If you have a 6 inch or larger casing it's possible to install both your existing AC pump and an efficient DC pump "piggyback" in the same well. The smaller drop pipe from the DC submersible will fit around the side of the AC pump above it. This gives you the best of both worlds: you get efficient DC water pumping for 90% of your needs, but still have the ability to fire up the generator and AC pump when you want lots of water fast.

Dear Dr. Doug:
It appears from the performance charts that your Real Pump will deliver the same amount of water as the Slowpumps with about two-thirds the power input at any given lift. But the Slowpump costs almost twice as much as the Real Pump. Can this be? Are you guys telling the truth about the Real Pump power consumption?

Answer: Honesty is the cornerstone of Real Goods' business. The Real Pump really *is* that much more efficient than the Slowpump. The Real Pump is also easy to rebuild in the field. When you need higher volumes you can parallel the Real Pump, or perhaps the Slowpump would be a better choice.

Dear Dr. Doug:
I need to pump my water almost 1,000 feet of vertical lift. Can I use several pumps like the Real Pump in series to accomplish this? Or do I have to use one monster pump to do the whole thing in one jump?

Answer: As long as we're talking about a surface pumping situation, the Real Pump is ideal for multiple series lifts. Pumps and panels can be spaced every 150 feet to 300 feet vertically depending on how many gallons per minute you want to produce. Be sure that all the PV panel arrays get equal sun exposure.

Dear Dr. Doug:
I've got a well that's cased with 6 inch PVC. The standing water level is about 40 feet down. Your new Real Pump looks like the most efficient one you carry, but you don't give the diameter for it. I'd like to lower a pump down the well and suspend it above the water.

Answer: Bad idea! You can fit some of the surface pumps down a 6 inch well casing, but it isn't recommended. First, none of these pumps are submersible (instant death there); second, the pump must be within 10' of the water level, if the level changes the pump either loses suction or gets dunked. Third, and most important, it's a 100% humidity environment down there and corrosion is a very serious problem. Life expectancy is severely reduced. Bite the bullet and buy a **real** submersible.

Total Lift or Total Head: The sum total of suction lift, static discharge head, and friction losses. With pressurized systems, multiply system pressure times 2.31 and add any vertical lift above water level for the total lift.

Dear Dr. Doug:
I'm contemplating drilling my own well with one of the inexpensive do-it-yourself units. Problem is, I'll have a 2 inch casing when finished, and I don't see anything in your catalog for 2 inch casings. Why not?

Answer: A 4 inch casing is the smallest that you'll find for submersible pumps. It just isn't technically possible to build a motor and water pump that will fit down a tiny 2 inch pipe. With a 2 inch casing you're limited to a jet pump system or possibly a jack pump. Jet pumps are the least efficient pump you can possibly use. They are **not** recommended for alternative energy systems, unless you're on hydro power and literally have power to burn. Jack pumps are highly reliable, but very expensive. Put in a 4 inch casing at the minimum.

Dear Dr. Doug:
I've got an installation where the pump is only lifting about 40 feet vertically, but I've got 2,500 feet of pipe run to do it. How much lift do I size the pump for?

Answer: You can size your pump for 40 feet of lift, so long as your 2,500 feet of pipe is large enough to prevent friction loss. For most of the small PV-powered pumps we sell, 3/4" or 1" pipe is fine. The pump doesn't "see" horizontal run unless the pipe is too small and is constricting the flow. (See the friction loss charts in the Appendix.)

Friction loss charts are on page 492.

Dear Dr. Doug:
I live way out in the Alaskan bush and whatever pump I buy needs to be field serviceable. Which of your pumps meet my needs?

Answer: These are pumps you might want to consider:

The Real Pump: Four screws disassemble the pump head for easy access to the diaphragm and check valves. We carry all repair parts.

Slowpump Series: This pump head unbolts easily from the motor, but must be returned to the manufacturer for service. Not a good choice unless you keep a spare pump head on hand.

Solar Force Piston Pump: Comes complete with an extensive spare parts and service kit. An excellent choice.

Solaram Surface Pump: This is a diaphragm-type pump which is easy to service in the field. A good choice.

Flowlight Booster Pump: Same as the Slowpump above.

Solar Submersible Series: These can be easily field-serviced and even come with the rebuild kit packed inside the pump casing. What a great idea!

Solarjack Jack Pump: Can be entirely field-serviced. Most of the working components are on the surface for easy access.

High Lifter: Very easy to field-service. All seals are O-ring type and simple to replace.

Hypro D19 and D30 Pumps: Diaphragm types again. Easy to field-service.

Shallow-Source Water Systems

Here is another in a series of articles by Jon Vara, a customer from Vermont. This edition of the Real Goods AE Sourcebook features his articles on "Shallow Well Pumping Systems," and "Deep Well Pumping Systems." We think you'll find the articles on water pumping extremely educational and an excellent prelude to our pumping products.

Consider the ultimate low-energy water system: A house sits on a flat bench in a sunny, south-facing slope. Behind it, a forested hillside rises and steepens. About an eighth of a mile from the house, and a hundred vertical feet above it, a clear, cold spring bubbles out of the ground. The homeowners have enclosed the spring in a concrete casing and covered it with a concrete cap to keep out frogs, snakes, mice, and other surprises. One-inch black polyethylene pipe carries the spring water down the hill to the house; ample water pressure is provided by the hundred-foot drop, with no need for a pump. The system is quiet, inexpensive, and as reliable as gravity.

Unfortunately, even in hilly, well-watered areas—such as my home state of Vermont—such ideal water systems are the exception. Most gravity-powered water systems lack sufficient vertical drop, or head, to dispense with pumping altogether.

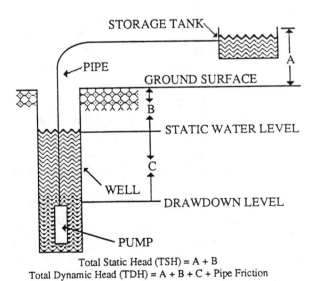

Total Static Head (TSH) = A + B
Total Dynamic Head (TDH) = A + B + C + Pipe Friction

Here are some significant numbers: Typical household water pressure runs between 20 and 50 psi. At 50 psi, water comes blasting out of the tap. Hold a glass loosely under the faucet, turn it on rapidly, and the force of the stream may knock it out of your hand. At 20 psi, the flow is much gentler; water pressure that falls much below 20 psi will seem sluggish and inadequate to most people.

The pressure tank enables the pump to remain off for much of the time.

For each psi of outlet pressure in a gravity system, 2.3 feet of head is necessary. The hypothetical water system described above, with its hundred-foot head, would have water pressure of 100 divided by 2.3, or about 43 psi. (For the sake of simplicity, I'm ignoring head loss, the reduction in effective head that results from friction between the piping and the water it contains. The smaller the pipe, the greater the velocity, and the longer the run, the more severe the head loss. If large enough pipe is used, however, head losses will be insignificant. Most books on plumbing contain head-loss tables for various sizes and types of pipe.)

Because relatively few homes can tap into a hundred feet of head, though, partial gravity systems are more common. Let's relocate our hypothetical spring further down the hill, so that it sits only 10 vertical feet above the house. That will reduce the pressure of the incoming water to about 4 psi—not enough to provide usable pressure at the tap. The most economical method of stepping up the pressure, in most cases, is to install some sort of booster pump.

In a booster pump system, the low-pressure, gravity-supplied water is piped into the inlet of a small pump, which forces it out under higher

Here are some comments from Windy Dankoff (Flowlight Solar Power) after reading Jon Vara's article: "Under 20 psi can deliver plenty of flow. It is flow that you observe and benefit from at the water spout, not pressure. But you must have oversized plumbing to deliver satisfactory flow. I recommend low-pressure gravity flow systems use one or two sizes larger than "normal" piping throughout the system, and possibly drilling out larger holes in some shower heads. (Do not use water conserving shower heads). Toilets and washers will fill slowly but will work fine. Most tankless water heaters will not work properly, without a booster pump. Some people use an inexpensive booster pump just on the line to the heater." - Windy Dankoff

Water pump powered by wind and PV modules in Mendocino County, California.
Photo by Sean Sprague.

pressure. This high-pressure water may be piped directly from the pump to the faucets and other outlets, but ordinarily it first enters a pressure tank.

As the high-pressure water is forced into an inlet fitting in the bottom of a steel tank—which may have a capacity of anywhere from 2 or 3 gallons to 80 gallons or more—the air trapped in the open space above is compressed into a smaller and smaller volume. When the pressure has risen to a pre-set level—typically 40 psi or so—an inexpensive pressure switch cuts off the power to the pump. A one-way check valve, which may be built into the pump or installed in the pressure pipe between the pump and the tank, prevents the stored water from flowing backward through the pump.

When you turn on a faucet to fill the tea kettle, the pressurized water emerges from the outlet (also located somewhere in the bottom of the tank) and, pushed by the expanding compressed air, travels through the household plumbing and out the open tap. The water will continue to flow, without any help from the pump, until the tank pressure falls to 20 psi or so, at which point the pressure switch causes the pump to kick on again, recharging the tank.

The pressure tank, in other words, enables the pump to remain off for much of the time, and, when it *does* cycle on, to run for at least as long as it takes to recharge the pressure tank. In the absence of a pressure tank, the pump would have to switch on every time you ran a glass of water, increasing wear and tear on the pump and pressure switch and wasting energy in the process, since electric pumps require a substantial starting surge each time they cycle on.

Several types of low-voltage pumps work well in booster pump-systems. The top-quality option is the Flowlight Booster Pump, available in either 12 or 24 volts. The Flowlight pumps are beautifully made and will last forever if not abused. They are very quiet an important consideration in a pump that is to be located indoors, as booster pumps usually are. Their only disadvantages are their relatively high cost, and their susceptibility to damage from particles of sand, grit, or sediment. The incoming water must be drawn through a cartridge-type sediment filter before it enters the pump, and the filters replaced several times a year.

A cheaper option is a marine or RV-type pump, such as the Shurflo. Both are diaphragm pumps, and far more tolerant of particulates than the vane-type Flowlight pumps. A filter is usually not necessary. Where large chunks of foreign matter may be present—if water is to be drawn directly from a stream, for example—a 40-mesh intake strainer is recommended.

The diaphragm pumps typically put out somewhat less than 2 gallons per minute, far less than the Flowlight Booster. When the pressure tank is fully charged, that makes little difference, but when the tank is depleted and the pump is supplying water directly to the house, as when someone is taking a long shower, turning on the kitchen faucet will cause the flow in the bathroom to fall to a dribble.

A second drawback to the diaphragm pumps is the loud, irritating buzz they emit when operating. If located indoors, they should be placed in a location that can be effectively soundproofed.

In the pressure-tank booster system just described, water is delivered to the booster pump by gravity. But what if (as is more than likely) your water source is located below the house, rather than above it?

If it is not too far below the house, no important changes are necessary. Both the vane-type Flowlight and the lighter-duty RV pumps have the ability to pull water uphill, as you would sip iced tea through a straw, and force it into a pressure tank at the same time. That makes it possible to draw water directly from a shallow well, or a buried rainwater cistern.

The catch is something called the "suction limit," which is the maximum height to which a given pump will pull water. The suction limit of the Flowlight Booster, for example, is about 17 feet. The diaphragm Shurflo pumps will pull water to about half that height. (Those are the figures at sea level; because air pressure falls with altitude the suction limit of any pump decreases by approximately one foot with each thousand-foot increase in elevation.)

In theory, the horizontal distance between the water source and the house is far less important, since water is much easier to move sideways than it is to lift (assuming, again, that the piping is large enough to keep head losses reasonably low). In practice, however, a pump cannot be expected to move water more than, say, a hundred feet or so by suction, even if the lift involved is well below its suction limit.

The problem is that any air bubbles in the intake line will stop a centrifugal pump dead in its tracks. Unless the suction line to the pump can be sloped uniformly uphill, enabling trapped air to rise out of the line of its own accord, the bubbles will congregate at any localized high points and cut off the flow. Where water need not be lifted far, but must be moved a great distance horizontally, it's best to locate the pump at the water supply, and have it force the water by pressure, rather than pulling it by suction, since air bubbles in a pressure line do no harm.

That, however, brings up a different problem, which is the difficulty of supplying low-voltage electricity to a remote pump. My own water, for example, comes from a shallow well only a few feet lower than the house, but nearly 800 feet west of it. Transmitting 12- or 24-volt power to a pump at the well would require thousands of dollars worth of heavy-gauge wire.

The solution is to run a 120-volt AC pump through an inverter. Standard 120-volt centrifugal pumps, however, are inefficient (that is, they move relatively little water for the amount of power they consume), and can only be run with a large, expensive inverter, such as the Trace 2012. Instead, I chose an inexpensive 120-volt pump that is easily powered by my small 300-watt Heart inverter. That made it possible to run ordinary, inexpensive #10 direct-burial wire from the house to the pump, in the same three-foot-deep trench that protects the plastic water line from the frost. The pump itself sits on a shelf within the well casing, a few feet below the surface of the ground, and hence safe from freezing, but well above the high-water level.

The higher-quality Flowlight pumps are also available in 120-volt. I chose the cheaper, less durable pump because I use relatively little water, and because I suspected that given the remoteness of the pump, I might not get around to changing the inlet filter on a Flowlight as often as necessary.

Those are the basics of shallow-source water pumping. If you are lucky enough to live where good water lies within easy reach of the surface, I hope this information will enable you to divert some of it into your kitchen and bathroom.

—*Jon Vara*

How Many Watts Do I Need for Solar Pumping?

Here is a handy formula for estimating watts required for a given pumping task. It allows you to predict power requirement even if you don't know what type of pump to use:

$$\text{Watts} = \frac{\text{Feet x GPM}}{.053 \text{ x \% Pump Efficiency}}$$

Pump efficiency (wire-to-water) is 30–50% for most solar pumps and generally highest for the higher volume varieties.

Windy Dankoff, of Flowlight responds: "It is not correct to generalize that a Flowlight Booster Pump or Solar Slowpump requires several filter changes per year. If there is no sediment, it will never require changing. I haven't changed my booster pump cartridge in over a year.

I would emphasize that filter maintenance is absolutely dependent on water quality! When the pump starts to make noise, it is time to change the cartridge, so it is not necessary to pull pump from a casing to check filter, either—just listen!"

Still have questions? See our Solar Water Pumping Booklet on page 295.

Solar Water Supply Questionnaire

In order for us to thoroughly and accurately recommend a water supply system for you, we need to know the following information about your system to the best of your knowledge.

To estimate your daily water needs, see the charts on page 493 of the appendix

NAME:

ADDRESS:

PHONE:

DESCRIBE YOUR WATER SOURCE:

Depth of well:
Depth to water surface: If water level varies, how much?
Estimated yield of well (gallons per minute):
Size of well casing (inside diameter):
Problems? (silty water, corrosive, etc.):

WATER REQUIREMENTS:

Irrigation: gal/day required: which months of year:
Is gravity flow/low pressure OK?

Domestic: gal/day required:
Is this your year-round full time home?
Is house already plumbed?   Conventional? - describe.

Livestock watering: gal/day required: which months of year?

DESCRIBE YOUR SITE:

Distance from well to point of use:
Elevation above sea level:
Vertical rise or drop from top of well to point of use:
Can you easily locate a storage tank higher than point of use?
    How much higher?   How far away?
Complex terrain, or multiple usage?--Enclose map to describe site.
Do you have utility power at site?   How far away?
Can you connect well pump to nearby PV home/battery system?
Is home PV system present or proposed?   Describe(voltage, etc.):
Distance from home system to well:

DESCRIBE EXISTING EQUIPMENT - Energy system, pumping, distribution, storage, etc.

How effective is your present system?
Do you have a specific budget in mind?
Do you have a deadline for completion?

The Automatic Multi-fixture (Diaphragm) Water Pump System

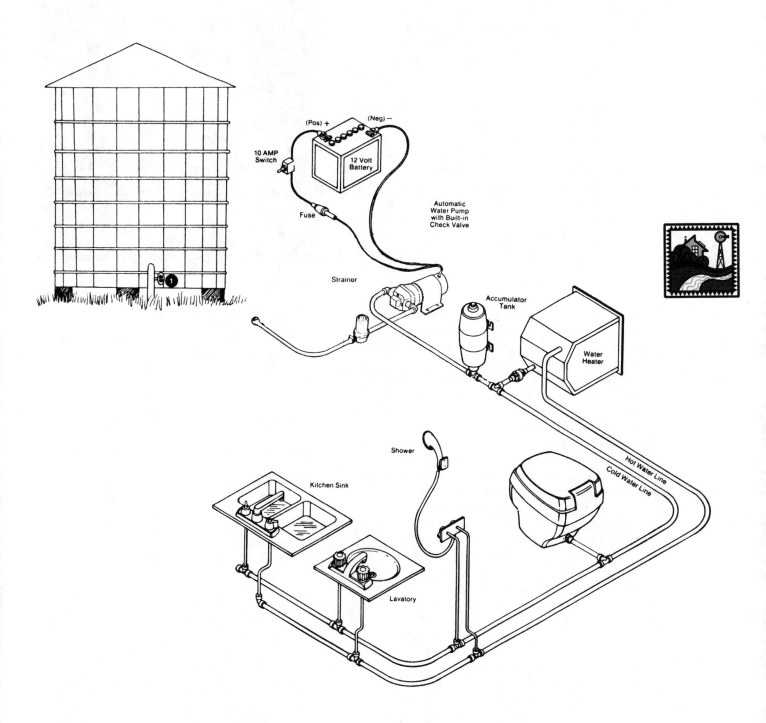

The automatic multi-fixture pump system (shown above) delivers water the instant a faucet is opened, just like home. Pump starts automatically when a faucet is opened or a toilet is flished. When all water outlets are closed, pressure in the discharge side of the pump rises to shut the pump off automatically. Pump draws water from a non-pressurized water tank. Standard household fixtures are used throughout. The heart of the system is the automatic water system pump.

The Real Pump

The Real Pump is a major breakthrough in one of the most challenging aspects of off-the-grid life; water logistics. It will pump 30% more water with the same wattage, at less than half the price of all its competitors! Co-developed by Real Goods and SunTronics, this pump is as versatile as it is heavy-duty. It will tolerate dirty or silty water, running dry, and is literally built to military specifications and capable of continuous duty.

It is designed for pumping surface water from springs, ponds, tanks, cisterns, rivers, streams, or shallow wells. It will pump up to 300' of vertical lift (130 psi), or can be used for pressurizing water systems. Suction lift capacity is 20' at sea level with open discharge. (Subtract 1 foot for every 1000 ft. above sea level.) The Real Pump is **not** submersible and not recommended to suspend in any well casing smaller than 24". (High humidity in well casings causes corrosion problems.)

The Real Pump will run on 12 or 24 volts (performance is best on 24V). The motor is a low wattage, high output, 30 volt military spec design that is more efficient than anything else available. The Real Pump can be powered either PV direct or from batteries. In PV direct applications a Linear Current Booster will allow earlier start-up and later shut-down each day, and will help the pump start in marginal cloudy conditions. In an emergency even a vehicle battery will give many hours of run time. Brush life is estimated at 4 to 5 years under **continuous** duty. If you ever manage to wear out a set they are externally serviceable and we carry them in stock (see the Accessories section).

The pump head is a positive displacement four-chamber diaphragm type that can pump dirty or silty water and will not be damaged by running dry. Simple filtration is recommended however since occasionally a grain of sand or strand or algae will stick under a check valve limiting or stopping pump output. This only takes 5 to 10 minutes to correct, but who needs the hassle? The pump is field serviceable with nothing more complicated than a Leatherman Pocket Tool. We carry both a Rebuild Kit which contains a new diaphragm and check valves, or a complete replacement Pump Head. (see the Accessories section) Either one can be done on site in less than an hour. A ¾-inch foot valve is included with the pump. All parts for a rebuild are kept in stock.

Multiply pump amperage draw by voltage, either 12 or 24, depending on your installation, to get wattage needed for PV panel sizing. Remember to oversize PV by 20 to 30%.

Compare the Real Pump to the competition for cost and efficiency and you will see why we had to develop it!

41-120 The Real Pump   **$399**

THE REAL PUMP

12 VDC FLOW CHART				24 VDC FLOW CHART		
LIFT	AMPERAGE					
HEAD (FEET)	DRAW	GPH		HEAD	DRAW	GPH
0	1.05	90		0	1.55	168
20	1.32	72		20	1.80	163
40	1.83	66		40	2.15	150
60	2.11	62		60	2.49	138
80	2.53	60		80	2.76	126
100	2.62	59		100	3.03	114
120	2.98	52		120	3.26	108
140	3.20	45		140	3.52	105
160	3.55	44		160	3.76	93
180	3.61	29		180	4.00	90
200	3.42	22		200	4.19	76
220	3.56	19		220	4.30	74
240	3.70	18		240	4.40	66
260	3.81	3		260	4.50	58
280	3.85	8		280	4.55	45
300	3.92	4		300	4.60	32

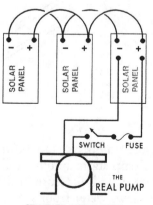

STANDARD INSTALLATION
12 VDC SOLAR PUMP

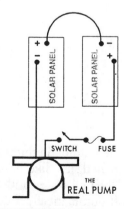

STANDARD INSTALLATION
24 VDC SOLAR PUMP

Accessories for the Real Pump

Pump Head Rebuild Kit

If you would like to rebuild your pump head, and not replace the whole thing, this kit contains all you need. Has 1 new diaphragm and 2 new check valves, 1 for inlet, and 1 for outlet. This is a good item to keep on your home workshop shelf. No special tools required.

41-129 Pump Rebuild Kit   **$15**

Port Kit

The Real Pump port kit is a pair of o-ringed port adapters with standard ¾-inch male iron pipe PVC ends. These are the same as supplied originally with the Real Pump.

41-119 Port Kit   **$5**

Pressure Gauges

Pressure gauges make system monitoring and troubleshooting much easier. Our gauges come in two sizes; 0–60 psi and 0–300 psi. Both gauges are fitted with a ¾-inch male bushing to ease installation.

41-121 Pressure Gauge 0–60 psi   **$9**
41-122 Pressure Gauge 0–300 psi   **$9**

Pump Controller with Built-in LCB and Remote Control

This controller is designed for PV direct systems and gives you everything necessary in one compact weather tight package. The controller has LCB technology for continued pump operation under marginal conditions. It will downconvert excess panel voltage into useable amperage. The pump will run slower, but without the LCB's help it would stop entirely. It also has a set of remote control terminals for connection to a float switch, pressure switch, or any other remote control. Connection wire for the remote control can be inexpensive 22 gauge telephone cable, and runs of several thousand feet will work. Works on 12 to 45 volts (open circuit voltage) and up to 10 amps.

41-196PV Direct Pump Controller   **$89**

Pump Controller Options for Versatility

For battery based pumping systems these two controllers are just what you need. Our day/night controller is versatile, doubling as a photo-sensor and a low-voltage disconnect. The photo sensor will turn the pump on at sunrise and turn it off at sunset in clear weather. In cloudy weather this controller will shut the pump off. The low voltage disconnect will prevent taking the batteries too low. Photosensor has 15-foot leads.

The water sensor controller also has a low-voltage disconnect and instead of a photosensor has a water level sensor. The water level sensor can be several hundred feet from the controller and will use inexpensive telephone wire to hook-up. The water level probes are gold plated for trouble-free service.

Low voltage disconnect set points are user adjustable on both controllers. Both units will operate on either 12 or 24 volts.

41-123 Controller w/photo sensor   **$75**
41-124 Controller w/water sensor   **$75**

Pump Head

Complete replacement pump head for the Real Pump. Installation takes only a few minutes.

41-128 Pump Head—Real Pump   **$69**

Replacement Brushes

We estimate that the brushes in the Real Pump will last 4–5 years in continuous duty (24 hours per day), (we told you this is a great pump!). However, when the time comes, here they are. These brushes replace externally with no special tools.

41-126 Replacement Brushes   **$16**

Heat Exchanger

If you find yourself in the situation of very high lift (200+ ft) combined with high temperatures (desert), you should invest in a water jacket for your Real Pump. This highly conductive aluminum water jacket fits around the Real Pump and routes a portion of the pumped water through the jacket to cool the pump. No water is wasted.

41-127 Real Pump Heat Exchanger   **$59**

Foot Valve

A foot valve prevents a pump from losing its prime, thus extending its life. Highly recommended for all pumps. Constructed of high-impact plastic to resist corrosion. This ¾" Foot Valve is *included* with the Real Pump.

41-125 Foot Valve - Plastic 3/4"   **$14**

See page 313 for a float switch to use with the PV Direct Controller. Pressure switches on page 296.

If water is dirty or silty, use the Inline Sediment Filter and Cartridges on the intake, page 294. Save yourself the hassle of pump cleaning!

Solar Slowpump

Since 1983, the Solar Slowpump has set the standard for efficiency and reliability in solar water pumping where water demand is in the range of 50 to 3,000 gallons per day. The Slowpump is a positive displacement pump that operates all through the solar day at the slow, varying speeds that result from variable light conditions. It is designed to draw water from shallow wells, springs, cisterns, tanks, ponds, rivers, or streams and to push it as high as 450 vertical feet (about 200 psi), for storage, pressurization, or irrigation. Suction lift capacity is 20 feet at sea level with free discharge (subtract 1 foot for every 1,000 feet of elevation).

Slowpump's positive displacement rotary vane pump has working parts of hard carbon-graphite and stainless steel—no plastic. The pump body is solid forged brass, so tough it survives most hard freezes. The pump starts easily even in low sunlight conditions and produces a smooth, nonpulsating flow, making it quiet and easy on your piping. They are self-priming and require no lubrication, and are rebuildable should damage or wear occur. Brushes last 5–10 years and are easy to inspect and replace.

The Solar Slowpump can be powered PV direct or from batteries. They are available in ¼-hp and ½-hp models. The ¼-hp models come in 12 or 24VDC and the ½-hp models come in 24 or 48VDC. They are also available in 120VAC upon special request.

Slowpumps must not be used without filtration. Their high-precision parts are damaged by sand, rust, or abrasive silt. If your water is consistently clear, a very fine intake strainer will provide sufficient protection. Otherwise, use an inline filter or our high-capacity intake filter. If in doubt, use a filter as the warranty does not cover damage from dirty water. Our filters use replaceable cartridges, which may last for months or years, depending on water quality.

Filters and accessories on page 294.

Power Sources and Storage

Slowpumps are powered by DC electrical current from solar panels or other power sources. They use a permanent magnet motor, which provides high efficiency even with varying voltages (typically encountered with direct solar power). They can also be powered by storage batteries including backup power from a vehicle battery. They are available in 120V DC or AC as well.

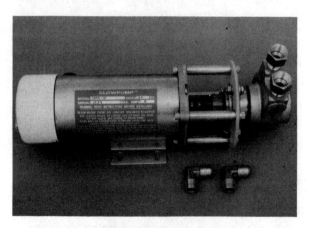

■41-108-12 1/4-hp Slowpump (12V) $515
■41-108-24 1/4-hp Slowpump (24V) $515
■41-112-24 1/2-hp Slowpump (24V) $649
■41-112-48 1/2-hp Slowpump (48V) $649
■41-114-24 1/2-hp Slowpump (24V) $649
■41-114-48 1/2-hp Slowpump (48V) $649
■41-116-24 1/2-hp Slowpump (24V) $649
■41-116-48 1/2-hp Slowpump (48V) $649
■41-118-24 1/2-hp Slowpump (24V) $649
■41-118-48 1/2-hp Slowpump (48V) $649

Please specify the complete model number from the Solar Slowpump chart. 120 volt AC models are also available for approximately $50 extra. Call for a quote. Send SASE for detailed spec sheet and illustrated sample installations.

Typical Slowpump installation from a spring

| | 1/4 HP Slowpumps | | 1/2 HP Slowpumps | | | | | | | | | |
| | 41-108 | | 41-112 | | 41-114 | | 41-116 | | 41-118 | |
lift in feet	gph	watts	gph	watts	gph	watts	gph	watts	gph	watts
20	258	74								
40	252	99								
60	244	133								
80	234	163								
100	225	199					120	136	201	193
120	216	228					118	153	198	209
140							116	168	195	228
160					60	116	114	184	192	246
180					60	123	112	198	189	263
200			21	103	59	130	110	215	186	283
240			21	118	59	147	107	250	181	324
280			20	132	58	162	103	286	176	375
320			19	147	57	179	100	328		
360			18	162	56	197	97	386		
400			18	178	56	214				
440			17	199	55	239				

Solar Slowpumps

Wattage listed is required power at pump. PV panels **must** be oversized 20 to 30% from manufacturers specifications. LCBs on page 214 are highly recommended to boost pump performance on PV based systems.

Flowlight Booster Pump

The Flowlight Booster Pump provides city water pressure for homes with 12 and 24-volt power systems. It represents a step up from "RV" pumps (like the Shurflo, and Jabsco). The Flowlight will far outlast the RV pumps. It uses the same forged brass, carbon graphite pump head as the Solar Slowpump providing efficient, quiet operation. It will use half the energy of an inverter-driven AC jet pump with one-fifth the starting surge! *A Booster pump pressurizing system is far cheaper than an elevated tank.*

The Flowlight Booster comes in the 5.5 gpm, 50 psi standard model or in the 3 gpm, 65 psi low flow model. If suction lift is greater than 10 vertical feet or frequent filter clogging is expected, the low-flow model is recommended. Maximum suction lift is 20 feet at sea level (subtract 1 ft for every 1,000 ft of elevation). Each model comes in 12 or 24 VDC. Other voltages and AC motors are available upon special request. *Filtration is required!* One year warranty.

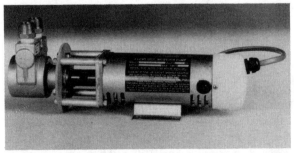

- ■41-141   Booster Pump Standard (12V)   **$565**
- ■41-142   Booster Pump Standard (24V)   **$565**
- ■41-145   **Booster Pump Low Flow (12V)**   **$555**
- ■41-146   **Booster Pump Low Flow (24V)**   **$555**

Installation kit on page 294.

Accessories for Flowlight Booster Pump and Slowpump

Adaptors
(included with the pump at **no charge**)
¼ hp pumps have a ½-inch threaded female adaptor.
½ hp pumps have a ¾-inch threaded male adaptor.

Dry Run Switch
Slowpumps can be damaged from overheating of the pump head due to running dry. There is now a thermal shut-off switch available that can be clamped onto any Slowpump, old or new. This provides better, simpler protection than float switch or low-pressure cut-off systems. The switch shuts off the motor when the pump begins to warm up and then must be manually reset.
■**41-135 Dry Run Switch (1300 models)** $79
■**41-144 Dry Run Switch (2500 or Booster)** $79
You must specify the pump model number!

Fine Intake Strainer/Foot Valve
A fine screen strainer is the minimum protection required for water that is free of suspended silt and rust. Water filtration is extremely important with the Slowpump as one grain of sand can damage the pump's precision surfaces. The new strainer has been upgraded and is made of all metal with an extra-fine 100 mesh metal screen.
■**41-136 Fine Intake Strainer/Foot Valve** $79

Inline Sediment Filter
Sometimes your water isn't dirty enough to mess with fancy and expensive filtration systems and all you need is a simple filter. Our inline sediment filters accept standard 10-inch filters with 1-inch center holes. They are designed for cold water lines only and meet National Sanitation Foundation (NSF) standards. Easily installed on any new or existing cold water line (don't forget the shutoff valve), they feature a sump head of fatigue resistant Celcon plastic. This head is equipped with a manually operated pressure release button to relieve internal pressure and simplify cartridge replacement. They're rated for 125 psi maximum and 100°F. They come with a ¾-inch FNPT inlet and outlet and measure 14"H x 4-9/16"Diameter. It accepts a 10-inch cartridge and comes with a 5-micron high-density fiber cartridge.
41-137 Inline Sediment Filter $55

Replacement Filter Cartridges
Our rust and dirt cartridge is made of white cellulose fibers with a graduated density. These filters collect particles as small as 5 microns (two ten-thousands of an inch). These are NSF listed components that take a maximum flow of 6 gpm. Our taste and odor filters are made with granular activated carbon. These filters effectively remove chlorine, sulfur, and iron taste and odor. These are NSF listed components. Maximum flow is 3 gpm. Note: filters should be replaced every six months to prevent bacterial growth or as needed. This is the cartridge to use with the inline sediment filter.
41-138 Rust & Dirt Cartridge (2) $14
41-436 Taste & Odor Filter (2) $35

30-Inch Intake Filter/Foot Valve
Replaces both the intake strainer and inline filter with a single high-capacity submersed cartridge unit. Good for silty streams, drilled well casings, and other problem applications. A spare micron cartridge is included. It also accepts three 10-inch cartridges. Comes with ¾-inch female fitting and ¾- to ½-inch reducer.
■**41-427 30" Intake Filter/Footvalve** $125

30-Inch Replacement Filter Cartridges
■**41-428 30" Cartridges (3-pack)** $79

Easy Installation Kit

All the small parts that you need to quickly install your booster pump system: accessory tee, pressure switch and gauge, check valve, drain valve, shut-off valve, pipe nipples, and two 18-inch flexible pipes with unions. All brass fittings. Often it's very hard to find these parts in out-of-the-way hamlets!

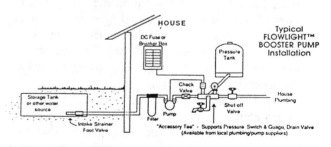

41-143 Easy Installation Kit $108

Solar Force Piston Pump

The Solar Force is a high-flow, medium to high lift positive displacement pump, designed for water lift or pressurizing. It utilizes solar-electric power to draw surface water from a shallow well, spring, pond, river, or tank. It can push water uphill and over long distances for home, village, livestock, irrigation, and fire protection. It is capable of producing from 4 to 9 gpm and it can lift up to 230 ft (or pressurize up to 100 psi). Its suction capacity is 25 vertical feet. (Subtract one foot per 1,000 feet of elevation above sea level.) *It is by far the most efficient pump available in its performance range,* coming in six models to meet your specific pumping needs.

It works efficiently at all speeds, even in low light conditions. This very durable cast iron and polished brass pump is designed to last decades. Features include a non-slip gear belt drive that allows *mechanical back-up or hand power,* a pressure relief valve, and a circuit breaker/switch enclosed in a weatherproof box. This is an extremely low maintenance pump designed for a 2–6 year maintenance interval. No special tools or skills are required. A crankcase viewing window allows easy oil inspection. A repair kit is included with the pump that contains all the seals and gaskets, valve parts, gearbelt, motor brushes, brass cylinder, oil additive, and assembly diagram. This is enough supplies to maintain the pump for 10–20 years of typical use. The pump is available in DC voltages of 12, 24, or 48 volts or in 120VAC. Two-year warranty.

Here is a chart for Solar Force pumps with performance measured at 14V (12V nominal), 28V (24V nominal) or 56V (48V nominal). Derate PV module wattage by 20–30% to calculate the number of solar modules necessary. For pressurizing: Total head = Vertical feet + (PSI x 2.31)

The dimensions of all Solar Force pumps are 22" x 13" x 16". The A1 and A3 weigh 70 lb and the B2 and B3 weigh 90 lb. On all pumps the inlet is 1¼-inches and the outlet is 1-inch pipe thread.

■41-181	Solar Force-A1—12V	$1,811
■41-182	Solar Force-A1—24V	$1,811
■41-183	Solar Force-A3—12V	$2,253
■41-184	Solar Force-A3—24V	$2,253
■41-185	Solar Force-B1—12V	$2,046
■41-186	Solar Force-B1—24V	$2,046
■41-187	Solar Force-B3—12V	$2,489
■41-188	Solar Force-B3—24V	$2,489
■41-189	Solar Force-B1—48V	$2,109
■41-190	Solar Force-B3—120VAC	$2,549*
■41-191	Solar Force-B1—120VAC	$2,109*
■41-193	Solar Forvce-B3—48V	$2,549

All Solar Force pumps are shipped UPS freight collect from New Mexico. The motor is packed separately.

*The 120VAC models use a DC motor with rectifier. They use AC power most efficiently, with 1/5 the starting surge of the common induction motor. This reduces energy use, wire cost and load on power inverters.

Performance chart on next page.

Solar Water Pumping Booklet

This booklet is a good place to start if you know nothing about solar water pumping. It's an eight-page booklet that introduces the concepts and the technology of solar pumping to potential users. It is written for general readership, and assumes little technical background.

80-613 Solar Water Pumping Booklet $1

Head in Feet	PSI	Model A1 GPM	Model A1 Watts	Model A3 GPM	Model A3 Watts	Model B1 GPM	Model B1 Watts	Model B3 GPM	Model B3 Watts
0-20	9	4.3	56	8.0	116	5.2	110	9.3	168
40	17	4.1	76	7.2	146	5.2	132	9.3	207
60	26	3.9	90	6.4	184	5.1	154	9.2	252
80	35	3.8	111	6.2	218	5.1	182	9.2	286
100	43	3.7	125			5.0	202	9.1	322
120	52	3.6	146			5.0	224	9.1	364
140	61	3.6	165			5.0	252	9.1	403
160	70	3.5	190			4.9	269		
180	78	3.5	190			4.9	280		
200	86					4.8	308		
220	95					4.7	314		

Accessories for Solar Force Pumps

Sediment Trap
Protects the pump from fine sand and debris. It is washable and comes with 1-inch male fittings. This is a 1-inch Arkal drip irrigation filter.

42-203 Sediment Trap for Solar Force   **$55**

Pressure Switch for Solar Force
These are DC-rated, adjustable pressure switches. Use the ¼-hp switch for the A-Series and the ½-hp switch for the B-series.

41-140 Pressure Switch—¼ HP   **$18**
41-194 Pressure Switch—½ HP   **$29**
41-195 Pressure Switch—¾ HP   **$98**

Here's an example of a typical system we sell for the Solar Force Piston Pump, to give you a better idea of total costs:

Sprinkler irrigation, domestic pressurizing, and fire protection: 9 gallons per minute at 30–50 psi for 7 hours per day (3,500 gallons per day, seven days per week). This system draws water from a storage tank or shallow water source. On-demand pressurizing necessitates a storage battery power system, which may be enlarged to power lights, tools, and appliances. The DC piston pump uses *far* less energy than conventional AC jet pumps, thus requiring a smaller, less expensive power system.

The Typical System
Solar Force Piston Pump (24V or 48V)
Eight 48 watt photovoltaic modules
Solar tracker/mounting rack
Eight storage batteries
Battery charge controller
System wiring
Pressure tank, etc.
System Price Estimate: $6,800

Solaram Surface Pump

The Solaram Surface Pump is the most efficient photovoltaic powered pump available for high head (in excess of 200 ft) and medium to high flows (3 to 9 gpm). It is used to pump from a surface water source such as a shallow well, spring, pond, river, reservoir, or holding tank. It is well suited to pressurizing as well as lifting. It may be used to lift water and pressurize a home water system at the same time. It may also be used for sprinkler and drip irrigation. A PV/battery system is needed for such on-demand pumping. With a PV/battery system, the pump's flow output and wattage requirement will be 20% lower than specifications.

The Solaram is a high-efficiency positive displacement diaphragm pump that is built like a tank for fulltime stock watering, home pressurizing, and irrigation use. It can pump as high as 960 vertical feet and produce flows as high as 9 gpm. It is a multiple diaphragm (two in 200 series and three in 400 series) industrial pump driven by a permanent magnet DC motor. It uses a cogged gear belt for high efficiency and long-term reliability with minimal attention. It comes with a Linear Current Booster (LCB), prewired in a weatherproof box. The LCB allows optimum pumping performance over a wide range of sunlight conditions. The entire unit is supported on a heavy galvanized steel

base and is protected from the weather by a galvanized steel hinged cover. A brass strainer/foot valve is included. The pump is protected from excessive pressure damage by a pressure relief valve. The motor is protected from overload by a DC circuit breaker and from overheating by an automatic reset, thermal cut-out switch.

The pump is filled with a nontoxic, vegetable base lubricating oil which should be replaced every 12 months of full time use (6–9 peak sun hours/day). The easy-to-replace pump diaphragms should also be replaced every 12

Surface Pump: This type of pump is designed for pumping from springs, water tanks, ponds, rivers or streams, or shallow wells. The pump is mounted in a dry, weatherproof location within 10 to 15 feet of the water surface. These pumps cannot be submersed (and be expected to survive).

PERFORMANCE CHART

HEAD FEET	MODEL 201 GPM	WATTS	MODEL 202 GPM	WATTS	MODEL 203 GPM	WATTS	MODEL 401 GPM	WATTS	MODEL 402 GPM	WATTS	MODEL 403 GPM	WATTS	MOTOR HP / VOLTS
0-80	3.0	170	3.7	207	4.6	285	6.2	258	7.5	339	9.4	465	
120	2.9	197	3.7	238	4.5	319	6.0	305	7.3	396	9.1	539	3/4 HP/24 V
160	2.9	225	3.6	268	4.5	352	5.8	354	7.2	453	8.9	619	
200	2.9	247	3.6	296	4.5	388	5.7	400	7.1	513	8.9	693	
240	2.8	265	3.6	327	4.5	427	5.6	453	7.0	572	8.6	724	
280	2.8	286	3.6	356	4.4	466	5.5	499	6.9	628	8.4	801	1 HP/24 V
320	2.8	315	3.5	388	4.4	496	5.4	548	6.8	686	8.3	869	
360	2.8	342	3.5	416	4.4	536	5.4	592	6.6	733	8.2	927	
400	2.7	363	3.4	450	4.4	572	5.3	649	6.5	782	8.7	1122	
480	2.7	416	3.4	505	4.3	649	5.3	717	6.5	900	8.5	1265	1.5 HP/180 V
560	2.7	456	3.3	570	4.3	693	5.2	800	6.6	1045	8.4	1397	
640	2.7	502	3.3	623	4.2	774	5.1	893	6.5	1166	8.2	1540	
720	2.6	551	3.2	690	4.0	856	5.1	1031	6.4	1287	8.1	1683	2 HP/180 V
800	2.6	589	3.2	715	4.0	931	5.1	1114	6.4	1408	8.0	1815	
880	2.6	647	3.2	774	4.1	1082	5.1	1206	6.3	1529	8.0	1958	3 HP/180 V
960	2.5	705	3.1	838	4.1	1190	5.0	1289	6.1	1650	8.0	2145	

GPM and **Wattage** levels based on "Array Direct" voltages. With PV/Battery systems, they will be 20% lower.

months of full-time use. A replacement diaphragm kit is included with the pump.

The pump may be turned on and off for the filling of a storage tank with Water Level Sensor 1 and turned off with the depletion of a water source with Water Level Sensor 2. The two sensors may be used in conjunction with each other. For a pressurizing system, use a pressure switch. A special pressure switch system can be used to switch the pump on and off for extreme long distance pumping.

The Solaram Surface Pump is designed for decades of use. It comes in four basic models, powered by a ¾, 1, 1½, 2, or 3 hp motor depending on your needs. One example: Model 403 at 1 hp will deliver 8.2 gpm at 360 vertical feet, using 927 watts at 24V. The pump is highly tolerant of dirty water and dry running, and it comes with a one-year warranty.

Installing of the Solaram is easy. Power is connected directly from the PV array to the terminals in the weatherproof LCB/breaker box. Water Level Sensors are easily connected with inexpensive telephone type wire. All pump models are 28"L x 16½"H x 16"W. The weight varies between 110 and 150 lb depending upon the model.

See page 214 for Water Level Sensors.

In a recent conversation with my father-in-law I was told that his brother, who moved to South Carolina about three years ago, was having difficulty with a pond that he had dug on the property when he moved there. It seems that he is using a gasoline powered pump to keep the pond at a reasonable level and that he is getting a bit discouraged with the never ending need to refuel and maintain the system. Without my ever having been to the property, or knowing any of his requirements, I was sure about the feasibility of solar power for him.
Richard Huck, E. Patchogue, NY

PV Array Sizing

Our technicians will be glad to help you size your system. If you'd prefer to calculate it yourself follow these directions: From the pump's performance chart, identify the pump *wattage* needed for the desired *flow* (gpm) at the required *head*, where *head* is calculated by vertical lift + pipe friction head + pressure head (if pressurizing system) in feet (feet = psi x 2.31). Also identify the appropriate pump motor *voltage*. Obtain specifications for the PV modules to be used. With this information, determine the total number of PV modules needed and the series/parallel module arrangement.

The Pressure System Kit includes all the parts you need to build a pressurizing system: accessory tee, adjustable pressure switch, pressure gauge, check valve, drain valve, and shut-off valve.

■41-286	Solaram #200, ¾ HP	$2,829
■41-287	Solaram #200, 1 HP	$2,897
■41-289	Solaram #400, ¾ HP	$2,939
■41-290	Solaram #400, 1 HP	$3,008
■41-294	Water Level Sensor 1	$33
■41-295	Water Level Sensor 2	$33
■41-296	Pulsation Damper	$231
■41-297	Pressure System Kit	$132
■41-298	Check Valve, Brass ¾"	$27
■41-299	Flow Gauge (1–10 gpm)	$107

All Solaram Pumps are shipped Freight Collect from New Mexico.
(Call for current prices on larger Solaram pumps; 1.5 through 3.0 hp.)

Low-Voltage Booster Pumps

Shurflo Pumps

Shurflo produces a high quality line of positive displacement, diaphragm pumps for RV and remote household pressurization, and for lifting surface water to holding tanks. These pumps will self prime up to 10' of suction with free discharge. The balanced three chamber design runs quieter than any other diaphragm pump and will tolerate silty or dirty water or running dry with no damage. (If water is known to have sand or debris it's best to use one of the inexpensive inline cartridge filters on the intake. 41-137 housing ($49) & 41-138 ($14) cartridge pair. Sand sometimes will lodge in the check valves necessitating disassembly for cleaning. No damage, but who needs the hassle? Put a filter on it!) Motors are all slow speed DC permanent magnet types for long life and the most efficient performance. All pumps are rated at 100% duty cycle and may be run continuously. There is no shaft seal to fail, the ball bearing pump head is separate from the motor. Pumps are easily field repaired and rebuild kits are available for all models with a factory 800# for Parts and Repair Kits. Built-in pressure switches are adjustable through the range listed for each pump. Pumps have 1/2" MIP inlets and outlets. All pumps carry a full one year warranty.

Low Flow Pump

For modest water requirements or PV direct applications. Available in 12 volt DC only, delivers a maximum of 60 psi or 135' of lift. Delivers 1.8 gpm at 5 psi using 2.9 amps; 1.5 gpm at 30 psi using 5.1 amps; and 1.25 gpm at 60 psi using 7.2 amps.

41-450 Shurflo Low Flow Pump $109

Shurflo Medium Flow Pump

The Medium Flow is available in 12 or 24 volt DC, performance specs are the same for both. Maximum pressure is 40 psi or 90' of lift for both models. Delivers 3.3 gpm at 5 psi using 5.4 amps; 3.1 gpm at 20 psi using 7.5 amps; and 2.8 gpm at 40 psi using 10.7 amps. (24 volt systems use 1/2 the amps)

41-451 Shurflo Medium Flow, 12 Volt $149
41-452 Shurflo Medium Flow, 24 Volt $165

Shurflo High Flow Pump

This is Shurflo's highest output model. Good for household pressurization systems, will keep up with all but the biggest water-hog fixtures. Available in 12 volt only. Maximum pressure is 50 psi or 115' of lift. Delivers 3.75 gpm at 5 psi using 6.2 amps; 3.1 gpm at 30 psi using 9.6 amps; 2.8 gpm at 50 psi using 12.5 amps.

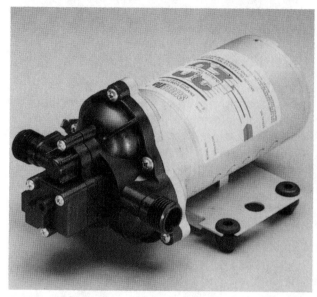

41-453 Shurflo High Flow Pump $189

Shurflo AC Pump - 115 V

For those that need to send power more than 150-200' to the pump this pump can save you big $$ on wire costs! It still uses an efficient permanent magnet slow speed DC motor, but has a rectifier pack to convert the power back to DC at the pump. Under heavy continuous use this motor is thermally protected and will shut off automatically to prevent overheating. This should only happen at maximum psi after 90 minutes of running. Pressure switch is adjustable from 25 to 45 psi. Maximum pressure is 45 psi. Delivers 2.6 gpm at 20 psi using 82 watts, 2.4 gpm at 30 psi using 96 watts, and 2.25 gpm at 40 psi using 108 watts. U.L. recognized. Weight 5.8#

41-454 Shurflo 115 v AC Pump $179

To order any of these products
or for more technical information
call us toll-free at
1-800-762-7325

Pressure Tanks

Unless you are lucky enough to have a gravity-fed water system, you probably need a pressure tank. Most people are dissatisfied with water pressure less than 15 psi: showers seem to dribble and it seems to take forever to fill the bathtub. Both Paloma and Aquastar instantaneous water heaters require at least 20 psi to work properly. Remember that city water pressure is 40 psi. So, if your water storage is not at least 45 vertical feet above your house, you'll probably be happier with a pressurized water system.

We design some of our water pump systems with the 2-gallon pressure tank. It prolongs the life of the pump by easing the demand on the pressure switch. Efficient tankless water heaters require a larger presure tank to stabilize water pressure. Twenty gallon capacity is the minimum recommended for this application.

Teel precharged pressure tanks have a permanent, factory pressurized air charge which is totally isolated from the water and cannot be absorbed as with standard style tanks. The interior of the tanks are epoxy coated for corrosion resistance, and the exterior is a baked green enamel. The 2-gallon tank has a ¾-inch MPT connection. The 2-gallon tank is shipped precharged at 20 psi and is easily adjustable. Maximum working pressure is 100 psi, maximum temperature is 120°F. Twenty gallon is the largest size that can be shipped by UPS. For larger systems you may save on shipping costs by purchasing locally.

If you need to raise your water pressure, see our Water Pressure Booster Kit on page 322.

41-401	2-Gallon Pressure Tank	$55
41-402	6-Gallon Pressure Tank	$135
41-403	12-Gallon Pressure Tank	$189
41-404	20-Gallon Pressure Tank	$195
41-405	36-Gallon Pressure Tank	$275

Gallon Capacity	Style	Maximum Drawdown	Diameter	Height
2	Vertical	0.8 gal	8.4"	12.6"
6	Horizontal	3 gal	10"	12.75"
12	Horizontal	6 gal	12"	15.5"
20	Vertical	9.5 gal	16"	32.5"
36	Vertical	18 gal	20"	37.75"

Pressure Switch

You must use a pressure switch for pressure tank systems. It will turn the pump off when the tank pressure reaches 40 psi, and then on again when the pressure drops to 20 psi. It also turns the pump off when pressure drops below 10 psi to prevent damage from running the pump dry. The settings are fully adjustable in the range of 5 to 65 psi. (This pressure switch is **included** in the Easy Installation Kit for the booster pump.)

41-140 Pressure Switch (6X535)   **$18**

Centrifugal Pumps

Bronze Centrifugal Pump—12V

This bronze-bodied and bronze-impelled pump is designed for applications where you have a lot of water to move at a minimum lift. The pump is not self-priming and must be mounted in a dry location where the motor is protected from dampness. Intake and outlet ports have ¾-inch internal pipe threads suitable for either brass or plastic pipe fittings. One year warranty.

Performance gpm at total feet of head

2 ft	6 ft	10 ft	12.5 ft
20 gpm	14.5 gpm	7.5 gpm	Shut-off
8.0A	6.7A	5.4A	4.5A

41-301 Bronze Centrifugal Pump—12V $149

March 12 & 24V Circulating Pumps

March circulating pumps are used primarily with hot tub, spa, solar thermal, and woodstove hot water systems. They use very little power at either 12 or 24 volts and keep needed water circulation happening. All pumps have a magnetic drive that eliminates the old-fashioned shaft seal. They are easy to service, only requiring a screwdriver, and the entire motor assembly can be replaced without draining the system. The 1/100 hp pump runs at 1,950 rpm and the 1/25 hp pump runs at 3,600 rpm. Six-month warranty. Will withstand temperatures to 200°F. Weight 7.1 lbs.

Specs @12V	1/100 hp	1/25 hp
Flow in gpm	5.5	7.5
Max head	7.1	15.5
Amps	1.5	3.8

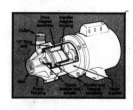 

41-501	**March Pump 1/100 HP 12V**	**$279**
41-503	**March Pump 1/25 HP 12V**	**$349**
41-504	**March Pump 1/25 HP 24V**	**$349**

Photovoltaic Pool Pumping System

After repeated requests for a solar powered pool pumping system, we have finally found one which offers a reliable and cost effective alternative to the standard pool pump.

A specially designed DC motor is matched with a proven swimming pool pump to provide the most efficient self-priming centrifugal solar pump available. Pump construction of glass reinforced thermoplastic ensures lifetime service with no corrosion. The strainer can be removed in seconds for cleaning without loss of prime.

The mechanical shaft seal is of the proven carbon on ceramic face type. All parts in contact with water are non-corrosive plastic, stainless steel or carbon/ceramic. The motor is of permanent magnet design with easily replaced brushes.

For solar pool filtration your system will consist of:

- solar pump and motor
- a PV array and support (either six or nine high power modules)
- a cartridge type filter (such as Jacuzzi CFT100)
- wiring & plumbing to complete the system

For pool water filtering, an annual electrical cost of hundreds of dollars can be replaced with a one time investment at installation. Tax credits may be applicable for the solar electric power system.

The cost of the system will depend on the size of the pool and the quality of your solar exposure. Prices range from $2,500 to $5,000. Call one of our technicians for an estimate.

Hartell Hot Water Pumps

The Hartell magnetic drive circulator pump series features a DC motor designed to circulate hot water using PV power or low-voltage battery systems. They are ideal for solar water heating or other closed-loop, low-flow pumping applications. This model will operate directly off of a solar panel—the brighter the sun, the faster the water is pumped! The pump has a 30,000-hour life expectancy.

Designed to be run directly off of a solar panel between 15 and 30 watts, it will pump to a maximum head of 11 feet with a maximum flow of 6.5 gpm. and will withstand temperatures to 200°F. Six-month warranty.

41-522 Hartell Pump (CP-10B-12) $259

Deep Well Pumping

Here is another of Jon Vara's installments, this time on deep well water pumping.

In the first installment of this article, I discussed pumping water from driven or dug wells, springs, cisterns, and lakes or streams. Such shallow water sources are vulnerable to contamination from surface runoff, and although it's possible to minimize the danger by careful site selection and proper development, the purity of shallow-source water cannot always be guaranteed. Moreover, in many areas there is simply no shallow water to be had, either because the local water table is uniformly deep, or because shallow sources that do exist are known to be contaminated.

In most cases, the only practical option is a drilled well. An experienced well driller will usually have a good idea of the depth to water and the probable yield, but there are no guarantees; and at a cost of $15 a foot or so, drilling a well can be an expensive, nerve-racking crap shoot.

If you are a lucky crap shooter, however, you may end up with a deep well that can be pumped in exactly the same way as a shallow well. Even where the source of the water lies hundreds of feet below the surface, a quirk of geology known as artesian pressure may force the water upward of its own accord. (Rarely will this come as a complete surprise, however; if artesian pressure commonly occurs at all in your area, your well driller will certainly know about it.)

In very rare instances, artesian pressure will force water from the ground in a steady stream —a freshwater gusher. More often, though, it will simply rise in the casing and stop somewhere below the surface of the ground. If it happens to rise to within 10 or 15 feet of the surface, and the supply is large enough that withdrawing a few gallons per minute won't cause the level to recede, you can simply drop a suction line into the casing, connect it to one of the shallow-well type pumps mentioned earlier, and you're in business.

In most cases, though, the water level in a drilled well will lie far below the suction limit of any pump, making it necessary to move the pump down, into the well itself. One approach is to lower a small nonsubmersible pump, such as one of the Flowlight Slowpumps, into the well casing until the pump itself is within 10 or 15 feet of the water level and the intake pipe is submerged. Provided that the suction limit is not exceeded, a correctly chosen Slowpump will **push** water to heads of over 400 feet.

At a cost of $15 a foot drilling a well can be an expensive, nerve-wracking crap shoot.

For that to work, however, two important conditions must be met. First, the pump must not draw water from the well more rapidly than the aquifer supplies it, or the level will drop as the pump runs, until the inlet begins to suck air rather than water. A Slowpump that is run dry will soon overheat and destroy itself, and although a dry-run switch is available to turn off an overheating pump, it's better to avoid the problem to begin with.

Secondly, the resting water level in the well casing—what well drillers call the static level—must remain constant, or nearly so, throughout the year. A water level that rises substantially in the spring of the year may flood the pump and short it out; receding water in late summer may leave the intake pipe high and dry.

Finally, bear in mind that when a Slowpump is lowered into a drilled well, out of sight doesn't mean out of mind. It will be necessary to pull the pump from time to time to change the foot valve/filter assembly, which protects the pump's inner workings from dirt and grit. That's not necessarily a difficult job, but it's one that must be attended to faithfully. If you're the sort of person who never changes the oil in your car, that may not be easy for you to remember.

Submersible pumps, by contrast, can be lowered directly into the water, which simplifies the problem of a fluctuating static level. One of the simplest ways to pump water from a deep well is to install a conventional 120-volt AC sub

comment from Windy Dankoff from Flowlight:
"Jack pumps are rarely used to deliver 'large' volumes (certainly not for large-scale irrigation!) Rather, they are used for high lifts (200 to over 1,000 feet) at low to medium volumes."

Oxen hauling water from deep well in India. Photo by Sean Sprague.

mersible—the standard deep-well residential pump—and power it through a large inverter. That requires only readily available, well-proven hardware of the sort that all plumbers, well drillers, and other service people are used to dealing with, which is no small advantage if you are not mechanically inclined yourself. If you already own an inverter with the capacity to start a 1/3-horsepower motor, it's a reasonable inexpensive way to go.

Unfortunately, it's also an energy-inefficient way to go. That's because conventional submersible well pumps, like so many electrically powered devices, are simply not engineered with the electric bill in mind.

Conventional deep well pumps operate on the centrifugal principle. In effect, the motor spins the water in the pump so rapidly that it is pushed against the inside of the housing—as coffee swirls against the sides of the cup when stirred with a spoon—and out and up the discharge pipe.

All truly efficient low-powered pumps, by contrast, rely on the general principle of positive displacement. In a positive displacement pump, a unit of water is drawn into a closed chamber and forced out by a rotating vane, oscillating diaphragm, or sliding piston. The Flowlight Slowpumps and Booster Pumps, for example, are vane pumps; the inexpensive RV-type Shurflo and the Solar Submersible are diaphragm pumps.

Piston pumps designed for deep well use are generally of the jack pump configuration, in which the pump motor sits on the surface atop the well casing, while the piston and cylinder may be located hundreds of feet below, beneath the static water level. A long "sucker rod", assembled in sections on the site, as you would screw together a chimney-cleaning rod, transmits the power from motor to piston. (Those on a tight energy budget may wish to consider a deep-well hand pump, which except for being powered by muscles instead of a motor is similar to a jack pump).

Jack pumps are simple, rugged, and capable of delivering large volumes of water. They are often used for large-scale irrigation and for similar demanding uses, but because they cost a minimum of several thousand dollars—not including batteries, PV panels, or the cost of the well itself—they are rarely used residentially, unless high water demand and great depth to water make less costly alternatives impractical.

One such alternative that may make sense in your application is a submersible diaphragm pump, such as the Solar Submersible. Submersible diaphragm pumps are comparable to the vane-type Flowlight pumps in terms of performance and overall efficiency, although they are substantially higher in cost. Being true submersible pumps, however, they are simpler to install—there is no danger of the water level rising and flooding the pump—and their diaphragm design enables them to pass particles of grit without damage, eliminating the need for troublesome filtration.

So far, so good. There is, however, one vital

aspect of deep-well pumping that I haven't addressed yet, and it's a far-reaching one. In discussing shallow-well pumping, we considered suction limits; now it's time to confront pressure limits, or the maximum head to which any given pump will force water.

Ultimately, suction is regulated by atmospheric pressure and the laws of physics. No matter how powerful a motor is harnessed to a pump, it's simply impossible to suck water more than about 30 feet uphill, and as we've seen, the practical limit is substantially less than that. Pressure, on the other hand, is an engineering problem. With the right pump, a sufficiently powerful motor, and strong enough piping, water can be pressurized to a head of thousands of feet. Theoretically, the sky is the limit.

The practical limit, however, has to do with the cost of a large motor, its physical size, and the amount of power it will require to run. Most efficient, low-voltage pumps will not force water to a total head of greater than three or four hundred feet, and often less.

Furthermore, just as a car with a small engine will have to slow down when climbing a steep hill, a pump that is forcing water to its maximum head will slow down, too. A pump that delivers several gallons per minute at a total head of 20 feet or so may deliver only a fraction of a gallon per minute at a head of 200 feet, and consume twice as much power to do it.

Still, a pump that delivers only ¼ gallon per minute will deliver 180 gallons in the course of 12 hours, an amount that should easily meet the water need of an ordinary household. The only difficulty is that few households *use* water at the rate of a fourth of a gallon per minute.

Imagine the water use in your own home. For hours on end—during the nighttime hours, for example—no water is used at all. Then, in a great surge of morning showering, afternoon clothes washing, or evening bathing, 40 or 50 gallons may be demanded in the space of half an hour or so, far more than the deep-well pump can deliver.

The solution is often a two-stage system of some sort, in which a deep-well pump brings up water at a slow, steady rate, and deposits it in a nonpressurized tank buried in the ground or safely stored in the cellar. A second, independent pump is used to pressurize the home water system from that convenient reservoir. That second stage, of course, is identical to the typical shallow-source system described in the first part of this article.

In areas with reliable sun, it is often possible to run the deep well pump directly from an independent array of PV panels. That will reduce the need for long wiring runs and eliminate battery losses. However, it will require a holding tank large enough to carry the household through cloudy spells, when little or no water is pumped from the deep well. Where prolonged cloudy spells are possible, however, it may be impractical to arrange that much storage capacity. In that case, the deep well pump, like the secondary pressuring pump, should be powered directly by the household battery bank. —*Jon Vara*

Solar Submersible Pump

Submersible Pump: This type of pump has a sealed motor/pump assembly. The entire assembly is installed below the water surface. Most commonly used when water level is more than 15 feet below the surface. Usually limited to 250 feet or less of lift.

This efficient, submersible, diaphragm pump will deliver up to 250 feet of vertical lift and is ideal for deep well applications. Remember, you're only lifting from the water *surface*, not from the pump. In a PV direct application it will yield from 300 to 1,000 gallons per day depending on lift and power applied. Solar submersible pumps are best used for slow steady water production into a holding tank, but may be used for direct pressurization applications as well. Flow rates range from .5 to 1.8 gallons per minute. The pump can be powered from either 12 or 24 volt sources. Output is best at 24 volt. The best feature of this pump is the fact that it is field serviceable, and the repair kit is in the bottom of the stainless steel pump housing, right where you'll need it years after you've forgotten all about it. Pump diameter is 3.8" (will fit 4" or larger casings), weight is 8 pounds making an easy one man installation possible. Warranty is one year.

In addition to the pump, you will also need:
• *PV panels & mount for direct systems*
• *A Controller for PV direct systems*
(see our PV Direct Controller, # 41-196)
• *¾" poly drop pipe, or equivalent*
• *#10 gauge submersible pump cable*
• *Poly safety rope*
• *Sanitary well seal*

The PV Direct controller is on page 291.

Drop pipe, pump cable, rope, and seal are all commonly available at any plumbing supply store. Splice kit is included.

41-150 Solar Submersible $895

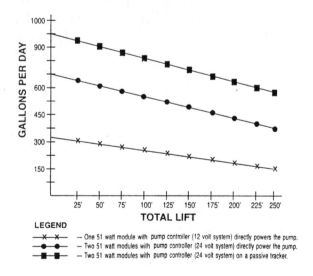

LVM-105 Submersible Pump

This 12-volt submersible pump measures only 1½-inches in diameter and 6½-inches long, and will pump a maximum of 4 gallons per minute. It will pump a maximum vertical head of 42 feet or 18 psi. At 25-feet of head the LVM-105 will deliver about 5/8 gallon per minute. A strainer is fitted over the input end to prevent large particles from entering the pump. The inlet and outlet sizes are ½-inch, it comes with 13 feet of cable with battery clips on the end, and it weighs only 1.2 pounds. Designed for intermittent use only. Three-month warranty

41-671 LVM-105 Submersible Pump $79

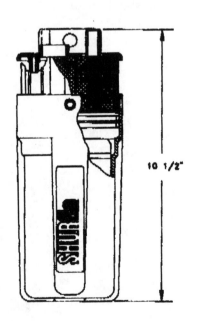

Solarjack Jack Pump

For your deep-well pumping needs (well depths to 1,000 feet and from ½ gallon to 10 gallons per minute) we offer the Solarjack Jack Pump water pumping systems; one of the most cost-effective and reliable photovoltaic solar pumping units available.

Solarjack has two Pump Jack models to choose from: the SJA (¾ hp) and the SJB (1 hp). Each system includes the Pump Jack, motor, controller, bridle assembly, and anchor bolts.

The SJA model is limited to 500-feet of total head and a maximum of 600 peak solar module watts while the SJB is capable of lifting water from 1,000-feet and limited to 1,080 peak module watts.

Installation

Each Pump Jack is mounted on a steel base which is then anchored to a concrete floor. (Anchors and bolts are included.) The solar modules are mounted nearby on a frame and anchored to a foundation. Drop pipe, sucker rod, polished rod, packing gland, and pump cylinder are necessary to complete the installation.

Maintenance

Very little supervision and maintenance is required. An occasional check of the oil level (SJA Jack) and lubrication of grease fittings is recommended, and it should operate for 20 years or more, according to the manufacturer, if maintained properly.

Pumps start at $3,000 for the basic unit. After adding the additional accessories, you're looking at a minimum of $5,000. Please write or call one of our technicians for an exact price quotation as these units are custom engineered for each application.

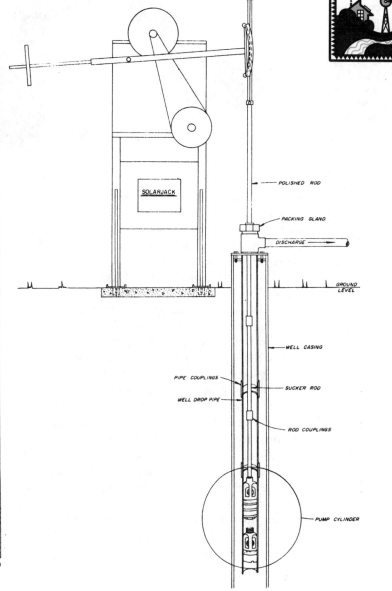

SJ PUMP SERIES
DEEP SET

FEET

U.S. GALLONS PER MINUTE

MODEL
A - SJ-24-47-JA
B - SJ-36-47-JA

Bowjon Wind-Powered Pumps

If you buy the efficiency first, the next best buy will be the appropriate renewable sources. And if you count — as our children surely will — the environmental damage and insecurity that fossil and nuclear fuels cause, then well-designed renewables look even better.
- Amory Lovins, in the Foreword

We had limited experience with Bowjon Pumps back in the late 70s and early 80s; a few raves and a few problems, then they disappeared from the market for several years. Now Bowjon is back promising a perfected product of rigid integrity with two years in redesign and development. So far, we're pleased with what we see.

Bowjon utilizes a unique "air injection" system that is ideal in areas with high wind velocities. You can place your Bowjon where the wind blows best, as far as ¼ mile away from your well. It uses five high-torque, heavy-duty aluminum blades. It is quickly installed on a single post tower. One person can assemble it without special rigs, tools, or equipment. Its minimal maintenance only requires checking the compressor oil every six months. There are no moving parts, leathers to replace, and there are no cylinders or plunger assembly, allowing the Bowjon to run dry without harm.

The Bowjon is available with an eight- or six-foot diameter single or twin in-line compressor with Timken bearings. The Bowjon's pumping system is capable of effectively lifting water up to 7 gallons per minute. More than one windmill can service the same well. The hub is a massive 17 lbs. of tempered aluminum that will withstand 3,200 psi. Bowjon's five blades are made of 6061-T6 tempered aluminum, each consisting of three layers for the maximum strength that's required in extremely high winds.

Bowjon Specifications:

Propeller: Tempered triple layer reinforced blades 6'8" or 8'8" across to minimize flex in high winds. Blades have a unique varied pitch to allow low-wind, high-torque start-up. They have high rpm ability.

Compressor: Industrial strength twin or single cylinder depending on air pressure needs for required water demand. Compressor is capable of 5 cubic feet of air volume per minute.

Start-Up: 5–8 mph.

Pump: Air injection. No moving parts, no cylinders, valves, rods, or leathers to wear out. Will run dry, accepting silt, sludge, and sand without harm. Length is 5 feet. Minimum well casing diameter is 2 inches.

Air Line: 3/8-inch polyethylene tubing (200 psi rated).

Water Line: 0–50' lift: 1-inch tubing or PVC Schedule 40 plastic pipe. 50–100-feet Lift: ¾-inch tubing or PVC Schedule 40 plastic pipe. 100-foot and over lift: ½-inch tubing or PVC Schedule 40 plastic pipe.

Submersion: At least 30% of the vertical lift distance from water level in the well to the highest point of delivery or storage. Where minimum recommended submersion is not possible, use of a collector tank and a regulator in the air line will prevent excessive air pressure and volume. Submersions less than 30-feet restrict vertical lift; but where water is only to be lifted a few feet, submersions of 5 or 10-feet with air tank and regulator are practical.

Lift/Submergence Ratios:

Vertical lift	Submergence below water	Optimum total well depth
50'	35'	90'
80'	56'	140'
120'	85'	210'
200'	100' (50% of lift)	350'

(If the submergence is too low for the amount of lift, the air will separate from the water. If the submergence is too high, the air will not lift the water. For these reasons, the submergence of the pump must be calculated carefully.)

Homesteader Model Bowjon Pump

The Homesteader model comes with 6½-foot diameter blades with a single cylinder compressor. It has a 3/8-inch air line 250 feet long and AL2 air injection pump. One-year warranty.

■**41-701 Homesteader Bowjon Pump $1,295**
Shipped freight collect from So. California

Rancher Model Bowjon Pump

The Rancher model comes with 8'8" diameter blades with a twin cylinder compressor. It has a 3/8-inch air line 250 feet long and an AL3 air injection pump. One-year warranty.

■**41-711 Rancher Model Bowjon Pump $1,595**

Ram Pumps

The ram pump works on a hydraulic principle using the liquid itself as a power source. The only moving parts are two valves. In operation, the liquid (usually water) flows from a source down a "drive" pipe to the ram. Once each cycle, a valve closes causing the liquid in the drive pipe to suddenly stop. This causes a water hammer effect and high pressure in a one-way valve leading to the "delivery" pipe of the ram, thus forcing a small amount of water up the pipe and into a holding tank. In essence, the ram uses the energy of a large amount of water falling a short distance to lift a small amount of water a greater distance. The ram itself is a highly efficient device; however, only 6% to 10% of the liquid is recoverable. Ram pumps will work on as little as 2 gpm supply flow. The maximum head or vertical lift of a ram is about 500 ft.

Selecting a Ram

Estimate Amount of Water Available to Operate the Ram — This can be determined by the rate the source will fill a container. Avoid selecting a ram that uses more water than available.
Estimate Amount of Fall Available — The fall is the vertical distance between the surface of the water source and the selected ram site. Be sure the ram site has suitable drainage for the tailing water. Often a small stream can be dammed to provide the 1.5 feet or more head required to operate the ram.
Estimate Amount of Lift Required — This is the vertical distance between the ram and the water storage tank or use point. The storage tank can be located on a hill or stand above the use point to provide pressurized water. Twenty or thirty feet water head will provide sufficient pressure for household or garden use.
Estimate Amount of Water Required at the Storage Tank — This is the water needed for your use in gallons per day. As examples, a normal household uses 100 to 300 gallons per day, much less with conservation. A 20-by-100 foot garden uses about 50 gallons per day. When supplying potable water, purity of the source must be considered.

Using these estimates, the ram can be selected from the following performance charts. The ram installation will also require a drive pipe five to ten times as long as the fall, an inlet strainer, and a delivery pipe to the storage tank or use point. These can be obtained from your local hardware or plumbing supply house. Further questions regarding suitability and selection of a ram for your application will be promptly answered by our engineering staff.

Aqua Environment Rams

We've sold these fine rams by Aqua Environment for over 10 years now with virtually no problems. Careful attention to design has resulted in extremely reliable rams with the best efficiencies and lift-to-fall ratio available. Construction is of all bronze with O-ring seal valves. The outlet gage and valve permit easy startup. Each unit comes with complete installation and operating instructions.

Typical Performance and Specifications

Vertical Fall (feet)	Vertical Lift (feet)	Pump Rate (Gallons/Day)			
		¾" Ram	1" Ram	1¼" Ram	1½" Ram
20	50	650	1350	2250	3200
20	100	325	670	1120	1600
20	200	150	320	530	750
10	50	300	650	1100	1600
10	100	150	320	530	750
10	150	100	220	340	460
5	30	200	430	690	960
5	50	100	220	340	460
5	100	40	90	150	210
1.5	30	40	80	130	190
1.5	50	20	40	70	100
1.5	100	6	12	18	25

Water Required to Operate Ram

¾" Ram - 2 gallons/minute	Maximum Fall - 25 feet
1" Ram - 4 gallons/minute	Minimum Fall - 1.5 feet
1¼" Ram - 6 gallons/minute	Maximum Lift - 250 feet
1½" Ram - 8 gallons/minute	

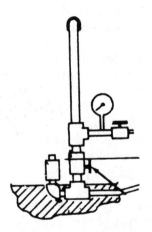

■41-811 3/4" Ram (#201) $195
■41-812 1" Ram (#202) $195
■41-813 1-1/4" Ram (#203) $245
■41-814 1-1/2" Ram (#204) $245

High Lifter Pressure Intensifier Pump

A new development in water pumping technology, the High Lifter Water Pump offers unique advantages for the rural user. Developed in Mendocino County, California expressly for mountainous terrain and low summertime water flow, this water-powered pump is capable of extremely high lifts in ratio to fall (over 1,000 feet) with just a trickle of inlet water. For example, assume a flow of 2 gallons per minute and a fall of 40-feet from a water source (spring, pond, creek, etc.) to the High Lifter pump with a 200-foot rise (head) from the pump up to a holding tank. The lift-fall ratio between the 40-foot fall and the 200-foot lift would be 6:1. The High Lifter pump would deliver 300 gallons of water per day from a 6:1 working ratio.

The High Lifter pump has many advantages over a ram (the only other water-powered pump). Instead of using a "water hammer" effect to lift water as a ram does, the High Lifter is a positive displacement pump that uses pistons to create a kind of hydraulic lever that converts a larger volume of low-pressure water into a smaller volume of high-pressure water. This means that the pump can operate over a broad range of flows and pressures with great mechanical efficiency. This efficiency means more recovered water. While water recovery with a ram is normally about 6%, up to 20% recovery may be achieved with the High Lifter Pump.

In addition, unlike the ram pump, no "start up tuning" or special drive lines are necessary. This pump is also quiet.

The High Lifter pressure intensifier pump is economical compared to gas and electric pumps, because no fuel is used and no extensive water source development is necessary.

There are two model High Lifter pumps available with a 4.5:1 and 9.1:1 working ratio. A kit to change the working ratio of either pump after purchase is available. A maintenance kit is available too. Choose your model High Lifter pump from the specifications and High Lifter performance curves. One year parts and labor warranty.

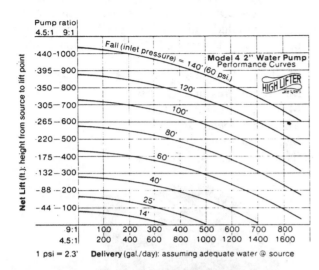

Pump ratio
4.5:1 9:1

Net Lift (ft): height from source to lift point

Fall (inlet pressure) = 140' (60 psi)

Model 4 2" Water Pump
Performance Curves

HIGH LIFTER

-440 –1000
-395 – 900
-350 – 800
-305 – 700
-265 – 600
-220 – 500
-175 – 400
-132 – 300
- 88 – 200
- 44 – 100

120'
100'
80'
60'
40'
25'
14'

9:1   100 200 300 400 500 600 700 800
4.5:1 200 400 600 800 1000 1200 1400 1600

1 psi = 2.3'   **Delivery** (gal./day): assuming adequate water @ source

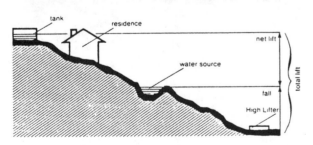

tank   residence

net lift

water source

fall   total lift

High Lifter

- ■41-801 High Lifter Pump—4.5:1 Ratio $895
- ■41-802 High Lifter Pump—9:1 Ratio $895
- ■41-803 Conversion kit (to other ratio) $97
- ■41-804 High Lifter Rebuild Kit $49

Tanaka Pressure Pumps

Tanaka makes great two-cycle gas pumps that are powerful, compact, easy to transport, and easy to use. Their only disadvantage is that they are very loud, like a chain saw. If noise isn't a problem they're hard to beat! Two-year warranty.

TCP-210

The smaller of the two Tanaka pumps weighs only 12 pounds and delivers a maximum of 1,900 gallons per hour (31.7 gpm) or a maximum lift of 98 feet, with a maximum suction of 23 feet. Fittings are 1-inch inlet and outlet, and the pump comes with 10-feet of suction hose.

41-901 Tanaka Pump TCP-210   **$395**

TCP-381

This Tanaka pump weighs only 21 pounds and pumps a maximum of 4,750 gallons per hour (79 gpm) or a maximum lift of 131 feet. The inlet and outlet are 1½-inches. At 50-feet of head this pump delivers 75 gpm, at 75-feet it delivers 50 gpm, at 100-feet it delivers 40 gpm, and at 120-feet it delivers 20 gpm.

41-902 Tanaka Pump TCP-381   **$495**

High-Pressure Gasoline Diaphragm Pumps

For high-pressure situations where a large vertical lift must be achieved, diaphragm pumps are the most reliable and most durable pumps available. These pump/motor combinations consist of the ultraquiet, overhead valve, Honda engines coupled with the highly efficient and durable Hypro pumps. They can be run dry without harm, and have built-in pulsation dampeners that act as shock absorbers. According to the manufacturer, the water pumped is not suitable for human or animal consumption. However, many of our customers have used these pumps for drinking water and we've never had a complaint in 14 years. Maintenance on the pumps is simple, and diaphragms are very easy to replace.

Hypro D19

The D19 comes with a 4-hp Honda engine with gear reduction and a control unit. This pump will deliver 5.3 gpm at 275 psi (635 feet of lift vertically). Simple assembly is required. Maximum suction lift is 10-feet.
Maximum flow: 6 gpm
Maximum pressure: 275 psi
Maximum lift: 635 feet
Maximum speed: 650 rpm
Maximum temp.: 140°F
(Not designed for potable water)

Hypro D30

The Hypro D30 diaphragm pump will deliver 8.5 gpm at 550 psi (1,270-feet vertical lift). It comes coupled with a 5-hp Honda engine. Simple assembly is required. Maximum suction lift is 10-feet.
Maximum flow: 9.5 gpm
Maximum pressure: 550 psi
Maximum lift: 1,270 feet
Maximum speed: 540 rpm
Maximum temp.: 140°F
(Not designed for potable water)

■41-912 Hypro D30 Gasoline Pump $1,495
Pump and engine shipped in separate boxes.

■41-911 Hypro D19 Gasoline Pump $1,095
Pump and engine shipped in separate boxes

Pump Accessories & Controls

Check Valves

A check valve must be installed in pressurized systems. Check valves seal the water system between the pump or pressure tank and the faucets in the home. It prevents reverse flow of water back into the pump and maintains pressure switch settings. All of our check valves come with accessory holes for a pressure gauge and pressure switch.

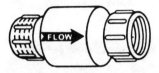

41-410	Check Valve, Bronze—¾"	$18
41-411	Check Valve, Bronze—1"	$19
41-412	Check Valve, Bronze—1¼"	$23
41-413	Check Valve, Bronze—1¼"	$28
41-414	Check Valve, Bronze—2"	$49

Foot Valves

Priming a pump is necessary to ensure that it isn't run dry. A foot valve prevents the need to do this every time that you start the pump. It is a one-way valve that lets water into the inlet line without letting it drain back into your water supply.

41-421	Foot Valve, Bronze—1"	$17
41-422	Foot Valve, Bronze—1¼"	$19
41-423	Foot Valve, Bronze—1½"	$35
41-424	Foot Valve, Bronze—2"	$49

Line Strainers

Roller pumps and most high-head pumps are particularly vulnerable to dirty water. A line strainer of at least 40 mesh should be used. We recommend 80 mesh for additional safety. These line strainers have clear bottoms so that you can see if they need to be cleaned.

42-508	Line Strainer, 80 Mesh—¾"	$29
42-509	Line Strainer, 80 Mesh—1"	$59

Float Switches

Water storage has its own special hassles and frustrations. These fully encapsulated mercury float switches make life a little easier for off-the-grid dwellers by providing automatic control of fluid level in the tank. You regulate the range of water levels by merely lengthening or shortening the power cable. Easily installed, the switches are rated to 5 amps at 12V. The "U" model will fill a tank (closed circuit when float is down, open circuit when up) and the "D" model will drain it (closed circuit when up, open circuit when down).

Note: if using these float switches with our PV Direct Controller on page 291, the U and D functions are reversed!

41-637	Float Switch "U"	$32
41-638	Float Switch "D"	$32

Water Storage

Polyethylene Storage Tanks

These poly tanks are approved for drinking water. They are slightly more durable than fiberglass and probably last about as long—some claim up to 30 years for either tank. The 1,400-gallon tank will nicely fit on any pickup truck. Tanks come with two 1¼-inch fittings at the top and one 1¼-inch fitting at the bottom. *All tanks are shipped freight collect from the manufacturer (Quadel) in Oregon.* The dimensions below are for the diameter on bottom x height.

Painting your water tank will protect it from UV rays and make either fiberglass or poly tanks last longer. (Might look nicer too!)

■47-201 1,400 gal 62"x97" 200 lbs. **$845**
■47-202 500 gal 48"x76" 125 lbs. **$595**

Kolaps-a-Tank

These handy and durable nylon tanks fold into a small package or expand into a very large storage tank. They are approved for drinking water and withstand temperatures to 140°F, and fit into the beds of several truck sizes. They will hold up under very rugged conditions, are self-supported, and can be tied down with D-rings. Our most popular size is the FDA98MT (525 gallons) which fits into a standard long-wide bed (5 x 8 ft) pickup.

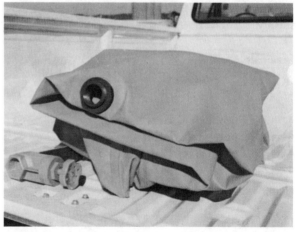

■47-401 FDA50MT 40"x50"x12" 73 gal $335
■47-402 FDA73MT 80"x73"x16" 275 gal $445
■47-403 FDA98MT 65"x98"x18" 525 gal $525
■47-404 FDA610MT 6'x10'x2' 800 gal $795
■47-405 FDA712MT 7'x12'x2' 1,140 gal $895
■47-406 FDA714MT 7'x14'x2' 1,340 gal $940

To order any of these products
or for more technical information
call us toll-free at
1-800-762-7325

Water Heating

Instantaneous Water Heaters

Would you keep your car running 24 hours a day, 7 days a week just in case you decided to go for a ride? Of course not. So why keep 30 to 50 gallons of water hot poised and ready just in case you jump up at 3 a.m. with the urge for a shower? Doesn't it make more sense to only heat the water when you need it?

This is the logic behind "tankless", "instantaneous," or "on-demand" water heaters. These heaters turn on only when a hot water tap is opened. They heat the water as it is passing through the appliance.

The benefits are numerous. There is no tankful of heated water to radiate heat to the surrounding area (often an unused cellar). Even with a well-insulated water tank, heat loss in an average household will be 15% to 20% of total energy input. In smaller households the loss is considerably higher because there is more "sitting" time.

Energy-saving tankless heaters have been commonplace in Japan and Europe for decades. The United States is one of the few civilized countries still using the archaic technology of storage tank-type water heaters.

We carry two brands of tankless heaters, the Paloma and the Aquastar. In both designs the water only touches noncorrosive brass, stainless steel, and copper. Unlike tank-type heaters that corrode, leak, and need replacement, all parts on tankless heaters are repairable or replaceable. With proper care, these are lifetime appliances. Because the water is always moving during operation, there is little or no scale build-up inside the heat exchanger. Tankless heaters use limited space and are designed for wall mounting. Venting must be vertical but may have a horizontal run (max. horizontal run 10 feet). Tankless heaters require a minimum of 20 psi water pressure to operate reliably.

Sizing Your Tankless Water Heater

The most important consideration is the flow rate in gallons per minute (gpm) for your hot water requirements. The following guidelines may be helpful:
Kitchen sinks: 1.5 to 2.5 gpm
Lavatories: .75 to 1.5 gpm
Low flow showerhead: 1.2 to 2.2 gpm
Energy hog guzzler showerhead: 2.5 to 3.5 gpm
Washing machine: 1.0 to 2.0 gpm
Dishwasher: 1.0 to 2.0 gpm

It takes about 50,000 BTUs/hr to reliably run a low-flow shower. The Paloma PH-5, at 38,100 BTUs/hr, and PH-6, at 44,800 BTUs/hr, are best used for sinks and kitchens. The PH-6 can be used for showers in summertime camp use or where input water isn't colder than 55°.

Due to a favorable foreign exchange situation, Aquastar units are presently the best buy. The model 80 (77,500 BTUs/hr), for one or two-person households, will supply any single fixture. The model 125 (125,000 BTUs/hr) will run multiple fixtures simultaneously for larger households. There is also an inexpensive recirculation kit available for Aquastars that allows them to be used for hot tub and recirculating water heater systems.

Aquastars have a thermostat on the output coils and will modulate the gas flame for steady output temperature regardless of changing flow or temperature rates (within the BTU limits of the heater). The standard units will modulate down to 20,000 BTUs/hour. The "S" option for solar or woodstove preheated water will modulate all the way down to zero. This makes it the only tankless heater to use when employing preheated water, as in solar domestic hot water systems and woodstove preheated water systems. Aquastar also has a toll-free customer and technical assistance number backed with two full time technicians and a fully stocked parts warehouse. These folks are committed to good customer back-up and service!

natural gas. This means that a heater rated at 91,500 BTUs/hr input will run for one hour on a gallon of propane. For those that buy propane by the pound, there are 4 pounds per gallon.

Water Heater Installation Tips

Gas piping: Most of these heaters require a larger gas supply line than the average tank-type heater. Just adapting a ½ inch supply line to ¾ inch at the heater won't work. If the heater says "¾-inch supply line," that means all the way back to the regulator. Most of these heaters also have a small regulator on the gas inlet. This is **in addition to** your standard regulator at the tank or gas entrance. Gas inlet is bottom center for all tankless heaters.

Pressure/temperature relief valve: Unlike conventional tank-type heaters, there is no P/T valve port built into the heater. This important safety valve must be plumbed into your hot water outlet during installation. Simply tee into the hot water outlet.

Venting: Tankless heaters **must** be vented to the outside. All models except the Paloma PH-5 come with draft diverter installed. The PH-5 can be used without the optional draft diverter only if installed in a protected outside location. All tankless heaters use conventional double-wall Type B vent pipe. This is the same type of vent as conventional water and space heaters. Vent pipe is **not** included with the heater. This vent pipe is easily available at plumbing, building supply, or hardware stores. The vent piping used with most of these heaters is larger than tank types use. Do not adapt **down** to existing vent size! Replace the vent and cut larger clearance holes as necessary if doing a retrofit. Venting may be run horizontally as much as 10 feet. Maintain 1 inch rise per foot of run. Ten feet of vertical vent is recommended at some point in the system to promote good flow.

Installation location: Don't install your tankless heater outside unless you are in a location where it **never** freezes. Freeze damage is the most common repair on these heaters. Don't tempt fate by trying to save a few bucks on vent pipe. If the heater is installed in a vacation cabin, put tee fittings and drain valves on both the hot and cold plumbing to ensure full drainage. The heat exchanger that works so well to get heat into the water works just as well to remove heat.

Customers frequently ask us about clearances for installing our Aquastars and Palomas. We've developed the following chart to help you plan your installation before you buy.

Water Heater Clearance Chart
(clearance from heater to combustibles)

Heater	BTU's/hr	Top	Bottom	Side	Front	Wall to flue centerline
Aquastar 80	77,500	12"	12"	1"	6"	5-1/8"
Aquastar 125	125,000	12"	12"	6"	36"	5-1/8"
Aquastar 170	165,000	12"	12"	6"	open only	7-3/4"
Paloma PH-5	38,100	16"	6"	6"	6"	5"
Paloma PH-6	43,800	16"	6"	6"	6"	5-1/8"
Paloma PH-12	89,300	16"	6"	6"	6"	6-1/2"

Dear Dr. Doug:
I don't have natural gas or propane in my house. Are there any tankless water heaters that use electricity?

Answer: There are some small under-counter electric units such as the Ariston that are ½ gpm tankless types. These heaters are designed for handwashing, but nothing bigger. Because whole-house tankless units heat the water "on the fly," energy input is **very** high when operating. A 240-volt electric heater of the same energy input as the Aquastar 125 would require considerably more energy than the average home electrical system is designed to handle (37,000 watts!).

Dear Dr. Doug:
Is it possible to use one of the tankless heaters on my hot tub?

Answer: It has often been done successfully. You need to bear in mind that these heaters are designed for installation into pressurized water systems. For safety reasons they won't turn on until a certain minimum flow rate is achieved. This is sensed by a pressure differential between the cold inlet and the hot outlet. Tankless heaters flow at rates of 5 gallons per minute and less, which is a much lower flow rate than hot tubs. A hot tub system is very low pressure, but high volume. What usually works is to tee the **heater inlet** into the **pump outlet**, and the **heater outlet** into the **pump inlet**. This diverts a portion of the pump output through the heater, and the pressure differential across the pump is usually sufficient to keep the heater happy. Aquastar has a very inexpensive "recirc kit" for its heaters that lowers the pressure differential required. This kit is mandatory! In addition you'll need an aquastat to regulate the temperature by turning the pump on and off (your tub may already have one if there was a heating system already installed). You'll also need a willingness to experiment and a good dose of ingenuity.

See diagram on page 321 for plumbing an Aquastar to a hot tub.

The Paloma PH Series

The Paloma is regulated by restricting the water flow to raise water temperature. This is a drawback for use as a solar backup, which really requires thermostatic gas control for automatic temperature regulation. The Paloma PH-6 is our best seller, providing 1.4 gpm at a 50°F temperature rise and is adequate for one tap at a time. Most customers are satisfied with the shower provided by the PH-6. The PH-5 is our most popular heater among RV users as the complete unit measures less than 16-inches high for easy installation (not code approved installation). The PH-5 produces nearly the volume of the PH-6 providing 1.2 gpm at a 50° temperature rise. The Paloma PH-12 will produce 2.9 gpm at a 50° rise. The Paloma carries a limited 5-year warranty on the heat exchanger and a 3-year warranty on all other parts and is available in white only. *Be sure to order either Natural Gas (NG) or Propane (LP).*

45-201 PH5-LP	$279
■45-201-N PH5-NG	$279
45-202 PH6-LP	$419
■45-202-N PH6-NG	$419
■45-203 PH12-LP	$795
■45-203-N PH12-NG	$795

All units shipped with draft hood except PH5

45-231 Draft hood for PH5	$55

Important note: All Palomas are available in either propane or natural gas models. Since we sell propane (LP) powered units by 30 to 1 over natural gas (NG) powered units, we will ship you a propane unit unless otherwise specified and *you will be responsible for the return freight if you order the wrong one.*

Aquastar and Paloma hot water heaters are not approved for use in RV's.

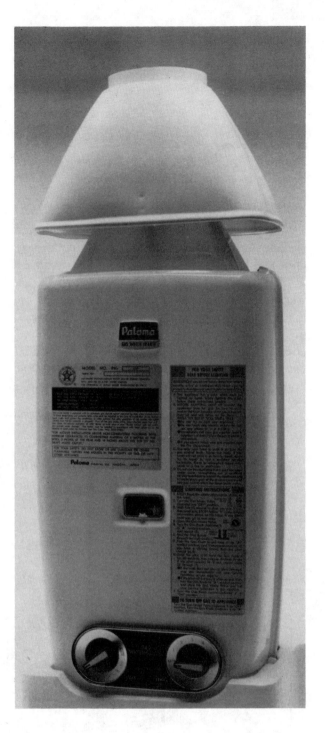

Paloma Tankless Water Heaters – Specifications

	PH-5-3F	PH-6D	PH-12MD	PH-16MD	PH-24MD
BTU/hr Gas input	38,100	43,800	30,000 (Min.) 89,300 (Max.)	30,000 (Min.) 121,500 (Max.)	37,700 (Min.) 178,500 (Max.)
Net weight in lbs.	13.7	20	36	66	76
Hot water output 50°F rise 100°F rise	g/m g/h 1.2 72 .6 36	g/m g/h 1.4 84 .7 42	g/m g/h (2.85) (171) 1.43 86	g/m g/h (3.8) (228) 1.9 114	g/m g/h (5.7) (342) 2.85 171
DIMENSIONS: Height in. Width in. Depth in.	15-13/16 11-17/32 9	29-5/8 10-1/4 9-7/8	35-3/4 13-3/4 11-3/4	40-3/4 18-3/4 13	40-3/4 18-3/4 14-7/8
Vent size in.	4	4	5	7	7
Gas connection in.	1/2 Male	1/2 Female	1/2 Female	1/2 Female	3/4 Female
Water connection in.	1/2 Male	1/2 Male	3/4 Male	3/4 Male	3/4 Male
Min water pressure	4.3 psi	4.3 psi	2.1 psi	2.1 psi	2.4 psi

TEMPERATURE RISE (°F)

( ) Mixing with cold water

Model	WATER FLOW RATE (Gallons per Minute)								
	0.5	1.0	1.5	2.0	2.5	3.0	3.5	4.0	4.5
PH-5-3F	119	59	40						
PH-6D	110/.6	71	48						
PH-12MD Hot Warm	100/.48 60/.79	100 60	100/1.43 60	90/1.59 60/1.59	(57)				
PH-16MD Hot Warm	100/.48 60/.79	100 60	100 60	100/1.9 60	90/2.11 60/2.11	(63)			
PH-24MD Hot Warm	100/.52 60/.87	100 60	100 60	100 60	100 60	100/2.85 60/3.17	90/3.17 60/3.17	(71)	(63)

WATER FLOW RATE AND TEMPERATURE RISE

( ) Mixing with cold water

Model		Temperature Adjusting Knob						
		Warm	1	2	3	4	5	Hot
PH-5-3F	Water flow (gpm) Max. Temp. Rise (°F)	1.5 40	1.5 40	1.5 40	1.3 46	1.1 54	0.8 75	0.6 100
PH-6D	Water flow (gpm) Max. Temp. Rise (°F)	2.0 35	1.8 39	1.5 47	1.0 70	0.8 88	0.7 102	0.6 117
		Warm		• • •			Hot	
PH-12MD	Water flow (gpm) Temp. Rise (°F)	.79 – 1.59 60		60 – 100			.48 – 1.43 – 1.59 100 90	
PH-16MD	Water flow (gpm) Temp. Rise (°F)	.79 – 2.11 60		60 – 100			.48 – 1.9 – 2.11 100 90	
PH-24MD	Water flow (gpm) Temp. Rise (°F)	.87 – 3.17 60		60 – 100			.52 – 2.85 – 3.17 100 90	

"Minimum water pressure" in this chart refers to pressure differential between cold inlet and hot outlet. All tankless heaters need about 20 psi system pressure to operate reliably.

Pressure booster kits are available for off-the-grid applications where sufficient pressure is a problem. See page 322.

To order any of these products
or for more technical information
call us toll-free at
1-800-762-7325

Aquastar Tankless Heaters

The French-made Aquastar tankless water heater performs much the same way as the Paloma except that it is thermostatically controlled, making it an excellent unit for a solar hot water or woodstove hot water system's backup. Aquastar was rated #1 by the leading consumer magazine. The Aquastar features a safety thermocouple at the burner and pilot, an overheat fuse, a manual burner control adjustment for finer temperature control, and built-in gas shut-off valves.

The Aquastar 80 is designed for use with one tap at a time and will produce 1.8 gallons per minute at a 60°F temperature rise. The Aquastar 125 will produce 3.25 gpm at a 60°F temperature rise. The Aquastar 170 will produce 4.5 gpm at a 60° temperature rise, and all 170 models are "zero modulated". The Aquastar is the only instantaneous water heater that should be used with preheated water systems. The "S" series is designed for use with solar or woodstove preheated water and is "zero modulated." What this means is that if the incoming water is hot enough, the Aquastar's burner will not come on at all. The Paloma, on the other hand, can be turned down to 20,000 Btu/hr, which means if the water is preheated the burners will fire at least 20,000 Btu/hr.

All Aquastars have a 10-year warranty on the heat exchanger and a 2-year warranty on all other parts. You must order either in propane (LP) or natural gas (NG).

45-102-P	Aquastar 80LP	$545
45-102-N	Aquastar 80NG	$545
45-105-P	Aquastar 125LP	$625
45-105-N	Aquastar 125NG	$625
45-106-P	Aquastar 80LPS	$595
45-106-N	Aquastar 80NGS	$595
45-107-P	Aquastar 125LPS	$695
45-107-N	Aquastar 125NGS	$695
■45-104-P	Aquastar 170LP	$995
■45-104-N	Aquastar 170NG	$995

Model 170 only shipped freight collect from Vermont or California.

Be sure to use correct item number to indicate propane (LP) or natural gas (NG).

The model 170 needs a 1 gpm flow rate to turn on and is usually not recommended for residential applications.

Aquastar Tankless Water Heater – Specifications

Aquastar and Paloma hot water heaters are not approved for use in RV's.

Specifications	Model 80	Model 125	Model 170
Power BTU input	77,500	125,000	165,000
Recovery time (gallons/hour)	78	126	174
Gallons/Minute			
60° rise to 115°	1.8	3.3	4.5
70° rise to 125°	1.7	2.8	3.8
90° rise to 145°	1.3	2.1	2.9
Minimum flow to activate burners (gallons/minute)	.75	.75	1.0
Vent Size	4"	5"	6"
connections			
Water	.5" Cu sweat	.5" Cu sweat	.75" NPT male
Gas	.75" NPT male	.75" NPT male	.75" NPT male
water pressure (psi)			
Minimum	15	15	15
Maximum	150	150	150
Height x Width x Depth	27.5x12x9.5	27.5x17x9.5	not given

Model 125 is the most popular heater for full households. It will support multiple fixtures simultaneously.

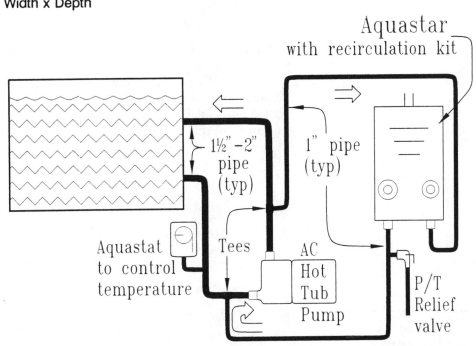

A typical hot tub heating system will look something like this. Remember! This type of heating system will require a willingness to experiment and a good dose of ingenuity, but the payback is simple inexpensive heating.

Tankless Water Heater Booster Kit

Both the Aquastar and Paloma water heaters require at least 15 psi to operate. Many of our rural customers have low water pressure and are unable to operate these heaters. For them we've developed our booster kit to allow a full pressure hot shower for even the most remote dwellers. The kit consists of three basic components: a Shurflo 12V pump (the highest flow model), a 20-gallon bladder-type pressure tank or equivalent to even out the pulses, and a pressure switch to control the system. The standard pipe fittings and 12V battery are not included.

45-235   Booster Kit $295

(Photo is only indicative. Actual kit differs)

Un-Clog-It Descaling Kit

The Un-Clog-It works great on tankless water heaters, as well as spa/hot-tub heaters, humidifiers, ice machines, water coolers, air conditioners, and cooling jackets. It will dissolve harmful mineral buildup from water lines. Minerals occurring in hard water will adhere to the sides of copper pipe, gradually choking the water path and eventually interfering with the normal operation of your water heating and cooling appliances. This kit dissolves the mineral buildup by circulating a safe hot acid solution. The kit contains a submersible pump, 1/2-inch inside diameter clear plastic hoses, heavy-duty brass swivel fittings, and the descaling agent consisting of 2 lbs of sulfamic acid, which will yield four 1-gallon treatments. The kit is available in three options: 1/2", 3/4", or Aquastar fittings for either 12V or 120V.

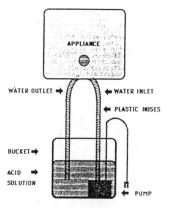

■**45-491 Un-Clog-It Kit (1/2")** $115
■**45-492 Un-Clog-It Kit (3/4")** $115
■**45-493 Un-Clog-It Kit (Aquastar)** $159
Please specify 12 or 120 volt
(Please note: there is a 25% restocking charge on returns of this item.)

To order any of these products
or for more technical information
call us toll-free at
1-800-762-7325

Solar Water Heating

Americans devote from 5 to 25% of their household energy budget to heating water. Often the water is heated to excessive temperatures due to ignorance or poorly calibrated thermostats. We can save energy by making sure that the water is not heated more than is necessary. The normal, everyday uses of hot water are referred to as **domestic hot water** (**DHW**) applications. Uses include heating swimming pools, spas and hot tubs, and some use hot water for space heating. All or part of the energy required can be provided by solar energy systems. Anyone who has been scalded by the water left in a garden hose knows that the sun has the power to do the job.

Heating water cost-effectively with energy from the sun has been practiced worldwide for nearly 100 years. A large number of designs have developed over the years, and we classify these types of solar systems as **solar thermal** systems, which take advantage of the heat available from the sun. **Solar electric** (photovoltaic) modules are not designed for the purpose of generating heat and are very different in design.

All solar thermal systems have two common elements: a solar collector and thermal storage. These two components may be integrated, as is the case in a "batch" heater (see drawing), or remote from one another. Energy is accumulated over the course of a sunny day to be used at the user's discretion. It is **critical** that every design incorporates an effective means to prevent freeze damage, as one freeze can cause expensive repairs that eliminate whatever economic gain had been previously achieved. Most properly sized DHW solar systems will reduce the heating required by conventional water heaters year round and eliminate the need for "conventional" energy input for several months of the year. One can save energy further by using a tankless, on-demand water heater in conjunction with a solar thermal DHW system.

The various designs of solar heating systems each have advantages and disadvantages. Batch heaters are usually the least expensive, however, they often have a limited storage capacity and they are heavy and therefore may overtax an unreinforced roof's load-carrying ability. Freeze damage is less likely with a batch heater, although supply and return lines should be well insulated. Batch heaters (as well as "thermo-syphon" and the Copper Cricket solar heating systems) are referred to as "passive" systems, meaning they do not employ pumps or controls, but rely on physical and chemical properties related to heat, such as conduction, convection, or phase change. One great advantage of passive systems is that, if properly designed, you can expect trouble-free operation for many years.

For large solar applications such as space heating, swimming pools, and industrial applications, active systems are preferable. Active systems are often classified by the type of freeze protection employed; antifreeze, drain-down, drain-back, or recirculation. Cost and effectiveness of each of these varies. (The Real Goods technical staff does not design active systems, although we will share our knowledge with interested customers.) If you are considering an active system we suggest soliciting several competitive bids and thoroughly check the contractors' references.

Want more info on Solar hot water? See the owners manuals on page 413.

Cold water in

Hot water to house

Hot water out

Blackened tank

Reflector

Glazing

Conventional water heater

Insulated box

Batch water heater – Passive solar batch heater preheats water for a conventional water heater; its darkened tank serves as both collector and storage.

Sun Family Solar Water Heater

We have found that in hot climates the PK-20 has just too much horsepower - in some cases, even with a family of four. In light of this we are now offering the PK-10 system. This is one four- tube 21 gallon collector. In hot areas the PK-10 can be cycled twice a day (totally drawn down and refilled). You will still need a PK-20 Sub System, which can handle up to 6 PK-10's. If you want to add one or more collectors later it is a simple task.

The Sun Family solar heater keeps water hot through the night, the same way a thermos bottle keeps your coffee at the right temperature—with a vacuum layer that drastically reduces heat loss. Made of super-strong, high-tech glass, that can withstand the blow of a 1¼" hailstone, the heater utilizes a double tube design that provides a full 360° of heat collecting surface.

Regardless of the sun's angle, the Sun Family unit soaks up the maximum possible amount of heat, morning, noon, and evening, in every season of the year. Even sunlight reflected from the roof is captured and absorbed. Combine a Sun Family with a supplemental instantaneous-demand water heater, and you'll always have all the hot water you need—at tremendous savings. Since this unit both heats *and stores* a generous 42 gallons of water, you need no seperate tank. And because it's directly connected to your water supply, you'll always have a high level of pressure, no matter at what level it's installed. It can be mounted at ground level, or against a wall. Modular design allows interconnecting as many units as required. This makes the system highly successful for large apartment houses, as well as small, individual installations. The Sun Family heater is the most exciting development in water heating to date—and the most cost-effective system on the market. We at Real Goods give it our highest recommendation.

Early tests show the Sun Family Water Heater providing approximately 35,000 BTU per day in the Sacramento climate.

The Subsystems contains an expansion tank, P/T valve, dribble valve, and miscellaneous parts required to complete the system. Although these parts can be obtained locally, we have bulk purchased the components and are offering them at our cost for your convienience.

■45-407 **PK-20 Solar Water Heater**	$1,995	■45-409 **PK-10 Solar Water Heater**	$995
■45-408 **Subsystem**	$165	■45-408 **Subsystem**	$165
■05-211 **Crating Charge**	$60	■05-216 **Crating Charge**	$30

Shipped freight collect from CA

Copper Cricket Solar Water Heater

The Copper Cricket has received attention for good reason; its designers learned from the failures of previous designs and came up with a new concept that assures performance and trouble-free operation for many years. The Cricket allows one to store a reasonable volume of water at ground level (not on the roof) with the special Cricket collector up to 36 feet above the tank. Heat is transferred with *absolutely no* controls or pumps! An antifreeze solution of 15% methyl alcohol and water is circulated via a "geyser effect" similar to a coffee percolator.

The Copper Cricket system specifies the use of an inexpensive 50 or 60-gallon electric water heater purchased locally instead of an expensive "solar" storage tank. The water is solar heated and stored in this tank. If the hot water has been heated sufficiently, it may then be directed to the end uses. If further heating is necessary, the preheated water is fed into the existing hot water heater where the temperature is raised to the final temperature. The complete system package includes the Copper Cricket collector, solar pad heat exchanger, and valve pack with mounting and connecting hardware.

The Copper Cricket, under ideal conditions, will provide 52 gallons per day of hot water heated to 160° F. enough to meet the normal needs of three or four people. A good handyperson can install the system in a weekend. System life expectancy is over 30 years. *Popular Science* selected this as one of 1989's most significant scientific breakthroughs.

The Copper Cricket is now available in a two panel system for larger families. Two standard Copper Cricket "geyser pumping" panels are plumbed to a single "Solar Pad" heat exchanger. An 80 gallon electric water heater tank is recommended as the minimum for this system, although one can use a tank as large as 120 gallons with this double panel system. Extra mounting hardware is included.

Installation kit includes drain valves, access valves, miscellaneous parts & pieces, and the hand vacuum pump that depressurizes the system and allows the "geyser effect" to start the Cricket chirping away! Note: $125 less the cost of any parts used will be refunded upon return of the Installation Kit. You will need the Installation Kit for proper installation!

Use either of these solar water heaters with the Aquastar "S" model for the most efficient water heating system possible.

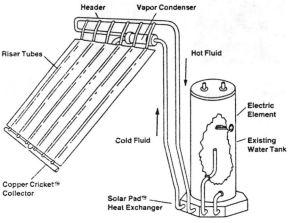

- ■45-405 Copper Cricket — $2,180
- ■05-230 Cricket Crating Charge — $50
- ■45-404 Copper Cricket (2 Panel) — $3,870
- ■05-231 Cricket Crating Charge (2 Panel) — $100
- ■45-415 Installation Kit — $150

Shipped Freight Collect from Oregon
Allow 4 weeks for delivery

Copper Cricket:"The most exciting, cost-effective solar hot water system to be developed in the last decade."
Worldwatch Institute

"It is what solar always should have been...it ought to replace pumped and controlled active systems... Personally I'd recommend it as the best system on the market."
Amory Lovins

Five-Gallon Solar Shower

This incredible low-tech invention uses solar energy to heat water for all your washing needs. The large 5 gallon capacity provides ample hot water for at least several hot showers. On a 70°F day the Solar Shower will heat 60° water to 108° in only 3 hours for a tingling-hot shower. This unit is built of four-ply construction for greatest durability and efficiency.

Solar showers feel better, because you're not stealing anything from your kids.
- Amory Lovins, in the Foreword

90-416 Solar Shower   **$12**

Solar Shower Enclosure

Our outdoor enclosure was designed specifically for the *Solar Shower*. It can also be used around portable toilets. The curtain measures 30" x 30" x 65" and is constructed of tough 4 mil vinyl. The inflatable frame sets up quickly with only a few puffs of air. The entire unit folds compactly for easy storage & weighs less than 4 pounds. Color: blue.

90-392 Shower Enclosure   **$26**

Power Shower Pump

Our new 12V pumping system greatly enhances our *Solar Shower's* performance and eliminates the need to hang it up. Plug the cigarette lighter adapter into your car or 12V home system, remove the *Solar Shower's* nozzle, and slide the power shower in. You can shower anywhere, wash dishes, clean fish, or clean equipment. Capacity is 200 gph at 1.7A on 12VDC. Also great as an emergency pump. *Solar Shower sold separately.*

90-391 Power Shower Pump   **$29**

Wood Burning Water Heaters

Chofu Wood-Fired Water Heater

The Chofu is a precision-built, high-efficiency, wood-fired water heater. It circulates water by thermosyphon without need for a pump. The Chofu can be used for hot tubs, domestic hot water systems, thermal-mass heating, or solar collector back-up (in unpressurized systems only.)

The stove body of the Chofu is 23 inches long and 16 inches across, constructed of durable 22 gauge stainless steel with a 1 1/8" thick water jacket surrounding the firebox. Heat transfers very rapidly through 9 square feet of heat exchange surface. The Chofu will heat 200 gallons of water from 55° to 105° F in 2½ hours. It takes wood up to 17-inches long and uses about two 5-gallon buckets of wood to heat a 200-gallon hot tub.

The Chofu is manufactured in Japan to a very high standard of quality. It was developed to maximize the heat output from the limited supply of wood available in Japan. (The Japanese have a passion for soaking in the "ofuro" or hot bath and use it daily.) The front of the stove is constructed of 1/8" steel, the circulation pipes are 1½", and the chimney stack is 4". A drain plug provides freeze protection.

Standard accessories provided include: through-wall tub ports, connecting hoses, ash rake, drill bit, and screws.

Many customers use galvanized "stock tanks" for hot tubs because of their low cost, wide availability, and ideal 24" depth. We have an enclosure kit available for converting a stock tank into a hot tub including foam insulation, cedar tongue and groove paneling (that makes the unit very attractive), stainless straps, and a drain system. A variety of accessories are available including vinyl covers, filters, stainless stove pipe, and more.

The photograph shows a 5-foot-diameter stock tank with the cedar enclosure kit.

■45-402 Chofu heater (½" tub wall) **$595**
■45-416 Chofo heater (½" plus tub wall) **$625**
■45-403 Cedar enclosure kit (Chofu) **$465**
You will need to measure the thickness of your tub wall, if it is greater than ½", please order item # 45-416.
Shipped freight collect from Washington State.

Send SASE for brochure on Chofu or Enclosure kit

"Since my introduction to the Japanese bath many years ago I use the Chofu to heat my small hot tub almost every day. The Chofu has served me as a faithful friend for 4 years. The grates, however, are now beginning to break and I will need to get some new ones. Thank you for the best purchase I have ever made!"
Eric Nelson, Vashon, WA

Here's a letter we received from Frazier Mann in Clinton, Washington:

"The Chofu heater and Snorkel stove are quite comparable in effectiveness as hot-tub heaters but differ greatly in the type and size of tub they are best suited for. The Chofu cannot practically heat over 350 gallons and therefore is best for heating stock tanks or other low-volume tubs. The Snorkel requires a larger tub to accommodate it inside and typically is used with a wooden tub of 450-900 gallons. From a survey of hot-tub users we found that a four-person stock tank heated by the Chofu has a similar heating time as a four-person wooden tub heated by the Snorkel. A complete system costs a little less with the Chofu. The choice is a matter of style, cost, and water and wood consumption."

The Snorkel Hot Tub Heater

The Snorkel Stove is a wood-burning hot-tub heater that brings the soothing, therapeutic benefits of the hot tub experience into the price range of the average individual. Besides being inexpensive, simple to install, and easy to use, it is extremely efficient and heats water quite rapidly. The stove may be used with or without conventional pumps, filters, and chemicals. The average tub with a 450-gallon capacity heats up at the rate of 30°F or more per hour. Once the tub reaches the 100° range, a small fire will maintain a steaming, luxurious hot bath.

Stoves are made of heavy-duty, marine-grade plate aluminum. This material is very light, corrosion resistant, and very strong. Aluminum is also a great conductor of heat, three times faster than steel. The Snorkel hot tub stove will heat up a hot tub 50% faster than an 85,000-Btu/hr gas heater and 3 times as fast as a 12-kilowatt electric heater!

■45-401 Snorkel Stove $595
Shipped freight collect from Washington State.

Real Goods Analysis of Wood-Burning Water Heaters

Now that we're featuring two different wood-burning water heaters we thought it might be time for some objective analysis. Below we list the weak and strong points, of each heater, with comparisons of each heaters' specifications:

Chofu heater
Heat exchange surface area – 9 sq ft
Approx. output – 32,000 Btu/hr
Firebox dimensions – 18"L x 14"W x 8.5"H
Wood length used – 17"
Time between refills at fast burn – 45 minutes
• **Does not take up space in the tub**
• **Can be used as a back-up hot water heater**
• **Slower heating, less efficent**

Snorkel Stove
Heat exchange surface area – 22 sq ft
Approx. output – 120,000 Btu/hr.
Firebox dimensions – 27"L x 12"W x 19"H
Wood length used – 26"
Time between refills at fast burn – 1 hour
• **Must be installed inside tub**
• **Tub heater only**
• **Fast, efficient heater**

Propane Space Heating

Ecotherm Direct Vent Heater

Direct venting makes the Ecotherm the ideal gas heater for bedrooms, bathrooms, and all other closed rooms, as well as super airtight homes. It draws in outside air for combustion and exhausts it through the same vent. Consequently, it doesn't deplete oxygen in the room and doesn't waste heated air.

The sleek, slimlined Ecotherm installs easily almost anywhere on an outside wall. Just cut a 4.75" hole and cut the direct vent tube to the thickness of the wall. Hang the heater on the wall and bring a ½" gas line to the valve. Since it heats so well without fans or blowers, the Ecotherm requires no electricity. Ignition is piezo-electric. A sensitive built-in thermostat accurately regulates heat for significantly lower gas consumption. This unit should not be located under windows which can be opened.

The special design insures that the front panel stays at a safe temperature. All components are of the highest quality. The steel heat exchanger is enamelled inside and out. The vent system is aluminum for corrosion resistance. Its strong outside terminal needs no additional protective grid, and the offset dual throat makes it waterproof. Will heat up to a 225 square feet or approximately a 13 x 13-foot room. Maximum Btu/hour output: 10,350. Size: 24.4"H x 27.09"W x 6.7"D. Weight: 55 lbs. Limited five-year warranty.

65-203 Ecotherm Direct Vent Heater $475
Please specify whether propane or natural gas

Catalytic Propane Heaters

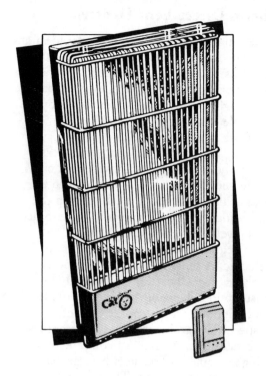

The "CAT" is made in Washington State and is the only vented propane catalytic heater on the market approved by the AGA (American Gas Association) for use in a home. It is a power vented heater that allows no combustion inside the room. It has a thermostat and has passed NFPA-501-C making it fully legal in California for either homes or RV's. They're quiet, easy to maintain, thermostatically controlled, and very economical with propane! Catalytic heaters work like the radiant heat of the sun.

CAT 6P-12A

The 6P-12A is a 6,000 Btu per hour heater with 12V automatic ignition, using less than ½ amp to power the fan. Its propane consumption is only ¼ lb. per hour. It will warm an area up to 175 square feet. Size is 21"H x 12"W x 5 3/8"-D. Mounting can be either surface or 2 3/4" recessed. Two-year full warranty.

65-201 6P-12A 12V Catalytic Heater $445

CAT 1500-XL

The 1500-XL is a 5,200 Btu per hour propane catalytic heater designed for use in homes. It has 110V automatic ignition. It is vented, thermostatically controlled, and fully approved by the AGA. It too will warm an area up to 175 square feet. Size is 28"H x 12 3/8"W x 5 3/8"D. Surface mounting. Two-year full warranty.

■65-202 1500-XL 110V Catalytic Heater $475

To order any of these products
or for more technical information
call us toll-free at
1-800-762-7325

Waterless Composting Toilets

All species must cope with the problems of waste disposal. In this nearly-twenty-first century, however, many of us (even the so-called enlightened ones) have no idea what happens to our waste after we hit the flush handle on our toilets.

In "wet," or waterborne, sanitation systems, excrement is flushed through a network of pipes along with other household wastes and effluent from business and industry. Raw or partially treated sewage is too often discharged into rivers, lakes, or coastal waters. Most cities deliver household sewage to central treatment plants, where physical, biological, and chemical actions eventually yield a purified wastewater effluent and the nutrient-rich substance called sludge.

Because of the trace heavy metal content in sludge (caused by industrial pollution), disposal options are increasingly limited and expensive. Landfilling is banned in many places, and where it is allowed it can cost as much as $150 per ton.

City dwellers worldwide generate between 100 and 150 million tons of nutrient-rich human waste every day, most passed into rivers, streams, and oceans without a thought toward reuse. The same basic nutrients (nitrogen, phosphorus, and potassium) that farmers purchase in chemical fertilizers are discharged wantonly into our waters. Not only does this foul the waters, it wastes the nutrient content. The Environmental Protection Agency estimates that American sludge has a total nutrient content equivalent of 10% of the chemical fertilizers farmers purchase—or a value of over $1 billion per year!

Contrast our waste disposal to third world countries without sophisticated "treatment plants," where soil fertility could not have been maintained over the last 4,000 years without human waste nutrient recycling. Chinese fish farmers fertilize their ponds with night soil as they have for thousands of years. Plants and microorganisms feed on the waste, and, in turn, become food for the fish.

While our methods of waste processing have some undeniable sanitary benefits, in a world of finite resources, closing nutrient cycles is a key to building a sustainable society. We can, and should, have the best of both worlds — sanitation and resource management.

Composting toilets make sense from numerous perspectives. They are environmentally and ecologically correct, and are economically efficient. Over 200,000 composting toilets have now been sold worldwide. Composting toilets are environmentally safe, require no septic system, no holding tank, use no chemicals, and produce no pollutants, while facilitating the work of nature.

Dear Dr. Doug:
I have a remote home that is powered primarily by PV panels. The outhouse routine is getting real old, especially in the winter. I'd like to move the bathroom back into the house. My questions are, do they smell, can I use my PV panels to power the electric model, what happens to all that stuff, and finally, are they legal?
Answer: "Do they smell?" In all honesty, they might occasionally smell, depending on loading, installation, and wind. The ventilation system on the toilet should ensure that no smell is detectable in the bathroom. If more ventilation is needed on non-electric models, (due to down draft from high surroundings etc.) a 12 volt fan can be added to the vent stack, to prevent odor.

What takes place in an outhouse is anaerobic decomposition: decay *without* oxygen. What takes place in a composting toilet is aerobic decomposition: decay in the presence of oxygen. Aerobic decay smells earthy, like a well-turned garden compost pile. (If an odor occurs, adding a fan to the vent stack usually provides a quick cure.)

"Can you use your PV panels?" Not unless you've got excess energy during the coldest months. We recommend the electric models whenever utility power is available. They feature a thermostatically controlled heating element to keep the compost at maximum activity level, and a fan in the stack to keep air circulating. The heater draws 250 watts, which is a substantial load for most PV systems. If you have hydro power, go for it!

"Where's it all go?" 90% of what goes in the top of a composting toilet is water, which is evaporated. This is why instead of flushing you add peat moss or similar material. The peat makes thousands of little wicks and air spaces to allow water to evaporate. The balance of the material composts down to a fluffy dry material that drops into the collection tray. If evaporation can't keep up with the load, then liquid will build up in the bottom and eventually drain into the overflow tube.

"Are they legal?" Yes and no. This is new technology, surprisingly, and the decision to accept or reject a composting toilet depends on the level of enlightenment of your local health official. The XL electric model with its **N**ational **S**anitation **F**oundation (NSF) seal of approval makes that decision easier for most county officials. The folks at SunMar have been very helpful and effective in convincing recalcitrant helath officials that these toilets are fine.

Sun-Mar Composting Toilets

We have now supplied well over 1,000 Sun-Mar toilets (formerly Bowli's) to our customers, and we are pleased to report that these toilets have exceeded our expectations and the manufacturer's specifications. We're convinced that this is the best small composting toilet system on the market. The old problem of resistance from doubting building inspectors has virtually been eliminated with the NSF approval (on the XL).

The heart of the Sun-Mar system lies in the revolutionary "Bio-Drum" composting process. The toilet's inventors are the same people who were involved in the original Swedish composting systems 27 years ago. The inventor was the recipient of the Gold Medal for the best invention at the International Environmental Exhibition in Geneva, Switzerland.

Sun-Mar composting toilets work like a compost pile. Human waste, peat moss, and kitchen scraps are introduced into the toilet; heat and oxygen transform this mixture into good fertilizing soil. The Bio-Drum ensures effective aeration and sterilization, killing anaerobic microbes and mixing the compost well. Turning the drum periodically maintains the aerobic composting process. Oxygen is provided by the ventilation system. The material entering the toilet is approximately 90% water, which is evaporated into water vapor and carried outside through the venting system. The remaining waste is transformed into an inoffensive earthlike substance (compost!).

Freezing temperatures *will not* damage the toilet or the compost; however, in temperatures below 50°F the composting action decreases. Toilet paper is composted along with the rest of the material. Composted material is removed one to four times per year, depending on use. Residential use may require removal slightly more often. This residual compost is the best garden fertilizer you can get.

How can the compost be dumped onto the garden when fresh waste is present? The device's bottom drawer serves the purpose: when the drum is approximately half to two-thirds full, some of the compost is cranked into the bottom drawer where final composting action occurs prior to use in the garden.

We have only rarely had a Sun-Mar customer complain about odor. The air flow provides a negative pressure which ensures no back draft. The air is admitted through ventilation holes in the front. The rotation and aeration by the Bio-

Drum or shaft-mixers along with the addition of organic material ensures a fast, odorless, aerobic breakdown of the compost.

All Sun-Mar toilets need a minimum of maintenance. All that needs to be added is a cup of peat moss per person per day, plus, if available, some other organic material such as vegetable cuttings, greens, and old bread. If the toilet is used continuously, once every third day the compost needs to be aerated and mixed. This is simply done by giving the handle a few turns.

All Sun-Mar toilets have a full 2 year parts warranty, and come with the vent stack and everything necessary for a Do-It Yourself installation. All units are certified by Canadian authorities, and the XL is fully tested and certified by NSF (the National Sanitation Foundation).

Sun-Mar N.E. (Non-Electric) Composting Toilet

The Non-Electric is our most popular selling composting toilet. It's perfect for many of our customers living off-the-grid and not wanting to be dependent on their inverters. The tremendous aeration and mixing action of the Bio-Drum, coupled with the help of a 4 inch vent pipe and the heat from the compost, creates a "chimney" effect which draws air through the system similar to a woodstove. The N.E. is designed for one to three people in residential use and four to six people in cottage use. In residential use a 12 volt fan is recommended to improve capacity, aeration, and evaporation. The N.E. is 22½"W x 28"H x 33"L.

44-101 N.E. Composting Toilet $1,149
05-212 Add for locations west of Rockies $50
Shipped Freight Collect from Buffalo, NY to locations east of the Rockies, and from our warehouse to locations west of the Rockies.

12V Toilet Exhaust Fan

At last, a powerful 12V fan designed to fit easily over a standard 4 inch ABS vent pipe! When used as a standard bathroom ventilation fan it can be wired in to your 12-volt lighting system via a switch.

It can also be used on a continuous daytime basis by connecting a 10 watt solar module (SM10 (#11-513) or Amorphous 10W (#11-555)). When used for ventilating composting toilets, such as the Sun-Mar N.E., it should be connected to a battery to provide around the clock ventilation. It is also a good system for greenhouse ventilation.

The bathroom fan uses only 9 watts (3/4 amp at 12 volts) and the fan motor is encapsulated to protect it from corrosion and locally fused to protect against motor burnout.

This exhaust fan is highly recommended for residential use and in locations where downdraft can occur, such as areas surrounded by mountains or high trees. It will greatly improve airflow and capacity.

44-802 12V Toilet Exhaust Fan $69

BioActivator Compost Helper

Our BioActivator greatly aids in the composting process. It greatly speeds the rotting of residue and works wonders when all else fails. This has been the perfect solution for many of our customers who have a difficult time maintaining an active composting environment. You can bring compost to life by using only one tablespoon per cubic foot of material or ½ pound per ton. Each 2 ounce packet contains over 50 billion beneficial bacteria in a nutrient-rich mix of kelp and bran that will turn 500 pounds (10 cubic feet) of organic waste into valuable compost.

44-250 BioActivator (2 oz.) $3
44-251 BioActivator (8 oz.) $6

What do you do with grey water? See page 413 for more information.

I'm writing to tell you that we have been using this Sunmar NE Composting Toilet and are very happy with it. His name is Roger (as in Roger Rabbit) and we feed him rabbit-appropriate food everyday. Melissa & Gregg Bach, Dover, NH.

If you have utility power available, the XL model or the electric version of the WCM are the way to go!

Dear Real Goods, Enclosed is a check for yet another Sun-Mar Non-Electric toilet. My neighbors love mine so much, they want one of their own. All of my original apprehensions have been completely answered. We've had our Sun-Mar installed 4 months here in Hawaii and there is no odor whatsoever. When we dump the compost it's a fine black humus that we can put right on our vegetable garden. How we could have ever contemplated spending $3,000 for a septic system is beyond me. Thanks for saving us so much money, and for making our lives simpler. Stan Dzura, Kealekekua, HI

The XL

The Sun-Mar XL has received full N.S.F. (National Sanitation Federation) certification after being put through rigorous testing at its maximum rated usage levels for a full 6 month period. For those with access to 120V electricity, the XL is a high-capacity unit ideal for year round or seasonal use, using a maximum of 280 watts for the thermostatically controlled heater and fan. It is designed for five to seven people in cottage use and two to four in residential use. For short periods the XL will safely handle double these numbers.

The Bio-Drum gives the XL (and other Sun-Mar models) an incredible ease of maintenance. You simply add peat moss and turn the handle every third day if the toilet is used continuously or once before leaving if used on weekends only. Comes with 1-1/2" vent pipe.

Dimensions are 22½"W by 29½"H by 33"L.

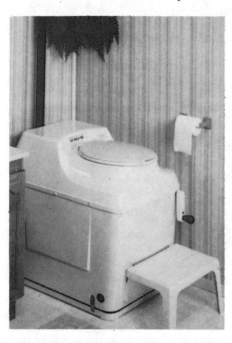

■44-102 **XL Composting Toilet** $1,295
Shipped Freight Collect from Buffalo, NY

The WCM

WCM stands for "Water Closet Multrum." The WCM is a high-capacity remote composter which operates in conjunction with a one-pint flush toilet. The toilet (or toilets) is connected to, and flushes into, the remote composter via a gravity-fed 3 inch ABS pipe. The WCM composter can be located up to 12 feet horizontally, and any distance vertically from the toilet, and placed either outside, or in a basement. For outside installation, a minimum of 31 inches (and preferably 3-4 feet) is required under the dwelling to permit installation.

The WCM has the Bio-Drum contained within and a thermostatically controlled heating element and electric fan. This unit is ideal for a larger family or people who don't want the compost in the same room with the toilet. It is large enough for a family of three to five in residential use or six to eight in cottage use. (For a short period these numbers can be doubled.) The WCM's fiberglass and stainless steel construction make it resistant to exposure to the elements, but for winter use it may be necessary to protect the unit so that the compost can maintain a minimum temperature. The WCM is the only flush toilet system that does not require a septic field or holding tank for the handling of human waste.

We offer the "Aqua Magic" low-flush toilet, made of plastic, or the Sealand low-flush vitreous china toilet with the WCM— both use less than one pint of water to flush. The compact low-flush toilet is mounted on a standard 3 inch floor fitting.

Sealand has a very effective flushing mechanism with a 360-degree rim flush. Installation is fast and easy with four easy-to-reach closet bolts that secure to the floor, and the base cover snaps in place. There is usually no need to alter plumbing fittings or connections; the water valve is on the side where service and connection are easy. All Sealand toilets are covered by a 2-year warranty and are built in the USA. Recommended for WCM or RV use only. These toilets cannot be connected to a sewer system. The Sealand 910 projects 19¼ inches from the wall and is 15½ inches wide and designed for areas where space is limited. It has 3/8 inch water connections which require a minimum of water pressure.

As an option the WCM is also available in a Non-Electric version. However, it is supplied with a 4 inch vent pipe which has to be installed vertically (no elbows). In the example, the

WCM-NE is installed only partially under the dwelling, so the vent pipe can run on the outside wall of the cottage. The WCM-NE capacity is rated at five to seven people in cottage use and two to four people in residential use. For residential applications where no electricity is available, we strongly recommend using the WCM-NE instead of the standard NE. We feel so strongly about this that we have just lowered our price (below our normal profit margin) on the WCM-NE to encourage you to purchase it instead on the NE. Most people prefer to have the composting action occurring in a location other than their own bathroom! Because the N.E. lacks a heater like the larger models, it has a small evaporation capacity. It therefore requires a small drain connected to a small 1-foot square drain pit for occasional overloads, typically ½ gallon per day; 8 feet of ¾ inch pipe is included for this purpose.

For those operating with a generator, the WCM is also available in an AE version, which means it has a 4-inch WCM-NE vent in addition to the regular 1½-inch WCM vent stack. The WCM-AE provides the increased capacity of the WCM while the generator is running, but works as a WCM-NE when electricity is not available. For residential use, when running in the WCM-NE mode, we recommend a 12 volt fan be fitted into the WCM-NE vent stack.

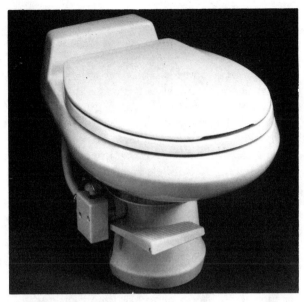

■44-405 **Sealand China toilet** $189
Shipped Freight Collect when ordered with Sun-Mar toilets; shipped UPS when ordered alone. **Send a business size SASE for a color brochure on the entire Sun-Mar line.**

Real Goods AE Hybrid

The Alternative Energy Hybrid was co-developed by Real Goods and Sun-Mar specifically for Real Goods' off-the-grid customer who derive part of their power from generators. The A.E. is identical to an XL, except that it is fitted with an N.E. drain, and with an additional 4" N.E. vent installed next to the XLs 1½" vent stack. The A.E. provides the increased capacity of an XL unit when a generator is running, but operates as a non-electric unit when 120 volt electricity is not available. For residential use, a 12 volt fan is recommended, when the A.E is running in a non-electric mode. System **does not** include 12 volt fan.

■44-201 **WCM Composting Toilet** $1,295
■44-202 **WCM-NE Composting Toilet** $1,149
■44-204 **WCM-AE Composting Toilet** $1,359
Shipped Freight Collect from Buffalo
■44-203 **Aqua Magic plastic toilet** $139

■44-103 **AE Hybrid** $1,359
44-803   12 Volt Fan $59
These are made to order and take 2 to 3 weeks. Shipped Freight Collect from Buffalo, NY

This quote comes from an article in Country Life about the Sunmar WCM: "The WCM is the only flush toilet system that does not require a septic field or holding tank. However, it gets 'hungry'. "That's why we call it Audrey. Actually the manufacturer does recommend that your consider the system a rabbit. It enjoys raw kitchen wastes (vegetables, green leafy material, bread or grass), but you wouldn't feed it coffee or other cooked waste. Like a rabbit, it benefits by using bedding material—peat moss, dead leaves and soil."

CTS 410 Composting Toilet

While our Sun-Mar toilets work great for small capacity systems, *Composting Toilet Systems* makes a great unit for larger capacities. This unit is nearly identical to the original Clivus Multrum. It is constructed of double-wall fiberglass with 1 inch of insulation (R-8) between the walls. The CTS uses absolutely no water, no chemicals, and has no adverse impact upon the environment. CTS toilets are in use by the Army Corps of Engineers, the National Park Service and has lots of residential users.

Aerobic decomposition takes place in the fully insulated fiberglass tank. This digester tank has a sloping floor upon which the composting waste pile is built. Organic materials such as waste, tissues, toilet paper, and wood fibers are accumulated over a period of time. Baffle walls and air channels are part of the digester tank, creating an atmosphere rich with oxygen, which is the ideal environment for microorganisms to digest the organic waste. The natural air flow in the stack is assisted with either an AC or DC solar fan, which creates a vacuum inside the digester tank. The end product is fertile organic humus, has no odor, and is easily removed.

The CTS 410 is designed for full-time long-term use for up to four people. Larger models are available for public facilities and large groups. The CTS can accommodate up to 40,000 uses per year. While the Sunmar is generally dumped six to twelve times per year, the CTS 410 is dumped only once a year after the second year of use. Only fully digested humus is removed, about two bushels worth annually. The small fan (specify 12V or 120V) is used only when the toilet is being used so only very minimal amounts of power are used. All CTS products are warranted for 3 years, except 1 year for the fan. **Write for free brochure.**

■**44-601 CTS 410 Composting Toilet $3,650**
Shipped Freight Collect from Washington State

TANK DIAGRAM LETTERING LIST:
1. Vent Pipe
2. Turbine Fan
3. Electric Fan
4. Kitchen Garbage Chute
5. Toilet
6. Emergency Access Door
7. Compost Storage Chamber
8. Air In
9. Air Ducts
10. Compost Access Door
11. Height 72"
12. Easy Drain System
13. Length 98"

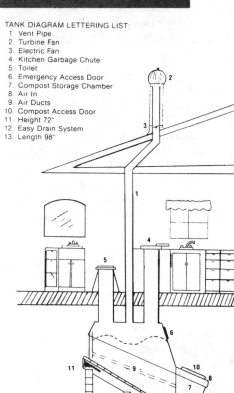

This is the basic installation, two toilets, either Back to Back or Upstairs & Down can be connected to one CTS tank.

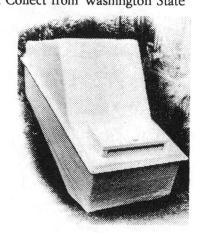

Tools

The use of tools is often cited in scientific circles as the behavior that distinguishes the human species from the rest of the animal world. Sure, the occasional ape has been found fishing for termites with a straw, but this hardly compares to the off-the-gridder in Maine using a 486 computer to communicate via the EcoNet satellite network with a South African animal rights activist about legislative tactics to promote passage of an environmental bill in Washington.

What's the tool here? Is it the computer? The modem? The phone lines? The PV panels that generate the power? Or the Pocket Leatherman that is used to fix almost everything that goes wrong?

Real Goods has a Tools category. Everyone agrees that tools belong in Real Goods. No one, however, can agree what consitutes a tool, and therefore what belongs in the category. If we could recapture the time and energy that has gone into this debate at the company... Oh well, in the final analysis, the time spent on tools has been time well spent, as this is where we define what qualifies as a "real good."

We have never bothered to define on paper what makes something a "real good"; that would be too limiting. We sell toys, books, videos, power systems, mosquito repellents (nontoxic, of course), subscriptions... whatever is needed to make independent living achievable, even if it is in your suburban back yard.

The products must represent good value, be durable, fairly priced, and justify their existence economically.

Once we get enough of a certain type of product, and receive signals from our customers that this is something they need us to supply, we create a new category, such as we did recently under the tag "mobility."

What about sunglasses? Should Real Goods sell sunglasses? You can buy them at any convenience store, but only the chintzy ones with neon rims that look zippy at the beach but afford no real protection against ultraviolet rays. You can get UV protection with the designer glasses that cost a small fortune (to pay for all that advertising in *Vogue* and *Esquire*) and come with a handsome, durable, nonslip, braided string—worth at least 30 cents—so that you can show your friends at the cocktail party that you can afford Serengeti sunglasses, even if you have no intention of wearing them.

Isn't there an honest pair of sunglasses? Something durable, that doesn't have neon rims or the little string, but accomplishes an essential function—protecting your precious eyes from the sun—and does it at a fair price?

If such sunglasses exist, we have been unable to source them. If we do, however, you will likely find them in the "tools" category.

How about work gloves? Not the disposable kind from the discount store that don't fit well or are uncomfortable to begin with (but it doesn't matter because they are meant to be thrown out after one day in the garden anyway), but good, quality gloves.

Or boots... or how about sunscreen... lip balm? Undyed cotton clothing? Non-toxic paints? Recapped tires? Recycled timber frame homes? None of these products are currently offered by Real Goods, but all are possibilities. If we can source the right product at the right price, and obtain it reliably from a supplier who is honest and quality conscious, then it is a "real good." And you will see it first on the pages in the "tools" category.

At present the heart of our tools offering consists of manual products that substitute for powerized, fuel-guzzling counterparts. The push mower is a perfect example. In the fifties power mowers were a rare possession for the average homeowner. Then technology managed to put the equivalent of one-and-a-half to three horses underneath the metal housing. We immediately cut grass like crazed characters from a Looney Tunes cartoon, in the process converting peaceful Sunday afternoons into noisy, fume-filled affairs. "Hey, hon. See if the kids have cut off any fingers or toes yet, and while you're up, get me a Bud."

This is progress?

We are just beginning to emerge from a generational tunnel where the fascination with power was so complete that no application was considered too absurd. Electric can openers, the Water-Pik, the leaf blower, the weed whacker, the electric hedge clippers, the electric charcoal starter, the Dust Buster—hang on to these beauties, because they will be tomorrow's collectibles. Someday you will tell your grandchildren, "Yes, little Elvis, when I was your age, I really had an electric pencil sharpener in my own room. Sure, it cost a lot to buy and to operate, and when it broke, it was unfixable, but for a while I thought it was so neat that I'd buy new pencils and sharpen them right down to the nub!"

– J.S.

POWER DOWN

My Own Personal, Toxic Waste Dump
(Recycling Made a Slob Outta Me)

Let me tell you about my workshop.

It's a simple place. The walls are rough sawn, addled with nails and hooks that hold this and that. There is a jukebox that holds all my old 45s and plays them with scratchy reverence. There is a deep sink that is perfect for washing the bottles I use for home brewing.

There are lots of tools, in various states of disrepair and disorganization. Heavens knows, it is not called a work shop because any work gets done there. The tools are for caressing, and to make me feel appropriately masculine. I'm always buying new ones at yard sales. Occasionally, when I'm depressed, I buy something from Sears, Brookstone, or Real Goods. I have tools that have not been used in 15 years and won't be for the next 15 years.

Such tools can last a lifetime.

If you get the impression that this simple space is a sanctuary for me, an oasis from the turmoil of everyday life, the pressures of the Showdown in the Gulf, an escape from a family that wants more than I can give and a career that somehow never measured up to the potential that everyone thought I had...

If you get the idea that I like this workshop, you're right. If you think that I lose myself in its nooks and crannies, you're right. If you think that this is the only place in the world where a middle-aged guy can act like himself, you're right.

Maybe, then, you can understand my distress at the fact that my workshop has become a landfill. The environmental movement is partially to blame. Recycling means that we separate everything into separate bins, so instead of having two full trash cans, I now have eight, each one-quarter filled.

I don't resent the environmental incursions on my workshop; at least the sacrifice is for a good cause. It's the equipment morgue that bothers me. These are the alleged tools that I own that no longer function, can't be repaired, but are in too good shape to discard. I've got two weed whackers that fall into this category (one electric, one gas), a chain saw, an electric sander, a power drill, two boomboxes, a Walkman (oops, Walkperson), and a riding lawn mower.

This does not even include the kitchen gear in the basement morgue. That's a separate scrap pile.

Every year I dutifully haul the stuff out for a yard sale, only to bring it back at the end of the day, sad reminders of the cracks in the American Dream.

The lawn gear makes me maddest, because my precious workshop space is devoted to the dysfunctional implements designed to turn my property into a personal toxic waste dump.

The manicured suburban lawn that has always been a key component of the American Dream has now been massaged, lobotomized, sanitized, and airbrushed to the point it has the personality of an empty-headed starlet. Our personal patches of land are groomed, dosed, and pumped so aggressively that they are more flawless than Astroturf. The male warrior (the same one who, as a teenager, would go to any extreme to avoid mowing the lawn) now feels justified in the use of any poison, herbicide, and pesticide and the deployment of lawn control devices built to military specifications to bring nature under control on his 100 by 80 foot lot.

Have you ever flown over a suburban city, especially in the Southwest? Have you ever wondered why the lawns are greener than any color found in nature and the water (in the aboveground pools) is bluer than the Caribbean?

Our mastery of backyard nature has been achieved in much the same way that we "won" the war in the Persian Gulf—by blasting life to smithereens with weapons of immense destruction. Now, increasingly, we are finding that our overkill has been indiscriminate. By killing the beetles and the crabgrass, we are doing ourselves in, as well.

Things are so out of kilter, however, that it is unreasonable to expect overnight readjustment. According to an article in *Time* Magazine (6/3/91, "Can Lawns Be Justified?"), Americans spent $6.4 billion on lawn-care products in 1989. Statistics are not readily available, but the management at Real Goods estimates that this is more than 20 times the amount spent nationally on products that promote energy independence.

The best solution is probably the most simplistic. If the pursuit of the flawless lawn has led us down the path of perdition, let's try another path. Let's take Dad (who's probably in need of a little exercise anyway) off his F-4 Phantom Riding Lawnmower and put him behind the push mower that he loathed as a lad. Before long the lawn will have shrunk to a manageable size, and Dad can spend his new-found leisure time powered down in the hammock. As he sips a well-earned beer, he may discover that there is a new American Dream right in his own back yard—the dream of energy independence.

It promises to be a hard fight. We need a few brave volunteers.

— Stephen Morris

Notes and Workspace

Tools

Workbikes

Bicycles have a long and honorable history as work vehicles. Now that our cities are clogged with traffic and the air is seriously polluted, human-powered cargo transport makes more sense than ever. Workbikes are the first machines that cycling enthusiasts can readily appreciate and enjoy riding. Both the Express and the Long Haul models feature a comfortable upright operating position, with straight handlebars and soft, supportive seats. They have low standover height—one frame fits all. Frames are constructed of true temper 4130 cromoly straight gauge steel tubing with TIG welding and powdercoat finish. They're guaranteed for 1-year.

The bikes utilize 20-inch front and 26-inch rear alloy wheels with stainless steel spokes, and alloy three-piece cranks. There are three gearing options: one-speed Bendix foot-activated rear coaster brake; Sachs internal five-speed hub with lever-activated front and rear brakes; or Suntour 21-speed drivetrain with Sachs drum brakes front and rear. Both models are available with rack or container. The Express with rack weighs 42 lbs., has a 43-inch wheelbase and a capacity of 100 lbs. The 17" x 20" x 20" container weighs 15 lbs., carries 80 lbs. The Long Haul with rack weighs 60 lbs., has 73-inch wheelbase and 200 lb. capacity. The 24" x 24" x 36" container weighs 26 lbs., carries 170 lbs.

■91-835	Express 1 Speed w/Rack	$825
■91-836	Express 5 Speed w/Rack	$945
■91-837	Express 21 Speed w/Rack	$1,075
■91-838	Express 1 Spd w/Container	$925
■91-839	Express 5 Spd w/Container	$1,045
■91-840	Express 21 Spd w/Container	$1,175
■91-841	Long Haul 1 Speed w/Rack	$1,225
■91-842	Long Haul 5 Speed w/Rack	$1,345
■91-843	Long Haul 21 Speed w/Rack	$1,475
■91-844	Long Haul 1 Spd w/Container	$1,475
■91-845	Long Haul 5 Spd w/Container	$1,595
■91-846	Long Haul 21 Spd w/Container	$1,725

Glenn Portable 24V MIG Welder

This portable Glenn MIG wirefeed welder will make a welder out of anyone. Only 23 lbs., it can weld from 60 to 200 amps, 24VDC, for sheet metal gauges through ½-inch plate. With voltage and wire speed (current) control it uses 0.035-inch self-shielding wire—no gas bottle is required. It comes with a MIG wire feed gun, ground clamp, and 6 feet of input cable. Output: 60–200 amps at 15–20 volts, input: 24 VDC, Dimensions: 11"W x 10"H x 20"D. Warranty 1 year. Replacement 2 lb. wirefeed spools are available for mild steel only.

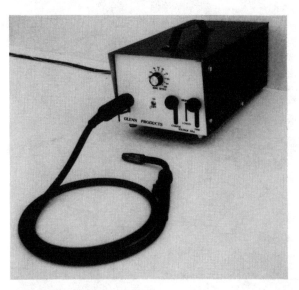

■63-120	Glenn 24V Welder	$795
■63-121	Wire Spool	$ 19

Butane Gas Match

This is the best "butane match" we've been able to find. It saves many a burnt finger in lighting gas cookstoves, hot water heaters, woodstoves, lanterns, barbecues, and campfires. It has an easy piezo-electric ignition so it never needs batteries or flints. It refills with a standard butane canister and is very easy to refill and adjust the flame. It has a 5-inch metal barrel.

63-310	Gas Match	$12

Gas Alarm

The gas alarm detects fumes from propane, butane, bottled gases, and natural gas. It detects gas concentrations at 10% of the lower explosion limits. A powerful alarm sounds at 84db (measured at 3 ft.) L.E.D. power indicator. It operates at temperatures from 50° to 120° F. and has a 2-year warranty. Available for standard 120V house current or 12VDC. Power draw is 3 watts at 12VDC and 6 watts at 120V-AC. Sensor lifespan is 8–10 years. Our tests indicate that the unit is quite sensitive and reliable. Mount above source for lighter-than-air natural gas detection, below the source for heavier-than-air propane or butane.

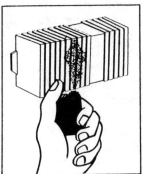

57-120	Gas Alarm(120V)	$89
57-121	Gas Alarm(12V)	$89

To order any of these products
or for more technical information
call us toll-free at
1-800-762-7325

The Sun Oven

The Sun Oven is a great solar cooker. We have been using it for over a year now, and love it. The Sun Oven is very portable—one piece—and weighs only 21 lbs. It is ruggedly built with a strong fiberglass case and tempered glass door. The reflector folds up and secures for easy portability on picnics, etc. It is completely adjustable, with a unique leveling device that keeps food level at all times. The interior oven dimensions are 14"W x 14"D x 9"H. It ranges in temperature from 360° to 400°F. This is a very easy oven to use and it will cook anything! After preheating, the Sun Oven will cook one cup of rice in 35-45 minutes.

63-421 Sun Oven $195

Pyromid Portable Barbecue

The Pyromid is a portable barbecue, stove, oven, smoker, and a roaster that's great for picnics, beaches, camping, or backyard use. It's easy to carry (folds to 1-inch thick by 12-inches square and comes with a tote bag), lightweight (6 lb) and sets up without tools ready to use in 15 or 20 seconds.

The Pyromid's surfaces are all stainless steel, making it rust-free and easy to clean. These surfaces cool down quickly, unlike hibachis or kettle grills, which take up to a half hour to cool. Standard accessories include a hood which lets you bake, regulate heat, smoke, or just shield from the wind.

All units come with a lifetime warranty. The Pyromid is a very ingenious and simple piece of technology.

Zip Stove

This little stove is remarkable. Great for camping, ideal for backpacking, the Zip Stove Sierra uses any fuel you can find. It has a built-in fan that blows on the fire like a blast furnace. In our test using a handful of oak twigs, 2 cups of water boiled in just 90 seconds — better than a gas stove. Zip stove operates on one AA battery, with storage space for a spare *(batteries not included)*. An optional windshield and grill assures fast cooking on windy days. Windshield-/grill is made of lightweight aluminum, folds flat, weighs only 16 oz.

63-422 Zip Stove $45
63-423 Windshield & Grill $22

63-418 Pyromid Portable Barbecue $75

A Radiation Stove! No moving parts, no electricity, and a half life of 50,000 years! Contact Three Mile Island for further details.

Our Grill-Stove slips neatly into a saddle bag for trips to the wilderness. Up to now we have never carried a stove.
Carl & Erla Kemp, Denver, CO

My wife cooked me a wonderful birthday cake on the beach when we were camping. The rest of the campers went nuts!
Jeff O, Potter Valley, CA.

Food Dehydrators

Preserve your excess produce the easy way, with this food dehydrator. Enjoy healthful snacks, such as apple rings, dried tomatoes, squash, carrots, and herbs for rich flavor in soups and sauces. Think of the money you'll save, not buying snacks and other foodstuffs! Within the dehydrator's attractive black case, a fan forces air horizontally across the plastic trays, removing moisture rapidly, so that flavor, color, and nutrients are locked in. And there's no need to rotate trays. The 120VAC unit has a wattage draw of 252 for the 4-tray, 430 for the 5-tray, and 650 for the 9-tray, but wattage is not continous because the unit's thermostat regulates temperature up to 145°F, by turning the power off and on. Preparation guide and recipes are included. One-year limited warranty.

■63-369	4-Tray Dehydrator	$109
■63-368	5-Tray Dehydrator	$155
■63-367	9-Tray Dehydrator	$229

Apple/Potato Peeler

A great time saver for busy cooks. This traditional, newly updated kitchen tool peels, cores, and slices apples in one operation, peels potatoes quickly and cleanly. It reduces the incidence of cut fingers, too. Just clamp it on a counter or table, attach the apple or potato, and turn the crank. The stainless steel blades neatly remove the skin while taking off very little of the fruit or vegetable. The frame is strong cast iron.

63-364 Apple/Potato Peeler   **$25**

Wheatgrass Juicer

Among the many claims for the benefits of wheatgrass juice is that it acts as a blood purifier and general detoxifier, due to the highly available, live chlorophyll it contains. It's touted as a high-energy, high-protein, health-promoting drink, rich in vitamins A, B-complex, C, E, and K. But most juicers are unable to extract the juice from wheatgrass. This manually operated, 15 inch high model is specially designed for the job, and can handle leafy vegetables, sprouts, and herbs as well. Made of fine cast iron, hot tin-dipped for a durable finish, it's precision machined for superior juice extraction and smooth operation. The strainer is stainless steel. There's a clamp attachment for portability, and a screw-slotted base for semipermanent mounting.

63-366 Wheatgrass Juicer   **$159**

Victorio Multi-Strainer

Enjoy the fruits of your garden year round by preserving food for the winter, as our grandparents did. The Victorio Multi-Strainer lets you create delicious tomato sauce, paste, and juice, applesauce, and other fruit sauces, for canning and freezing. Puree pumpkin for pies, or any other cooked vegetable. Make your own strained baby foods, too. It's much faster and easier than laborious, old-fashioned methods. No peeling or coring is necessary. As you turn the handle, juice and pulp are automatically separated from skin and seeds. A squirt guard catches splashes; a rubber seal prevents leakage around the handle. The Victorio is equipped with a 1/16 inch screen for tomatoes and apples; a 1/8 inch pumpkin screen; a 3/64 inch berry screen; a spiral cone for processing grapes; and a plunger. Hopper, squirt guard, spiral cone, and plunger are strong plastic that won't react with acid foods. The housing is sturdy cast aluminum. Handle and shaft are steel; screens are double-plated steel.

■**63-405 Victorio Multi-Strainer**   **$79**

Champion Juicers

The Champion juicer is the finest, most reliable, and most versatile juicer on the market. It works great on carrots, all vegetables, apples, and also makes nut butters. It is 120VAC only and runs great off of Trace inverters (812 and larger). Amp draw is 5.7A from a 120V source. Available in almond or white. All units come with a 1-year warranty on the motor and a 5-year warranty on parts. Specify color.

■**63-401 Champion Juicer**   **$269**
■**63-402 Grain Mill Attachment**   **$ 89**

Corona Stone Mill

The Corona is the deluxe corn and grain mill. It's a stone mill with interchangeable metal plates that's easy to convert back and forth. With the stones you can grind fine flour in one operation with no heating problems. You can easily adjust from coarse to fine grind for any dry grain. The stones are manufactured especially for the Corona hand mill and are bound with a special bonding that will allow no flaking off. The stones are made of a vitrified carbon material that will last a lifetime. When you need to crack cereal or grind moist items simply attach the metal plates. Comes with high hopper and two augers for easier changing of plates.

63-410 Corona Stone Mill   **$89**

Kirby Vacuum

Hokys are used extensively by large hotels and theaters for their quiet operation.

One of our long-time customers has a business reconditioning Kirby Vacuum cleaners. Kirbys, like lots of other American appliances, were made better in the Fifties and Sixties than today. Of even greater significance to Real Goods customers is that these older models are lots more efficient, using only 3 or 4 amps compared to today's electricity hogs that use up to 7 amps! That means they'll run on a 600 watt Trace inverter handily and be gentler on your batteries if you have a 12-volt system. For those of you not familiar with Kirby, it is quite simply the best vacuum ever made, and *it uses no replaceable bags!* Two replacement belts and a spare guide light are included as well as an instruction booklet and a six month warranty. (reconditioned)

The Kirby Tool Kit includes a crevice tool, duster brush, utility air nozzle, portable handle, shoulder strap, suction coupler, curved extension tube, 5-1/2' flexible hose, & two straight extension tubes equalling 36".

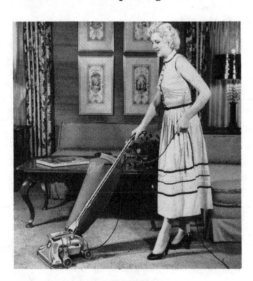

■63-320 **Kirby Vacuum** $219
■63-318 **Kirby Tool Kit** $49

Hoky Carpet Sweepers

Meet the undisputed king of manual carpet and floor sweepers — the incomparable Hoky. Operating almost noiselessly, it cleans astonishingly well. In fact, many Hoky owners seldom or never use their electric vacuum cleaners anymore. Why pay the power company just so you can pick up some dirt? Or if your home is powered by alternative energy, why use current that you could save for appliances that can't function without it? We offer two Hoky models: the economical, efficient 23T, and the heavy-duty, commercial N/T. The spiral rubberized blades scoop up lint, feathers, pet hair, nails, broken glass, string, sand, gravel, and dirt from thick shag carpet, bare wood, vinyl tile - just about any surface in the home or office. Horsehair brushes get right up to the wall and push dirt into the 12.5"-wide path of the blades. The effort you expend is minimal, since the sweeper is so lightweight and maneuverable — though very well made and durable. About the only maintenance you'll have to do is occasionally wipe clean, or comb the rotor brush. And there's no awkward bag to dump out or change — only easy-emptying, hinged-lid receptacles. The handle is 43" long. Fifty years ago and more, most homes had a carpet sweeper. Now it's back, newly engineered and planning to dethrone your vacuum cleaner.

63-413 **Hoky 23T Sweeper** $45
63-449 **23T Replacement Brush** $11
63-451 **Hoky Commercial** $75

Leatherman Pocket Tool

The Leatherman "pliers in the pocket" tool was developed and perfected over a 7 year period by a machinist who realized his Swiss Army knife was severely limited for many tasks. The pocket survival tool is constructed of 100% stainless steel, with the optimum grade/hardness for each tool/blade. Its main attribute is its full-sized, full-strength "fan handle" pliers combo (regular and needlenose) and wire cutters inside. The tool is designed to military specs and all parts are interconnected. The stainless steel file/saw cuts wood and metal and sharpens fish hooks. It has a full range of screwdrivers (small, medium, large, and Phillips). It also features a knife blade, 8-inch ruler, can/bottle opener, and an awl/punch. Closed, it's only 4-inches long and 5 ounces, and comes with its own leather carrying case. Guaranteed for 25 years!

63-201 Leatherman Pocket Pliers Tool $39

Mini-Leatherman

Leatherman has introduced a second tool for those who want the full-size, full-strength pliers, but like a lighter tool in their pocket. It features needlenose pliers, regular pliers, wire cutters, knife blade, ruler, can opener, bottle opener, 1/4-inch tip screwdriver, and a metal file. It is constructed of 100% stainless steel, guaranteed for 25 years, measures 2-5/8" x 1" x 1/2" and weighs only 4 ounces.

63-202 Mini-Leatherman Pocket Tool $34

Uncle Bill's Tweezers

The "Sliver Grippers" made by Uncle Bill's Tweezer Company are quite simply the finest tweezers you'll ever use. Made of spring-tempered stainless steel, the precision points are accurately ground and hand-dressed. With these tweezers it's easy to find and grip even the tiniest splinter or stinger. No pocket, purse, first aid kit, or tool box should be without a pair! All tweezers come with a lifetime moneyback guarantee and a convenient holder that fits on your keychain! Our local Lyme Disease Control Center is now recommending Uncle Bill's Tweezers for removing ticks.

63-428 "Sliver Gripper" (set of 3) $12

Lyme Disease Protection Kit

The menace of Lyme disease is spreading. Now the inventive people who brought us Uncle Bill's tweezers are doing their part to safeguard your health. The Tick Removal Kit contains a pair of Uncle Bill's tweezers, a magnifying lens, two sealable plastic pouches to put the ticks into for later lab analysis, two antiseptic swabs, instructions on tick removal, and a photograph of a deer tick (the common carrier of Lyme disease), and a list of Lyme disease symptoms. If you live in an area where Lyme disease is known to be established, or any tick-infested region, this outfit belongs in your first aid kit.

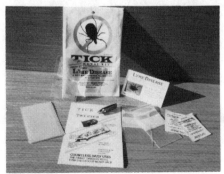

63-429 Tick Kit $8

Celestron Sports Binocular

Here's a product that was a natural for our product review committee. Our entire group of five were so impressed that each ordered a pair immediately. This new 10x50 "Sport" binocular from Celestron, famous for quality optics, is an exceptional value featuring fully coated optics, a Porro prism, center focus, and a 7-degree field of view with a near focus of 30-feet. Weighing only 29 oz., the incredible clarity is unmatched in its price range. A thin, soft rubber covering prevents damage from rough handling and provides a sure grip. Tripod adaptable, carrying case and neck strap included.

67-203 Sport Binocular $119

8x25 Mini Monocular

The 8x25 is Celestron's smallest and lightest monocular, weighing a scant 4 oz. It's an ideal traveling companion and built to last. It has 8-power magnification, a 25-mm objective lens, and is only 3.7-inches long. A carrying case, neck strap and lens cap are standard.

67-201 Mini Monocular $59

10x25 Mini Binocular

This is a lightweight but powerful binocular; 4-inches in height, it weighs 9 oz. and has 10-power magnification and a 25mm objective lens.

67-202 Mini Binocular $149

Economy Handwarmer

Winter is unforgettable, with its gift of numb, aching, icy fingers. But now your hands can stay warm and comfortable as you battle the frost giants. Take the handwarmer hiking, skiing, or camping. Use it around the house, in your car, in your office. New, non-toxic heat storage technology replaces electrical cords and dangerous chemicals. You get safe, moist heat whenever you need it: simply push the button, and the solution pad heats instantly. Stays hot nearly twice as long as microwavable gel packs and the like. Recharge the 3" x 4" pad by boiling it in water. Made from food-grade sodium acetate and water.

63-361 Heat Solution (Pack of 3) $15

Indoor/Outdoor Thermometer/Clock

This great thermometer features a huge LCD display (3/4-inch numbers) that can be read across the room, and a 10-foot cable with a weatherproof remote temperature probe. This can be used to monitor a fishtank, freezer, doghouse, or outside temperature. A simple switch toggles from inside to outside temperature. It reads from –58° to 122°F. and comes with a 1-year warranty. Also included is an accurate quartz clock. Battery included.

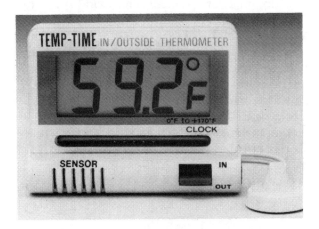

63-501 Indoor/Outdoor Thermometer   **$25**

Stick-Up Thermometer

This is an ideal simple thermometer for inside refrigerators, RVs, houses, or cars. It comes with a magnet and adhesive disk to stick anywhere. It measures accurately from –20° to 120°F. One-year warranty. Two-inch round dial.

63-507 Thermometer(set of 3)   **$12**

Perception II Indoor Climate Monitor

The Perception II measures temperature, barometric pressure, and humidity simultaneously. Not merely interesting, informative, and a signaler of weather shifts, the Perception II can be helpful in maintaining the proper environment for computers, specialized electronic equipment, delicate plants — not to mention people. A barometric trend arrow shows at a glance if pressure is changing. Features: temperature from 32°F to 140°F; high and low temperatures; barometric pressure (with memory recall); humidity; high and low humidity; time and date recorded with highs and lows; alarms for temperature, humidity, and time; barometric trend alarm for change greater than .02, .04, or .06 inches of mercury in an hour; 12/24 hr. clock; date. Scans selected functions; reads in metric or U.S. units. Designed for desk, shelf, or wall; measures 5.25" X 5.9" X 3". Works on AC with included adapter or 12V with optional lighter cord (listed below); has battery backup.

63-380 Perception II   **$195**

Weather Wizard II

The Weather Wizard II is a professional-quality station that monitors indoor and outdoor weather at the touch of a button. Everyone lives and works in microclimates that may vary in a short distance from "frost pocket" to "banana belt," and where conditions may differ greatly from those reported for the general area. So Weather Wizard is a useful tool for farming, for the dedicated home gardener, and for any business affected by weather. Additionally, it's fascinating for anyone with a scientific bent or a closeness to Nature. Functions include inside temperature from 32°F to 140°F; outside temperature from -50°F to 140°F; highs and lows; wind direction in 1° or 10° increments; wind speed to 126 mph; high wind speed; wind chill to -134°F; low wind chill; highs and lows with time and date; temperature, wind speed, wind chill and time alarms; 12/24 hr. clock; date. Measures 5.25" x 5.9" x 3"; can be wall mounted. AC operation with battery backup or 12V operation with optional lighter cord (listed below). Comes with anemometer with 40 feet of cable; external temperature sensor with 25 feet of cable; junction box with 8 feet of cable; AC power adapter.

63-381 Weather Wizard II   **$265**

Weather Monitor II

This top-of-the-line model combines all the operational abilities of Weather Wizard II and Perception II, without any increase in size: it, too, is just 5.25" X 5.9" X 3". Glance at the display, and see wind direction and wind speed on the compass rose. Check the barometric trend arrow to see if pressure is rising or falling. The unit provides inside temperature readout from 32°F to 140°F; outside temperature from -50°F to 140°F; high and low temperature; wind direction in 1° or 10° increments; wind speed to 126 mph; high wind speed; wind chill to -134°F; low wind chill; barometric pressure (with memory recall); inside humidity; high and low humidity; timed and dated highs and lows; alarms for temperature, wind speed, wind chill, humidity, and time; barometric trend alarm; 12/24 hour clock; date. Comes with anemometer with 40 feet of cable; external temperature sensor with 25 feet of cable, junction box with 8 feet of cable; AC power adapter. Optional 12V DC lighter cord (listed below).

63-382 Weather Monitor II   **$395**

Extension Cable

Gives more latitude in the placement of the anemometer, external temperature sensor, Rain Collector, or Weatherlink, by adding 40 feet to the length of the standard cables. Depending on the electromagnetic and radio frequency interference in the area, cables may be linked together for runs up to 80-160 feet.

63-387 4 Conductor 40 Foot Cable   **$19**
63-388 6 Conductor 40 Foot Cable   **$24**

Weatherlink

For in-depth weather studies, Weatherlink teams the Perception II, Weather Wizard II, or Weather Monitor with an IBM-compatible computer, to create graphs, calculate average weather conditions, generate summaries, analyze trends, and more. For example, over time one might be able to trace weather effects from global warming, fluctuations in atmospheric and oceanic currents, or sunspot activity. Weatherlink stores data until it is transferred into the PC. Data may be exported to Lotus 1-2-3 or dBase III compatible spreadsheet or database software. Software Features: instant weather bulletin displays the weather on one screen. Graph any function on a daily, weekly, monthly, or yearly basis. Graph two days, weeks, months, or years on the same screen. Display two different functions on the same graph. Track information from two or more weather stations (one Weatherlink required for each station).

Installation: Fit Weatherlink inside the mounting base of the weather station and plug it in. Then run the cable to a serial port on the PC. To monitor weather conditions in remote locations, use the Modem Adapter with a Hayes or compatible modem.

Specifications: For IBM PC, XT, AT, PS/2 or compatible personal computers with 512K conventional memory. Requires Hercules monochrome, CGA, EGA, VGA or compatible video graphics adapter and monitor, MS-DOS or PC-DOS 2.1 or higher, and one serial port. Supports RS-232 serial ports 1, 2, 3, or 4 and most dot matrix and laser jet printers. Comes with 9-pin and 25-pin RS-232 serial port adapters, 8 feet of cable, and 5.25" 360K floppy disk.

63-383 Weatherlink Module   **$175**
63-384 Modem Adapter   **$7**

Rain Collector

Since rainfall is one of the major components of climate, Weather Wizard II and Weather Monitor II achieve their maximum potential when joined with Rain Collector. It allows reading both daily and accumulated rainfall totals. The self-emptying receptacle measures precipitation in 0.1" (3mm) increments, with exceptional accuracy. The ruggedly built unit has an easy plug-in connection; comes with 40 feet of cable and mounting hardware.

63-385 Rain Collector   **$59**

DC Lighter Cord

(For climate monitors)
For greater flexibility in weather research. For travelers by car, truck, RV, or boat who like to keep track of weather conditions. And for those who live and/or work off the grid. This cord with cigarette lighter plug replaces the standard AC adapter, allowing all the weather stations to operate on 12V DC power.

63-386 Weather Station 12V Cord   **$9**

Weathercaster Barometer

Also known as a "storm glass," this is a very interesting and unusual weather instrument. It consists of a beautifully crafted hand-blown glass flask, and brass hanger with drip tray. This centuries-old barometer is used by simply filling with water, (colored if you like) and hanging on the wall. Changes in barometric pressure force the water up or down the spout, indicating the approach or recession of a storm. This is a conversation piece that is reliable, accurate, and fun to use for all ages. The flask measures about 9-inches tall, and the hanger is 12-inches.

63-346 Weathercaster Barometer   **$29**

Colonial Sundial

The ultimate in solar clocks, our beautiful brass-plated reproduction of an antique New England sundial is buffed to a rich luster. It features the wording, "Let others tell of storms and showers; I'll only mark your sunny hours." A dragon supports the gnomon (pointer) and it measures 8¾-inches in diameter and 4½-inches tall. Sundials have been in use for thousands of years and have never required winding or batteries. This will make an elegant addition to any garden or yard.

63-347 Colonial Sundial   **$39**

The Sou'wester Anemometer

The Sou'wester is a low-cost instrument perfect for those who want basic information about their current windspeed. Easy to install, and powered by a self-generating anemometer, the Sou'wester displays current windspeed on an easy-to-read dial. Comes complete with a wind sensor, 60' of cable, stub mast and instructions.

■63-353 The Sou'wester   **$139**

Hand-Held Windspeed Meter

This is an inexpensive and very accurate windspeed indicator. It features two ranges, 2–10 mph and 4–66 mph. With included chart, conversions to knots can be made. Speed is indicated by a "floating" ball viewed through a clear tube. Many of our customers have been curious about the wind potential of various locations and elevations of their property, but are reluctant to spend big money for the bulk of the anemometers on the market. Here is an economical solution. Helpful tip: to find the wind speed at higher elevations above ground, try taping the meter to the side of a long pipe (a piece of tape over the finger hole for high range reading) and have a friend hold it up while you stand back and read with binoculars. Complete with protective carrying case and cleaning kit.

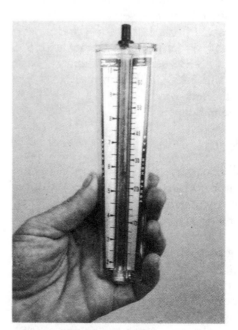

63-205 Windspeed Meter   **$19**

The 2100 Totalizer Anemometer

The 2100 Totalizer is a moderately-priced instrument that accurately determines your average windspeed. The Totalizer is an odometer that counts the amount of wind passing through the anemometer to accurately indicate average windspeed in miles per hour; simply divide miles of wind by elapsed time. A 9-volt alkaline battery (included) provides 1-year of operation. The readout can be mounted in any protected environment. Complete kit includes a Maximum #40 Anemometer, 60-feet of sensor cable, battery, stub mast for mounting, and instructions.

■63-354 The 2100 Totalizer $219

8000 Series Wind Hawk Anemometer

The 8000 Series Wind Hawk is ideal for wind site surveys for home, farm, and small business wind power applications. Wind turbine owners find it invaluable for monitoring turbine generation, power production, and wind variations. The 8000 continuously computes and displays eight functions: present windspeed, peak gusts, average windspeed, power density, elapsed time, hour of peak gust, hours above cut in and wind status. Comes complete with sensor, 60 feet of cable, stub mast, AC adapter, instructions.

■63-355 The 8000 Series Wind Hawk $395

The Totalizer or the Wind Hawk are ideal for monitoring a wind site for future generation potential.

To order any of these products
or for more technical information
call us toll-free at
1-800-762-7325

After more than a decade of hard use (homesteading grime, diapers, kid dirt), I can report that the James Washer is a solid alternative. Like any other move away from the easy (but often environmentally costly) gadgetry of our times, it takes a little longer, and certainly more involvement, but it's not outrageous. (Three full sized loads takes me about 2 hours from first fill to final clothespin - less than 1/4 of that time spent actually agitating.) It's been a valuable tool.
J. Foote, Copper Hill, VA.

"I thought you might be interested that we have had a James washer for almost a year now. I do about two loads a week and I find it's not very difficult work at all. I do all my son's diapers on it and they get very clean and soft, without using any bleach or softener. It's nice to be able to do laundry at home, and even nicer to do it outside in nice weather."
Sue Calhan, Waldoboro, ME

Solar Pathfinder—Site Analyzer

One part of designing a system is picking the best place to mount the panels. Also, you may want to know if there are any good sites at all on your property. This simple-to-use tool eliminates all the guess work. Place it anywhere and it will instantly show you the sun path for every day of the year. If there is a telephone pole or a tree blocking the way, the Pathfinder will tell you what time the sun will be blocked by it and for how long. Put it on your kitchen table and it will tell you when the sun will shine on a plant there all year round. We have investigated several of these instruments and found this one to be the simplest to operate and by far the most versatile. It even works great in cloudy conditions, and even in moonlight! When you have taken a reading, you can record it on paper and compare results from other sites. Includes data for your specific latitude; data for other latitudes are available. No solar technician or installer should be without this valuable and finely made tool. Useful for photovoltaic, solar hot water, and passive solar applications. Please specify latitude at which Pathfinder will be used.

63-348 Solar Pathfinder   **$189**

James Washers

The James hand-washing machine is made of high-grade stainless steel with a galvanized lid. It uses a pendulum agitator that sweeps in an arc around the bottom of the tub and prevents clothes from lodging in the corner or floating on the surface. This ensures that hot suds are thoroughly mixed with the clothes.

The James is sturdily built. The corners are electrically spot-welded. All moving parts slide on nylon surfaces, reducing wear. The faucet at the bottom permits easy drainage. *Wringer attachment pictured with the washer is available at an additional charge.*

63-411 James Washer   **$195**

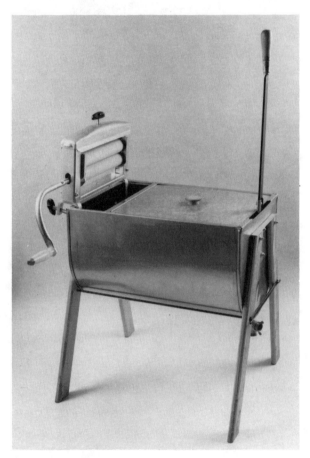

Hand Wringer

The hand wringer will remove 90% of the water, while automatic washers remove only 45%. It has a rustproof, all-steel frame and a very strong handle. Hard maple bearings never need oil. Pressure is balanced over the entire length of the roller by a single adjustable screw. We've sold these wringers without a problem for over 13 years.

63-412 Hand Wringer   **$119**

Solar Clothes Dryers

Electric and gas clothes dryers are big-time energy hogs. Even worse, the rough spin ages clothes prematurely. Our solar dryer uses sun and wind to gently dry clothes, leaving them with a naturally fresh scent that can't be duplicated by chemical additives. The large accordion laundry rack, 48" x 36" x 27", gives you over 40 feet of drying space. It's made of poplar and birch; the hardwood and extra-thick dowels provide superior strength and durability. We've found that, even with two children in cloth diapers, this rack can do the job.

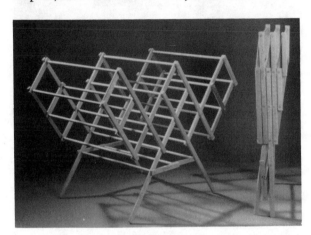

■63-435 **Classic Wooden Rack** $79

Space Saving Clothes Dryers

Here is the old-fashioned clothes line with a modern twist. These compact dryers mount on a wall and are ready to quickly pull out when you need them. When you are done with the washing, they reel right back in.

The **Sunline's** five parallel lines extend to 34 feet providing 170 feet of straight usable line. Tightening knob keeps lines taut and tangle-free. Can be mounted on outdoor post, wall, in garage, or basement. Easily moved to new locations, comes with lightweight case.

The **Minidryer's** five parallel lines extend 12-1/4 feet each, providing a total drying area of over 61 feet. Lines retract to protect from dust, dirt, and stains. Tension knob keeps lines taut.

The **Reeldryer's** 40 feet of vinyl-coated single line neatly stores (tangle-free) when not in use. Line end has a spring hook for quick attachment to the provided screw hook. Easily moveable.

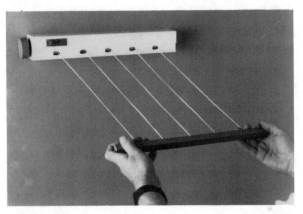

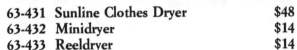

63-431	Sunline Clothes Dryer	$48
63-432	Minidryer	$14
63-433	Reeldryer	$14

In most cases, the best buy is to get the most efficient end-use device you can, then just enough renewable energy supply to meet that greatly reduced demand.
- Amory Lovins, in the Foreword

See page 263
for our 12V
Bed Warmer

Heated Mattress Pads

Here are 120VAC bedwarmers truly designed for maximum comfort. They feature a zone system that puts more warmth at your feet, where you need it most; medium heat in the center; and no heat at your head, where the pillow rests. The pads do a better job of keeping you cozy than electric blankets and are more energy efficient, because the heat rises from below and is held in by the covers. The silent, lighted thermostatic control keeps the pad at the setting you prefer. At the highest setting it reaches 75°F, more than enough for most people. With other AC bedwarmers, unhealthful electromagnetic fields have been a cause for concern. These pads greatly reduce EMF radiation through a new wiring design that allows electrical current to flow in two opposite directions. Made of 100% *Trevira* polyester, the pads are machine washable and dryable, with less than 2% shrinkage. They're fully warranted for 5 years.

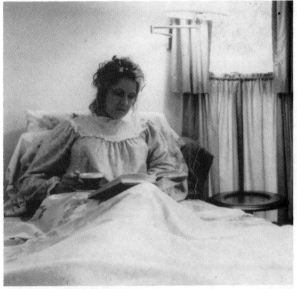

63-370	Twin 39" x 75"	$45
63-371	Full 54" x 75"	$49
63-372	Queen 60" x 80"	$59
63-373	Dual King 78" x 80"	$89
63-374	Calif. King 72" x 84"	$89

To order any of these products
or for more technical information
call us toll-free at
1-800-762-7325

Wood Heating

The Heartland Cookstove

At the turn of the century, before the advent of central heating, the kitchen was the energy center of the house. The pinnacle of design was achieved with the wood cookstove, a powerhouse of form and function that was the family's source of warmth, food, and even hot water. The Oval Cookstove was originally designed in 1906 and has been the focal point of energy independent homes ever since.

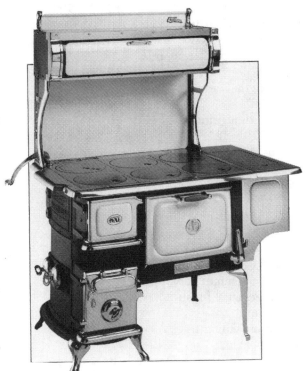

The Oval of today (and its smaller sister the Sweetheart, both produced by the Heartland Company of Canada) shares the same stunning appearance and multi-functional purposefulness of its ancestor, but its interior has been steadily improved over the years, to make wood-fired cooking easier than it was in great-grandmother's day.

The Oval can produce up to 55,000 BTU/hr, enough to heat a maximum of 1,500 square feet. Its airtight firebox can hold a generous 35-pound load of wood, enough to burn through the night. Its proficiency as a heater is matched by its cooking versatility. The oven can handle a 25-pound turkey, while the 6-square-foot cooktop, with six separate temperature zones, gives you flexibility to handle all the needs of an elaborate holiday dinner. There's even a warming cabinet on top for the dessert pies and two ways of heating water. First is a traditional copper-lined reservoir with spigot, second is the optional stainless steel water jacket inside the firebox to be plumbed into your hot water storage. Both the Oval and the Sweetheart have durable, firebrick-lined fireboxes, totally capable of full-time operation with wood or coal. They are solidly constructed of cast iron, with porcelain finish and easy-to-clean nickel fixtures.

Thousands of these classic stoves have been sold worldwide. Now, discover how simple, effective—and heartwarming—energy independence can be. These stoves demonstrate some principles of economy, efficiency, and beauty that the age of oil has made us forget.

All stoves are FOB Ontario, Canada. Call for freight quote; charges will be added to your invoice. Freight is quite reasonable since the manufacturer has a 60% discount with the freight company.

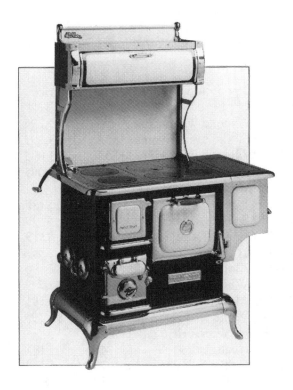

Heartland Specifications

	Sweetheart	Oval
Heating capacity – Sq. Ft.	800-1,500	1,200-1,500
Heat output(BTU/hr)	12,000-35,000	18,000-55,000
Firebox capacity (hardwood)	22 lb.	35 lb.
Wood length	16"	16"
Stove top dimensions	22" x 32"	24-1/4" x 34"
Cooking elements or lids	6	6
Flue size	7"	7"
Weight (without reservoir)	385 lb.	500 lb.

Standard Features
Adjustable summer cooking firebox
Porcelain-enamel finish
Nickel trim
Ash pan on sliding track
Warming cabinet
Certification: Warnock Hersey Laboratories

Optional Extras
Coal grate
Water Reservoir (5 gal./Sweetheart, 7 gal./Oval)
Water jacket

Item #	Description	Weight (lb.)	Price
Oval			
■61-110	Almond w/reservoir	560	$3,295
■61-111	White w/reservoir	560	$3,295
■61-112	Almond w/o reservoir	520	$3,095
■61-113	White w/o reservoir	520	$3,095
Oval Accessories			
■61-118	Heatshield	48	$185
■61-119	Stainless Water Jacket	15	$199
■61-120	Coal Grate Package	36	$119
Sweetheart			
■61-114	Almond w/reservoir	430	$2,595
■61-115	White w/reservoir	430	$2,595
■61-116	Almond w/o reservoir	390	$2,395
■61-117	White w/o reservoir	390	$2,395
Sweetheart Accessories			
■61-121	Heatshield	48	$185
■61-122	Stainless Water Jacket	18	$170
■61-123	Coal Grate	22	$89

Freight Additional, Call for quote

Kettle Humidifier

This traditional cast iron humidifier looks great on any woodstove and provides a moistening steam to mitigate the dry wood heat. It holds 3 quarts of water and has a brass handle.

63-436 Kettle Humidifier $24

Dragon Humidifier

Fill this cast iron or brass medieval sculpture with water and set it on top of your woodstove. As the stove heats up, the dragon breathes steam out its nostrils creating not only a dazzling sensation but humidifying the air as well.

63-415 Cast Iron Dragon $119
63-417 Polished Brass Dragon $179

Farm and Garden

Mighty Mule Solar Gate Opener

This is a heavy-duty, durable automatic gate opener. The Mighty Mule will open gates up to 16-feet wide or up to 250 pounds. It is very easy to install and doesn't require an electrician. It comes with an adjustable timer for automatically closing the gate, and will automatically latch the gate shut. A 12V, 6.5 amp-hour battery is included that can be easily charged with the optional 5-watt solar module. The unit can also be powered by 110V with a step-down transformer if you don't need the solar panel option. All units come with a wireless remote transmitter so that you can open the gate without getting out of your vehicle. Additional transmitters are often purchased so that you can keep one in each car. It takes 18 seconds for the gate to open and 18 seconds to close.

Mighty Mule's electronics are state-of-the-art. The gears are all metal, not nylon or plastic. The highly efficient motor provides enough power to operate ornamental iron and commercial chain link gates up to 250 pounds. Even in high cycle applications the Mighty Mule will not overheat. Solar panel mount not included. Does not include mounting hardware.

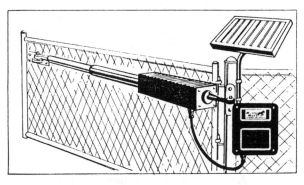

∎63-126	Mighty Mule	$735
∎63-127	5-watt solar panel option	$95
∎63-142	Horizontal Gate Latch	$119
∎63-144	Extra Transmitter	$39
∎63-146	Digital Keypad	$69

Send SASE for free brochure.

Parmak Solar Fence Charger

The 6 volt Parmak will operate for 21 days in total darkness and will charge up to 25 miles of fence. It comes complete with a solar panel and a 6-volt sealed, leakproof, low internal resistance gel battery. It's made of 100% solid-state construction with no moving parts. Fully weatherproof, it has a full 2-year warranty.

63-128	Parmak Solar Fence Charger	$195
63-138	Replacement 6V battery for Parmak	$35

Rainmatic 2500

The Rainmatic is great for summer vacations and peace of mind as it will automatically water your garden—up to four times per day and from one to seven days per week. On and off watering times can be easily programmed to last from one minute to 24 hours. It can be switched to manual with one touch, and it uses your existing hose or faucet. It's ideal for automating drip systems and is simple to install. It uses 4 "C" batteries (not included) Alkalines will last one full season, or Golden Power Nicads about three months.

43-601	Rainmatic 2500	$59

The Curmudgeon joins with those gentle readers that think it shocking to turn the benevolent rays of the sun to this purpose.

Rainmatic 3000

The following letter from Pam Wilkenson says it all: *"Dear Real Goods, I just ordered a Rainmatic 2000 water timer from you, & would like to suggest that you carry the Rainmatic 3000 instead of or in addition to the 2000. The 3000 model is far more versatile, offering eight waterings per day (instead of four) and an additional misting option, which can be programmed to water in any interval from 1 minute to 24 hours, for time periods from seconds to hours. The 3000 is wonderful, and is the only battery-powered water timer with a misting feature on the market!"*

Pam Wilkinson, Santa Barbara, CA

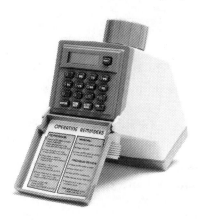

43-602 Rainmatic 3000   **$79**

Inline Sediment Filter

Sometimes your water isn't dirty enough to mess with fancy and expensive filtration systems and all you need is a simple filter. Our inline sediment filters accept standard 10-inch filters with one-inch center holes. They are designed for cold water lines only and meet National Sanitation Foundation (NSF) standards. Easily installed on any new or existing cold water line (don't forget the shutoff valve), they feature a sump head of fatigue-resistant Celcon plastic. This head is equipped with a manually operated pressure release button to relieve internal pressure and simplify cartridge replacement. They're rated for 125 psi maximum and 100°F. They come with a ¾-inch FNPT inlet and outlet and measure 14" high by 4-9/16"in diameter. It accepts a 10-inch cartridge and comes with a 5-micron high-density fiber cartridge.

41-137 Inline Sediment Filter   **$49**

Arkal Filters

The Arkal filter is probably the best designed filter for drip irrigation systems. It also works famously for all filtration systems. Instead of using the traditional cartridge or fine mesh screen for filtering, the Arkal uses an assembly of thin rings. A spring holds the rings tightly together when the filter cover is on and lets them separate when the cover is removed. This makes cleaning the filter very quick and easy, as well as avoiding the expense of replacing cartridges.

Another advantage of this system is that it can maintain a higher gph and psi than with a cartridge or mesh system.

The filters have built-in valves that can be used both to turn your water flow off and to regulate pressure up to 120 psi. They have a 140 screen mesh.

	3/4" Filter	1" Filter
Max. flow rate:	18 gpm	27 gpm
Filtering vol.:	6 cu. in.	27 cu. in.
Filtering area:	25 sq. in.	47 sq.in.

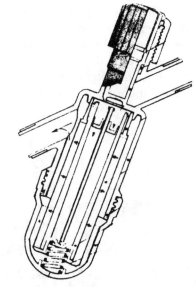

43-201 Arkal Filter with shutoff (3/4")   **$39**
43-202 3/4" Arkal w/o shutoff   **$22**
43-203 1"Arkal Filter   **$69**

Water Filters

Our rust and dirt cartridge is made of white cellulose fibers with a graduated density. These filters collect particles as small as 5 microns (2 ten-thousands of an inch). These are NSF-listed components that take a maximum flow of 6 gpm. Our taste and odor filters are made with granular activated carbon. These filters effectively remove chlorine, sulfur, and iron taste and odor. Maximum flow is 3 gpm. Note: filters should be replaced every 6 months to prevent bacterial growth or as needed. This is the cartridge to use with the inline sediment filter.

41-138	**Rust & Dirt Cartridge (2)**	$14
41-436	**Taste & Odor Filter (2)**	$34

Cedar Composter

The Cedar Composter is a handsome 3-foot square bin made from aromatic cedar (from tree farms — the spotted owls are safe). Cleverly constructed (no tools required), this is the compost organizer for the productive garden. Units may be combined to tackle any size composting task, and handy options help fine-tune compost production. The standard unit measures 36 by 36 by 31-inches or 23.25 cubic feet. Double or triple your composting volume, by ordering 36-inch *bin extenders*. Improve efficiency with the optional *Super-Charger*, a perforated air flow panel made with recycled polyethylene plastic, 36 by 18 by 2-inches that accelerates decomposition by enhancing convective air flow.

54-303	**Cedar Composter (36")**	$59
54-304	**Cedar Bin Extender**	$49
54-305	**Compost Super Charger**	$19

Soil Saver Composter

There are over 150,000 SoilSaver recycling composting systems in use in North America today. Backed by a 10-year warranty and 14 years of research and development, this is the best contained composting system we've ever found. Manufactured from 50% recycled polyethylene resin, the SoilSaver holds 9.6 bushels (12 cu. ft) of compost in its 26" square by 32" high bin. The SoilSaver's design allows for optimum heat retention, moisture, and aeration with an aerobic, odor-free process. Two doors allow for removal of finished compost. Optional composter base bars pests while keeping the bin square. Free composting guide included.

54-310	**SoilSaver Composter**	$99
54-312	**Composter Base**	$14

Compost Tool

Compost needs to be aerated and turned so it doesn't stagnate and get anaerobic. Our rust-resistant lacquered steel tool makes compost mixing easy. 29-inches long.

54-311	**Compost Tool**	$14

Take the time you used to put into working to pay your electric bill and put it instead into your garden, your compost pile, a walk, a fishing trip. Take the time you used to work to pay your medical bills and build a greenhouse.
- Amory Lovins, in the Foreword

To order any of these products
or for more technical information
call us toll-free at
1-800-762-7325

Manual Lawn Mower

After numerous requests from customers, we decided to research the manual lawn mower market. We've found a great manual lawn mower made by the oldest lawn mower manufacturer in the U.S. This mower is safe, lightweight, and very east to push. It's perfect for small lawns and hard-to-cut landscaping. The reel mower provides a better cut than power mowers, keeping lawns healthy and green, and it doesn't create harmful fumes or noise pollution. The short grass clippings from the mower can be left on the lawn as natural fertilizer, or you can purchase the optional grass catcher and add the grass to your compost pile. Cutting width is 16-inches. It has 10-inch adjustable wheels, five blades and a ball bearing reel.

| 63-505 | Lawn Mower | $109 |
| 63-506 | Grass Catcher | $19 |

Hydrosource Polymer

Ever lost a beloved plant to your own neglect? Never again. The Rolls-Royce of water-absorbing polymers is now available for home use. Originally designed for landscaping and professional agriculture applications, Hydrosource can absorb and store enormous amounts of fluid (up to 400 times its weight!), so that you can significantly decrease water usage and extend the time between waterings. Plant roots grow right through the polymer "reservoirs" and tap the nourishment, resulting in better health and increased yields. Many customers have reported watering their houseplants only once per month after using Hydrosource. For gardens and lawns use 2–20 pounds per 1000 square feet. For potted plants plan 1–2 ounces per cubic foot of soil. Lasts for 8 years or more. Completely safe, decomposes into carbon dioxide, water, nitrate, and lactic acid. Call for quotes on 1 ton or more.

46-142	Hydrosource (1#)	$9
46-143	Hydrosource (5#)	$39
46-144	Hydrosource (50#)	$275

Mobility

In eight short years we enter a new millenium. Will we still be fighting wars for oil? Will people in our inner cities continue to choke on fumes spewed by archaic internal-combustion engines? Will global warming caused by the unchecked emission of carbon dioxide plague the future of the planet? The answers to these questions will be directly related to our attitudes toward burning gasoline on the highway.

First, a look at the current scorecard:
- Number of vehicles in America: 186 million
- Total mileage: 1.96 trillion
- Percent of total energy consumption attributable to vehicles: 27%
- Percent of petroleum consumption attributable to vehicles: 62%
- Pounds of carbon released per gallon of gas: 23

But there are human costs as well. Gasoline consumption exacts a cost on our health. There is the price of defending it (the Gulf War) and maintaining open shipping lanes. There is lost productivity from traffic jams and the economic and environmental costs of oil spills. The Union of Concerned Scientists has estimated external costs of $2.53 per gallon of fuel. Whether or not one accepts this figure, the days of the internal-combustion engine appear numbered. Additionally, new legislation mandates car manufacturers to produce zero or low emission, alternatively fueled cars, including electric vehicles, by the year 2000.

With this in mind, we recently traveled east to attend the world's largest Solar and Electric Car Show sponsored by the Northeast Sustainable Energy Association. We came away extremely bullish on the potential for electric vehicles in America.

The professional caliber of the attendees was proof that electric vehicles are coming of age. (There were equal representations of three-piece suits and long hair.) The opening speech was given by Robert Stempel, the CEO of General Motors, who outlined GM's commitment to develop the Impact, a sporty yet affordable model that promises a top speed of 100 mph and a range of 120 miles, by the mid-1990s.

The second day's session featured Dr. Roger Billings, who has for years been an advocate of hydrogen fuel. Hydrogen can be synthesized from water, safely stored in powdered form, and emits only water vapor and very small amounts of nitric oxide as by-products of combustion. A hydrogen fuel cell is up to 60% more efficient than electric, with a range comparable to a gasoline-powered vehicle. A prototype photovoltaic-powered hydrogen plant using water electrolysis is under construction in Riverside, California. The plant will refuel internal-combustion engine cars that have been modified to run on hydrogen.

The typical American family owns two cars, one for commuting and short town trips and the other for out-of-town travel. Ninety percent of American commuters drive less than 25 miles per day, well within the range of electric vehicles.

What became apparent to attendees of the conference is that technology now exists to make it possible to abandon a fossil fuel-based transportation economy. Electric vehicles (which have been in existence as long as the internal combustion engine) are practical for 80% of the nation's short-range commuting needs.

To put this in quantitative perspective, if we could limit half of our urban driving to electric vehicles, we could save 20 billion gallons of gasoline every year while eliminating the emission of 240 million tons of carbon dioxide into the atmosphere.

There are other benefits, too. Electric vehicles are less complex than gasoline-powered vehicles and require less maintenance. They do not need tune-ups, oil changes, mufflers, fuel pumps, or carburetors. They produce no exhaust fumes and no waste oil. They consume no gas or antifreeze. They waste no energy when sitting at idle. (In Los Angeles alone, 72 million gallons of gas are wasted each year by vehicles sitting in traffic.) The City of Los Angeles is so convinced that electric vehicles are the answer that the L.A. Electric Vehicle Initiative was passed that requires 10% of all vehicles on the road by the year 2003 to be zero-emission vehicles. Massachusetts and New York followed suit shortly thereafter.

The economic argument for electric vehicles is just as compelling. An electric vehicle costs approximately $0.05 per mile for fuel (regardless of traffic) while gasoline, even at $1 per gallon, can cost up to 40% more. Maintenance on a traditional vehicle averages $300 to $500 per year, while an electric vehicle is almost maintenance-free. Moreover, the lifespan of an electric engine is much greater than its gasoline counterpart. Total expenses over 100,000 miles are about $0.08 per mile for an electric vehicle and $0.22 per mile for a gas car of comparable use.

With such compelling evidence, why are electric vehicles only hovering on the energy horizon, rather than sitting in our garages? Their main limitations have been electricity storage (batteries) and driving range. Most EVs can travel only 50 to 75 miles on a single charge. Although this is within the needs of the average commuter, this will need to be extended before EVs enjoy widespread acceptance. The second limiting factor is the high weight and limited lifetime of deep-cycle lead-acid batteries. A typical small EV requires a minimum of 72 V (12 ea. 105 ah batteries at 67 lbs ea.) or an optimum of 96V (16 ea. 105 ah batteries at 67 lbs). Battery companies are working feverishly to develop technology with nickel-cadmium, nickel-zinc, and nickel metal-hydride systems.

While the Big Three automobile manufacturers are pursuing electric vehicle marketing strategies for the latter part of the decade, there is no reason to delay your commitment. Each electric car that replaces a gasoline vehicle saves 500 gallons of gasoline and eliminates 10,000 pounds of carbon dioxide each year.

The most viable immediate solution for the eco-car buff is to convert a gasoline car to electric. Most electric cars in the U.S. today are conversions of conventional cars. Taking a standard "econo-box," like a VW Rabbit, Geo Storm, Geo Metro, or Honda, and converting the engine to an electric motor (usually minimum 10 HP) and a battery storage system is a relatively simple process.

At Real Goods we are educating our staff to the world of electric vehicle conversions. Several employees have already made commitments to commute to work in EVs, and we intend to set up a charging station outside our showroom in Ukiah. Although our EV offering is limited, there are several time-tested conversion kits in this edition of the *AE Sourcebook*. Check the "Knowledge" section for a great compendium of books on the subject and access to electric vehicle manufacturers.

This is only the beginning. We intend to put major efforts into convincing America of the sanity and wisdom of alternative forms of transportation. The technology, motivation, and resources are here now. But technology is only part of the answer. Behavior modification will be critical as we need to train ourselves to travel less, and when we do, to extract every bit of energy from our fossil fuels and to convert it into useful mileage. We have to learn to substitute electronic communication for physical transportation, cruising at the keyboard for pleasure, perhaps, rather than at the wheel.

In the end we will find ourselves more mobile than ever. And mobility is no more than another form of freedom.

– J.S.

POWER DOWN

Cruise control

Interstate 89. A peaceful summer evening. Midweek, no traffic. No crazy Canadians heading for Old Orchard Beach. It's just me and the road. Oh yes, and the car. It's not even a fancy car. Four years old. American made. Got all the whistles and bells, though. Comfortable enough.

Well, not quite enough. Needs a little fine tuning.

As I swing from the access ramp to the main highway, I accelerate to cruising speed (in Vermont 69.5 miles per hour). I snap on the cruise control and take off my shoes. I will now adjust myself into another world, a world of slothfulness and decadence, a world known as the comfort zone.

First, I set the odometer. Since I will be writing a column about this little slice of life, I will be writing off the mileage. Twenty-seven point five cents per mile. Thanks, Uncle Sam. Heh-heh-heh.

Now, for the temperature. When I was young and foolish my comfort zone was between 50 and 85 degrees. As long as my feet didn't freeze or I could stick my arm out the window, I could survive. As an old, but foolish, guy my comfort zone has narrowed to between 72 and 72.5 degrees, and not only does the temperature have to be right, but so does the air flow.

When I am driving barefoot (because the cruise control is on), I prefer a floor to ceiling, front to rear air flush of fresh air that fully aerates my tootsies. But then, don't we all?

On this particular day the comfort zone can be reached without air conditioning, which I am loath to use because of its deleterious impact on the planet's ozone layer. My environmental conscience has its tolerance limit of 2 degrees plus or minus. That's when the little guy with the forked tail appears on my shoulder and says, "What, are you nuts? No one's going to see you depleting the ozone. This is your car,

your castle."

Next, it's time to think about my butt. I remember when I first bought this car. The salesman said that power seats would pay for themselves at re-sale time. The same for power windows. That's good news, because these suckers have cost me about two grand in repair bills, and frankly I can use the dough.

As long as my feet didn't freeze or I could stick my arm out the window, I could survive.

The seats adjust up and down, front and back, side to side, left to right, inside and out. There are over 1,500 possible positions, yet I can never find the perfect one.

When I get finished with my butt, I spend a half hour playing with the power side mirrors with built-in heaters which, unfortunately, are not adjustable.

Once I am settled in, with the neck rest adjusted so that I can see all three mirrors by just moving my eyes, it's time for—you guessed it—**tunes**!

You know you have made it in life when you can afford a music system that is twice as smart as you are. Mine is so complicated that even my nine-year-old can't figure it out. There is a setting for type of tape—metallic or Dolby or retrograde or molybdenum. I owned this car for two years before I learned that these designations did not refer to the music on the tape.

Finally, there is the piece de resistance, the graphic equalizer. All of my friends own graphic equalizers, and not one of them know what the device does, or how to operate it. Every once in a while they slide the levers up and down to make sure that the thing-that-they-don't-know-what-it-does still does it. In my case, it's a moot point, as all my cassettes are missing from their boxes anyway. (This is another of life's mysteries. How can there be so many cassette tape boxes and so few tapes? Oh well.)

This is status, the completely adjustable life. As I cruise on down Interstate 89, the only thing that could make me happier would be a bank of toggle switches... and maybe a rheostat or two... and some of those digital gauges. They do not need to be hooked up to anything, just so long as they permit diddling.

My adjustment is now complete. I even adjust my attitude with one of those nonalcoholic beers in the foil-necked bottles that costs a **dollar seventy-five**!

Comfortable and cruisin'. Passed my exit 20 minutes ago, and don't even care.

When you think of it, the car is modern man's castle. In some ways a vehicle is a working laboratory of off-the-grid life. Our light, our heat, our tunes, our cellular phones all come to us without being plugged in to the utilities. We can control this environment—loud, soft, hot, cold. We can even buy one of these new, little inverters from Real Goods and bring our TVs and Macintoshes into the car! It's all there at fingertip control.

The automobile would be a perfect cell of energy efficiency, but for one tragic flaw. They move. And to move, they consume nonreplaceable fossil fuels. In gobbling the fuel they spew noxious gases into the atmosphere that threaten life. Important, note that the last sentence did not read "threaten our way of life," but simply "threaten life."

It is only if we adjust our "way of life" as nimbly as we adjust our cruise control, by developing more efficient, less toxic means of locomotion, that we can realize the independence that the automobile has promised us.

— Stephen Morris

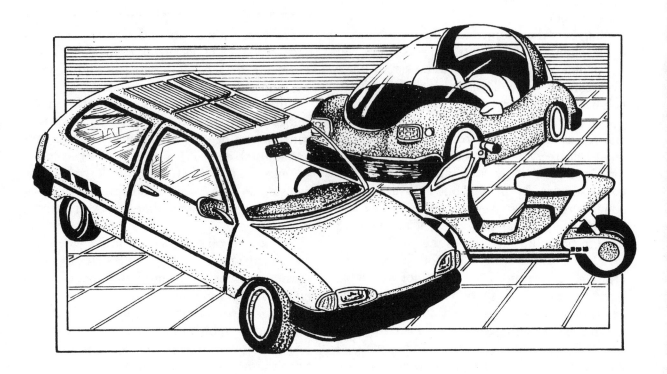

Electric Vehicles

What is an electric vehicle, or EV? An electric vehicle is one that uses an electric motor instead of an internal-combustion engine, and batteries instead of a fuel tank and gasoline. The motor is a larger version of the kind that powers your hair-dryer, or the refrigerator in your kitchen, or the tape player in your car. The batteries are similar in size and shape to the one used to start your car's engine, only there are many more of them.

The energy of the battery pack is routed to the motor through an electronic controller. Housed in a small black box, this works like a light-dimmer switch (or the speed control on an electric drill), smoothly delivering power to the motor and controlling its speed.

Driving a car that has been converted to electric propulsion is virtually identical to driving one that has a gas-powered engine. The same operator controls are used—accelerator and brake pedals. Your foot pushes on the accelerator to control the speed of the vehicle, and pushes on the brake pedal to stop it. You will notice differences, though. You don't have to start an electric car like you must start an engine. The accelerator starts up the electric motor, and you are on your way. You don't have to warm up the motor, either. Unlike gas engines, electric motors work well without any warm-up period.

Stopped at a signal light or stuck in traffic, there's another important difference. If your foot is not on the accelerator pedal, the electric motor is not running. That will seem strange at first, but you will get used to it quickly. Since the electric motor is not running, it is not consuming or wasting power.

There is a long list of other advantages to be found in the electric vehicle. For example, since no fuel is consumed, there is no exhaust pipe. That's right— there is zero pollution when an electric vehicle is operated.

A very early electric vehicle.

The best technologies on the market can save about three-quarters of all electricity now used in the United States, while providing unchanged or improved services.
- Amory Lovins, in the Foreword

GMCs Sunraycer

The gasoline-powered internal-combustion engine car now reigns supreme. Transportation now accounts for more than 70% of oil consumption in the USA annually. The private automobile, at 41%, is the biggest single offender.

Our dependence on foreign oil has lately been underscored by the Persian Gulf war of 1991. America currently imports about 50% of its oil. Based on current trends, imports will grow to 75% by the year 2010. These are dollars flowing out of our country and economy. People in our inner cities continue to choke on the fumes spewed by archaic internal-combustion engines. The global warming problem and the emission of carbon dioxide continue to plague the future of our planet. And the national trade deficit continues to rise. All of these ills are the direct result of burning gasoline on our highways.

Oil and gasoline are subsidized so heavily that we pay only one-tenth of its real cost to us at the pumps. Alternative technologies and energy-efficient consumption get no subsidies, and so appear impractical. In addition to the direct and measurable costs of gasoline, we must also pay attention to health costs and other external costs, such as the military costs to keep oil flowing from the Middle East, lost economic productivity due to traffic congestion, oil spill cleanup, mitigation of greenhouse gas emissions, and death and injuries from traffic accidents. The Union of Concerned Scientists estimates these external costs to be $2.53 per gallon of fuel.

Our own sun shows us the way it ought to be. More energy than we have used through our entire tenure on this planet strikes the earth every day in the form of solar energy. Solar cells right now can convert better than 10% of the incoming solar rays into useful electrical form. Solar-thermal devices use the sun's energy directly, efficiently and inexpensively providing water and space heating.

Transportation today is a matter of convenience. We want what energy does for us—heat, light, power, water, food, and fast wheels. We haven't paid much attention to how it was done, and at what price.

It's time to give the alternatives a chance. Fast, efficient mass transit. Solar power plants. Telecommuting. Needing less. Living closer to work. Wouldn't it be nice to wake up one morning to a world that did it this way?

It's time to give the alternatives a chance.

At long last, even the government has accepted the fact that the internal combustion engine's days are numbered. The Clean Air Act, legislation in California for air quality, and the new proposed National Energy Policy all mandate car manufacturers to produce significant numbers of zero or low-emission cars, including electric vehicles, by the year 2000.

In Third World cities there is a grocery store within easy walking distance of everyone's house. The suburban lifestyle with its dependence on the automobile for every errand is turning out to be a major ecological blunder. Short of doing an instant rebuild of all our cities, electric powered vehicles offer a sane, cost-effective technology that can bridge the gap to a softer, gentler future.

Electric vehicles are infinitely less complex than gasoline-powered vehicles requiring far less maintenance. Electric vehicles do not need tune-ups, oil changes, mufflers, fuel pumps, or carburetors. That means no exhaust fumes, no gas, no coolant, and no waste oil. Greenpeace estimates that do-it-yourself mechanics dump an Exxon Valdez worth of used motor oil down storm sewers and drains every 2½ weeks! Electric vehicles are also quite energy-efficient. For each barrel of oil, 3 to 5 electric vehicles can be powered over the same distance as one gasoline-powered car when the oil is burned in a utility-size power plant to produce electricity. It is much easier to monitor and control the pollution of a single smokestack than tens of thousands of tailpipes. Utility scale power plants operate 2 to 3 times more efficiently than car engines.

Although there are limits to the range of a practical electric vehicle, current technology permits travel at legal speeds as far as 50 to 120 miles with a fully charged battery pack. This is well within the average trip length for cars daily. The daytime range of the electric car is easily doubled, too. An onboard battery charger permits worksite recharging from a standard wall socket. So, the overnight charge gets you to work, and the recharge during work hours gets you home. A 100 to 120-mile range will meet the needs of more than 90% of the driving public, and 80% of the needs of the remaining 10% of the population.

Solar-Powered Cars

by Michael Hackleman

It excites the imagination: Wheels spinning from light. A vehicle achieving speeds of 35 mph on the sun's energy alone. Design so good that, with careful attention to aerodynamics, weight, and efficiency, one horsepower goes a long, long way. Combined with its zero-pollution nature, no other system of propulsion offers this unique characteristic.

The past six years have seen a flurry of races and rallies to demonstrate the potential of both electric and solar-powered vehicles. (See *Solar Cars Are Here!* below.)

Anyone who has worked on a solar-powered car knows that they are not very practical. The combination of electric propulsion and solar technology is an unlikely marriage, expensive and fragile when the panel is actually mated to the vehicle. However, a solar-powered car does turn people's heads. Anyone who walks up to a solar car for the first time will find that their inner child is alive and well!

Solar cars, then, demonstrate the practicality and simplicity of electric vehicles in three ways. Spectators are frequently amazed by the acceleration and speed these vehicles achieve working

with less than 2 horsepower. Also, the solar component of the car drives home the idea that solar is a power source that works anywhere under the sun.

Finally, a solar car project is a fun way to design solutions that consider a scope beyond mere machinery. Coupled with lightweight, aerodynamic bodies, electric propulsion is an elegant solution to the dilemma of smog-ridden basins throughout the United States, declining oil reserves, and other issues that affect the quality of life.

At the university level, a solar car project sensitizes students to environmental issues by exposing them to technologies based on a non--fossil fuel. In this way, they can bring classroom engineering to bear in designing solutions to today's problems.

At this point, racers are the only electric vehicles that use solar energy as a primary source of propulsive power. EVs designed to operate on the road may have solar cells attached to them, but it would be more accurate to say that they are solar-assisted. This definition will become clearer as the reader delves into the world of solar-electric cars.

Solar Cars Are Here!

World awareness of the potential of electric vehicles, particularly ones that could be powered from the sun, started in the mid-1980s. Here's a brief summary of this history to date:

1985—Tour de Sol. Switzerland starts the popular Tour de Sol, a race event that challenges the vehicles over the course of five days, including a stretch over the Alps! The annual event attracts entrants worldwide. The first race drew a crowd of 500 spectators, the second saw a gathering of 3,000, and the third drew a crowd estimated at 20,000. Today's races draw many tens of thousands of people.

1987—World Solar Challenge. Australia expands the solar racing circuit in a grand scale with the World Solar Challenge, challenging its entrants to a transcontinental race, north to south, of over 2,000 miles across the starkness and heat of the Outback.

1988—Solar Cup USA. California hosted the Solar Cup in Visalia, launching the first solar car race in the USA.

1989—Race veteran James Worden piloted his Solectria 5 on a 22-day transcontinental (west-to-east) USA run.

1989—**Tour de Sol USA** held its first American event, and provides organizational support to similar interests all over the country each year.

1990—**Sunrayce,** July. The GM-sponsored Sunrayce pits vehicles from 32 colleges and universities against each other and the elements in an 11-day, 1,800-mile transcontinental run through seven states from Florida to Michigan.

1990—**SEER (Solar Energy Expo and Rally)**, August. A gathering of electric vehicles and alternative energy technology and products from the Western USA. The first annual Tour de Mendo race was won by the Real Goods-sponsored SUnSUrfer from Stanford University.

1990—**World Solar Challenge**, November, the second transcontinental Australian race. Over 40 entrants participate, with shorter completion times recorded, and record crowds.

1991—**Phoenix 500 Race**, April. Professional race at the Phoenix International Raceway to publicize and promote the development of electric vehicles.

1992—**Tour de Sol**, May. From Albany, NY to Boston, MA. Real Goods is a major sponsor of this event which exposes thousands of new people to the potential of non-polluting vehicles.

1992—**Sun Day Challenge Race**, June. Cape Canaveral, FL. Solar racing comes to the space center of the United States.

How Does a Solar-Powered Car Work?

In principle, solar-electric cars are fairly simple. Sunlight striking solar-electric modules on the vehicle is converted directly into electricity and routed to the electric motor. Typically, a small set of batteries is carried onboard to help with high loads such as acceleration, hill climbing and higher rates of speed. The batteries also assist with propulsion when the panel can't receive much sunlight, during the early morning and late afternoon hours, and through overcast and shaded conditions. In a racing format, of course, the energy that comes out of the battery must be replenished with electricity generated from the sun's energy.

There is a maximum of 1,000 watts of energy in a square meter of area on the earth's surface when exposed to sunlight on a clear day at high noon. Commercial-grade cells yield efficiencies of 12–14%, yielding approximately 100 watts of electricity per meter after losses. Racing electric vehicles get a 25% overall increase in power from their panels by using more expensive solar cells. With a 17% conversion efficiency, space-quality cells produce 120 watts or more per square meter.

Drivetrain efficiency—including solar panel, power conditioners, motor and controller, and transmission—is essential when the power source is so restrictive. Aerodynamics, or slicing through the wind without expending much energy to do it, is also vital. Winning solar-powered cars are lightweight. Accelerating the vehicle to some speed, particularly from a standstill, consumes much more energy than it takes to maintain a steady speed. Climbing a hill, even at a constant speed, consumes power in big gulps.

Successful solar-powered cars utilize a peak-power tracker, like the Maximizer and LCB. High temperatures, varying light intensity, and changes in the panel's orientation all affect panel output. The peak-power tracker's job is to maintain the best match between solar input and the vehicle's battery or motor bus voltage.

Commercial Electrics

The oil crisis in the 1970s brought about many electric vehicle prototypes. Advanced primarily by the automotive manufacturers through Department of Energy funding, their promises of production died when the price of gasoline dropped.

The steady rise of gasoline prices has again prompted a hard look at electric propulsion. Currently, Japanese auto manufacturers appear to be the most convinced that there is a market in electric vehicles. This is also a good market for entrepreneurs and industry that is already manufacturing components for the automotive trade. Two of the most visible electric vehicles are the Impact and the Horlacher Electric.

The Impact

It was a good sign for the '90s that General Motors opened 1990 by unveiling its bold new prototype, Impact. This is a car that most environmentalists would enjoy: it's pollution-free, quiet, uses one-fifth as much energy from fossil fuels as regular cars, and has a recyclable battery pack.

The Impact boasts a respectable 120-mile range — whether you're cruising along at freeway speed or in the stop-and-go urban driving cycle. Here, the aerodynamists and stylists have finally gotten their act together. The Impact has a competitive look, yet boasts the lowest drag coefficient in the history of commuter vehicles. (Low drag means the vehicle is slippery moving through the air, consuming less energy to go the same speed.) The battery pack is basically off-the-shelf technology — lead-acid batteries that pack a lot of energy, are low in weight, and recharge quickly (2 hours for 60% of capacity, 4 hours for 95%). Since the batteries are sealed (recombinant), no watering maintenance is required.

---

*The Impact blew away
a Miata and a Nissan 300ZX
in the quarter mile.*

The Impact prototype is a concept advanced and built primarily by Paul MacCready and the gang at AeroVironment. This is the crew responsible for the Gossamer series of human-powered aircraft.

The Impact is a high-performance electric vehicle. The footage shown during its introduction proves that. From a standstill, the Impact *blew away* a Miata and a Nissan 300ZX in the quarter mile. Obviously, the design was intended to blow holes through the perception that electric vehicles are golf carts with pretty bodies over them.

The big question on GM's mind seems to be: Is there a market for electric-powered cars? That GM is unsure about this is revealed in the way they announced it, immediately underselling both the vehicle and technology. GM claims that operating expenses for the Impact are

The Impact

double that of a standard internal-combustion engine car, that the Impact is handicapped with periodic battery replacement, and that it is not a get-in-and-drive-anywhere car. The Impact also needs an infrastructure to recharge and service this new type of car. GM appeared genuinely surprised at the excitement and warm reception given by industry, government, and the general public. They probably thought that everyone was going to laugh. Nobody did!

Still, GM faces an interesting dilemma: Its marketing effort must tiptoe through any comparisons that could hurt its greater investment in internal-combustion technology. Still, this is potentially the decade of the environment, and no company is better poised to commit to the effort. Will GM hold out for market share and delay aggressive tactics until other car companies have tested the waters?

Horlacher Electric

With gasoline in Europe at $3-5 a gallon, Boris Horlacher's electric is a welcome sight at the annual Tour de Sol events, sprinting up alpine grades with the best of them. While Boris refuses to go into mass production until he's satisfied with the design, folks who have driven one of the early prototypes say that the Horlacher meets ALL of their expectations. Propulsive power comes from a 12-hp Brusa drive — the motor-controller used in all the winning Tour de Sol electrics — and it pushes this peppy three-wheel twin-seater at speeds up to 45 mph for an hour or more. There's even a four-wheel version of the Horlacher now. Department of Transportation regulations and federal safety standards will play havoc with introducing these small, lightweight vehicles into the USA. How do we get 2-ton monster vehicles off the road to make it safe for environmentally sound cars like the Horlacher?

Are Solar Cars Practical?

At this time, there is nothing inexpensive or practical about solar-powered transportation. Add up the cost of a racing solar-powered car and you may wonder why I fume when I hear people claim that a few solar modules will let you zing down the road at 55 mph to work. Putting solar modules on an electric vehicle doesn't make the car a solar-powered vehicle. Solar-assisted, yes. Solar-powered, unlikely. Currently, a 48-watt module costing $300 and charging in perfect weather for 7 days will only store enough electricity to help a lightweight EV *about 5 miles* down the road at 45 mph. The addition of solar panels on electric vehicles makes them fragile, bulky, and prone to theft. To be solar-powered, a car does not need the panel mounted on the car. There ARE alternatives.

- Plug into the grid, particularly if your regional power comes from hydro-generated electricity.
- Install a grid interface of a home-mounted solar array. This gives you credit for daylight solar charging at home, cashed during the overnight charging of your EV.
- Recharge your EV from a homepower system (pack-to-pack charging); this is particularly easy when the home pack is a higher voltage than the EV's pack. (This is highly unlikely, but we won't go into it).
- Utilize battery exchange, switching a daytime charged pack for a depleted one when you get home.

Hybrid Electric Vehicles

The battery pack in an electric vehicle stores a finite amount of power. However large this amount is or how efficiently it is converted into propulsion, the time will come when the vehicle rolls to a stop and requires a recharge.

Facilities (infrastructure) that will fast-charge or exchange a depleted battery pack with another are needed to make the "pure" EV as workable as cars with gas engines. However, until this infrastructure is set in place, there is an alternative: the hybrid EV.

An electric vehicle that uses two or more power sources (one of which is a battery pack) is referred to as a "hybrid" EV. The power produced by the extra energy source varies considerably with the type and design, but it is designed to augment the power from the battery pack, adding to overall range, performance, and other operational characteristics.

90% of American commuters drive less than 25 miles per day, well within the range of electric vehicles.

Within this broad definition of hybrid EVs, many alternative energy or power sources qualify as an additional energy source. This includes solar panels, a wind arch, and an engine-generator unit. Even a regenerative braking circuit might permit a hybrid classification, since it produces electricity from the vehicle's momentum, and extends range.

Let's look at the hybrid EV that uses a genset, or engine-driven generator. The term genset is an abbreviation for "generator set". Folks in the construction trade know gensets as stand-by generators. The 120-volt, 60-cycle AC generator is a typical genset. If none of these terms is familiar to you, imagine hooking up the small engine (like one that runs a lawnmower) to the alternator or generator in your car through a V-belt. This would be a low-power unit, but when placed in an EV, it becomes an onboard charging system. The engine itself can be fueled with gasoline, propane, alcohol, hydrogen, or diesel.

In the hybrid EV, the ACU (auxiliary charger unit) is a small engine (5–8 hp) coupled directly to an alternator. The alternator's output is connected to the batteries. A power cable connects both the ACU's output and the batteries to the motor controller. The motor itself is connected to more wheels through a standard transmission and differential. Note that the ACU is not coupled to the drive train mechanically.

A hybrid EV combines the best features of electric motors with the best features of engines. The electric motor contributes its flat torque/rpm and variable-load characteristics, short-term high-power endurance, and its light weight. The engine contributes its high power density and fuel availability. In the process, each offsets the disadvantages inherent in the other in a mobile propulsive environment.

Hybrid EVs address two fears held by the driving public concerning electric vehicles in general: low performance and getting stuck somewhere with a dead battery pack. Then, too, all energy sources have inherent advantages and disadvantages. Using two or more sources frequently adds the good features of each source and offsets the shortcomings inherent in any one source. City driving favors use of the battery pack; freeway driving favors ACU operation. Short distance suggests battery-only operation, while long-distance driving requires the ACU. Presumably, you know where you're going, how far it is, and how fast you'll drive. This lets you select battery only, or ACU operation. The hybrid EV makes it possible to use the existing infrastructure for fossil-fueled vehicles without actual dependence on it.

Use of an ACU increases overall system complexity and initial costs. It may increase overall vehicle weight, too. Hybrid EVs are really a transition technology, until we are fully able to outlaw any combustion technology, since this is at the root of increased carbon in the atmosphere.

How does the ACU work? When you're stopped at a light or stop sign, all of the ACU's power is going into the batteries. When you're traveling down the road at 15 mph, some of the ACU's power goes to the electric motor, and the remainder goes into the battery pack. At some speed, say 35 mph, all of the ACU's output goes into the motors. At 50 mph, the batteries supply the additional power (above the ACU's output) needed to reach and hold that speed. Note that the engine is relieved of the task of producing *propulsion* and assigned the task of producing *power* toward the propulsive effort, battery storage, or both. Thus, when the ACU is operational, the power it produces is never wasted. It's used or stored. Compare that to a gasoline car stuck in a traffic jam or waiting for a signal light!—**MH**

How to Take Action

How can you make what you do matter? Politicians and successful corporations long ago learned that the **real** vote each individual holds is the dollar. Spend a dollar on gasoline and you vote for it, and for whatever policies the people who sell it to you choose. Don't feel guilty about it. Instead, move toward a mode that decreases or eliminates your consumption of oil. Increase your awareness of the issues and technology in transportation. Buy, construct, use, or support electric vehicles and other alternatives!

Publications

Alternative Transportation News

Alternative Transportation News is a 36-page magazine that has the ambitious goal of changing the transportation habits of the world in this decade! It is about transportation issues, alternatives to our current systems, and the means to achieve them. It unites a historical perspective with state-of-the-art technologies to suggest ways we can achieve transportation that is less abusive to life on this planet.

ATN describes how these products and technologies represent a fertile but virtually untapped market to individuals, entrepreneurs, and industry in general. It is written for the layperson, translating otherwise technical information into terms anyone can understand. It is a networking tool, helping organizations and corporations worldwide to become aware of each other's efforts in fuel and vehicle technologies and to work together to integrate these efforts. It represents a vision of the future without making promises on how easy, fast, or inexpensive it will be to extract ourselves from our dilemmas.

ATN has departments on human-powered vehicles, air and water transport, fuels and cells, design, basic and advanced projects, and history. It reviews books and products, notes calendar events, and provides for reader input in many ways.

An ATN subscription is $12 annually (6 issues). Send subscriptions for ATN to:

Earthmind
P.O. Box 743-RG
Mariposa, CA 95338.
(Checks payable to Earthmind.)

Solar Mind

Joe Stevenson is an excellent writer and we highly recommend you subscribe to his publication, *Solar Mind*, which is full of vision, insight, and practical solutions to the internal-combustion engine. *Solar Mind* is a unique merging of spirit, mind, technology, and environmental concerns. It is a place to share and discuss ideas, insights, and solutions to earthly problems. To subscribe, send $3 for a sample issue, or $25 for six issues ($12 for students, nonprofit groups, and retirees), to:

Joe Stevenson
Solar Mind
759-RG South State Street #81
Ukiah, CA 95482

Convert It

This book by Mike Brown is a step-by-step manual for converting a gas car to an electric powered car. It's a very readable and practical manual for the do-it-yourselfer wanting the fun and educational experience of converting a conventional automobile into an electric vehicle that will be both practical and economical to operate. Included are a generous number of illustrations of an actual conversion, along with instructions for testing and operating the completed vehicle. Mike Brown shares a wealth of his own personal experience in making conversions and includes many practical tips for both safety and ease of construction that will be useful for the novice or the experienced home mechanic alike. Brown does a great job of explaining the cost of the conversion components. This book has been strongly recommended by the Electric Auto Association, Electric Vehicle Progress, and Alternate Energy Transportation. The book is 56 pages long. We realize the price seems a bit steep—we have pleaded with the author to republish in quantity to reduce his cost. However, for the time being this is the *only* book available on the subject, and it seems worth the price for that reason.

80-404   Convert It $35

Electric Vehicle Video

Electro Automotive Information Services. Yes, electric cars are simple. But should you use a series DC motor, brushless DC, or an AC? What about aircraft generators? In this new video, Mike Brown, author of *Convert It*, sorts out all the choices. It's a good starting point if you are considering building an electric vehicle, or just want to know more about how they work. Brown's business, Electro AutoMotive, has been supplying components for electric vehicles since 1979, and is known for equipping the winners in countless races and rallies. The tape, with footage of dozens of cars and components, runs approx. 90 min.

80-125   Electric Vehicle Video $39

Electric Vehicles Unplugged!

An enthusiastic resource from Douglas Marsh. It covers the whys and hows of electric vehicles and then gets specific, with the best list of makers, converters, clubs, kits, plans, and parts we've seen. 57 pages.

80-408 Electric Vehicles Unplugged!   **$9**

Do-It-Yourself Automotive LPG Conversion

A 36 page primer written by Tom Jennings. Of all major fuels, only liquified propane gas (LPG) can be produced and consumed without polluting the earth. Moreover, it can be stored and transported safely. Cars run better and last longer when powered with LPG. The author assumes you understand how the major systems work, but instructions, even though generic, are clear and doable. The author estimates the conversion cost at under a grand!

80-406 Do-It-Yourself Auto LPG Conversion $5

Electric Vehicle Directory

Philip Terpstra. This book shows over 20 vehicles on the market complete with specifications. Lists sources of other electric vehicles as well as components, associations, and newsletters. 80 pages.

80-403 Electric Vehicle Directory   **$11**

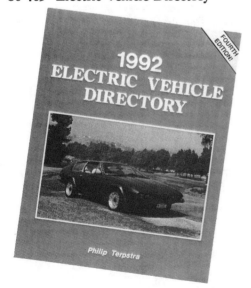

Deluxe Electric Auto Conversion Kit

Attention, small car owners: the electric parade starts here. Using this kit, anyone with mechanical aptitude can convert most small gas or diesel cars to efficient electric power drive. Your vehicle should weigh under 2,800 pounds before conversion, have a manual transmission, and no power steering.

The kit was developed by Mike Brown, who has had over 27 years of automotive experience, and more than 12 years of electric car experience. In short, Mike knows what he's doing: his system works in the real world, not just in computer simulations. Vehicle range is approximately 50 to 70 miles on a full charge.

Kit includes:
10-hp advanced DC Motor, Motor/transmission adaptor, Noalox anticorrosion paste, Gauges and shunt, Cooling fan, Belleville washers, PMC 1221 controller, Shrink tube, PMC PB-6 potbox, Crimping tool, Cable shears, Charger, Cable, Lugs, Main contactor, Battery filler, Circuit breaker, Hydrometer, Fusible link, *Convert It* book.

Please specify make and model of car, production date (year and month), engine size, and type of transmission (four or five speed) to determine proper adaptor plate.

Customers desiring extra power, or driving in hilly areas, should take the optional 19-hp motor.

This kit is certified by the state of California to be exempt from the state portion of sales tax, and to qualify for a $1,000 credit on state personal income tax. Instructions for claiming the tax credit are included with the kit.

Note: Kit does not include batteries. Requires 16 "golf cart" batteries. Batteries are available locally for $60–$80 each depending on the outlet.

■91-820	Elec Auto Conversion Kit	$4800
■91-821	19-hp Option	$620
■05-225	Shipping Cost	$175

For another
alternative in
transporatation
be sure to see
the Workbikes
on page 341.

Clean Power for the Long Haul: Electric Truck-Conversion Kit

Sure, Detroit and Japan are gearing up (in a small way) for electric vehicle production. But why wait years for them to get it together? If you own a Chevy or GMC S-10 or S-15, you can make the switch over to electric transportation *now*. The truck is made to carry heavy loads, so can handle the weight of the batteries. The batteries are installed outside of the passenger compartment. And the truck offers better crash protection. On the down side, trucks, in general, have inferior aerodynamics. However, this can be partially offset by a ground effects add-on. Our pre-engineered kit, from Solar Car Corp., allows you to simply bolt and plug the system together; no welding needed, this super-simple system is completely modular. It's made for standard transmissions and utilizes regular golf-cart-type batteries. (Kits for power-steering-equipped trucks are not available). The conversion is a pleasurable project that you can do in your spare time, in approximately 25–50 person-hours, depending on mechanical experience. And we can guarantee you'll feel great about breaking your truck's addiction to the gas pump. Range is approximately 50 to 70 miles on a full charge. It has a top speed of over 70 mph. The regenerative brake kit is recommended only for hilly terrain. It will recover about 15% of the power it took to get up the hill.
Kit includes:
9 inch Advanced DC Motor (28 hp at 120V)
Motor mounts compatible with engine mounts (Please specify 4- or 6-cylinder engine)

Flywheel and bell housing adaptor
Molded plastic battery boxes
Front battery tray and controller mount
Power brake kit
Modular control package with ammeter
DC-DC converter with sub fuse assembly
On-board 120V charger (1,500 W)
Main disconnect
Battery cables with ends & quick disconnects
Molded plastic, dash-mounted meter holder
Fuel gauge (voltmeter)
Rear booster springs
Simple instructions
This kit does not include 20 "golf cart" batteries. Shipping costs for batteries are very expensive. Batteries are available locally for $60–$80 each depending on outlet.
Please specify standard or extended-cab truck.

■**91-810 Electric Truck Conversion $5,590**
Available options:
■**91-811 240V fast charger (4,000 watts) $250**
Substitute for onboard charger
■**91-812 Solar Panels (each) $495**
48W solar panels for hood, roof, or camper top
■**91-813   4.5 kW regen. brake kit $595**

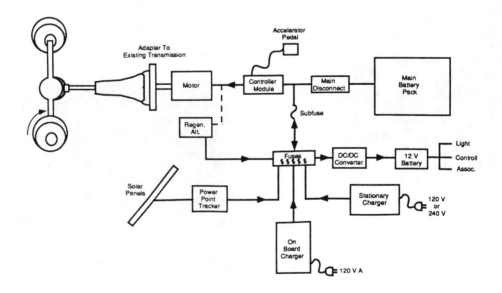

EcoScoot Electric Scooter

Electric scooters are a great form of electric transporation. We expect to see a lot of new additions to this field over the next few years and even are working to develop our own scooter. Stay tuned to the Real Goods News for updates.

While we all wait not so patiently for Detroit to get its act together with electric cars, or alternative is to make our own conversions for the time being. One shining star of a product that is now available is the EcoScoot Electric Scooter. It's hard for many to take the first step toward electric vehicles with a $10,000 to $15,000 investment in a car, but the EcoScoot allows you to enter the EV world for less than $2,000 with a very practical vehicle.

The EcoScoot can do everything that its dirty gas cousin can do, except it doesn't pollute the environment, make noise, drip oil, or foul spark plugs. It can travel over 20 miles and up to 30 miles per hour on a single 120-volt overnight charge costing only about 12 cents per charge. You can use it to scoot to school or work, or to run errands around town. You'll avoid traffic jams, parking problems, gas lines, and smog checks. The EcoScoot is powered by a smooth and reliable electric motor with an electronic speed controller. With just a twist of the handle

grip, the EcoScoot smoothly accelerates to your desired speed without gear changes or clutching. To stop, just back off the throttle and apply the brake. Anyone that can ride a bicycle can drive an EcoScoot. The EcoScoot is licensed as a motorized bicycle. A $6 DMV fee (in California) purchases the license plate and never has to be renewed. The EcoScoot requires two Trojan 22-NF 12V, 30 amp-hour batteries wired in series to produce 24V that are not included in the price. If you want to purchase the batteries locally to avoid freight, compatible batteries are available from KMart, Sears, Price Club, Costco, and Kragen.

■91-808	EcoScoot	$1,995
■05-213	Pallet Charge	$75
■15-106	Trojan 22-NF	$55

Scooter Shipped Freight Collect from San Luis Obispo, CA
Batteries Shipped Freight Collect from Santa Rosa, CA

The Doran Three-Wheel Electric Vehicle

After 6 years of professional engineering development and thousands of miles of testing, these sleek lightweight vehicles are nearly ready for the road (expected delivery date: spring 1993). Their design objectives are minimized weight, aerodynamic efficiency, sports car performance, and practicality. An incredibly strong foam-core composite body and the lack of a fourth wheel minimizes weight. A natural teardrop shape, minimal frontal area and flush-mounted windshield maximize aerodynamic efficiency. The unique three-wheel design lowers rolling resistance by 25% without sacrificing handling. Front wheel drive and steering with double wishbone front suspensions provide remarkable roadholding agility and overturn resistance matching many modern sports sedans. Capable of 0 .8 g acceleration on the skidpad, it also exhibits very stable braking from 60 mph in less than 137 feet. You can lower the side windows and stow the t-tops and rear hatch to enjoy convertible-like driving. Whether commuting to work or taking a drive through the countryside, these vehicles are always sporty and fun to drive.

The heavy-duty series wound DC motor propels the vehicle to 60 mph in 10.9 seconds and climbs hills with ease. A pulsewidth-modulated controller keeps efficiency high with solid-state reliability. The inexpensive lead-acid batteries propel the vehicle for about 60 miles around town. Best of all, the batteries charge overnight on your 120-volt outlet costing only 3 to 7 cents per mile depending on where you live. To maximize performance and range, the battery-to-vehicle weight ratio has been optimized. With a cruising speed of 60 mph and a top speed of over 80 mph, it has plenty of acceleration for the freeway and for quiet, pollution-free commuting around town.

Available in spring of 1993
Approximate Price: $15,000

Doran Construction Manual

This 97-page book walks you through all the steps necessary to build your own Doran three-wheeler, either powered by electricity or gasoline, and tells you where to purchase the necessary parts. The Doran Motor Company decided to offer plans for these vehicles so that others can duplicate their efforts. The manual includes background information on every subject so that you can see why they designed the vehicle the way they did, and it gives you enough insight to evaluate the technical feasibility of your own ideas so that you can customize your vehicle if you choose. The book answers all of the common questions asked about the vehicles and discusses the driving characteristics.

Example: What Will It Cost? This vehicle will cost from $5,000 to over $10,000 depending on the amount of labor you have done by others and how much you do yourself. The price of parts will also vary greatly depending on whether you buy them new, rebuilt, or salvaged.

80-410 Doran Construction Manual   **$39**

Just for Fun

Just because the planet has problems does not require that we wear the grim faces of impending doom. We can learn, we can solve, and we can have a great time. Recognizing that play is an essential component of life (as important as energy or shelter) is the first step. The problems of the planet will not be solved overnight, and a little laughter between now and then can only make the situation better.

Our product selection committee uses the San Francisco Science Exploratorium for inspiration. Displayed are over 650 natural wonders that you can see, touch, feel, and experience.

Our gift selection of solar educational toys is designed to demystify solar while teaching the benefits of sensible energy. Most importantly, these toys will provide hours of enjoyment and fun.

But providing "real goods" for kids is only part of our program. This year we decided to try an experiment to enlist kid power directly in our company mission. The program is called "Real Goods for Real Kids on Real Planets."

The program originated at a solar fair in the summer of 1991. Real Goods tries to participate in as many of these events as we can, but the limiting factors are always staff energy and time. This particular event was close to home, so everyone was rallied for the weekend effort. This meant a disruption to family life which we tried to minimize by making our kids part of the show. They helped us set up, then while we were preaching the gospels of renewable energy and recycling, they set up a booth to preach the gospel of healthy junk food. Salmon sticks, taro chips, and solar iced tea were flying out of the "Real Kids" booth.

There were no bored children tugging at Mom's or Dad's sleeve. No whining for special attention; the kids had a ball, and at the end of weekend they collected a nice pile of money that they donated to a local charity. (Of course, we parents were so proud that we then showered the children with praise and baubles that they would have purchased themselves, had they kept the money.)

The point is this: everyone had a good time, and everyone benefited from the experience. The idea for a more widespread program was born.

The first tenet of our "Real Goods for Real Kids on Real Planets" is that it is *not* a profit-making venture for the company. We don't need our cynics to think we are lining our pockets off the efforts of little nippers. Tenet number two is that we want to reward enterprise; number three is that we want the planet to be a better place because of the effort.

In a nutshell, the program works like this. We will provide non-profit organizations with quantities of a modified Real Goods catalog carrying planet-friendly products. Qualifying non-profits can take orders for this merchandise, submitting orders to the company according to a specified set of guidelines. Orders are fulfilled directly to the consumer, and a "commission" is rebated to the participating organization, the proceeds to be used for whatever purpose the group wants. Doesn't this sound better than selling candy or cookies?

Will it work? Time will tell. We think that by giving groups a means of helping themselves is the best way for Real Goods to help the most worthy organizations in the most local communities.

We see a lot of environmental organizations focusing their efforts on children, as if it is the kids' responsibility to clean up the messes *we* have created. Sometimes we become cynical about businesses propagandizing youngsters. We hope we have not fallen into that trap, but we're certain we can count on our vocal consumership to let us know if we have. Our goal is to treat children not as adults, but as full-fledged people who have as much stake in this planet as anyone.   – J.S.

POWER DOWN

The Power Down War

Jimmy Carter called our need for energy conservation "the moral equivalent of war." Nice turn of a phrase, but not exactly the rallying cry to inspire the masses. Most of us would prefer to avoid war in any size, shape, or form, even its "moral equivalent."

What we want to do is to have fun.

In the dim recesses of the average, adult American mind is a memory of when we were happiest and most free. In most cases (this, obviously, is the result of a highly empirical survey) the memory takes place when we are young, free, warm, active, and close to naked. Perhaps we are playing in the surf, running barefoot on a summer eve, or staring at a campfire. The common denominator is that these memories often take place in a total void of power, other than that supplied by our own bodies.

Turn the tables on this thought, and it still has validity. How many ideal memories take place in traffic jams, or sitting on the riding mower.

Power insulates us from our bodies. We enjoy running when we are small, because it is the only way we have to move fast. Later we succumb to go-carts or motorcycles, because we can move faster. The wind blows harder through our hair, and we feel even more free, our emancipation heightened by the roar of the engine and the adrenaline surge of knowing that a misstep can cost us our lives.

The thrill of power is irresistible, and the more you use, the bigger the thrill becomes. Ask the pilot of Stealth Bomber, or a driver at the Indy 500. Now, after a lifetime of increasingly dependency, some ya-hoo comes along and tells us to give it up? This is a hard sell.

In 1991 Real Goods sponsored the first-ever Off the Grid Day, the purpose of which was to try to re-create, if only for a day what life was like before power. What we hoped to accomplish was to get people to recognize what an essential role power plays in their lives, but also to develop an appreciation of the many activities that can be independent of power (at least, power from the grid). Among the activities we heard about were canoeing, sailing, biking, picnicking, walking, volleyball, golf..... the list was infinite.

Now the country is involved in a "War on Drugs." Again, we appreciate the need for belligerence to get people's attention for a worthy cause. The fallacy of the logic, however, is that none of the people signing up to fight the war will be the ones most affected by drugs. It is unlikely that any addicts or pushers will be in the ranks, nor any of their victims. Perhaps a more appropriate war would be one fought against hopelessness and true powerlessness that leads people to the escape promised by drugs.

Similarly, the soldiers in the trenches in the war against energy dependence are the people who are already most enlightened. While they can probably be successful in furthering their own sense of "empowerment" (currently, the word most in favor amongst the privileged elite), the entire movement is destined to become yet another temporary blip on the national consciousness unless we are able to enlist the support of the people who currently think there is no problem.

Instead of watching adventures, let's live them.

This can be accomplished either by screaming hysterically that there is a problem (and if these people have not heard the warnings about global warming, acid rain, nuclear meltdown, and ozone depletion by now, they never will), or by devious psychological means.

We choose the latter. Here is the Power Down Plan—we declare war, a Power Down War. We're going to have so much fun, that powering down will seem like a lifelong trip to Disneyland.

The Power Down War begins with a healthy body, because being anything less than 100% is no fun at all. We'll pay attention to what we put on and in our bodies, then we'll use our bodies to their full capacity and treat them with pampering affection that Dad used to give the '53 Buick Roadster.

The second phase of the Power Down War involves people, because doing anything is more fun if you share it. Instead of watching adventures, let's live them. (Frankly, I can do a better job of acting out my role than can Kevin Costner.) Wars need soldiers, and we will do our best to recruit, because we want this movie to have thousands of extras.

There is a financial element to the war (as with all wars). The equipment (from PV panels to windsurfers, from compact fluorescents to down comforters) will be purchased from the money we save by reducing power consumption, and from self-imposed taxes that we pay for what we do consume. Best of all, the money gets spent on us, us, US! (Maybe we have finally grown out of the "me generation.")

We will need allies, and will find them in the sun, wind, and water. Only by aligning ourselves with these forces and by harnessing their immense powers can we hope for success.

The campaign promises to be long and arduous, marked by constant pampering of the body, stimulating bouts of conversation, grueling expansion of the mind, and hopefully the camaraderie of many fellows.

This is the Power Down War of endless challenge and endless stimulation. It won't be easy, but it *will* be fun. Come on board and join the pursuit for the new American Dream.

— Stephen Morris

Notes and Workspace

Solar Toys, Maps, & Gifts

Gift Certificates

For those of you enamored by our products but unable to decide what to give, we offer **Real Goods Gift Certificates.** Many of our customers have used these for wedding presents for would-be independently powered newlywed home steads. We'll be happy to send a gift certificate to the person of your choice, in any amount you choose, along with the our latest catalog.

00-001 Gift Certificate   Specify $ Amount

From the Rainforest

This irresistable Rainforest Munchies mix of dried bananas, coconut, papaya, pineapple, cashew, and Brazil nuts is packaged in a new reusable decorator tin. Pack toys, collections, or odds and ends in the tin after you've devoured the treats. A percentage of the profits from this mixture go to Cultural Survival, an orgainization dedicated to preserving the Rainforests of the world. Also, available packed in a can. Net weight 30 ounces each.

03-805 Rain Forest Munchies Tin $19
03-804 Rain Forest Munchies Can $14

Rainforest Crunch

Rainforest Crunch is an incredibly delicious, all-natural, highly-addictive cashew and Brazil nut buttercrunch that, best of all, helps preserve the rainforest. The creation of Ben Cohen (Ben & Jerry's Ice Cream), this is part of a larger project to show that the rainforest can be used more profitably as a living rainforest than by cutting, burning, and transforming it into plantations and ranches. Forty percent of the profit goes to rainforest preservation organizations and inter-national environmental projects. This very tasty candy provides an ideal vehicle for putting your money where your mouth is! The Crunch comes both in an attractive and colorful 1-pound reus-able tin, or a colorful box. *Available October thru April only, due to summer meltdown.*

03-801 Rainforest Crunch (1 lb. can) $15
03-802 Rainforest Crunch (8 oz. box) $5

To order any of these products
or for more technical information
call us toll-free at
1-800-762-7325

Reversible Earth/ Monkey and Elephant

Two toys in one! Unzip the animal's back and turn it inside out, and you have a globe showing where endangered species still cling. Everybody loves this toy. Made in China of child-safe materials. For each animal purchased, 5 square feet of rainforest is saved by the maker, Friends of the Forest.

| 90-427 | Earth Monkey | $19 |
| 90-467 | Earth Elephant | $19 |

Bumper Stickers

Stop Solar Energy

Guaranteed to crank heads and provide lots of double-takes. With the words: "Stop Solar Energy; Mutants for Nuclear Power, Village Idiots for a Toxic Environment," this bumper sticker will go great on the bumper of your electric vehicle!

STOP SOLAR ENERGY
MUTANTS FOR NUCLEAR POWER ☢
VILLAGE IDIOTS FOR A TOXIC ENVIRONMENT

90-741 Stop Solar... Bumper Sticker $2

I Get My Electricity from the Sun

Black letters on a yellow background. Flaunt your energy independence!

I GET MY ELECTRICITY FROM THE SUN
REAL GOODS TRADING CO. ☼ 966 MAZZONI ST. UKIAH, CA. 95482 (707) 468-9214

90-742 Electricity from...Bumper Sticker $1

Solar Watch

Casio's solar watch is powered by an amorphous solar cell, eliminating the need and the expense of a battery. It comes with a daily alarm, can be used as a stopwatch, can be set for a 12 or 24 hour format, and has an automatic calendar that determines and sets the number of odd or even days in the month. The watch is also water resistant.

68-405 Solar Watch (Casio) $39

Solar Tea Jar

The small solar panel on the top of this 1-gallon plastic unbreakable tea jar will stir your tea with the paddle and keep your tea warm in the sun. A built-in spigot is provided at the bottom for easy pouring. This brewing method reduces the acidity and bitterness of the tea. The unit measures 10-1/4 inches high, 6 inches diameter.

90-202 Solar Tea Jar $18

Solar Wooden Model Kits

Our most popular toy. Our wooden model kits have been incorporated into many other popular product catalogs. These easy-to-assemble (about 1 hour) models are a perfect demonstration to children (and adults) of the wonder of solar power. The parts snap together easily; glue is included. They make great science projects. We've sold over 4,000 of these over the last two years.

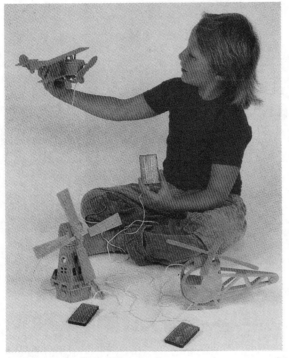

90-402	Airplane Kit	$18
90-403	Helicopter Kit	$18
90-404	Windmill Kit	$18

Solar Construction Kit

This popular four-in-one construction kit is entertaining and very educational for kids from 5 years old. Children can construct a helicopter, windmill, airplane, or water wheel—each with a solar-powered moving part. The kit is complete with an electric motor, a small solar panel, and over 100 plastic pieces.

Build a Sun Runner - the Solar Model Car Kit

Tamiya's sleek, futuristic solar car is more than just an engrossing toy: it's a miniature harbinger of the future. Equipped with an adjustable solar panel, the car will even run indoors under a strong light - so it's good for lots of winter amusement, as well as summer. The steerable front end lets you direct the car in the direction you desire. All the components are designed for running efficiency. Ultra-thin tires are matched to moondisc wheels for the "experimental-vehicle" look. The side-winder gears transmit power directly and effectively to the rear wheels. There's a choice of three different gear ratios to meet your trail requirements. The kit contains body parts, wheels, high-efficiency motor, and a 0.5V-1200mA solar panel with a protective case. A screwdriver is all you'll need for quick assembly. Measures about 2"x10".

90-464 Solar Model Car Kit $45

90-400 Solar Construction Kit $22

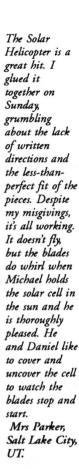

The Solar Helicopter is a great hit. I glued it together on Sunday, grumbling about the lack of written directions and the less-than-perfect fit of the pieces. Despite my misgivings, it's all working. It doesn't fly, but the blades do whirl when Michael holds the solar cell in the sun and he is thoroughly pleased. He and Daniel like to cover and uncover the cell to watch the blades stop and start.
Mrs Parker, Salt Lake City, UT.

Solar Speedboat

Whenever the sun shines on the boat the tiny motor propels it across the water or around in circles depending on where you set the rudder. It's the closest thing ever invented to a perpetual motion machine! A great educational toy.

90-406 Solar Speedboat $24

Naturalist Microscope

The portable Naturalist Microscope brings you close to the hidden beauty and wonder of nature—crystals, leaf veins, and insects. Its "easy-focus" glass lens provides 60 power magnification. Equipped with built-in case, two specimen vials, tweezer, probe, two glass slides, and instruction book. Designed for ages 5 through adult, this elegant but inexpensive instrument is made in Russia. Most inexpensive microscopes never seem to come into focus; the Naturalist offers a crisp and very sharp image.

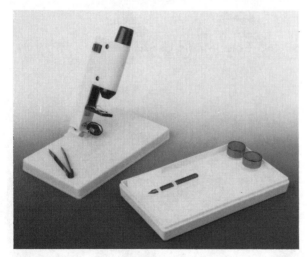

90-465 Naturalist Microscope $19

150 Solar Experiments

With this kit, you can build a solar furnace, a stroboscope, moiré patterns, an electronic thermometer, electroplating, photosynthesis, and solar energy mobiles. Components include a DC electric motor, a 600 milliamp solar cell, a 50-microamp voltmeter, a solar water heater, a parabolic reflector, a strobe disk, a magnifying glass, a thermometer, diodes, LEDs, and a test tube. It's the most comprehensive solar experimentation kit that we've been able to find.

90-401 150 Solar Experiments $39

Solar Educational Kit

Here is a simple and inexpensive educational kit that will get any child (or adult) started with the graphic basics of solar power. A small solar (PV) panel is included along with a motor, a small fan, and a pinwheel for making a variety of objects that are really functional.

90-405 Solar Educational Kit $9

Educational Solar Energy Kit

This kit helps prepare your child for the coming world society based on renewable energy. Young people 8 years old and up will learn how solar energy works in a practical way. Comes with eight mini-photovoltaic modules that can be wired in endless combinations of series and parallel circuits. They'll get a sense of accomplishment from assembling a solar circuit themselves, and seeing how electricity functions, plus the excitement of actually powering a radio, calculator, battery charger, or the fan that comes with the kit. Our favorite solar experiments kit.

90-470 Educational Solar Energy Kit $19

High-Power Solar Project Kit

This educational kit comes complete with a 6 by 5 inch amorphous solar panel with an 8-volt 100-mA nominal output. The small solar panel works great for nicad battery charging, running small fans and motors, and countless other solar experiments and hobbies—limited only by your imagination! Instruction sheet, motor, propeller, and wiring clips are included. Some soldering required.

90-422 High-Power Solar Project Kit $11

Science Lab

Our new Science Lab puts together three science laboratories in one package. Solar Energy, Chemistry, and Microscopy are combined to give hours of fascinating entertainment, education, and astonishment for budding scientists of all ages. The kit comes complete—no batteries or soldering are needed. There are suggestions for more than 75 exciting experiments and activities. Recommended for ages 8 through adult.

90-423 Science Lab $25

Electrics Education Kit

This kit contains 80 electrical experiments. Learn about static electricity, magnetism, and electro-chemistry. Make batteries from household items, or even your tongue! Components include electromagnet, compass, Morse key, LEDs, wire coils, magnet, electrodes, resistors, plastic circuit board, battery case. There are many hours of fun and learning in this kit. Ages 8 to adult. Four AA batteries not included.

90-417 Electrics Education Kit $39

Solar Energy Lab

Our new mini-lab is a great starting point for solar energy experimenters. Included are a parabolic solar concentrator, test tube and holder, thermometer, solar water heater, focusing lens, and book with over 25 experiments. Easy assembly. Ages 8 through adult.

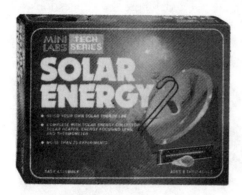

90-420 Solar Energy Lab   **$12**

Solar-Powered Music Box

There are no batteries or main-springs to break in our Solar Music Box. It is 100% powered by light (either the sun or a 75-watt bulb) through its solar cell. Place the music box in a sunny window and watch the melody happen. Mounted in a 4½-inch-square black base, 1-7/8 inch high, the song played is "The Entertainer."

90-431 Solar Music Box   **$35**

Solar-Powered Sun Chime

Our new Sun Chime will play in the heat of summer when your wind chime is dormant. Hang it in direct sunlight for a melodious harmony from the six brushed-brass tubes. The motor circulates a small chain, which strikes the chimes. The chime can be suspended from the ceiling or hung with the included suction cup in a sunlit window. It is 9-inches tall.

90-428 Sun Chime   **$21**

Solar Musical Keychain

The Solar Musical Keychain is a functional keychain with a small LED light for finding your keyhole in the dark. When the PV panel is directed toward the sun or a bright light it plays a song. A great PV educator.

90-407 Solar Musical Keychain   **$6**

AM/FM Solar Radio

Our solar-powered AM/FM radio folds out to recharge the nicad batteries in the sun. It gives good quality sound out of the built-in high fidelity cobalt mylar speaker, and it accepts 3.5-mm headphones as well (not included). The unit measures 4.5 cm wide by 11 cm tall. It works directly from the sun. No house current is needed, no disposable alkaline batteries that go dead when you're at the beach or in the wilds. One of our all time greatest sellers.

90-409 AM/FM Solar Radio   **$25**

Dynamo & Solar AM/FM Receiver

The dynamo radio needs no electricity or disposable batteries other than two AA nicads. It has a small solar panel for solar charging and a hand-cranking dynamo for hand charging. The dynamo charges at a 10:1 ratio—one minute of cranking gives you 10 minutes of radio operation.

90-418 Dynamo Receiver   **$29**

Solar Pocket Radio

Of all the solar radios we sell this new AM/FM radio has the best sound quality. It is compact at only 2.3 by 3.5 by 0.5 inches, and weighs less than 3 ounces! It is true stereo, with high-quality earphones included (there are no speakers). The integral solar panel charges the built-in nicad battery, allowing 8 hours of play in total darkness. Features include auto-power on-off and detachable two-position carrying clip. The unit produces rich, high-volume sound, with superb FM reception. Take it anywhere; it will never need batteries.

90-468 Solar Pocket Radio   **$39**

Solar Visor AM/FM Speaker Radio

This solar visor AM/FM radio is great for jogging or working in the garden. The radio can be used at night or indoors with the included battery pack. One hour of sun allows for approximately 6 hours of indoor use.

90-421 Solar AM/FM Visor Radio   **$24**

Solar Mosquito Guard

Our solar mosquito guard really works to repel mosquitoes. It puts out a high-frequency wave that will repel mosquitoes in a 12 foot radius. The battery will recharge in 3 hours of sun and it comes with an on-off switch. Be forewarned that the Solar Mosquito Guard's high-pitched sound is definitely audible. With no other noise it can be annoying, but with a little background music or commotion it works great, and is certainly less annoying than the music of a mosquito buzzing in your ears! We must be honest: of the 5,000 units we've sold since our Spring 1990 catalog was released, we have had around 100 returns from folks who say they do not work—these cases may be due to deaf mosquitoes! We have gotten many, many raves as well.

I am a resident of the former Mosquito County in Florida, and I am impressed with the Mosquito Guard...the relief from exploratory drilling provided by the device is most welcome. Hugh E Webber, Orlando, FL

90-419 Solar Mosquito Guard　　　　**$8**

Solar Safari Hat

Our safari-style solar cooling hat is similar to the Solar Cool Cap but more attractive and a better conversation piece. Switch from solar power to battery power for those humid, steamy nights in the African savannah. A real eye grabber, and they really do work wonderfully! Specify Tan (90-411-T) or White (90-411-W)

90-411 Solar Safari Hat　　　　**$45**

Solar Cool Caps

These caps work as miniature evaporative coolers. The cool breeze blowing on your moist forehead quickly acts to cool down your entire body. A switch allows you to select solar or battery and a battery compartment is included for two AA cells (not included). Definitely more than a gadget or conversation piece—they really work! We equipped an entire roofing company with them for the hot Ukiah summers.
Specify your first and second choice of colors: Red 90-410-R, Blue 90-410-B, or White 90-410-W.

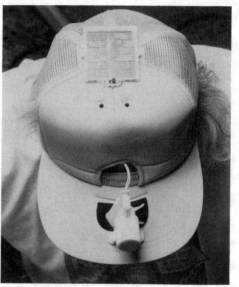

90-410 Solar Cool Cap　　　　**$24**

Earthalert – Environmental Game

Here is a board game that actively involves its players in environmental issues. Learn about the environment while having fun. The game itself, of course, is made entirely of recycled materials. This game involves skill and role playing and is suitable for ages 8 to adult.

90-438 Earth Alert Game　　　　**$29**

Birding Game

We put this great new game into our catalog by popular demand from staff children Ashley and Sara, who as new product testers played for 16 hours in two days! Incorporating the field identification system of Roger Tory Peterson, the inventor of modern natural history field guides, this game is recommended by the American Birding Association. Up to six players compete to raise money for their chosen conservation cause by identifying birds and answering questions about the natural history, behavior, and conservation of over 200 North American birds. Using any of four birding skill levels, the game is highly entertaining and incredibly visually attractive, with enormous educational value. Did you know 60 million Americans are casual bird watchers? Four years of testing went into perfecting this game.

90-439 Birding Game   **$39**

Night Sky Star Stencil

The Night Sky Star Stencil is one of the most fun and exciting new products we've found. It's like camping out in your own home. The stars are not visible in normal light, but when your room is darkened, the ceiling becomes a spectacular star display: more than 350 stars become instantly visible with completely accurate position and brightness.

Application is simple (1 or 2 hours): put up the reusable stencil with the provided adhesive, paint the stencil holes, take down the stencil, turn off the lights, and the stars come out. Stars will glow 30 minutes to 2 hours depending on light exposure before room darkening, and the phosphorescent glow paint lasts for years.

It's available in either summer sky or winter sky stellar configurations and in two sizes: 8 by 7½ feet and 12 by 12½ feet. Having tried all the various paste-on star kits, we find the Night Sky Star Stencil to be the queen of home planetaria. It's ideal for helping kids conquer their fear of the dark! Glow paint is a water-based nontoxic, nonradioactive material.

90-341	Winter sky (8 ft)	$25
90-342	Winter sky (12 ft)	$30
90-343	Summer sky (8 ft)	$25
90-344	Summer sky (12 ft)	$30

Tyvek World Map

This is probably the nicest map of the world ever made. The colors are brilliant, and the countries and mountains seem to jump out at you. It appears 3-D even though it isn't. It's printed on virtually untearable Tyvek. The size is 30 x 53 inches, the scale is 1:28 million, and the projection is Van der Grinten. Fully updated as of April 1992.

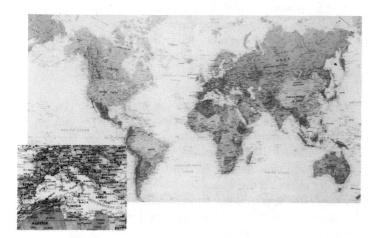

| 90-301 | Tyvek World Map | $27 |

Celestial Planisphere

Our Celestial Planisphere is by far the easiest way to learn the stars and constellations of the sky. Simply dial in the time of night and the day of the year, orient toward the North Star, and you're provided with a detailed map of the heavens. 10½" Planisphere is available in three latitudes.

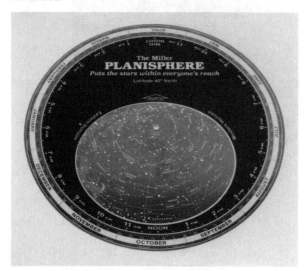

90-346	Planisphere—30°	$10
90-347	Planisphere—40°	$10
90-348	Planisphere—50°	$10

The Earth Flag

The Earth Flag's popularity gained a lot of momentum from Earth Day 1990, and can now be seen flying everywhere. It continues to be an inspiring symbol to peace, environmental justice, and global vision.

The Earth Flag is now available in three different sizes. Each flag consists of a four-color, photolike image of the Earth, based directly on the now-famous NASA photograph, printed on a dark blue background. The **Large Earth Flag** is 3 x 5 feet and is printed on nylon, with a canvas header and brass grommets. The **Medium Earth Flag** is 2 x 3 feet and is printed on cotton, with a canvas header and brass grommets. It is recommended for parades, indoor display, and classroom use. The **Small Earth Flag** is 6 x 9 inches, is printed on cotton, and mounted on an unfinished white birch stick.

90-731	Earth Flag—Large (3 x 5 ft)	$39
90-732	Earth Flag—Medium (2 x 3 ft)	$19
90-733	Earth Flag—Small (6 x 9 in)	$4

Whole Earth Globe

This extraordinary 16-inch globe depicts Earth as seen from space, without borders or boundaries. It can be used by educators, peacemakers, as a beach ball, or as an allpurpose toy for kids of all ages everywhere. It comes complete with the Global Handbook and makes a wonderful gift. We've found a second version of the globe—this one (the "Wildlife Version") has native animals depicted beautifully in their natural habitats.

90-415 Earth Ball (Nasa Photo) $8

90-430 Earth Ball (Wildlife Version) $9

Raven Landforms Map

Introducing the real America! This large, highly detailed map gives both children and adults a much better sense of the overall geography of the United States (except Alaska and Hawaii). It identifies great mountain ranges and plains, showing elevation by color. State lines, county seats, major cities and highways are also depicted. For superior protection and longer life, the 37 x 58 inch map is also available with vinyl lamination.

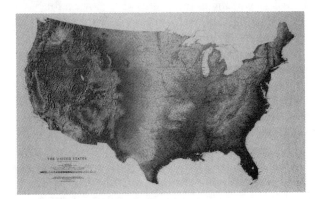

90-309 Landforms Map $39
90-310 Landforms Map Laminated $64

Raven Landforms and Drainage Map

As if you are an astronaut, you view the lay of the land and the patterns of river drainage all across the 48 states. In starling three-dimensional imaging on a flat surface, this 37 by 58 inch black-and-white map shows us what a dramatic and fascinating continent we live on. It's educational and inspiring. Again, you have the choice of lamination. Everyone who has seen this map has been stunned by its realism. It is truly exquisite.

90-311 Landforms & Drainage $39
90-312 Landforms & Drainage Laminated $64

To order any of these products
or for more technical information
call us toll-free at
1-800-762-7325

Raven State Maps

These are the best state maps we've ever seen anywhere of each of the 11 Western states, Hawaii, Alaska, New York, Ohio, Pennsylvania-New Jersey, and now Vermont/New Hampshire. Raven maps give you the startling illusion of three-dimensional imaging right on a flat surface. They offer remarkably clear views of physical features, and a wealth of detail on towns, roads, and railroads. Maps vary slightly in size but are approximately 42 x 55 inches.

All Raven maps retain the accuracy of the U.S. Geological Survey materials from which they are built. Intended for wall display, all maps are printed in fade-resistant inks on sturdy 70# paper.

Because of the startling realism of these maps, people are highly encouraged to touch them. For this reason, Raven also makes a lamination available of 1.5 mil vinyl coating on both sides. It is lightweight enough so that the map is easily rolled for shipping in a sturdy tube. *Rather than add an additional map tube shipping charge, we have simply included the $4 in our maps price.*

90-302 State Maps $24
90-303 Laminated State Maps $49
Specify state (90-302- AK, AZ, CA, CO, HI, ID, MA-CT-RI, ME, MT, NM, NV, NY, OH, OR, PA-NJ, UT, VT-NH, WA, WY)

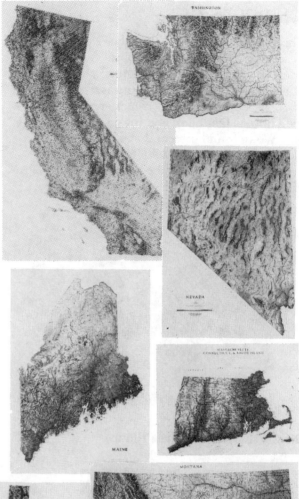

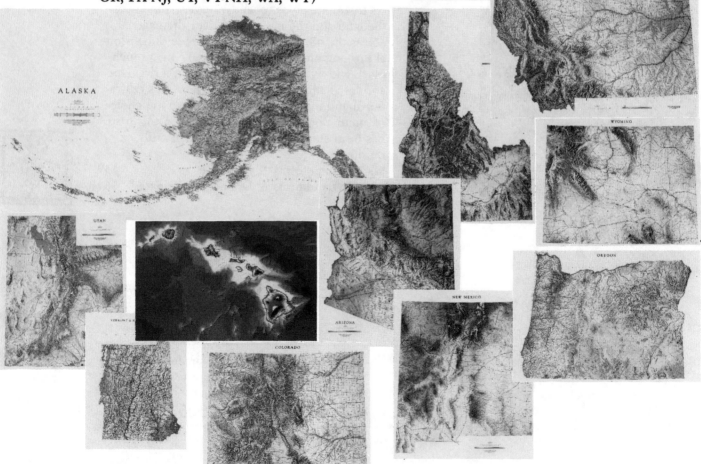

Real Goods T-Shirts

Marathon Man T-Shirt

This T-shirt was the hit of Earth Day booths everywhere. It features cover artist Winston Smith's rendition of a marathon man leaping over the earth carrying a solar panel, with the words: Real Goods, We Can Change the World. Our local T-shirt printer Bob Perkowski said it was the nicest T-shirt he's done in 10 years! Very vivid colors. Specify size (S,M,L,XL).

Cavemen in Tool Heaven T-Shirt

Featuring another cover gem by Winston Smith, our ever-popular Spring '89 "Real Goods News" cover has the enlightened Neanderthals in tool heaven with the words: "Real Goods Brings You Out of the Stone Age Into the Solar Age." Printed on 100% cotton. Specify size (S,M,L,XL).

90-605 Spring '90 Real Goods T-Shirt $12

90-600 Real Goods Caveman T-Shirt $12

Pocket Falcon T-Shirt

A falcon is perched on your pocket! A great illusion, this shirt is lots of fun to wear. Also known as a sparrow hawk, the kestrel is America's smallest and most colorful falcon. Our best all-time seller in T-shirts! 100% cotton—specify size (S,M,L,XL).
90-631 Pocket Falcon $15

Pocket Iguana T-Shirt

A green iguana has scaled your T-shirt! Wear this design only if you don't mind drawing attention to yourself. Our model, Stymie, was one of our most interesting subjects. Found in Mexico, Central America, and tropical south America, *Iguana iguana's* favorite perch is on a tree limb overhanging water. 100% cotton—specify size (S,M,L,XL).

90-633 Pocket Iguana   **$15**

The Rainforest T-Shirt

Our new Rainforest T-Shirt is stunning. Several of our new products staff thought it was the most beautiful shirt they'd ever seen! It's made of green cotton which is an untreated, unbleached, and uncombed natural cotton, with an ecru (tan) background. Specify size (M,L,XL-Not available in small).

Nuclear Fear T-Shirt

This image was created by German artist Carl W. Rohrig for the cover of *Stern* magazine, right after the Chernobyl Nuclear accident. The editor who OK'd this picture lost his job as a result. 100% cotton—specify size (M,L,XL).

90-683 Nuclear Fear T-Shirt   **$17**

90-614 Rainforest T-Shirt   **$17**

Earth—Love It or Lose It T-Shirt

Our new Earth—Love It or Lose It T-shirt is a striking image of the earth as a delicate balloon with a child perched on top wondering why the air is coming out. An innovative artist in Colorado makes this beautiful airbrushed design for us. This first appeared on our winter '92 *Real Goods News*. 100% cotton. Specify size (S,M,L,-XL).

90-672 Love It or Lose It T-Shirt   **$16**
90-710 Love It or Lose It Poster   **$16**

Pocket Snake T-Shirt

Slithering from your pocket is a real surprise—North America's most beautiful snake, the corn snake. Nonvenomous, *Elaphe guttata* ranges over much of the central and eastern United States, and preys on rodents, bats, and birds. It is named for its colorfully patterned similarity to kernels of Indian corn. The corn snake is primarily nocturnal and is a talented tree climber. Specify size (S,M,L,XL).

90-635 Pocket Snake   **$15**

Pocket Gator T-Shirt

Clinging precariously to your shirt, this year-old gator is startling! *Alligator mississippiensis* is found in the aqueous environments of the Gulf states and up the Atlantic coast to the Carolinas. Once hunted relentlessly for its meat and skin, its population has recovered since being designated a "threatened" species. It can reach over 19 feet, making it the largest reptile in North America. Specify size (S,M,L,XL).

90-636 Pocket Gator   **$15**

Satellite Images of Earth Revolutionize Map Making

This spectacular image of the World is a color mosaic of actual satellite images revealing the earth, in natural color, unobstructed by clouds for the first time. The image was created by electronically assembling thousands of individual satellite images on the Stardent computer. The final result is an astoundingly clear, sharp and colorful image of the planet Earth. This map will profoundly change the way you view our planet. Available in 24" x 36" on paper or laminated.

90-306 Earth From Space Map   **$15**
90-307 Laminated Map   **$20**

Knowledge

The reader of Real Goods materials has heard us say, many times over, "knowledge is our most important product."

But what does that statement mean? To us it means that we are in a learning phase, and even more important than the "real goods" that we need for independent living is the knowledge that lets us know the hows, whens, and whys of putting the tools to use.

Ever heard the story of the man with the broken TV set? He tried everything he could think of to fix it, enduring hours of frustration piqued by the whines of bored children. Finally, he called the TV repairman, who came and replaced one screw, thereby fixing the set.

"That will be forty dollars," the repairman said.

"Forty dollars!" the man protested. "That screw couldn't have cost more than five cents."

"Yep," answered the repairman, "It's five cents for the screw, and $39.95 for knowing which one to replace." The story is quaintly anachronistic now. Imagine the absurdity that a television set could actually be repaired, let alone that someone would come to your house to do it! The point is still valid: tools without knowledge are worthless.

As much as any company we know, Real Goods uses its publications to disseminate knowledge. Most businesses would never devote the space we do to the Readers' Forum in the "Real Goods News," or devote the staff time and energy to the *Alternative Energy Sourcebook*. We think it is an essential part of our formula, and we are always seeking new ways to push the frontiers of independent living.

A new forum for knowledge is the Institute for Independent Living. The purpose of the institute is to provide a face-to-face, interactive environment where people can come and learn about independent living from a group that collectively has a wealth of experience that belies their years.

In time, we plan the Institute to be a hallowed physical reality, with a completely off-the-grid campus, that demonstrates the products, technologies, and techniques of the independent lifestyle. Landscaped with ecologically appropriate plants, needing minimal amounts of precious water or fertilization, the campus will be at once beautiful while so environmentally compatible as to be nearly invisible. The professors will learn as much from the students as the students from the teachers. The products of Real Goods will be omnipresent, unobtrusive, yet essential in their functionality.

Of course, it's a pipe dream! But it has to start somewhere.

Initially, the Institute of Independent Living will offer weekend seminars and special interest forums related to the topics of energy independence and environmentally sane living. Eventually the Institute will encompass our publishing wing, and perhaps our product testing facility.

This is where we will *apply* our knowledge.

Truthfully, we can't be sure how it will evolve. Perhaps, in a few years, we will look back on this idea, as we now look back on the electric toothbrush we bought in 1979, and laugh. What we can feel certain of is that our creation of the Institute is a physical commitment to our statement: "Knowledge is our most important product."

– J.S.

POWER DOWN

Horsepower or Maybe Just Horse Manure

The first car I ever owned was a 1960 Pontiac Catalina. Quite rightly, it holds a place in my heart right up there with my first love and the time I actually hit one out in a Little League game. Increasingly, however, I am more amused by the gross folly of this vehicle than warmed by the soft images of its memory.

The Catalina was one of the largest cars Pontiac ever made. It had eight cylinders, something like 356 horsepower, got eight miles to a gallon (on high test!), and cruised comfortably at 120 miles per hour. Believe me, I know.

Even though gas was cheap in those days, the voracious appetite of the vehicle kept me constantly impoverished, but I considered it a small price to pay for the ability to hit one-twenty on the straightaways. After all, isn't this an American birthright? It was only years later, as I grew wise and gray, that I developed perspective on this vehicle.

The occasion was a local fair that included displays of animal husbandry. I took my small son into the barn that housed the work horses. We stood in awe of these gentle beasts chewing hay in their stalls. We patted their heavily muscled flanks (as high as we could reach, anyway), keeping a respectful distance. We were fascinated by the control that the horse's owners demonstrated over these towers of power. With sharp gee's and haw's and well-timed hand slaps, they could actually get these beasts to do their bidding. It was easy to see why "horsepower" became the standard unit of measurement in matters of work.

The role of the work horse in early rural life was explained to us. This warm-blooded machine pulled the plow, dragged the wood from the forest, pulled the sled that carried the sap, and carried the family to church on Sundays. For its pay the work horse required only fresh water and enough grass or hay to feed it. Not that this last requirement was to be taken lightly. Depending on the quality of the grassland, up to 20 acres might be required to feed a horse year round.

The horse required little maintenance—no waxing, no brake jobs, no STP Oil Treatment. For residue, the animal left only organic substances that might be put on a garden to enhance growth. In time a work horse (the right kind, anyway) could reproduce itself.

As I heard these tales, holding my son's small hand, I thought about an 18 year-old driving a Pontiac Catalina that boasted **three hundred and fifty-odd** horses beneath its hood. Imagine a wagon holding one lad holding the reins connected to a line of 350 beasts that stretched beyond the horizon? One young man, I don't care how many part-time jobs he held, could never earn the money to buy the hay for such a herd. How big would a barn have to be to feed them? One would have to own a ranch or farm the size of Mendocino County!

How could someone intimidated by the

snorting mass of a single work horse be comfortable commanding 350 of them at the turn of a key? And how could something that makes as much sense as a work horse be replaced by something as illogical as a Pontiac Catalina?

How could someone intimidated by the snorting mass of a single work horse be comfortable commanding 350 of them at the turn of a key?

It was to answer this question that I decided to purchase a bicycle generator from Real Goods. Here would be an interactive demonstration of the consequences of power consumption in our lives. The concept is simple: if you want power, make it. If you want to watch television, start pedaling. By logging 15 minutes on the machine you will earn yourself a sitcom. You will have to decide whether or not "Roseanne" is worth the effort.

No longer, I realized, would I need to fret about the mind-rotting effect of video games. My children's young minds would rot only insofar as their bodies' cardiovascular endurance permitted.

Best of all, my sons would not grow up as I had, totally disconnected to the ramifications. The mental image of the wagon pulled by 350

horses, or a water skier towed by a veritable cavalry is enough to make one realize how potent is the energy contained within our fossil fuels. We don't want to turn our backs on this power, but we need to manage it sanely, and to use it more productively than did the 18 year-old hurtling along at the wheel of his Catalina.

As for why the work horse was replaced by the internal-combustion engine, the reason is fairly obvious. Have you ever considered how much work it would be to clean up the shit from 350 horses? At the turn of the century, when the horse population of our burgeoning industrial cities was at an all-time high, the waste problem was out of control. Gutters were filled with horsecrap. The waste problem was obvious to anyone with eyes and a nose. One of the miracles of the internal-combustion engine was that it made the waste problem disappear.

Or so it seemed.

Now we are learning that we simply took the shit from the streets and put it into the air. No problem was solved, but for a while it was swept under the rug. This is the second half of the lesson that my children will learn from their bicycle generator: there is no free lunch, and there is no free TV.

The tools to teach these critical lessons are now at hand. Only one thing prevents me from teaching my children the essential life-lesson on the precious nature of power that I learned too late in life. Courage, or lack thereof.

— Stephen Morris

Something New Under the Sun. It's the Bell Solar Battery, made of thin discs of specially treated silicon, an ingredient of common sand. It converts the sun's rays directly into usable amounts of electricity. Simple and trouble-free. (The storage batteries beside the solar battery store up its electricity for night use.)

Bell System Solar Battery Converts Sun's Rays into Electricity!

Bell Telephone Laboratories invention has great possibilities for telephone service and for all mankind

Ever since Archimedes, men have been searching for the secret of the sun.

For it is known that the same kindly rays that help the flowers and the grains and the fruits to grow also send us almost limitless power. It is nearly as much every three days as in all known reserves of coal, oil and uranium.

If this energy could be put to use — there would be enough to turn every wheel and light every lamp that mankind would ever need.

The dream of ages has been brought closer by the Bell System Solar Battery. It was invented at the Bell Telephone Laboratories after long research and first announced in 1954. Since then its efficiency has been doubled and its usefulness extended.

There's still much to be done before the battery's possibilities in telephony and for other uses are fully developed. But a good and pioneering start has been made.

The progress so far is like the opening of a door through which we can glimpse exciting new things for the future. Great benefits for telephone users and for all mankind may come from this forward step in putting the energy of the sun to practical use.

BELL TELEPHONE SYSTEM

Winston Smith came across this gem while searching through some 1950s National Geographic *magazines.*

The Efficient Home

Preaching to the converted has become the Real Goods curse. One way you can help: endow your local public or high school library with some of these fine books. The truth will make us free.

Efficient House Sourcebook

Robert Sardinsky and the Rocky Mountain Institute. "Most of the entries are excerpted deftly enough to be considered information sources themselves. You'd have to subscribe to a truckload of periodicals to keep up with what's presented here in one book." — J. Baldwin, *Whole Earth Review*, Summer 1988. This directory lists and critiques—complete with addresses, prices, and contact names—the periodicals, books, schools, organizations, and agencies that deal with all aspects of resource-efficient house design, construction, retrofit, and much more. This is the fully updated and revised 1991 edition (formerly called *Resource Efficient Housing*.) 160 pages. **80-108 $17**

Nontoxic, Natural, & Earthwise

Debra Lynn Dadd. This book helps you identify products that are safe for your home and shows you how to protect the environment as you shop. It contains the most comprehensive listing of healthful products available and uses a rating system that indicates both safety and environmental impact. It evaluates air and water filters, organic foods, biodegradable cleaners, cosmetics, pest controls, energy-saving appliances, clothing, gardening supplies, baby products and more. *Recommended by The Green Consumer, 50 Simple Things..., and the Whole Earth Ecolog.* 360 pages. **80-817 $13**

Practical Home Energy Savings

David Bill and the staff of the Rocky Mountain Institute. Concise, factual, and well-organized, this slim booklet contains as much intelligence about energy sensibility as most of the fat greenploitation books on the market. How can it do that in a mere 48 pages? Easy: the authors assume there's intelligence on the reader's end, and simply suggest what to do instead of belaboring the point. Well done, RMI. **80-123 $8**

Real Goods takes great pleasure in offering the works about alternative technologies - off the grid, off the map, off the wall projects that make sense enough to someone that it's become a book. Part of the joy comes, we suppose, from being publishers ourselves. Our best-selling Alternative Energy Sourcebook is the bible of the alternative energy industry, and our Remote Home Kit Manual is an excellent primer on getting a comfortable energy lifestyle from the sun.

How do frogs feel about the way we treat the planet? – Hopping Mad

Hot Water and Warm Homes from Sunlight

A teacher's guide for grades 4-8, part of the Greater Explorations in Math and Science series produced by Lawrence Hall of Science, University of California at Berkeley. Parents as well as teachers can use it to involve children in comprehending and experiencing the benefits of "going solar." Hands-on experiments involve building small paper houses to study solar heating, and using aluminum pie pans as model water heaters. There's also a section on making solar cookers. Suggestions for discussing the results of the experiments are included. The book supplies summary outlines and student data sheets. 56 pages.   **80-171 $8**

Non Toxic Household Books

America's Neighborhood Bats

Merlin Tuttle. A wonderful companion to your bat house. It provides a wealth of helpful information on bat behavior, biology, and habitat. It includes range maps, a source list, and easy-to-understand text. Spectacular color photos! 128 pages.   **80-830 $12**

The Green Consumer

Elkington, Hailes, and Makower. This is a brand-specific handbook of attractive, cost-competitive, and easily available products that make great environmental sense. It includes baby products (diapers too), long-lasting light bulbs, biodegradable detergents, soaps, garbage bags, wood products that won't destroy the rainforests, groceries that aren't overpackaged and won't overburden our waste disposal sites, and much more. Printed on recycled paper. Our only complaint is that they don't list Real Goods. Write to them! 342 pages.   **80-809 $10**

Least Toxic Home Pest Control

Dan Stein who is an exterminator who believes in controlling critters by outsmarting them. Intelligently written and beautifully illustrated, this book explains life cycles and vulnerabilities of pests, and shows ways to exploit those weaknesses without overkilling or polluting our homes. A masterpiece of environmental sensitivity. 87 pages   **80-826 $9**

Common Sense Pest Control

Olkowski, Daar, and Olkowski. This weighty tome provides a thorough review of the literature of least-toxic pest control in 715 large-format small print pages: everything you ever wanted to know about controlling pests, and much, much more. The authors advocate a close-the-barn-door-before-the-horse-escapes approach rather than quick-fix chemical overkill, the much beloved panacea of the petrochemical poison industry. There is an excellent short chapter on diversity, trap-crops, and companion planting in the garden.   **80-837 $40**

The Curmudgeon here again. There's an explosion of green books. (This is the bandwagon publishers have jumped on now Yuppie self-improvement books aren't selling so well.) Again, we've tried to find the best and most useful, but while we all need to do 50 simple things to save the earth, let's remember to tackle one or two difficult tasks, too.

Clean & Green

Annie Berthold Bond. An encyclopedic source of solutions to 485 yucky household problems —non-toxic cleanser for every cleaning challenge you can think of, from toilet bowls to laundry, from ovens to fireplaces. Mostly grandma's home wisdom, and a wonder we have forgotten so much! 162 pages.   **80-820 $9**

Shopping for a Better Environment

Laurence Tasaday with Katherine Stevenson. Just by participating in a modern industrial society we damage the environment to some degree, even with the best of intentions. But we can minimize our impact on the earth. Tasaday gives us the information to make that possible. He evaluates the environmental impacts of over 100 categories of consumer products—from beef to air fresheners to lawn mowers—and recommends specific types and brands. The reader learns how to choose, use, recycle, and dispose of items purchased. Helpful tables range from "Tropical Hardwoods to Avoid" to "Finding the Right Water Filter". This is the most complete, up-to-date, and easy-to-use guide to green consuming. 333 pages. **80-839 $10**

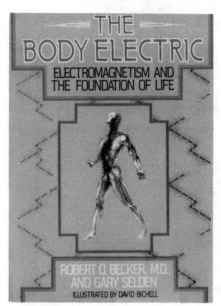

The Body Electric

Dr. Robert Becker and Gary Seldon. Electromagnetic radiation (EMR) can affect the human immune system, the rate of cancer growth, even thoughts and emotions. Dr. Becker discusses the hazards of EMR produced by everything from ordinary household appliances to utility power lines, radar, and microwave transmitters. This information has been suppressed for years. Our present AC electrical distribution network is putting the entire planet at risk. *"Changes would have been made years ago but for the opposition of power companies concerned with their short-term profits, and a government unwilling to challenge them...The entire power supply could be decentralized by using wind, flowing water, sunlight...greatly reducing the voltages and amperages required to transmit power over long distances."* This book is fascinating reading and will give you background information and motivation to clean up the magnetic pollution in your local environment, live longer, and avoid cancer! 364 pages. **80-610 $11**

To order any of these products
or for more technical information
call us toll-free at
1-800-762-7325

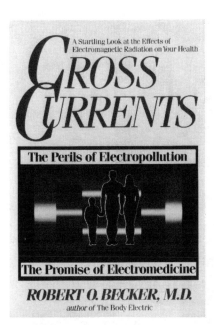

Cross Currents

Dr. Robert Becker. This is the definitive work on energy medicine (electro-medicine), the rapidly emerging new science that promises to unlock the true secrets of healing. Beginning with our historic link to the earth's natural pulsating magnetic fields, we are led through experiments and discoveries that have produced our present new knowledge of electromagnetic medicine and its effect on the healing process. In the last 30 years electro-pollution and cancer have (not coincidentally) both risen at the same alarming rate. The book does an excellent job of getting vitally important information out of the scientific priesthood and into the hands of the general public—who should be making the important environmental decisions! Dr. Becker is *the authority* on the biological effects of magnetism, a pioneering researcher in biological electricity and regeneration, and a professor at two medical universities. 336 pages. **80-612** **$22**

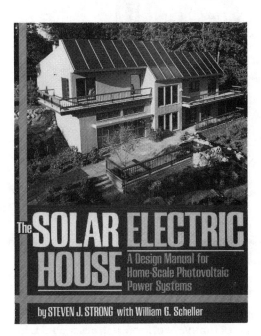

The Solar Electric House

Steven Strong. The author has designed more than 75 PV systems. This fine book covers all aspects of PV from the history and economics of solar power to the nuts and bolts of panels, balance of systems equipment, system sizing, installation, utility-intertie, stand-alone PV systems, and wiring instruction. A great starter book for the beginner. New edition. 276 pages.

80-800 **$20**

The New Solar Electric Home

Joel Davidson. Gives you all the information you need to set up a first-time PV system, whether it be remote site, grid connect, marine, or mobile, stand-alone, or auxiliary. Good photos, charts, graphics and tables. Written by one of the pioneers. Perhaps the best all-around book for getting started with alternative energy. 408 pages. **80-101** **$19**

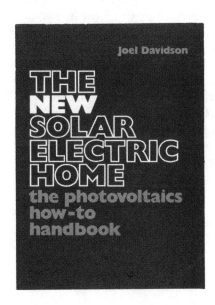

The same Administration that touted the virtues of the free market pressed home its strenuous efforts to deny citizens the information they needed to make intelligent choices. And Federal tax credits meant to help offset the generally much larger subsidies —totaling at least $50 billion per year—given to renewable's competitors were generally abolished, while most of the subsidies to depletable and harmful energy technologies were maintained or increased, tilting the unlevel playing field even further.
- Amory Lovins, in the Foreword

Living on 12 Volts With Ample Power

David Smead and Ruth Ishihara. This book has become the definitive work on batteries and DC refrigeration systems. It thoroughly covers all aspects of power systems from the workings of solar panels to the optimization of a balanced energy system. This book will show you why the battery charger you may be using is killing your batteries! The best book available for marine applications. Highly recommended for serious AE users. 344 pages.   **80-103** **$25**

Wiring 12 Volts For Ample Power

David Smead and Ruth Ishihara. The most comprehensive book on DC wiring to date, written by the authors of the popular book *Living on 12 Volts with Ample Power*. This book presents system schematics, wiring details, and troubleshooting information not found in other publications. Chapters cover the history of electricity from 600 B.C. to the modern age, DC electricity, AC electricity, electric loads, electric sources, wiring practices, system components, tools, and troubleshooting. 240 pages.   **80-111** **$19**

The Remote Home Kit Manual provides specific help with the Real Goods Remote Home Kits but it's also an thorough overview of the nuts-and-bolts installation of an photo-electric system. We won't be offended if you use it as a template for designing your own.

Remote Home Kit Owner's Manual

Real Goods Staff. We wondered in our weaker moments whether our Remote Home Kit Owner's Manual would really be worth all the hours we spent on it. We're happy to report it was! Our 65-page installation manual gives step-by-step instructions for the complete novice to electrify a homestead with one of our Remote Home Kits. There are lots of valuable charts, diagrams, and reference materials contained making it indispensable for any AE library. We encourage all of our new customers to check one out before purchasing one of our turnkey packages—we're getting lots of compliments on this one! 65 pages.   **80-400** **$5**

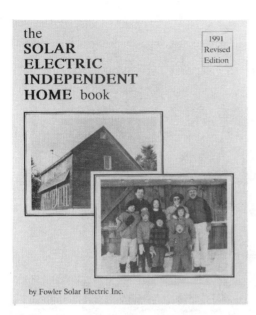

by Fowler Solar Electric Inc.

The Solar Electric Independent Home Book

Paul Jeffrey Fowler. A good, very basic primer for getting started with PV, written for the layman. Lots of good charts, a good glossary and appendix. It has recently been updated and includes 25 additional CAD diagrams (making 75 total) that are more detailed and much improved. The text has been updated to reflect changes in PV technology. Great charts and graphics. This is one of the best all-around books on wiring your PV system. 200+ pages.

80-102 $19

Practical Photovoltaics

Richard Komp. This book presents the theory and practice of photovoltaics in a nontechnical manner. It runs the gamut from explaining the physics of PV to offering hands-on instruction for the assembly of your own PV modules from salvaged cells. It's a unique combination of technical discussion and practical advice. 215 pages.

80-107 $17

RVers Guide to Solar Battery Charging

Noel and Barbara Kirkby. The authors have been RVing for over 20 years and have applied photovoltaics to their independent lifestyle. This book includes numerous example systems, illustrations, and easy-to-understand instructions. The *Whole Earth Catalog* calls it "A finely detailed guide to installing PV systems in your motorhome, trailer, boat, or cabin...this book has what you need to know." 176 pages. **80-105 $13**

PRACTICAL PHOTOVOLTAICS

Electricity from Solar Cells

Richard J. Komp, PhD
Skyheat Associates
English, Indiana

Second Edition

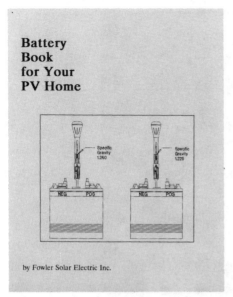

Battery Book for Your PV Home

This booklet by Fowler Electric Inc. gives concise information on lead-acid batteries. Topics covered include battery theory, maintenance, specific gravity, voltage, wiring, and equalization. Very easy to read and provides the essential information to understand and get the most from your batteries. 22 pages.   **80-104 $8**

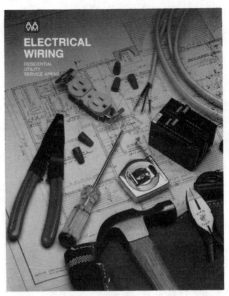

Electrical Wiring

The AAVIM staff. Thoroughly covers standard electrical wiring principles and procedures. This book has become an industry standard in training students, teachers, and professionals. It uses over 350 step-by-step color illustrations, covering circuits, receptacles and switches, installing service entrance equipment, and more. Revised in 1991 to include changes made in the 1990 National Electrical Code. 188 pages. **80-302 $23**

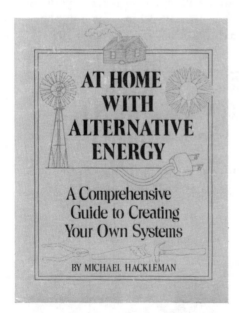

At Home with Alternative Energy

Michael Hackleman. The title is accurate. Emphasis on putting together your own "renewable" energy system from solar (no PV) to domestic hot water heating and cooling. Lots on wind energy; fun to read, exciting language. 145 pages.   **80-126 $12**

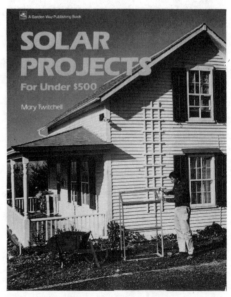

Solar Projects for Under $500

Mary Twitchell. Provides plans and step-by-step instructions for gardening, heating, and the homestead. Any do-it-yourselfer will enjoy building these simple, well-conceived projects. 130 pages.   **80-141 $12**

The Consumer Guide to Solar Energy

Scott Sklar and Ken Sheinkopf, explains solar energy technology and how to put it to use in any household. Neither too technical nor simplistic, it's the best primer we've seen: thorough and practical. *"This book gives every consumers the tools to provide permanent individual solutions to increasing energy prices and disturbances in the volatile Persian Gulf ... this book must be read!"*—Jeff Genzer (National Association of State Energy Officials). 344 pages. **80-828** **$10**

The 12V Doctor's Troubleshooting Book

Edgar J. Beyn. A simple, precise guide to tracking down problems in 12V electrical systems. Written for boat owners, it will make anyone who has to deal with a 12V setup more self-sufficient, by familiarizing them with tools, test equipment, and procedures for zeroing in on the trouble spot with as few tests as possible. Topics include volt-ohm meters, solving alternator problems, testing batteries, checking for electrolytic corrosion, and much more. 58 pages.

80-128 **$15**

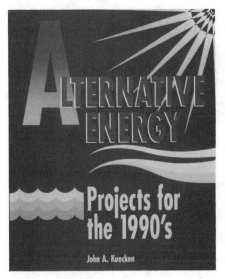

Alternative Energy — Projects for the 1990s

John A. Kuecken, an electromagnetic engineer whose expertise is evident in the carefully engineered plans for energy independence. This book is meant for the sophisticated do-it-yourselfer, with a strong background in electronics and computers. After a review of domestic, economic, and energy considerations, it proposes projects: inverters, power control, windmills, water engines, solar collection, and alternate energy storage. Scattered throughout are BASIC programs, excellent drawings, charts, and illustrations. Excellent technology. 254 pages.

80-122 **$15**

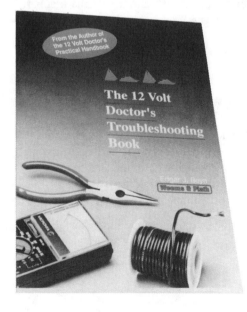

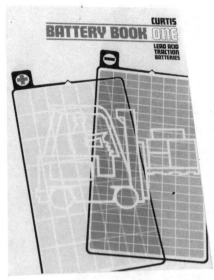

The 12V Doctor's Project Book

Edgar J. Beyn. Going beyond basic equipment, to suggest a wealth of useful accessories that may not be available for purchase. It is boat oriented, with broader applications. All designs are intended for use with 12V or are related to 12V battery systems and direct current. Learn about adding a third battery bank without adding another main switch; making an ammeter for high current; battery charger cutoff by voltage; high and low voltage indicator lights; and more than 20 other projects. The book lays out the necessary circuits and components, with brief descriptions of how they work. Requires some previous experience with electrical and mechanical work. 57 pages.   **80-127 $15**

Battery Book One:
Lead Acid Traction Batteries

A dependable manual on the care and feeding of "wet cells." Though designed with the electric forklift truck user in mind, it contains much information that applies to lead acid batteries in general. Electric vehicle and 12V home owners will find out how to minimize their energy costs by correctly selecting batteries, controlling the use of electrical energy in recharging batteries, and avoiding damaged batteries and equipment from over-discharging. 68 pages.   **80-129 $8**

Wind

Wind and Windspinners

Michael Hackleman. A do-it-yourself manual on building working wind-electric units, simply and inexpensively. Includes an extensive chapter on S-Rotor construction, one of the most versatile backyard aeroturbines. 140 pages. **80-121 $12**

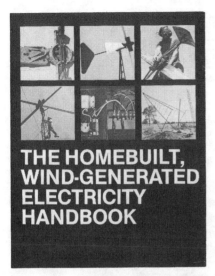

The Homebuilt, Wind-Generated, Electricity Handbook

Michael Hackleman, author of perennial top-seller *Wind and Windspinners*. This book is crammed with diagrams, explanations, instructions and advice about handmade windpower. This book and an abundance of motivation is all you need to harvest power from the wind. The diagrams and drawings are complete and accurate, and contain an ingredient rare in do-it-yourself energy books: a sense of whimsy, and even a hint that the author might be fallible—pictures of fallen windspinners—to remind the reader to take care. This is an extraordinarily useful book for anyone thinking of spinning their own power. 194 pages.   **80-116 $10**

An Owner's Manual for the Solar Hot Water System

This trio of books can help demystify that spaghetti of tanks, valves, pumps, and pipes which constitutes that solar hot water heating system you inherited or never understood in the first place. These well written books can help you evaluate the health of the system or even help design one for you. If your solar contractor never got around to providing you with an explanation, then these can be invaluable. A little knowledge can go a long way in keeping the system working smoothly and avoiding expensive repairs.

Manual for Closed Loop\Anti-freeze
Solar Systems   **80-133 $19**
Manual for Drainback Solar Systems
                                        80-134 $19
Manual for Draindown Solar Systems
                                        80-135 $19

Water

Planning for an Individual Water System

American Association of Vocational Instructional Materials (AAVIM). This is *the* definitive book on water systems and one of our all time favorites. It was a surprise and complete delight when one of our customers pointed it out to us. Incredible graphics and charts. Discusses water purity, hardness, chemicals; presents a thorough discourse on all forms of water pumps, discusses water pressure, pipe sizing, windmills, freeze protection, and fire protection. If you have a new piece of land or are thinking of developing a water source, you need this book! 160 pages.
                                        80-201 $19

Greywater Information Booklet

This is an excellent small booklet on just about anything you ever wanted to know about greywater disposal and usage. It contains all the information you need to construct and use a greywater system. Written by the folks at Oasis, it includes the guidelines to the approved use of greywater from the Goleta, California Water District in the appendix. 17 pages.   **80-203 $5**

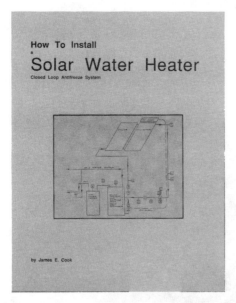

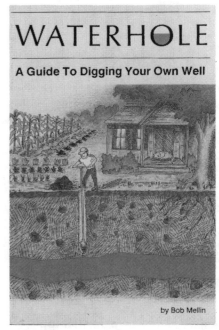

How to Install a Solar Water Heater, Closed Loop Antifreeze System

James Cook. This solar hot-water book is written for the average do-it-yourself homeowner who has basic tools and skills. Illustrations are provided in the manual that guide you step-by-step through the installation, maintenance, and troubleshooting phases. 96 pages. **80-109 $13**

Build Your Own Water Tank

An informative booklet by Donnie Schatzberg, gives you all the details you need to build your own ferro-cement (iron-reinforced cement) water storage tank. No special tools or skills required. The information given in this book is accurate and easy to follow, with no loose ends. The author has considerable experience building these tanks, and has gotten all the "bugs" out. 35 pages. **80-204 $10**

Earth Ponds

Tim Matson. This is a thorough and attractive treatment of a small subject. The first edition of this book appeared a decade ago, and was instrumental in the sculpting of thousands of ponds around the world. This new edition is much larger, more professionally illustrated, but, most important, it has grown through the author's experience, a wealth of anecdote and research back to the seventeenth century, and attentive love for his subject. This book is utterly indispensable for all pond dreamers, builders, and maintainers everywhere. 150 pages.

80-156 $17

Waterhole: A Guide to Digging Your Own Well

Bob Mellin, who has hand-dug several wells using the technique he recommends. This small but thorough book deals with all aspects of one of the simplest of tasks–digging a hole in the ground: site selection, where not to dig, how to dig, how to keep the hole uncontaminated, digging and pumping equipment. This book treats only one type of well, the small-bore auger dug well, with humor and balance. Very much in the tradition of great Real Goods do-it-yourself books. 72 pages. **80-614 $9**

Shelter

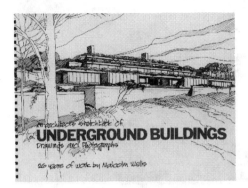

Underground Buildings

Malcolm Wells. An architect's sketchbook of 26 years of work. This is a delightful, irreverent, and very personal account of the struggle against the current of the architectural mainstream. Wells includes hundreds of sketches to illustrate the potential, the successes, and the failures of solar energy efficient, underground buildings. Handwritten, easy-to-read, 200 pages.

80-146 $15

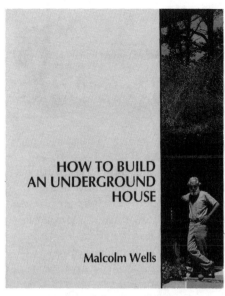

How to Build an Underground House

Malcolm Wells. Wells' fourth book about underground architecture. He has followed and pioneered underground architecture since 1964, his earlier books having sold 120,000 copies. The United States only boasts 3,000-4,000 earth dwellings due to inertia. This is a very elegantly and simply written book that talks about not only the underground house, but the appliances and energy systems needed to sustain it. Wonderful graphics, drawings, and plans, all hand-done in Wells' inimicable style. A must for undergrounders! 96 pages.   **80-155 $12**

Shelter

This is the classic book put out by Shelter Publications originally in 1973, when it sold 185,000 copies. Out of print since 1978, the book has recently come back. It features 176 pages on architecture— from bailiwicks to zomes. You won't find any palaces, pyramids, temples, cathedrals, skyscapers, or Pentagons. This is instead a brilliant book on homes and habitations for human beings in all their infinite variety. It is lavishly illustrated, with over 1,000 photographs, numerous drawings, and 250,000 words of text. It is a real piece of environmental drama and a great tool for the person just starting to fantasize a dream living space.   **80-145 $17**

Terra-Dome Video

Underground homes and commercial buildings dispel the stereotype of underground living as dark and damp. Terra-Dome designs emphasize openness and esthetic flexibility, at a price tag that rivals conventional construction. Modular design is the key to the low price (approx. $18 per square foot for the basic structure), while energy savings are extraordinary—up to 90% savings over traditional designs. Modules are available in versatile 24-foot and 28-foot configurations, connected by 16-foot arches. Walls are 10-inches thick, and the dome itself will support 8-feet of earth. Architectural styles range from Colonial to Tudor, Contemporary to Mediterranean. Terra-Dome builds homes that are exceptional in every way and at an incredible price. *Call for more data and an informative brochure, or order our 30-minute video showing houses, commercial structures, and building processes.*

80-106 $20

Earthship

Professionally produced video by Dennis Weaver of his independent sustainable living space, Earthship. Narrated with the same enthusiasm and down-home twang he brings to those Great Western Savings commercials. The video is a solid prospectus for a responsible way to build, using surplus material—tires and cans. The book is a builder's guide for an innovative construction technique integrating recycled waste with conventional building materials. The primary construction material—used tires—are a post-consumer disposal nightmare, and can be recycled into a building, **free**. A single tire filled with tamped earth weighs 300 pounds and is stable, immovable, and cheap: the houses in the book cost from $20 per square foot. Interior walls made with adobe-covered cans are strong, light, and can be sculpted to any shape. An exciting concept. Just as small problems lead to big problems, so do small solutions lead to big solutions. Weaver grinds his favorite axe—the greenhouse effect—and emphasizes the 3 Rs: Reduce, Reuse, Recycle. The companion books, Earthship, Vol.1 (230 pages) and Vol. 2 (260 pages), by Michael Reynolds, are a competently homegrown how-to manuals showing a builder how to use these new building materials. They provide detailed information from design to finish with Vol.2 giving interior details and including a chapter about getting permits.

80-117   Earthship Video $29
80-118   Earthship, Vol. 1 $29
80-131   Earthship, Vol. 2 $29

**To order any of these products
or for more technical information
call us toll-free at
1-800-762-7325**

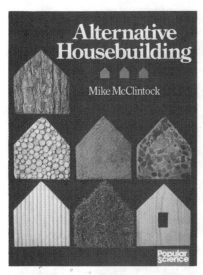

Alternative Housebuilding

Mike McClintock, syndicated how-to columnist for the *Washington Post*. Beautiful duotone illustrations carefully detail the planning and construction of log, timber-frame, pole, cord-wood-stone-earth masonry, and earth-sheltered houses. Full of ideas, step-by-step construction advice, and elegant examples that can help raise the level of owner-built homes to architecture. Sound techniques, excellent source listings. 367 pages. **80-119 $19**

The Natural House Book

David Pearson, an architect, planner, and London-based eco- and healthy-housing consultant. This book modestly bills itself as "a comprehensive handbook to show how to turn any house or apartment into a sanctuary for enhancing your well-being", and delivers much of what it promises. This handsome, color-illustrated book borrows historical and ethological successes in shelter and applies them to modern housing design. It considers every aspect of healthy living space by analyzing the whole house—healthy as contrasted to dangerous—and the elements that go into houses, the living systems and materials used in constructing the space, and then considering each of the kinds of rooms we use for living, sleeping, cooking, bathing, exercising, and growing plants. This book succeeds because it takes a whole-systems approach. 288 pages. **80-120 $20**

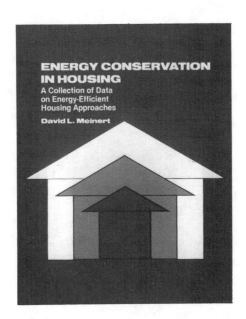

Energy Conservation in Housing

David Meinhert. Provides designers, builders and remodelers a collection of useful tables, clear plans, and well-thought-out ideas about the efficient use of energy in the living space. The book avoids the trap of jargon in presenting designs that efficiently manage heat loss and gain in tightly constructed homes while maintaining an adequate supply of fresh air. 150 pages. **80-143 $14**

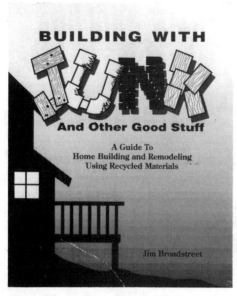

Building with Junk and other Good Stuff

A master scrounger's guide to the millions of dollars of building supplies thrown away every day. For dedicated recyclers and builders on a budget this hardbound book covers every aspect of home building from floors to roofing and from bankers to building inspectors. Readable and informative. The *Whole Earth Ecolog* says it well: 'detailed, true, and full of wit.' 162 pages.

80-142 $20

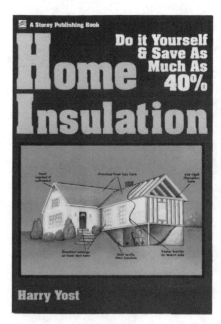

Home Insulation

Harry Yost, who has been insulating things for 30 years and presently lives in Palmer, Alaska. The book's thesis is that you can save big bucks installing your own insulation if you know what to do—and this book tells you precisely, simply, and encyclopedically. 138 pages.   **80-124 $12**

*Efficiency, during 1973-86, came to represent an annual energy source two-fifths bigger than the entire domestic oil industry, which had taken a century to build; yet oil had rising costs, falling output, and dwindling reserves, while efficiency had falling costs, rising output and expanding reserves...
- Amory Lovins, in the Foreword*

The Smart Kitchen

David Golbeck. This book caused a stir in architecture circles when it hit the stands 2 years ago, but until now, it's been hard to find. Excellent chapters on in-kitchen recycling, lighting, and refrigeration will help you save money, but it is the wealth of thought and research—a two-page bibliography on refrigerators alone—that has won this book such praise from Fine Home-building and Amory Lovins's Rocky Mountain Institute. If you're building or remodeling a kitchen, don't start without this book. 134 pages.   **80-144 $16**

Tools

Heaven's Flame

Joseph Radabaugh, who has been designing, building, and using solar cookers for 16 years. An inspiring initiation into the delights of building and cooking with one's own solar oven. Following the book's clear plans and diagrams, it's easy to construct an inexpensive, efficient SunStar solar cooker in just a few hours, using aluminum foil, glass, and cardboard boxes. Time after time, the SunStar design has proved its excellence. Radabaugh also goes into the history and theory of solar ovens, looks at several other cookers, examines general design and building concepts, and offers cooking tips. 68 pages.

80-701 $10

Backyard Composting

Harmonious Technologies, summarizes the theory and practice of turning organic garbage back into something useful. Good pictures of tools and techniques and lots of great reasons to be composting. Teaches you how to start and build your own composting bins from scrap materials, how to maintain a compost with recipes, and how to use the compost on your garden. If you aren't already composting, you oughta be ashamed—it's so simple! 96 pages.

80-151 $7

The Solar Cookery Book

Beth and Dan Halacy, who use solar energy almost every day to prepare their food. Just less than half the book provides plans and directions for making several solar ovens, and is well worth the price of the book. The last half of the book provides hints for adapting to the sun as heat source and a selection of recipes that will disappoint vegetarians and sugar-phobics. 108 pages.

80-138 $8

The Hydroponic Hot House: Low-Cost, High-Yield Greenhouse Gardening

James B. DeKorne. A well-illustrated guide to alternative-energy greenhouse gardening. Includes directions for building several different greenhouses, practical advice on harnessing solar energy, and many good suggestions for increasing plant yield. The first easy-to-use manual on home hydroponics. 178 pages.　　**80-838　$17**

Converted Machines

James Forgette. A guidebook and catalog —maybe it's competition, but welcome aboard! This booklet analyzes commercial appliances from the off-the-grid perspective and the pros and cons of converting to lower voltage. A narrow subject, and the author is clearly experienced. 36 pages.　　**80-306　$7**

To order any of these products
or for more technical information
call us toll-free at
1-800-762-7325

Electric Vehicles

Convert It

Mike Brown. A step-by-step manual for converting a gas car to electric power. It's a very readable and practical manual for the do-it-yourselfer wanting the fun and educational experience of converting a conventional automobile. Amply illustrated, with instructions for testing and operating the completed vehicle. Brown does a great job of explaining the conversion costs, using his own experience as a guide. Recommended by the Electric Auto Association, Electric Vehicle Progress, and Alternate Energy Transportation. 56 pages long. The price is steep, reflecting the high costs of providing such specialized information. Considered in light of the benefits, however, it's cheap at twice the price.　　**80-404　$35**

Electric Vehicle Video

Electro Automotive Information Services. Yes, electric cars are simple. But should you use a series DC motor, brushless DC, or an AC? What about aircraft generators? In this new video, Mike Brown, author of *Convert It*, sorts out all the choices. It's a good starting point if you are considering building an electric vehicle, or just want to know more about how they work. Brown's business, Electro AutoMotive, has been supplying components for electric vehicles since 1979, and is known for equipping the winners in countless races and rallies. The tape, with footage of dozens of cars and components, runs approx. 90 min.　　**80-125　$39**

Transportation 2000, Moving Beyond Auto America

From Transportation 2000, Boulder, Colorado. 29 minute video featurs transportation experts from across the United States (including Ralph Nader) Great visuals of the future of transportation, including wonderful shots of a magnetic levitation train; a good look at ingenuity, policies and statistics. It is inspiring, showing light at the end of the tunnel for the domination of the internal combustion engine. The information is solid, and encouraging. Great classroom discussion.   **80-113 $49**

Electric Vehicles Unplugged!

An enthusiastic resource from Douglas Marsh. It covers the whys and hows of electric vehicles and then gets specific, with the best list of makers, converters, clubs, kits, plans, and parts we've seen. 57 pages.   **80-408 $9**

Do-It-Yourself Automotive LPG Conversion

A 36 page primer written by Tom Jennings. Of all major fuels, only liquified propane gas (LPG) can be produced and consumed without polluting the earth. Moreover, it can be stored and transported safely. Cars run better and last longer when powered with LPG. The author assumes you understand how the major systems work, but instructions, even though generic, are clear and doable. The author estimates the conversion cost at under a grand!   **80-406 $5**

Why Wait for Detroit?

A collection of articles by experts on electric vehicles: buying and maintenance hints, a remarkably candid discussion of the alternatives, pros and cons, a directory of resources, and a look to the future. Real substance here, including mug shots of politicos telling us why we've got to get on the bandwagon, in case you need converting. 209 pages. **80-405 $12**

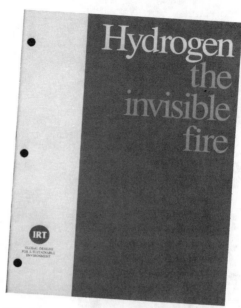

Hydrogen, the Invisible Fire

Patrick Kiernan, IRT Environment. A very up-to-date summary of hydrogen use; a welcome coverage of a subject pitifully ignored by the press. Includes information on all current research projects, current and potential uses, safety issues, storage and transportation problems, and pollution concerns. This book is worth the price just for the extensive resource lists and bibliography. A very honest assessment of both the positive and negative points. 48 pages. **80-150 $15**

Electric Vehicle Directory

Philip Terpstra. This book shows over 20 vehicles on the market complete with specifications. Lists sources of other electric vehicles as well as components, associations, and newsletters. 80 pages. **80-403 $11**

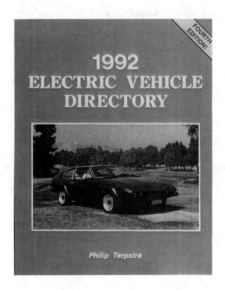

The Fuel Savers

Bruce N. Anderson, presents fuel-saving ideas, from simple to complicated, then rates them for relative cost-effectiveness using an oil barrel scale. The amount of oil left in the barrel represents the amount of energy saved relative to the amount spent. Subjects include energy conservation, window treatments, sunrooms, wall applications, solar water heaters, roof applications, and warm air collectors. Half of the royalties from this book will be donated to the Scholarship Fund of the American Solar Energy Society, and the other half will be donated to the National Foundation on Poverty and the Environment. 88 pages.   **80-148 $5**

Steering a New Course: Transportation, Energy and the Environment

Deborah Gordon. Highlights the coming transportation crisis and offers solutions. In the next 20 years, road congestion is projected to triple, with U.S. dependence on imported oil rising to 70% and smog-forming emissions increasing 30%. The book's examination of transportation/environment interaction is the most complete to date. Reducing hidden costs is also covered. The author is a senior transportation and energy policy analyst with the Union of Concerned Scientists and previously worked for Department of Energy and Chevron. 244 pages.
80-402 $20

Just for Fun

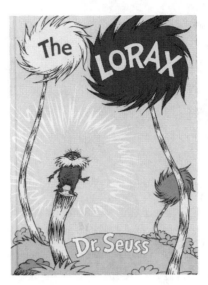

The Lorax

Dr. Seuss. Who would have thought Dr. Seuss could be controversial? This book was actually banned by one of our local school districts as being too hard on loggers! This great story about the deforestation of the Truffula trees is helping to awaken an entire generation of young environmentalists. 61 pages.   **80-815 $12**

The Green Activity Book

A slim (32 page) volume — just the right size — filled with ideas, games, and activities for good readers aged 7 to 11 years. It has the right tone ("It's not easy being green,' says our host the frog") and enough zaniness so that we could tell the authors hadn't lost track of what it's like to be young.   **80-822 $5**

Public Therapy Buses

Steven Johnson. Sub-titled *Information Specialty Bums, Solar Cook-A-Mats and other Visions of the 21st Century*, this very tongue-in-cheek book of cartoons offers up whimsical and slightly cracked visions of how technologies will intersect, how everyday items like automobiles, clothes, and houses will evolve, and how we will work, travel, play, and simply survive in the future. 128 pages.   **80-611 $8**

Rads, Ergs, and Cheeseburgers

A stylish and informative book that will interest good readers aged 9 to 14. Our protagonists are changed to points of light by an entity named Ergon and whisked away to the planets Barren and Noxious for a whirlwind tour of energy awareness. Along the way, politically unpopular issues neglected in school science classes are discussed with honesty and humor. 108 pages.

80-823 $13

Whole Earth Ecolog

Edited by J. Baldwin, et al. Introduction by Stewart Brand. Big and floppy, like the Whole Earth Catalogs used to be, the Ecolog focuses solely on the environment. Whole Earth looked at the plethora of Earth-saving books and made sure that the Ecolog was not just another version of *99 Recyclable Bottles of Beer on the Wall!* The Ecolog celebrates all groups doing something to change the world. Among the features: *Village Homes,* a parklike solar neighborhood now 15 years old where energy use is 50% less than the U.S. average; *The best sources of safe, environmentally righteous goods,* tools and hardware; *CoHousing,* attractive and affordable especially for single parents and seniors; *Exceptional books and videos* for getting the kids off to a good start. 126 oversized pages.

80-803 $16

Pocket Ref

Thomas J. Glover. This amazing book, measuring just 3.2 by 5.4 inches is like a set of encyclopedias in your shirt pocket! Here is a very small sampling of the hundreds of tables, maps, and charts within: battery charging, lumber sizes and grades, floor joist span limits, insulation R values, periodic table, computer ASCII codes, IBM PC error codes, printer control codes, electric wire size vs. load, resistor color codes, U.S. holidays, Morse code, telephone area codes, time zones, sun and planet data, earthquake scales, nail sizes, geometry formulas, currency exchange rates, saw blades, water friction losses, and a detailed index! Have we said enough? Indispensable! 480 pages. **80-506 $10**

50 Simple Things
You Can Do to Save the Earth

The Earthworks Group. This is a wonderful book that should be part of everyone's Earth Day library. We have quoted liberally from this book throughout our catalogs. It's chock full of amazing facts regarding the greenhouse effect, air pollution, ozone depletion, hazardous waste, acid rain, vanishing wildlife, groundwater pollution, and garbage. It offers numerous concrete steps that can be taken to make each one of us a contributor to solutions to the energy problem. This original comes highly recommended! 96 pages. **80-801 $5**

Eco-Tech Conference Video

The Eco-Tech Conference occurred in Monterey, California, in November 1991 and featured the world's finest minds brainstorming on the future of the planet. Amory Lovins, Stewart Brand, Peter Warshall, Peter Schwartz, Denis Hayes, Paul Saffo, Chellis Glendinning, Payson Stevens, Michael Rothchild, Peter Calthorpe, Eric Drexler, Fritjof Capra, Godfrey Reggio, and many others discuss the state of the planet, the extent of hopefulness for the future, and the urgent need for immediate action if we are to survive. 90 minutes. **80-307 $69**

Ecopreneuring

Steven J. Bennett. This is the first book written on how to start an environmental business. It is a start-up guide offering business opportunites in ecologically sound packaging, child and baby care products, household cleaning products, and foods & beverages, education and training and more. It goes through many case histories of successful eco-businesses (including Real Goods.) 320 pages. **80-507 $18**

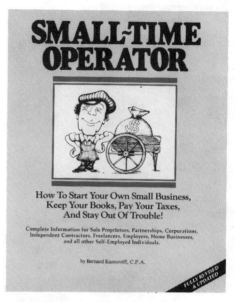

Small-Time Operator

Bernard Kamaroff, C.P.A. This is probably the best book ever written on starting up a small business. In fact, it's the book we at Real Goods used to start our first store in 1978. It's the book that nurtures your first ideals of going into business yet sets your reality firmly in concrete. Written in a personal, nontechnical style, it gives you everything you need to start up and stay in business for the first year, providing a giant boost to avoid the pitfalls of the 90% of new businesses that belly-up in year one! Over 330,000 in print! 192 pages. **80-502 $13**

50 Simple Things You Can Do to PAVE the Earth

Darryl Henriques. A delight and will bring timely grins for the depressed eco-activist. *"Without Junk Mail, Millions of Lonely People Get No Mail At All!"* Randy Hayes, of the Rainforest Action Network, says "if humor heals, this book will help preserve the earth." Darryl Henriques is a wonderfully funny person whose demented mind should be supported! 96 pages. **80-831 $5**

Global Warming —the Greenpeace Report

After reading this book, you'll wonder just which drugs the Reagan-Bush administration is **not** saying *no* to in its ongoing refusal to admit that global warming might be a problem, and that Americans are the most offensive perpetrators on the planet. Sadly, this book will preach only to the converted, but it **will** provide converts with an encyclopedic and authoritative compendium of facts, trends, and projections to show those who still have their heads in the sand. This book should be the bible of energy policymakers everywhere. 554 pages.

80-208 $11

Stewart's Green Line, The Environmental Directory

North America's most complete listing of environmentally friendly products and services, including developers of new technology and organizations and institutions for environmental education. This handy reference guide offers smart choices that will help speed up the cleaning up of the environment. It includes thousands of listings with telephone numbers, addresses, and quality buyer information. This is the most complete listing we've yet seen. 122 pages. **80-140 $9**

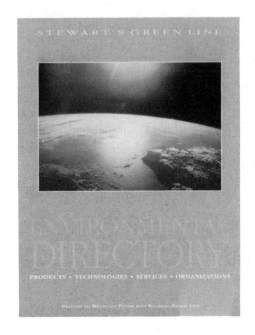

Beyond Oil: the Threat to Food and Fuel in the Coming Decades

Gever, Kaufmann, Skole, and Vorosmarty. The most thorough blueprint available for our energy future. *Beyond Oil* punctures the myth that the U.S. can increase its per-capita material standard of living and its population ad infinitum, through continuing depletion and degradation of nonrenewable and ecological resources. Using one of the most ambitious computerized assessments of future U.S. energy supplies ever conducted, it projects probable energy, economic, and agricultural outcomes well into the next century. *Beyond Oil* should be required reading for every politician, business leader, and informed citizen in America. 258 pages. **80-409** **$18**

The Harper Collins Dictionary of Environmental Science

An invaluable reference tool for students and professionals alike—and fascinating just for browsing. It's really more of an encyclopedia than a dictionary: the over 2,000 entries go far beyond basic definitions to provide in-depth explanations and examples. Subjects include geology, climatology, botany, zoology, human geography, and planning, in a full spectrum from Acid Rain to Zero Population Growth. Diagrams and charts illustrate numerous concepts such as radiocarbon dating and plate tectonics. Anyone with love of wonder at the earth's infinite variety will find much pleasure here. 453 pages. **80-842** **$13**

Real Goods—the Electronic Edition

EcoNet, a non-profit network for environmental information, is our partner in an electronic publishing breakthrough: Real Goods Online. We are particularly proud to share disk space with the best collection of planetary information available, and to make it available to our readers.

To get into Econet call the Institute for Global Communications at 415/442-0220 (fax 415/546-1794) and give them your name, address, phone number, and credit card number.

If you have a computer but no modem, we have laid in a supply of external modems—modulator-demodulators—that let computers talk to each other. Our special modem package includes a top-of-the-line Supra 2400 baud modem, a communications program needed to get on line, a cable to connect to your IBM-compatible's or Mac's serial port, and a 5-year warranty. (To order for an IBM-compatible, you need to know how many pins in your serial port – they come in two flavors, 9-pin and 25-pin.)

09-911	**DOS 25-pin Modem Kit**	**$110**
09-912	**DOS 9-pin Modem Kit**	**$110**
09-913	**Mac Modem Kit**	**$110**

Appendix

GLOSSARY

A

Activated Stand Life • The period of time, at a specified temperature, that a battery can be stored in the charged condition before its capacity falls.

Alternating Current (ac) • An electric current that reverses its direction at a constant rate.

Ambient Temperature • The temperature of the surroundings.

Amorphous Silicon • A type of PV silicon cell having no crystalline structure. See also Single-Crystal Silicon, Polycrystalline Silicon.

Ampere (Amp) (A) • Unit of electric current measuring the flow of electrons. The rate of flow of charge in a conductor of one coulomb per second.

Ampere-Hour (Ah) • The quantity of electricity equal to the flow of a current of one ampere for one hour.

Angle of Incidence • The angle that a light ray striking a surface makes with a line perpendicular to the reflecting surface.

Anode • The positive electrode in an electrochemical cell (battery) toward which current flows. Also, the earth ground in a cathodic protection system.

Array • A collection of photovoltaic (PV) modules, electrically wired together and mechanically installed in their working environment.

Array Current • Current produced by the array when exposed to sunlight. See Rated Module Current.

Array Operating Voltage • The voltage provided by the photovoltaic array under load. See Battery Voltage.

Availability • The quality or condition of being available. PV system availability is the amount of time a PV system is 100 percent operational.

Azimuth • Horizontal angle measured clockwise from true north; 180° is due south.

Base Load • The average amount of electric power that a utility must supply in any period.

Battery • A device that converts the chemical energy contained in its active materials directly into electrical energy by means of an electrochemical oxidation-reduction (redox) reaction. This type of reaction involves the transfer of electrons from one material to another through an electrical circuit.

Battery Capacity • The total number of ampere-hours that can be withdrawn from a fully charged battery. See Ampere-Hour, Rated Battery Capacity.

Battery Cell • The smallest unit or section of a battery that can store an electrical charge and is capable of furnishing a current.

Battery Cycle Life • The number of cycles that a battery can undergo before failing.

Battery Self-Discharge • Loss of chemical energy by a battery that is not under load.

Battery State of Charge • Percentage of full charge.

Battery Terminology

    Captive Electrolyte Battery Type • A battery type having an immobilized (gelled or absorbed in the separator) electrolyte.

    Lead-Acid Battery Type • A general category that includes batteries with pure lead, lead-antimony, or lead-calcium plates and an acid electrolyte.

    Liquid Electrolyte Battery Type • A battery type containing free liquid as an electrolyte.

    Nickel Cadmium Battery • A battery type containing nickel and cadmium plates and an alkaline electrolyte.

    Sealed Battery Type • A battery type with a captive electrolyte and a resealing vent cap, also called a valve-regulated, sealed battery.

    Vented Battery Type • Free liquid electrolyte type battery with a vent cap for free escape of gasses during charging.

Blocking Diode • A diode used to prevent current flow within a PV array or from the battery to the array during periods of darkness or of low current production.

British Thermal Unit (Btu) • A unit of heat. The quantity of heat required to raise the temperature of one pound of water one degree Fahrenheit.

Bypass Diode • A diode connected in parallel with a block of parallel modules to provide an alternate current path in case of module shading or failure.

Capacity (C) • The total number of ampere-hours that can be withdrawn from a fully charged battery. See Battery Capacity.

Cathode • The negative electrode in an electrochemical cell.

Charge Controller • A device that controls the charging rate and state of charge for batteries. See Charge Rate.

Charge Rate • The rate at which a battery is recharged. Expressed as a ratio of battery capacity to charge current flow, for instance, C/5.

Cloud Enhancement • The increase in solar insolation due to direct beam insolation plus reflected insolation from partial cloud cover.

Concentrator • A photovoltaic module that uses optical elements to increase the amount of sunlight incident on a PV cell.

Controller Terminology

    Adjustable Set Point • A feature allowing adjustment of voltage disconnect levels.

    High Voltage Disconnect • The battery voltage at which the charge controller will disconnect the batteries from the array to prevent overcharging.

    Low Voltage Disconnect • The battery voltage at which the charge controller will disconnect the batteries from the load to prevent over discharging.

    Low Voltage Warning • A warning buzzer or light that indicates low battery voltage.

    Maximum Power Tracking • A circuit that keeps the array operating at the peak power point of the I-V curve where maximum power is obtained.

    Multistage Controller • Unit that allows multilevel control of battery charging or load.

    Reverse Current Protection • A method of preventing current flow from the battery to the array. See Blocking Diode.

    Single-Stage Controller • A unit with one activation level for battery charging or load control.

Temperature Compensation • A circuit that adjusts the battery high or low voltage disconnect points as ambient battery cell temperature changes. See Temperature Correction.

Conversion Efficiency • The ratio of the electrical energy produced by a photovoltaic cell to the solar energy received by the cell.

Converter • A unit that changes and conditions dc voltage levels.

Crystalline Silicon • A type of PV cell made from a single crystal or polycrystalline slice of silicon.

Current • The flow of electric charge in a conductor between two points having a difference in potential (voltage), generally expressed in amperes.

Cutoff Voltage • The voltage at which the charge controller disconnects the array from the battery. See Charge Controller.

Cycle • The discharge and subsequent charge of a battery.

D

Days of Storage • The number of consecutive days the stand-alone system will meet a defined load. This term is related to system availability.

Deep Cycle • Battery type that can be discharged to a large fraction of capacity. See Depth of Discharge.

Design Month • The month having the combination of insolation and load that requires the maximum power out of the array.

Depth of Discharge (DOD) • The percent of the rated battery capacity that has been withdrawn.

Diffuse Radiation • Radiation received from the sun after reflection and scattering by the atmosphere.

Diode • Electronic component that allows current flow in one direction only. See Blocking Diode, Bypass Diode.

Direct Current (dc) • Electric current flowing in only one direction.

Discharge • The withdrawal of electrical energy from a battery.

Discharge Rate or C Rate • The rate at which current is withdrawn from a battery. Expressed as a ratio of battery capacity to discharge current rate. See Charge Rate.

Disconnect • Switch gear used to enable or disable components in a PV system.

Dry Cell • A cell with a captive electrolyte. A primary battery cell.

Duty Cycle • The ratio of active time to total time. Used to describe the operating regime of appliances or loads in PV systems.

Duty Rating • The amount of time an appliance can produce at full rated power.

E

Efficiency • The ratio of output power (or energy) to input power (or energy). Expressed in percent.

Electrolyte • The medium that provides the ion transport mechanism between the positive and negative electrodes of a battery.

Energy Density • The ratio of the energy available from a battery to its volume (Wh/L) or weight (Wh/kg).

Equalization • The process of restoring all cells in a battery to an equal state of charge.

F

Fill Factor • For an I-V curve: the ratio of the maximum power to the product of the open-circuit voltage and the short-circuit current. Fill factor is a measure of the "squareness" of the I-V curve shape.

Fixed Tilt Array • A PV array set in a fixed position.

Flat-Plate Array • A PV array that consists of nonconcentrating PV modules.

Float Charge • The charge to a battery having a current equal to or slightly greater than the self discharge rate.

Frequency • The number of reptitions per unit time of a complete waveform, as of an electric current, usually expresssed in Hertz.

G

Gassing • Gas by-products produced when charging a battery. Also, termed out-gassing. See Vented Battery Type.

Grid • Term used to describe an electrical utility distribution network.

I

Insolation • The solar radiation incident on an area over time. Usually expressed in kilowatt-hours per square meter. See also Solar Resource.

Inverter • In a PV system, an inverter converts dc power from the PV array to ac power compatible with the utility and house loads. An inverter is also called a power conditioner or power-conditioning subsystem.

    Square Wave • A waveform that contains a large number of harmonics. A waveform that can be generated by opening and closing a switch.

    Modified Sine Wave • A distorted sine wave consisting basically of one single periodic oscillation.

    Sine Wave • A waveform corresponding to a single-frequency, periodic oscillation, which can be shown as a function of amplitude against angle and in which the value of the curve at any point is a function of the sine of that angle.

Irradiance • The instantaneous solar radiation incident on a surface. Usually expressed in kilowatts per square meter.

I-V Curve • The plot of the current versus voltage characteristics of a photovoltaic cell, module, or array.

K

Kilowatt (kW) • One thousand watts.

Kilowatt Hour (kWh) • One thousand watt hours.

L

Life • The period during which a system is capable of operating above a specified performance level.

Life-Cycle Cost • The estimated cost of owning and operating a system for the period of its useful life.

Load • The amount of electric power used by any electrical unit or appliance at any given moment.

Load Circuit • The current path that supplies the load. See also Load.

Load Current (amps) • The current required by the electrical unit during operation. See also Ampere.

Load Resistance • See Resistance.

Langley • Unit of solar irradiance. One gram calorie per square centimeter.

Low Voltage Cutoff (LVC) • Battery voltage level at which a controller will disconnect the load.

M

Maintenance-Free Battery • A battery to which water cannot be added to maintain electrolyte volume. All batteries require inspection and maintenance.

Maximum Power Point • A mode of operation for a power conditioner, whereby it continuously controls the PV source voltage in order to operate the PV source at its maximum power point.

Module • The smallest replaceable unit in a PV array. An integral encapsulated unit containing a number of PV cells.

Modularity • The concept of using identical complete subunits to produce a large system.

Module Derate Factor • A factor that lowers the module current to account for normal operating conditions.

N

NEC • National Electrical Code, which contains safety guidelines for all types of electrical installations. The 1984 and later editions of the NEC contain Article 690, "Solar Photovoltaic Systems."

Normal Operating Cell Temperature (NOCT) • The temperature of a PV module when operating under 800 W/m² irradiance, 20°C ambient temperature and wind speed of 1 meter per second. NOCT is used to estimate the nominal operating temperature of a module in its working environment.

Nominal Voltage • The terminal voltage of a cell or battery discharging at a specified rate and at a specified temperature.

N-Type Silicon • Silicon having a crystalline structure that contains negatively charged impurities.

Ohm • The unit of electrical resistance equal to the resistance of a circuit in which an electromotive force of 1 volt maintains a current of 1 ampere.

Open Circuit Voltage • The maximum voltage produced by a photovoltaic cell, module, or array without a load applied.

Operating Point • The current and voltage that a module or array produces under load. See I-V Curve.

Orientation • Placement with respect to the cardinal directions, N, S, E, W; azimuth is the measure of orientation.

Outgas • See Gassing.

Overcharge • Forcing current into a fully charged battery.

P

Panel • A designation for a number of PV modules assembled in a single mechanical frame.

Parallel Connection • Term used to describe the interconnecting of PV modules or batteries in which like terminals are connected together.

Peak Load • The maximum load demand on a system.

Peak Power Current • Amperage produced by a module operating at the "knee" of the I-V (current-voltage) curve. See I-V Curve.

Peak Sun Hours • The equivalent number of hours per day when solar irradiance averages 1,000 W/m². Six peak sun hours means that the energy received during total daylight hours equals the energy that would have been received had the sun shone for six hours at 1,000 W/m².

Peak Watt • A manufacturer's unit indicating the amount of power a photovoltaic module will produce at standard test conditions (normally 1,000 W/m² and 25° cell temperature).

Photovoltaic Cell • The treated semiconductor material that converts solar irradiance to electricity. See Cell.

Photovoltaic System (PV System) • An installation of PV modules and other components designed to produce power from sunlight.

Plates • A thin piece of metal or other material used to collect electrical energy in a battery.

Pocket Plate • A plate for a battery in which active materials are held in a perforated metal pocket on a support strip.

Polycrystalline Silicon • Material used to make PV cells which consist of many crystal structures.

Power (Watts) • A basic unit of electrical energy, measured in watts. See also Watts.

Power Conditioning System (PCS) • See Inverter or Converter.

Power Density • The ratio of the rated power available from a battery to its volume (W/liter) or weight (W/kg).

Power Factor • The cosine of the phase angle between the voltage and the current waveforms in an ac circuit. Used as a designator for inverter performance.

Power Loss • Power reduction due to wire resistance.

Primary Battery • A battery whose initial capacity cannot be restored by charging.

Pyranometer • An instrument used for measuring solar irradiance received from a whole hemisphere.

R

Rated Battery Capacity • Term used by battery manufacturers to indicate the maximum amount of energy that can be withdrawn from a battery at a specified rate. See Battery Capacity.

Rated Module Current • Module current measured at standard test conditions.

Remote Site • Site not serviced by an electrical utility grid.

Resistance (R) • The property of a conductor by which it opposes the flow of an electric current resulting in the generation of heat in the conducting material. The measure of the resistance of a given conductor is the electromotive force needed for a unit current usually expressed in ohms.

S

Seasonal Depth of Discharge • An adjustment factor providing for long-term seasonal battery discharge. This factor results in a smaller array size by planning to use battery capacity to fully meet long-term load requirements during the low insolation season.

Secondary Battery • A battery which after discharge can be recharged to a fully charged state.

Self-Discharge • The loss of useful capacity of a battery due to internal chemical action.

Semiconductor • A material that has a limited capacity for conducting electricity.

Series Connection • The interconnecting of PV modules or batteries so that the voltage is additive.

Shallow Cycle Battery • Battery type that should not be discharged greater than 25 percent. See Depth of Discharge.

Shelf Life • The period of time that a device can be stored and still retain a specified performance.

Short Circuit Current (Isc) • Current produced by a PV cell, module, or array when its output terminals are shorted.

Silicon • A semiconductor material commonly used to make photovoltaic cells.

Single-Crystal Silicon • A type of PV cell formed from a single silicon ingot.

Solar Cell • See Photovoltaic Cell.

Solar Insolation • See Insolation.

Solar Irradiance • See Irradiance.

Solar Resource • The amount of solar insolation a site receives, usually measured in kWh/m^2/day. See Insolation.

Specific Gravity • The ratio of the weight of the solution to the weight of an equal volume of water at a specified temperature. Used as an indicator of battery state of charge.

Stand-Alone • A photovoltaic system that operates independent of the utility grid.

Starved Electrolyte Cell • A battery containing little or no free fluid electrolyte.

State of Charge (SOC) • The capacity of a battery expressed at a percentage of rated capacity.

Stratification • A condition that occurs in a deep-cycled liquid-electrolyte lead-acid battery when the acid concentration varies from top to bottom in the battery. Periodic controlled overcharging tends to destratify or equalize the battery.

String • A number of modules or panels interconnected electrically to obtain the operating voltage of the array.

Subsystem • Any one of several components in a PV system (i.e., array, controller, batteries, inverter, load).

Sulfating • The formation of lead-sulfate crystals on the plates of a lead-acid battery. Can cause permanent damage to the battery.

Surge Capacity • The requirement of an inverter to tolerate a momentary current surge imposed by starting ac motors or transformers.

System Availability • The probability or percentage of time a PV system will fully meet the load demand.

System Operating Voltage • The PV system voltage.

System Storage • See Battery Capacity.

T

Temperature Compensation • Allowance made for changing battery temperatures in charge controllers.

Temperature Correction • A correction factor that adjusts the nameplate battery capacity when operating a battery at lower than 20°C. See Temperature Compensation.

Thin Film PV Module • See Amorphous Silicon.

Tilt Angle • Angle of inclination of collector as measured from the horizontal.

Total ac Load Demand • The sum of the ac loads, used when selecting an inverter.

Tracking Array • A PV array that follows the daily path of the sun. This can mean one axis or two axis tracking.

Trickle Charge • A low charge current intended to maintain a battery in a fully charged condition.

U

Uninterrupted Power Supply (UPS) • Designation of a power supply providing continuous uninterrupted service.

V

Varistor • A voltage-dependent variable resistor. Normally used to protect sensitive equipment from sharp power spikes (i.e., lightning strikes) by shunting the energy to ground.

Vented Cell • A battery cell designed with a vent mechanism to expel gases generated during charging. See Vented Battery Type.

Volt (V) • The practical unit of electromotive force or difference in potential between two points in an electric field.

Voltage Drop • Voltage reduction due to wire resistance.

Watt (W) • The unit of electrical power. The power developed in a circuit by a current of one ampere flowing through a potential difference of one volt; 1/746 of a horsepower.

Watt Hour (Wh) • A unit of energy measurement; 1 watt for one hour.

Waveform • Characteristic shape of the ac output from an inverter.

Wet Shelf Life • The period of time that a battery can remain unused in the charged condition before dropping below a specified level of performance when filled with electrolyte.

Water Pumping Terminology

    Centrifugal Pump • A class of water pump using a rotary or screw to move water. The faster the rotation, the greater the flow. This type of pump is used in shallow well applications.

    Dynamic Head • Vertical distance from center of pump to the point of free discharge including pipe friction.

    Friction Head • The energy that must be overcome by the pump to offset the friction losses of the water moving through a pipe.

    Positive Displacement • See Volumetric Pump.

    Static Head • Vertical distance from the top of the static water level to the point of free discharge.

    Storage • This term has dual meaning for water pumping systems. Storage can be achieved by pumping water to a storage tank, or storing energy in a battery subsystem. See Battery Capacity.

    Suction Head • Vertical distance from surface of free water source to center of pump (when pump is located above water level).

    Volumetric Pump • A class of water pump utilizing a piston to volumetrically displace the water. Volumetric pumps are typically used for deep well applications.

Glossary courtesy of the National Technical Information Service; U.S. Department of Commerce

PHOTOVOLTAIC POWER SYSTEMS
AND
THE *NATIONAL ELECTRICAL CODE*

SUGGESTED PRACTICES

March 1992

Compiled and Distributed by

Southwest Region Experiment Station
Southwest Technology Development Institute
New Mexico State University
P.O. Box 30001/Dept. 3SOL
Las Cruces, New Mexico 88003-0001

National Electrical Code® and *NEC*® are registered trademarks of the National Fire Protection Association, Inc., Quincy, Massachusetts 02269

The original drafts of this guide were prepared under U. S. Department of Energy Contract DE-AS04-90AL57510.

REAL GOODS

PURPOSE

This guide provides information on the *National Electrical Code® (NEC)* and how it relates to photovoltaic (PV) systems. It is not intended to interpret, or replace the *NEC*. It merely paraphrases the NEC and aligns information contained in the NEC with PV subsystems. Any PV system designer, equipment manufacturer, or installer should have a thorough knowledge of the *NEC* and a full understanding of the engineering principles and hazards associated with photovoltaic power systems. This material is not intended to be a design guide nor an instruction manual for an untrained person. Furthermore, this guide is not intended to cover all aspects of the *NEC* or PV systems--it must be used in conjunction with the full text of the *National Electrical Code*. This guide will be revised and updated as needed. Suggestions should be sent to the address on the front cover.

The *National Electrical Code* including the 1990 *National Electrical Code* is published and updated every three years by the National Fire Protection Association (NFPA), Batterymarch Park, Quincy, Massachusetts 02269. The *National Electrical Code* and the term *NEC* are registered trademarks of the National Fire Protection Association and may not be used without their permission. Copies of the 1990 *National Electrical Code* are available from the NFPA at the above address, most electrical supply distributors, and many bookstores.

In many locations, all electrical wiring including photovoltaic power systems must be accomplished by a licensed electrician and inspected by a designated local authority. Some municipalities have more stringent codes that supplement or replace the *NEC*. The local inspector has the final say on what is acceptable. In some areas, compliance with codes is not required.

DISCLAIMER

Neither the authors, the Southwest Region Experiment Station, the Southwest Technology Development Institute, New Mexico State University, Sandia National Laboratories, the U. S. Department of Energy, nor the National Fire Protection Association assume any liability resulting from the use of the information presented in this manual. This information is believed to be the best available at the time of publication and is believed to be technically accurate. Any application of this information and results obtained from the use of this information are solely the responsibility of the reader.

ACKNOWLEDGEMENTS

Numerous persons throughout the photovoltaic industry reviewed the drafts of this manual and provided comments which are incorporated in this version. Particular thanks go to Joel Davidson, Heliopower/Hoxan; Mike McGoey, Remote Power; Harlen Chapman, Siemens; Paul Garvison, Integrated Power; Floyd Rose, Salt River Project; John Wohlgemuth, Solarex; Tim Ball, Solar Engineering; Bob Nicholson, Solar Cells Inc.; Carl Lovette, State of New Mexico; Mike Thomas, Sandia National Laboratories; Steve Willey, Backwoods Solar; and all those who provided useful information at seminars on the subject.

SUGGESTED PRACTICES

OBJECTIVE

- SAFE, RELIABLE, DURABLE PHOTOVOLTAIC POWER SYSTEMS

- KNOWLEDGEABLE MANUFACTURERS, DEALERS, INSTALLERS, CONSUMERS, AND INSPECTORS

METHOD

- WIDE DISSEMINATION OF THESE SUGGESTIONS

- TECHNICAL INTERCHANGE BETWEEN INTERESTED PARTIES

INTRODUCTION

The National Fire Protection Association has acted as sponsor of the *National Electrical Code (NEC)* since 1911. The original Code document was developed in 1897. With some exceptions, electrical power systems installed in the United States in this century have had to comply with a legally mandated local code or the *NEC*. This includes many photovoltaic (PV) power systems. In 1984, Article 690, which addresses safety standards for the installation of PV systems, was added to the Code. This article has been revised and expanded in the 1987 and 1990 editions.

Many of the PV systems in use and being installed today may not be in compliance with the *NEC* or other local codes. There are several contributing factors to this situation:

- The PV industry has a strong "grass roots," do-it-yourself faction that may not be fully aware of the dangers associated with low-voltage, dc, PV power systems.

- Some people in the PV community may believe that PV systems below 50 volts are not covered by the *NEC*.

- Some electrical inspectors have not had significant experience with direct-current portions of the code or PV power systems.

- The electrical equipment industries do not advertise or widely distribute equipment suitable for direct current (dc) use that meets *NEC* requirements.

- Popular publications are presenting information to the public that implies that PV systems are easily installed, modified, and maintained by untrained homeowners.

- Photovoltaic equipment manufacturers have been generally unable to afford the costs associated with testing and listing by national testing organizations like Underwriters Laboratories.

- Photovoltaic installers and dealers in many cases have not had significant experience in installing ac residential and commercial power systems.

Not all systems are unsafe. Some PV installers in the United States are licensed electrical contractors and are familiar with all sections of the *NEC*. These installer/contractors are installing reliable PV systems that meet the *National Electrical Code* and minimize the hazards associated with electrical power systems. However, many PV installations may have numerous defects and may not meet the 1990 Code. Some of the more prominent problems are listed below.

- Improper ampacity of conductors
- Improper insulation on conductors
- Unsafe wiring methods
- No overcurrent protection on many conductors
- Inadequate number and placement of disconnects
- Improper application of equipment having Underwriters Laboratories or other Nationally Recognized Testing Laboratory listing
- No short-circuit current protection on battery systems
- Use of nonapproved components when approved components are available
- Improper system grounding
- Lack of equipment grounding
- Use of underrated components
- Unsafe use of batteries
- Use of ac components (fuses and switches) in dc applications

The Code may apply to many PV systems regardless of size or location. A single PV module may not present a hazard, and small systems in remote locations may not present many safety hazards because people are seldom in the area. On the other hand, two or three modules connected to a battery may be lethal if not installed and operated properly. A single deep-cycle storage battery (6 volts, 220 amp-hours) can discharge about 8,000 amps into a short-circuit. Systems with voltages of 50 volts or higher present shock hazards as do inductive surge currents on lower voltage systems. Storage batteries can be dangerous; hydrogen gas and acid residue from lead-acid batteries must be dealt with safely.

The problems are compounded because unlike ac equipment and systems there are few *UL*-listed components that can be easily "plugged" together to make a PV system. Connectors and devices do not have mating inputs or outputs, and the knowledge and understanding of what works with what is not second nature to the installer. The dc "cookbook" of knowledge does not yet exist.

To meet the objective of safe, reliable, durable photovoltaic power systems, the following suggestions are made:

- Dealer-installers of PV systems become familiar with the methods of wiring residential and commercial ac power systems.

- All PV installations be inspected, where required, by the local inspection authority in the same manner as other equivalent electrical systems.

- Photovoltaic equipment manufacturers build equipment to *UL* or other recognized standards and have equipment tested and approved when practical.

- Listed or recognized subcomponents be used in assembled equipment where formal testing and listing is not practical.

- Electrical equipment manufacturers produce, distribute, and advertise listed, reasonably-priced, dc-rated components.

- Electrical inspectors become familiar with dc and PV systems.

- The PV industry educate the public, modify advertising, and encourage all installations to comply with the *NEC* or other codes.

- All persons installing PV systems obtain and study the current *National Electrical Code*.

- Existing PV installations be upgraded to comply with the *NEC* or modified to meet minimum safety standards.

SUGGESTED PRACTICES
Scope and Purpose of the *NEC*

Some local inspection authorities use regional electrical codes, but most jurisdictions use the *National Electrical Code*–sometimes with slight modification. The *NEC* states that adherence to the recommendations made will reduce the hazards associated with electrical installations. The *NEC* also says these installations may not necessarily be efficient, convenient, or adequate for good service or future expansion of electrical use [90-1]. (Numbers in brackets refer to articles in the *NEC*.)

The *National Electrical Code* addresses nearly all PV power installations, even those with voltages less than 50 volts. It covers stand-alone and grid-connected systems. It covers billboards, other remote applications, floating buildings, and recreational vehicles (RV) [90-2a, 690, 720]. The Code deals with any PV system that produces power and has external wiring or electrical components or contacts accessible to the untrained and unqualified person.

There are some exceptions. The *National Electrical Code* does not cover installations in automobiles, railway cars, boats, or on utility company properties used for power generation [90-2b]. It also does not cover micropower systems used in watches, calculators, or self-contained electronic equipment that has no external electrical wiring or contacts.

Article 690 of the *NEC* specifically deals with PV systems, but many other sections of the *NEC* contain requirements for any electrical system including PV systems [90-2, 720]. When there is a conflict between Article 690 of the *NEC* and any other article, Article 690 takes precedence [690-3].

The *NEC* suggests, and most inspection officials require, that equipment identified, listed, labeled, or tested by a Nationally Recognized Testing Laboratory (NRTL) be used when available. Two of the several NRTL's are the *Underwriters Laboratories (UL)* and Factory Mutual Research (FM) [90-6,100,110-3]. *Underwriters Laboratories* and *UL* are registered trademarks of Underwriters Laboratories Inc., 333 Pfingsten Road, Northbrook, IL 60062.

Most building and electrical inspectors expect to see *UL* on electrical products used in electrical systems in the United States. This is a problem for the PV industry because obtaining *UL* approval is not free, and the market is often small. Some manufacturers claim their product specifications exceed those required by the testing organizations, but inspectors readily admit to not having the expertise, time, or funding to validate these unlabeled items.

THE PLAN

The suggested installation practices contained in this guide progress from the photovoltaic modules to the electrical outlets. For each component, *NEC* requirements are addressed and the appropriate Code articles are referenced in brackets. A sentence, phrase, or paragraph followed by a *NEC* reference refers to a requirement established by the *NEC*. The words "will," "shall," or "must" also refer to *NEC* requirements. Suggestions based on field experience with PV systems are worded as such and will use the word "should." The availability of approved components is noted, and alternatives are discussed.

Appendix A presents diagrams for PV systems of varying sizes showing suggested connection and wiring methods.

PHOTOVOLTAIC MODULES

Three manufacturers, Siemens/ARCO (*UL*), Tideland Signal (*UL*), and Solarex (FM), offer listed modules at the present time. Kyocera has redesigned its junction box to meet *UL* standards, and this module is being marketed in Europe. Introduction into the United States is planned in the future. Also, Kyocera and Hoxan are reviewing the *UL* standard to determine what will be required for *UL* labeling.

Methods of connecting wiring to the modules vary from manufacturer to manufacturer. The Code requires strain relief be provided for connecting wires. If the module has a closed weatherproof junction box, strain relief and moisture-tight clamps should be used in any knockouts provided for field wiring. Where the weather-resistant gaskets are a part of the junction box, the manufacturer's instructions must be followed to ensure proper strain relief and weatherproofing [*UL* Standard 1703]. Figure 1 shows various types of strain reliefs. The one on the left is a basic cable clamp for interior use with nonmetallic sheathed cable. The clamps in the center and on the right are watertight and can be used with either single or multiconductor cable--depending on the insert. The plastic unit on the right is made by Heyco.

Figure 1. Strain Reliefs.

Module Marking

Certain electrical information must appear on each module. If modules are not factory marked (required by the listing agency--*UL*), then they should be marked at the site to facilitate inspection and to allow the inspector to determine the requirements for conductor ampacity and rating of overcurrent devices. The information supplied by the manufacturer **will** include the following items:

- Polarity of output terminals or leads
- Maximum overcurrent device rating for module protection
- Rated open-circuit voltage
- Rated operating voltage
- Rated operating current
- Rated short-circuit current
- Rated maximum power
- Maximum permissible system voltage [690-51]

Figure 2 shows a typical label that appears on the back of a module.

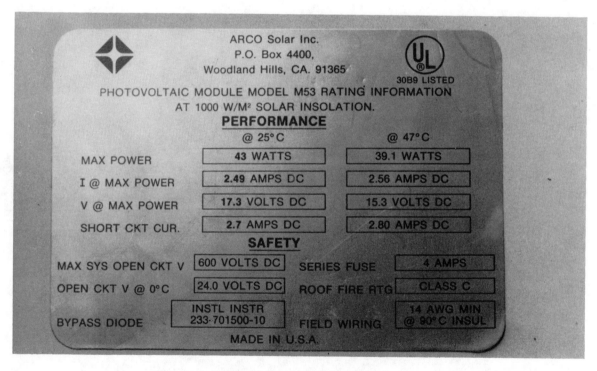

Figure 2. Label on Typical PV Module.

Module Interconnections

Copper conductors are recommended for almost all photovoltaic system wiring. Copper conductors have lower voltage drops and maximum resistance to corrosion. Aluminum or copper-clad aluminum wires can be used in certain applications but their use is not recommended--particularly in dwellings. All wire sizes presented in this guide refer to copper conductors.

The *NEC* requires No. 12 American Wire Gage (AWG) or larger conductors to be used with systems under 50 volts [720-4]. Single-conductor, type-UF (Underground Feeder) cable identified as sunlight resistant is permitted for module interconnect wiring [690-31b]. Stranded wire is suggested to ease servicing of the modules after installation [690-34]. Unfortunately, single-conductor, stranded, UF sunlight-resistant cable is not readily available. The limited amount that was available had a gray or red insulation rated at 60°C. This insulation was not suitable for long-term exposure to direct sunlight at temperatures likely to occur on roofs near PV modules. Such wire has shown signs of deterioration after four years of exposure.

A widely available and acceptable substitute is black, single-conductor cable identified as Underground Service Entrance Cable (USE). When made to the *UL* standard, it has a 90°C temperature rating and is sunlight resistant even though not commonly marked as such. It is acceptable to most electrical inspectors [Table 310-13 and 16].

Where No. 10 AWG meets ampacity considerations, it is a good compromise between ease of installation and minimizing the voltage drop in the array wiring. Where modules are connected in parallel, the ampacity of the conductors will have to be adjusted accordingly. Ampacity of conductors at any point must be at least 125 percent of the module (or array of parallel modules) rated short-circuit current at that point [690-8a, b1]. If flexible two-conductor cable is needed, electrical tray cable (TC) is available, but must be supported in a specific manner as outlined in the *NEC* [318]. It is sunlight resistant and is generally marked as such. Although frequently used for module interconnections, SO, SOJ, and similar flexible, portable cables and cordage are not sunlight resistant and are not approved for fixed (nonportable) installations [400-7, 8].

Crimped ring terminals are recommended for use in the module junction box to ensure all strands of the conductor are connected to the screw terminal. If captive screws are used, then fork-type crimped terminals could be used, but no more than two should be used on any one screw. Crimping and soldering the ring or fork terminal to the wire is recommended--particularly in areas of high humidity.

Crimping tools designed for crimping smaller wires used in electronic components usually do not provide sufficient force to make long-lasting crimps on connectors for PV installations even though they may be sized for No. 12-10 AWG. Insulated terminals crimped with these light-duty crimpers frequently develop high resistance in a short time and may even fail as the wire pulls out of the terminal under light pressure. It is strongly suggested that only heavy-duty industrial-type crimpers be used for PV system wiring. Figure 3 shows four styles of crimpers. On the far left is a stripper/crimper used for electronics work that will crimp only insulated terminals. Second from the left is a stripper/crimper that can make crimps on both insulated and uninsulated terminals. The pen points to the dies used for uninsulated terminals. With some care, this crimper can be used to crimp uninsulated terminals on PV systems if the terminals are soldered after the crimp. The two crimpers on the right are heavy-duty industrial designs with ratcheting jaws and interchangeable dies that will provide the highest quality connections. They are usually available from electrical supply houses.

Figure 4 shows some examples of insulated and uninsulated terminals. In general, uninsulated terminals are preferred (with insulation applied later if required), but care must be exercised to obtain the heavier, more reliable *UL*-listed terminals and not unlisted electronic or automotive grades. Again, an electrical supply house rather than an electronic or automotive parts store is the place to find the required items.

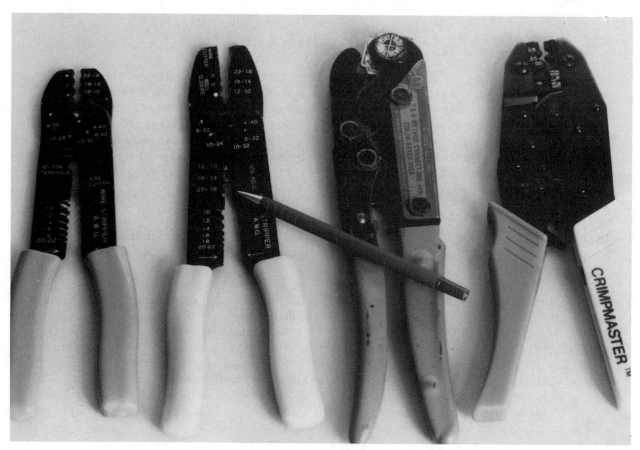

Figure 3. Terminal Crimpers.

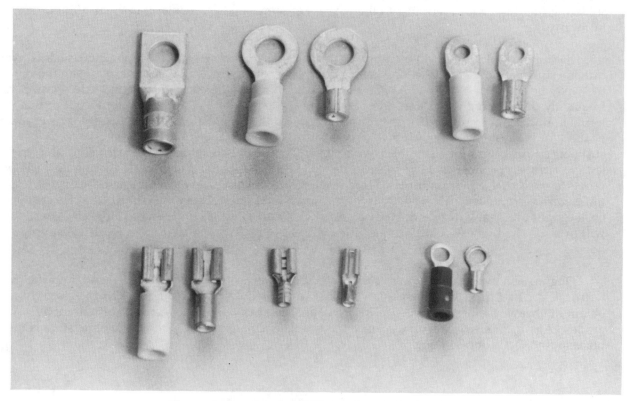

Figure 4. Insulated and Uninsulated Terminals.

If the junction box provides clamping-type terminals, it is not necessary to use the crimped terminals. Although time consuming, the crimping and soldering technique should be considered to ensure the connections last as long as the modules themselves. Because of the relatively high cost of USE and TC cables and wire, they are usually connected to less expensive cable at the earliest possible opportunity. Other than module interconnections, all other PV system wiring must be made using one of the methods included in the *NEC* [690-31, Chapter 3]. Single-conductor, exposed wiring is not permitted except with special permission [Chapter 3]. The most common methods used for PV systems are individual conductors in rigid metallic and nonmetallic conduit and nonmetallic sheathed cable.

Where individual conductors are used in conduit, they should be conductors with at least 90°C insulation such as THHN. The conduit can be either thick-wall or thin-wall electrical metallic tubing (EMT). If nonmetallic conduit is used, electrical (gray) PVC rather than plumbing (white) PVC tubing must be used [346, 347].

Two-conductor with ground, UF cable that is marked sunlight resistant is frequently used between the module interconnect wiring and the PV disconnect device. Black is the preferred color because of higher resistance to ultraviolet light, but the gray color seems durable because of the thicker insulation associated with the cable. Splices from the stranded wire to this wire must be protected in rainproof junction boxes such as NEMA style 3R. Cable clamps must also be used. Figure 5 shows a rainproof box with a pressure connector terminal strip installed for module wiring connections.

Interior exposed cable runs can be made only with sheathed cable types such as NM, NMC, and UF. The cable should not be subjected to physical abuse. If abuse is possible, physical protection must be provided [300-4, 336 B, 339]. Single conductor cable (commonly used between batteries and inverters) **shall** not be used in exposed locations--except as module interconnect conductors [300-3a].

WIRING
Module Connectors

Concealed module connectors must be able to resist the environment, be polarized, and be able to handle the short-circuit current. They **shall** also be of a latching design with the terminals guarded. The grounding member **shall** make first and break last [690-32, 33]. The *UL* standard also requires that the connectors for positive and negative conductors **shall** not be interchangeable. Even if the connection is concealed, they **shall** be able to resist the environment.
Module Connection Access

All junction boxes and other locations where module wiring connections are made **shall** be accessible. Removable modules and flexible wiring will allow accessibility [690-34]. This means modules should not be permanently fixed (welded) to mounting frames, and solid wire that could break when modules are moved to service the junction boxes should not be used.

Splices

All splices **must** be made in approved junction boxes with an approved splicing method. There are, however, some *UL*-listed devices that can be used for taps on nonmetallic sheathed cable outside junction boxes. Conductors **must** be twisted firmly to make a good electrical and mechanical connection, then brazed, welded, or soldered, and then taped [110-14b]. Although solder has a higher resistivity than copper, a rosin-fluxed, soldered splice will have slightly lower electrical resistance, and potentially higher resistance to corrosion than an unsoldered splice. Mechanical splicing devices such as split bolt connectors or terminal strips are also acceptable. Crimped splicing terminals may also be used if heavy-duty crimpers are used.

Figure 5. Rainproof Junction Box. Shown with Custom
Terminal Strip for Module Connections.

If the highest reliability is needed, then ultrasonic welding should be used for splices. Also, properly used pressure connectors give high reliability. Fuse blocks, fused disconnects, and circuit breakers are available with these pressure connectors.

Twist-on wire connectors (approved for splicing wires) have not proved adequate when used on low-voltage (12-50 volts), high-current PV systems because of thermal stress and oxidation of the contacts.

Where several modules are connected in parallel, a terminal block or bus bar arrangement must be used so that one module can be disconnected without disconnecting the grounded (on grounded systems) conductor of other modules [690-4c]. On grounded systems, this indicates that the popular "Daisy Chain" method of connecting modules is not acceptable because removing one module in the chain will disconnect the grounded conductor for all of those modules further out the chain. This becomes more critical on larger systems where paralleled sets of long series strings of modules are used. Figure 6 shows unacceptable and acceptable methods.

Several different types of terminal blocks and strips are shown in Figure 7.

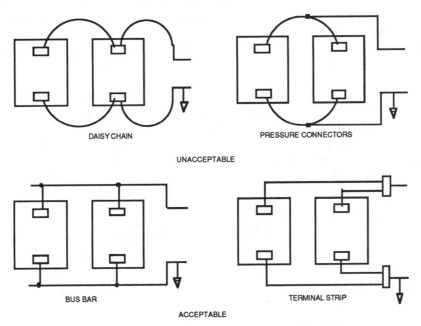

Figure 6. Module Interconnect Methods.

Figure 7. Power Splicing Blocks and Terminal Strips.

Conductor Color Codes

The *NEC* established color codes for electrical power systems many years before either the automobile or electronics industries were started. PV systems are being installed in the arena covered by the *NEC* and therefore **must** comply with those standards that apply to both ac and dc power systems. In a system where one conductor is grounded, the insulation on all grounded conductors **must** be white or natural gray (not much UV light resistance) or be any color but green and marked with white plastic tape or paint at each end. Conductors used for module frame grounding and other exposed metal equipment grounding **must** be bare (no insulation) or have green or green with yellow-striped insulation or identification [200-6, 7; 210-5].

The *NEC* requirements specify that the grounded conductor be white. In most PV-powered systems that are grounded, the grounded conductor turns out to be the negative conductor. A prominent exception is the telephone system, which uses a positive ground. In a PV system where the array is center tapped, the center tap or neutral **must** be grounded [690-41] and this becomes the white conductor. There is no *NEC* requirement designating the color of the ungrounded conductor, but the convention in power wiring is that the first ungrounded conductor is colored black and the second ungrounded conductor is colored red. This suggests that in two-wire, negative-grounded PV systems the positive conductor be red or any color with a red marking but green or white and the negative grounded conductor be white. In a three-wire, center-tapped system, the positive conductor could be red, the grounded center tap conductor **must** be white and the negative conductor could be black. Article 200-6 currently allows only No. 6 AWG and larger conductors to be marked.

GROUND-FAULT PROTECTION AND ARRAY DISABLEMENT
Ground-Faults

Article 690-5 of the *NEC* requires a ground-fault detection, interruption, and array disablement (GFID) system for fire protection if PV arrays are mounted on the roofs of dwellings. Ground-mounted arrays are not required to have this device. A group of devices to meet this requirement is under development but is not currently available. These particular devices will require that the system grounding conductor be routed through the device. To keep costs to a minimum, the device under development will most probably replace the PV disconnect switch and will serve multiple functions, such as for

- Manual PV disconnect switch
- Ground-fault detection
- Ground-fault interruption
- Array disablement
- Array wiring overcurrent protection

If a revised version of the *NEC* specifies equipment that is not available, the preceding edition of the code may be used with the approval of the inspecting authority. In this case, the 1987 *NEC* did not require a GFID device [90-4].

Array Disablement

Article 690-18 requires that a mechanism be provided to disable portions of the array or the entire array. The term disable has several meanings, and the *NEC* is not clear on what is intended. The *NEC* Handbook does elaborate. Disable can be defined several ways:

- Prevent the PV system from producing any output
- Reduce the output voltage to zero
- Reduce the output current to zero

The output could be measured at either the PV source terminals or at the load terminals.

Fire fighters are reluctant to fight a fire in a high-voltage battery room because there is no way to turn off a battery bank unless you can somehow remove the electrolyte. In a similar manner, the only way a PV system can have zero output at the array terminals is by preventing light from illuminating the modules or by removing the terminals on the modules. The output voltage may be reduced to zero by shorting the PV module or array terminals. When this is done, short-circuit current will flow through the shorting conductor, which in a properly wired system with bypass diodes, does no harm. The output current may be reduced to zero by disconnecting the PV system from any load. The PV disconnect switch would accomplish this action, but open-circuit voltages would still be present on the array wiring and in the disconnect box.

During PV module installations, the individual PV modules can be covered to disable them. For a system in use, the PV disconnect switch is opened and the array is either short circuited or left open circuit depending on the circumstances. In practical terms, some provision should be made to disconnect portions of the array from other sections for servicing. As individual modules or sets of modules are serviced, they may be covered and/or isolated and shorted to reduce the potential for electrical shock.

Ground-fault detection, interruption, and array disablement devices might accomplish the following actions automatically:

- Sense ground-fault currents exceeding a specified value
- Interrupt or significantly reduce the fault currents
- Open the circuit between the array and the load
- Short the array output terminals

These actions would reduce the array voltages to nearly zero (minimizing human shock hazards and equipment damage) and would serve to force the fault currents away from the fault path and back into the normal conductors. For fault location and repair, the array shorting device would have to be opened.

GROUNDING

The subject of grounding is one of the most confusing issues in electrical installations. A few definitions from Article 100 of the *NEC* address the situation.

Grounded:	Connected to the earth.
Grounded Conductor:	A system conductor that normally carries current and is intentionally grounded. In PV systems, one conductor (normally the negative) of a two-conductor system or the center-tapped wire of a bipolar system is grounded.
Grounding Conductor:	A conductor not meant to carry current used to: (1) connect the exposed metal portions of equipment to the ground electrode system or the grounded conductor, or (2) connect the grounded conductor to the grounding electrode or grounding electrode system.
Equipment Grounding Conductor:	See Grounding Conductor 1.
Grounding Electrode Conductor:	See Grounding Conductor 2.

Grounding--System

For a two-wire PV system over 50 volts (open-circuit PV output voltage), one dc conductor **shall** be grounded. In a three-wire system, the neutral or center tap of the dc system **shall** be grounded [690-7, 41]. These requirements apply to both stand-alone and grid-tied systems. Such system grounding will enhance personnel safety and minimize the effects of lightning and other induced surges on equipment. Also, grounding of all PV systems will minimize radio frequency noise from dc-operated fluorescent lights and inverters.

The system grounding electrode conductor for the direct current portion of a PV system **shall** be connected to the PV source circuits as close to the modules as possible but still on the load side of the PV disconnect switch [690-42, 250-22]. In grid-connected systems in which the inverter output is grounded, the grounding point could be on the array side of the PV disconnect switch, and even with grounded conductor switching, most of the system would be grounded. In a stand-alone system, if the grounding point were on the PV side of the array disconnect switch and the grounded conductor were opened, most of the system including the battery and the load would be ungrounded. The *NEC* Handbook clarifies the issue somewhat by stating that one of the purposes of opening the grounded conductor with the ground-fault device is to purposely unground the array and open the ground path for ground faults. This implies that the grounding point **must** be on the load side of the PV disconnect switch.

In grid-tied systems, for which the inverter design allows a dc grounded system, it is suggested that the grounded conductor not be switched by the PV disconnect switch. In this case, the grounding point for the dc portion of the system would be near or at the inverter input, and the dc grounding conductor would be tied to the same ground rod as the ac grounding conductor. Inverters that do not allow for dc grounding of the array on systems over 50 volts do not meet the requirements of the *NEC* unless some other method of grounding has been approved.

The direct-current system-grounding electrode conductor **shall** not be smaller than No. 8 AWG or the largest conductor supplied by the system [250-93]. This may present a problem in some installations in which the battery-to-inverter cables are AWG 1/0 or 2/0 or larger. The conductor from the battery to charge controller and then to the PV disconnect device may be as small as No. 10 or 12 AWG (depending on charging current), and this is where the system grounding conductor could be connected. According to the *NEC*, the larger conductor (i.e., 1/0) will have to be used for the system grounding electrode conductor, and in this example it would be connected to the much smaller wire between the PV disconnect device and the battery. Figure 8 illustrates this problem for a particular system (the conductor sizes are representative only). To be conservative, the larger wire size could be used as the grounded conductor from the attachment point for the grounding electrode conductor all the way to the battery.

Since the *NEC* does not provide a clear definition of where the PV output circuits end, the single grounding point may be on the battery side of the charge controller or even the negative terminal on the battery or the inverter. The negative dc input to the inverter is grounded in some designs, and this would be an appropriate place to connect the grounding electrode conductor and other equipment grounding conductors. Connection of the grounding electrode conductor to either the negative battery terminal or negative inverter terminal would avoid the "large wire/small wire" problem outlined above.

It is imperative that there be no more than one grounding connection to the negative conductor of a PV system. Failure to limit the connections will allow currents to flow in uninsulated conductors and will create unintentional multiple ground faults in the grounded conductor [250-21]. Also keep in mind that future ground-fault interrupter systems may require that this single grounding connection be made at a specific location.

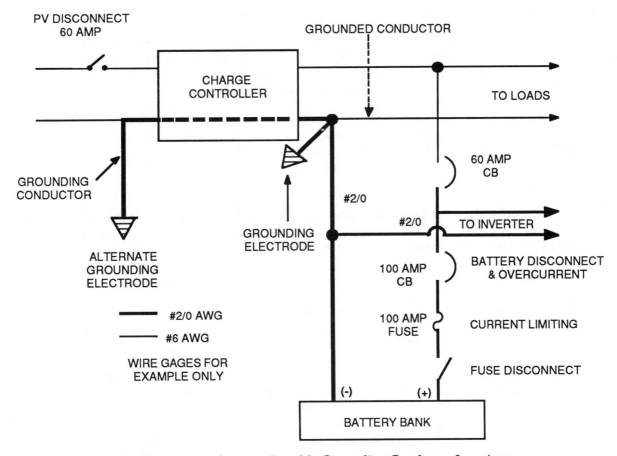

Figure 8. Typical System: Possible Grounding Conductor Locations.

Some inverter designs have the entire chassis used as part of the negative circuit. Also the same situation exists in certain radios--automobile and shortwave. These designs will not pass the current *UL* standards for consumer electrical equipment and will probably require modification in the future since they do not provide electrical isolation between the exterior metal surfaces and the current carrying conductors. Also they create the very real potential for multiple grounding conductor connections.

A few charge controllers break the negative lead internally, and others use a current shunt in the negative lead. It appears that breaking the negative lead will unground some part of the system on negative-grounded systems and thus violate the provisions of the code. This, of course, depends on the location of the grounding point, but in any case some part of the system becomes ungrounded when the controller opens the circuit. A current shunt installed in the negative lead of a charge controller or elsewhere in a properly designed system will have the same or greater ampacity as the negative conductor from the PV source circuits and poses no problems. Some telephone systems ground the positive conductor, and this may cause problems for the PV system. An isolated-ground, dc to dc converter may be used to power subsystems that have different grounding polarities from the main system. In the ac realm, an isolation transformer will serve the same purpose.

In larger utility-tied systems and some stand-alone systems, high impedance grounding systems might be used in lieu of or in addition to the required hard ground. The discussion and design of these systems are beyond the scope of this guide because they are very site specific.

Grounding--Equipment

All noncurrent-carrying exposed metal parts of junction boxes, equipment, and appliances in the entire PV and dc load system **shall** be grounded [690-43, 250-42, 720-1 & 10]. All PV systems, regardless of voltage, **must** have an equipment grounding system for exposed metal surfaces (e.g., module frames and inverter cases). The grounding conductor **shall** be sized as required by Article 690-43 or 250-95. Generally this will mean a grounding conductor size equal to the size of the current-carrying conductors. If the array can provide short-circuit currents that are more than twice the rating of a particular overcurrent device, then grounding conductors **must** be used that are based on the rating of the overcurrent device [690-43, 250-95].

Grounding Electrode

The dc system grounding electrode **shall** be common with or bonded to the ac grounding electrode (if any). It **shall** be located as close to the system grounding-conductor connection point as practical [690-44, 250-26c]. The system grounding conductor and the equipment grounding conductor **shall** be tied to the same ground electrode or ground electrode system. Even if the PV <u>system</u> is ungrounded (less than 50 volts), the equipment grounding conductor **must** be connected to a grounding electrode [250-50]. The grounding electrode **shall** be a manufactured device, 5/8 inch in diameter with at least 8 feet driven into the soil at an angle no larger than 45 degrees from the vertical [250-83]. Metal water pipes and other metallic structures as well as concrete encased electrodes may also be used in some circumstances [250-26c, 250-81, 250-83].

A bare-metal well casing makes a good grounding electrode. If it is distant from the PV array or the main disconnect, it should be part of a grounding electrode system. The central pipe to the well should not be used for grounding because it is sometimes removed for servicing.

For maximum protection against lightning-induced surges, it is suggested that a grounding electrode **system** be used with at least two grounding electrodes bonded together. One electrode would be the main system grounding electrode as described above. The other would be a supplemental grounding electrode located as close to the PV array as practical. The module frames and array frames would be connected directly to this electrode to provide as short a path as possible for lightning-induced surges to reach the earth. This electrode **must** be bonded with a conductor to the main system grounding electrode [250-81]. The size of the bonding or jumper cable **must** be related to the ampacity of the overcurrent device protecting the PV source circuits. This bonding jumper is an auxiliary to the module frame grounding that is required to be grounded with an equipment grounding conductor. Table 250-95 gives the requirements. Generally, this **must** be no smaller than No. 8 AWG to comply with bonding jumper requirements. Grounding conductors are allowed to be smaller than circuit conductors when the circuit conductors become very large. Article 250 of the *NEC* elaborates on these requirements.

Do not connect the negative conductor to the grounding electrode, the grounding conductor, or the frame at the modules. There should be one and only one point in the system where the grounding electrode conductor is attached to the system-grounded conductor. See Figure 9 for clarification. The wire sizes shown are for illustration only and will vary depending on system size. Chapter 3 of the *NEC* specifies the ampacity of various types and sizes of conductors.

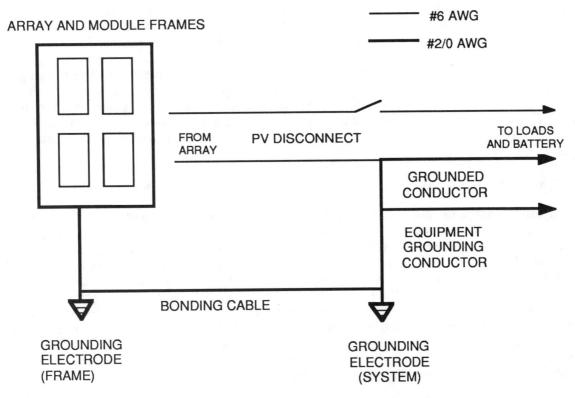

ARRAY AND MODULE FRAMES

—————— #6 AWG

—————— #2/0 AWG

FROM ARRAY   PV DISCONNECT   TO LOADS AND BATTERY

GROUNDED CONDUCTOR

EQUIPMENT GROUNDING CONDUCTOR

BONDING CABLE

GROUNDING ELECTRODE (FRAME)

GROUNDING ELECTRODE (SYSTEM)

Figure 9. Example Grounding Electrode System.

CONDUCTOR AMPACITY

Photovoltaic modules are limited in their ability to deliver current. Their short-circuit currents are nominally 10 to 15 percent higher than their operating currents. With reflective ground cover such as sand or snow and with reflections from clouds, PV output may reach 125 percent of rated output for short periods of time (minutes). Therefore, common overcurrent trip devices cannot be used to disconnect a PV array under shorted conditions.

Another problem for PV systems is that the conductors may operate at temperatures as high as 65°C when the modules are mounted close to a structure, there are no winds, and the ambient temperatures are high. Temperatures in module junction boxes frequently are in this range. This may require that the ampacity of the conductors be derated or corrected with factors given in *NEC* Table 310-16. For example, a No. 10 AWG USE single conductor cable used for module interconnections has a 90°C insulation and an ampacity of 40 amps in an ambient temperature of 26-30°C. When it is used in ambient temperatures of 61-70°C, the ampacity of this cable is reduced to 23.2 amps.

The ampacity of conductors in PV source circuits **shall** be at least 125 percent of the rated module or parallel-connected modules <u>short-circuit current</u> [690-8]. The ampacity of the PV output circuit conductors **shall** be at least 125 percent of the <u>normal rated output current</u> [690-8]. Rated operating current refers to the manufacturer's rating of peak power voltage and current at standard conditions of 1,000 watts per square meter of irradiance and a cell temperature of 25°C. Conservative design dictates using 125 percent of the parallel module <u>short-circuit current</u>, which will be higher than the <u>normal rated output current</u>. The ampacity of conductors to and from an inverter or power conditioning system **shall** be 125 percent of the rated operating current for that device [690-8]. In a similar manner, other conductors in the system should have an ampacity of 125 percent of the rated operating current to allow for long duration operation at full power. Operation when snow or cloud enhancement increases the PV output above normal may require additional ampacity.

A 1989 revision to the *UL* Standard 1703 for PV modules requires that the module marking and installation instructions be based on 150 percent of the 25°C ratings. Conservative design practices require oversizing wire and increasing the ratings of overcurrent devices on PV source and output circuits. The ampacity of conductors and the sizing of overcurrent devices is an area that demands careful attention by the PV system designer/installer. Temperatures and wiring methods **must** be addressed for each site.

Stand-Alone Systems

In stand-alone systems, inverters are frequently used to change the direct current (dc) from a battery bank to 120-volt or 240-volt, 60-Hertz (Hz) alternating current (ac). The conductors between the inverter and the battery **must** have properly rated overcurrent protection and disconnect mechanisms. These inverters frequently have short duration (tens of seconds) surge capabilities that are as high as four times their rated output. For example, a 2,500-watt inverter might be required to surge to 10,000 watts for 10 seconds when a motor load **must** be started. The *NEC* requires the ampacity of the conductors between the battery and the inverter to be sized by the rated 2,500 watt output of the inverter. For example, in a 24-volt system, a 2,500-watt inverter would draw 105 amps at full load and 420 amps for motor-starting surges. To minimize steady-state voltage drops, account for surge-induced voltage drops, and increase system efficiency, most well-designed systems have conductors several sizes larger than required by the *NEC*.

When the battery bank is tapped to provide multiple voltages (i.e., 12 and 24 volts from a 24-volt battery bank), the common negative conductor will carry the **sum** of all of the simultaneous load currents. The negative conductor **must** have an **ampacity at least equal to the sum** of all the amp ratings of the overcurrent devices protecting the positive conductors or have an ampacity equal to the sum of the ampacities of the positive conductors.

The *NEC* does not allow paralleling conductors for added ampacity, except that cables 1/0 AWG or larger may be paralleled under certain conditions [310-4]. DC-rated switchgear, overcurrent devices and conductors cost significantly more when rated to carry more than 100 amps. It is suggested that large PV arrays be broken down into subarrays, each having a short-circuit output of less than 80 amps. This will allow use of 100-amp-rated equipment (125 percent of 80 amps) on each source circuit.

OVERCURRENT PROTECTION

The *NEC* requires that every ungrounded conductor be protected by an overcurrent device [240-20]. In a PV system with multiple sources of power (PV modules, batteries, battery chargers, generators, power conditioning systems, etc.), the overcurrent device **must** protect the conductor from overcurrent from any source connected to that conductor [690-9]. If the PV system is directly connected to the load without battery storage or other source of power, then no overcurrent protection is required if the conductors are sized at 125 percent of the short-circuit current [690-8].

When circuits are opened in dc systems, arcs can be sustained much longer than they are in ac systems. This presents additional burdens on overcurrent-protection devices rated for dc operation. Such devices **must** carry the rated load current and sense overcurrent situations as well as be able to safely interrupt dc currents. AC overcurrent devices have the same requirements, but the interrupt function is considerably easier.

Ampere Rating

The PV source circuits **shall** have overcurrent devices rated at least 125 percent of the parallel module short-circuit current. The PV output circuit overcurrent devices **shall** be rated at least 125 percent of the rated PV currents [690-8]. Since some installations have experienced the blowing of fuses for unknown reasons, conservative practice might call for increasing the rating of these overcurrent devices and the ampacity of the conductors they protect to 150 percent of the short-circuit current or more. This agrees with the recent *UL* requirements mentioned above. Time delay fuses or circuit breakers would minimize nuisance tripping or blowing. In all cases, dc-rated devices having the appropriate voltage rating **must** be used and adequate ventilation **must** be provided.

Both conductors from the PV array **shall** be protected with overcurrent devices when both positive and negative conductors of the PV output circuit are opened by the disconnect switch [Diagram 690-1]. This requirement also makes PV disconnects resemble service entrance disconnects. Since PV module outputs are current limited, these overcurrent devices are actually protecting the array wiring from battery or power conditioning system short circuits.

Often PV modules or series strings of modules are connected in parallel. As the conductor size used in the array wiring increases to accommodate the higher short-circuit currents of paralleled modules, each conductor size **must** be protected by an appropriately sized overcurrent device. This device **must** be placed nearest the source of the largest potential overcurrent for that conductor [240-21]. Figure 10 shows an example of array conductor overcurrent protection for a medium-size array broken into subarrays.

Either fuses or circuit breakers are acceptable for overcurrent devices provided they are rated for their intended uses--i.e., they have dc ratings when used in dc circuits, the ampacity is correct, and they can interrupt the necessary currents when short-circuit currents occur [240 E, F, G]. Figure 11 shows dc-rated, *UL*-listed circuit breakers being used for PV source circuit disconnects.

Since PV systems may have transients--lightning and motor starting as well as others--inverse time circuit breakers (the standard type) or time-delay fuses should be used in most cases. In circuits where no transients are anticipated, fast-acting fuses can be used. They should be used if relays and other switchgear in dc systems are to be protected.

Branch Circuits

Fuses used to protect dc or ac branch circuits **must** be tested and rated for that use. They **must** also be of different sizes and markings for each amperage and voltage group to prevent unintentional interchange. However, dc-rated fuses that meet the requirements of the *NEC* are difficult to find [240F]. Figure 12 shows *UL*-listed, dc-rated, time-delay fuses on the left that are listed for branch circuit use, which would include the battery fuse. Acceptable dc-rated, *UL*-recognized fast-acting supplementary fuses are shown on the right and can be used in the PV source circuits. The fuses shown are made by Littelfuse, and the fuse holders are by Marathon.

These particular requirements eliminate the use of glass, ceramic, and plastic automotive fuses as branch circuit overcurrent devices because they are neither tested nor rated for this application.

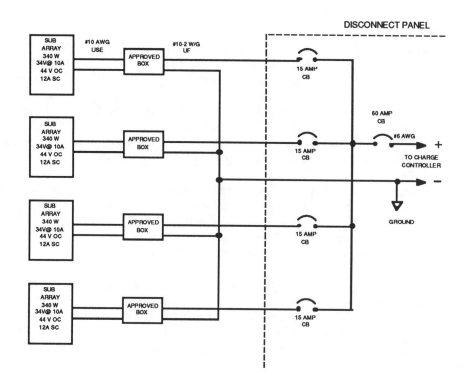

Figure 10. Typical Array Conductor Overcurrent Protection
(with Optional Subarray Disconnects).

Figure 11. Approved Circuit Breakers.

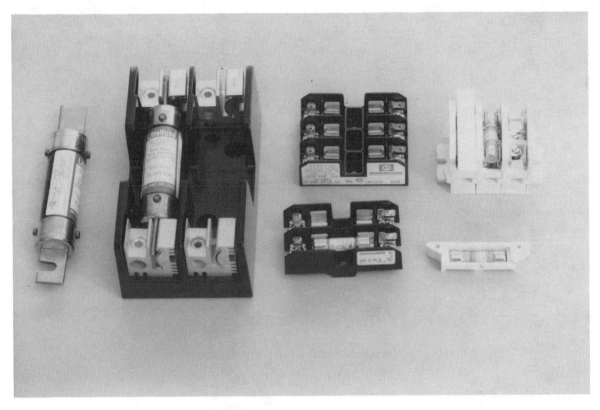

Figure 12. Listed and Recognized Fuses.

Automotive fuses have no dc rating by the fuse industry or the testing laboratories. When rated by the manufacturer, they have only a 32-volt maximum rating, which is less than the open-circuit voltage from a 24-volt PV array. Furthermore, these fuses have no rating for interrupt current, nor are they generally marked with all of the information required for branch circuit fuses. They are not considered supplemental fuses under the *UL* component recognition program and **must** not be used anywhere in the PV system. Figure 13 shows unacceptable automotive fuses on the left and recognized supplemental fuses on the right. Unfortunately, even the supplemental fuses frequently have no dc rating.

Circuit breakers also have specific requirements when used in branch circuits, but they are generally available with the needed dc ratings [240 G]. Figure 14 shows examples of dc-rated, *UL*-recognized circuit breakers on the left. They may be used in the PV source circuits for disconnects and overcurrent protection. The larger units are dc-rated, *UL*-listed circuit breakers that can be used in dc load centers for branch circuit protection. The breakers shown are produced by Square D and Heinemann. Airpax also produces a dc *UL*-listed circuit breaker, and Potter Brumfield and others produce dc-rated, *UL*-recognized breakers.

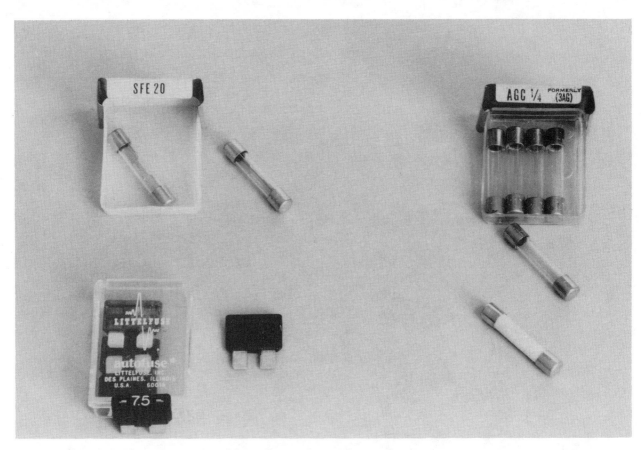

Figure 13. Acceptable and Nonacceptable Fuses.

Figure 14. *UL*-Recognized and Listed Circuit Breakers.

Ampere of Interrupt Current (AIC)--Short-Circuit Conditions

Overcurrent devices--both fuses and circuit breakers--**must** be able to safely open circuits with short-circuit currents flowing in them. Since PV arrays are inherently current limited, high short-circuit currents are not a problem, and the normal ampacity of the conductors is sufficient. In stand-alone systems with storage batteries, however, the short-circuit problem is very severe. A single 220 amp-hour, 6-volt, deep-discharge, lead-acid battery may produce short-circuit currents as high as 8,000 amps for a fraction of a second and as much as 6,000 amps for a few seconds in a direct terminal-to-terminal short circuit. Such high currents are likely to create an arc in an underrated overcurrent device causing that device to burn or blow apart. Two paralleled batteries would generate twice as much current, and larger capacity batteries would be able to deliver proportionately more current into a short circuit. In dc systems, particularly stand-alone systems with batteries, the interrupt capability of every overcurrent device is important. This interrupt capability is specified as amps of interrupt capability or AIC.

Most dc-rated, *UL*-listed, branch circuit breakers that can be used in PV systems have an AIC of 5,000 amps, however, Heinemann Electric makes one with an AIC of 25,000 amps. Some dc-rated, *UL*-recognized circuit breakers have an AIC of only 3,000 amps. DC-rated fuses normally have an AIC of a few thousand amps, up to 10,000 amps if they are of the current-limiting variety.

Fusing of PV Source Circuits

The *NEC* allows supplementary fuses to be used in PV source circuits [690-9c]. A supplementary fuse is one that is designed for use inside a piece of equipment. These fuses supplement the main branch fuse and do not have to comply with all of the requirements of branch fuses. They should, however, be dc rated and able to handle the short-circuit currents they may be subjected to. Unfortunately, many supplemental fuses are not dc rated, and if they are, the AIC (when available) is usually less than 5,000 amps. The use of ac-rated supplementary fuses **is not** recommended for the dc circuits of PV systems.

Current Limiting Fuses--Stand-Alone Systems

A current-limiting fuse or high AIC circuit breaker should be used in the positive conductor near the battery to limit the current that a battery bank can supply to a short-circuit and to reduce the short-circuit currents to levels that are within the AIC of downstream overcurrent devices. These fuses are available with *UL* ratings of 125 and 300 volts dc, currents of 0.1 to 600 amps, and a dc AIC of 20,000 amps. They are classed as RK5 or RK1 current-limiting fuses and should be mounted in class-R rejecting fuse holders or dc-rated, fused disconnects. For reasons mentioned previously, time delay fuses should be specified. One of these fuses and the associated disconnect switch should be used in **each** bank of batteries with a paralleled amp-hour capacity up to 1,000 amp-hours. Batteries with single cell amp-hour capacities higher than 1,000 amp-hours will require special design considerations because these batteries may be able to generate short-circuit currents in excess of the 20,000 AIC rating of the current-limiting fuse. When calculating short-circuit currents, the resistances of all connections, terminals, wire, fuse holders, circuit breakers, and switches **must** be considered. These resistances serve to reduce the magnitude of the available short-circuit currents at any particular point. The suggestion of one fuse per 1,000 amp-hours of battery size is only a general estimate, and the calculations are site specific. The fuses shown in Figure 12 are current limiting.

For systems less than 65 volts (open circuit), Heinemann Electric 25,000 AIC circuit breakers may be used or APT 400 amp fused disconnect. These circuit breakers are not current limiting even with the high interrupt rating so they cannot be used to protect other fuses or circuit breakers. An appropriate use would be in the conductor between the battery bank and the inverter. This single device would minimize voltage drop and provide the necessary disconnect and overcurrent features.

Current-Limiting Fuses--Grid Connected Systems

Normal electrical installation practice requires that service entrance equipment have fault-current protection devices that can interrupt the available short-circuit currents [230-208]. This requirement applies to the utility side of any power conditioning system in a PV installation.

Whenever a fuse is used for an overcurrent device and is accessible to other than qualified persons, it **must** be in a circuit where all power can be removed from both sides of the fuse for servicing. It is not sufficient to reduce the current to zero before changing the fuse; there **must** be no voltage present on either end of the fuse prior to service. This may require the addition of switches on both sides of the fuse location--a complication that increases the voltage drop and reduces the reliability of the system [690-16, Diagram 690-1]. Because of this requirement and the complications it causes and the need for disconnects, it is recommended that a current-limiting fuse be used at the battery and that circuit breakers be used for all other overcurrent devices.

DISCONNECTING MEANS

There are many considerations in configuring the disconnect switches for a PV system. The National Electrical Code deals with safety first and other requirements last--if at all. The PV designer **must** also consider equipment damage from over voltage, performance options, equipment limitations, and cost.

A photovoltaic system is a power generation system, and a specific minimum number of disconnects are necessary to deal with that power. Untrained persons will be operating the systems, and the disconnect system **must** be designed to provide safe, reliable, and understandable operation.

Disconnects may range from nonexistent in a self-contained PV-powered light for a sidewalk to the space-shuttle-like control room in a large, multi-megawatt, utility-tied PV power station. Generally, local inspectors will not require disconnects on totally enclosed, self-contained PV systems like the sidewalk illumination system or a pre-wired attic ventilation fan. This would be particularly true if the entire assembly were *UL*-listed as a unit and there were no external contacts or user serviceable parts. However, as the complexity of the device increases and separate modules, batteries, and charge controllers having external contacts **must** be be wired together and possibly operated and serviced by unqualified persons, the situation changes.

Photovoltaic Array Disconnects

Article 690 requires all current-carrying conductors from the PV power source to have a disconnect provision. This includes the grounded conductor, if any [690-13, 14; 230 F]. If a grounded conductor is opened by the PV disconnect switch, all other conductors **must** open simultaneously. This requires a multipole switch. Diagram 690-1 and Article 690-13 indicate and imply that the negative conductor disconnect **must be a switch** and no further elaboration is given--even in the *NEC* Handbook. However, Article 690-17 does discuss using switches or circuit breakers for disconnects of the ungrounded conductors, and Article 230 F does say that ac service-entrance grounded conductors may be disconnected with a bolted connection. The *NEC* Handbook provides further insight when discussing array disablement where the system grounding conductor is attached to the load side of the PV disconnect switch. In this case, the switched grounded conductor is used to interrupt the ground fault path by ungrounding the array.

In normal operations (with no ground faults) on a system where the grounded conductor is opened by the PV disconnect switch, some part of the system becomes ungrounded, and this presents a safety problem. If the array were ungrounded, static voltages and inductive surges could soar on the conductors and cause conductor insulation breakdown and spark hazards. Ground faults are rare in Block V PV modules manufactured since 1985. For these reasons, it is suggested that the grounded conductor remain <u>unswitched</u> to stabilize the system voltage with respect to ground and enhance system safety rather than be configured to remove ground faults that rarely happen. A bolted disconnect **must**, however, be used to comply with the *NEC*. The local inspection authority may have specific requirements on this issue. Of course in an unground, 12- or 24-volt PV system, both positive and negative conductors **must** be switched.

Equipment Disconnects

Each piece of equipment in the PV system **shall** have disconnect switches to disconnect it from all sources of power. The disconnects **shall** be circuit breakers or switches and **shall** comply with all of the provisions of Article 690-17. DC-rated switches and fuses are expensive and difficult to locate. The ready availability of moderately priced dc-rated circuit breakers with ratings up to 48 volts and 70 amps would seem to encourage their use in all 12- and 24-volt systems. When properly located and used within their approved ratings, circuit breakers can serve as both the disconnect and overcurrent device. In simple systems, one switch or circuit breaker disconnecting the PV array and another disconnecting the battery may be all that is required.

A 2,000-watt inverter on a 12-volt system can draw nearly 200 amps at full load. Disconnect switches **must** be rated to carry this load and be protected by current-limiting fuses, since the battery-to-inverter wiring is usually a very large gage wire to minimize voltage drops. Again a dc-rated, *UL*-listed circuit breaker may prove less costly than a switch with the same ratings.

Battery Disconnect

When the battery is disconnected from the stand-alone system, either manually or through the action of a fuse or circuit breaker, care **must** be taken that the PV system not be allowed to remain connected to the load. Low-level loads will allow the PV array voltage to increase from the normal battery charging levels to the open-circuit voltage, which will shorten lamp life and possibly damage electronic components.

This potential problem can be avoided by using multipole circuit breaker or fuse disconnects as shown in Figure 15. This figure shows two ways of making the connection. Separate circuits, including disconnects and fuses between the charge controller and the battery and the battery and the load as shown in Figure 16, may be used if it is desired to operate the loads without the PV array being connected. If the design requires that the entire system shut down with a minimum number of switch actions, the switches and circuit breakers could be ganged multipole units.

Charge Controller Disconnects

Some charge controllers are fussy about the sequence in which they are connected and disconnected from the system. Most charge controllers do not respond well to being connected to the PV array and not being connected to the battery. The sensed battery voltage (or lack thereof) would tend to rapidly cycle between the array open-circuit voltage and zero as the controller tried to regulate the nonexistent charge process. This problem will be particularly acute in self-contained charge controllers with no external battery sensing.

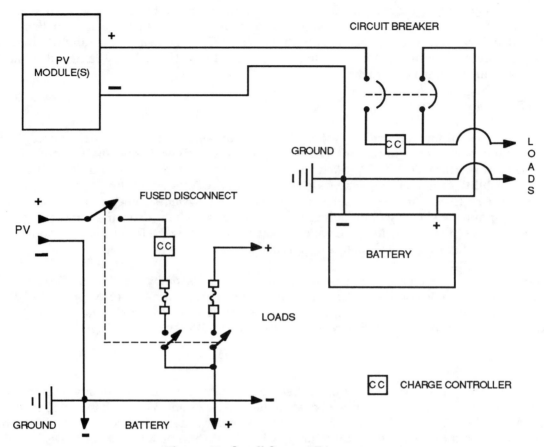

Figure 15. Small System Disconnects.

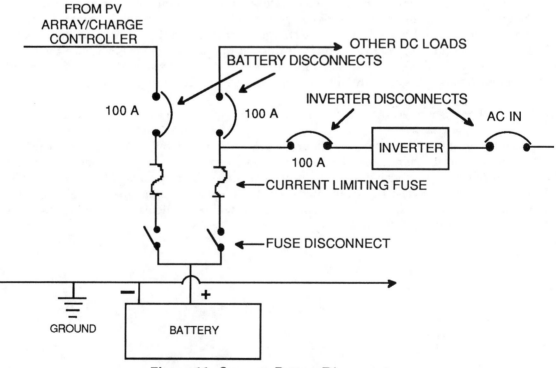

Figure 16. Separate Battery Disconnects.

Again, the multipole switch or circuit breaker can be used to disconnect not only the battery from the charge controller, but the charge controller from the array. Probably the safest method for self-contained charge controllers is to have the PV disconnect switch disconnect both the input and the output of the charge controller from the system. Larger systems with separate charge control electronics and switching elements will require a case-by-case analysis--at least until the controller manufacturers make their products more tolerant. Figure 17 shows two methods of disconnecting the charge controller.

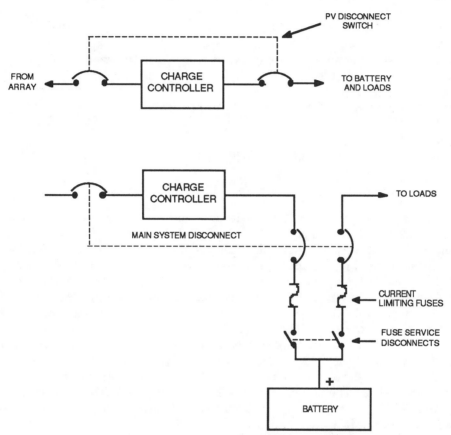

Figure 17. Charge Controller Disconnects.

Multiple Power Sources

When multiple sources of power are involved, the disconnect switches **shall** be grouped and identified [230-72, 690-15]. No more than six motions of the hand **will** be required to operate all of the disconnect switches required to remove all power from the system [230-71]. These disconnects include those for the PV output, the battery system, any generator, and any other source of power. Multipole disconnects or handle ties should be used to keep the number of motions to six or fewer.

PANEL BOARDS, ENCLOSURES, AND BOXES

Disconnect and overcurrent devices **shall** be mounted in approved enclosures, panel boards, or boxes [240-30]. Wiring between these enclosures **must** be by an *NEC*-approved method [690-13a]. Appropriate cable clamps, strain-relief methods, or conduit **shall** be used. All openings not used **shall** be closed with the same or similar material to that of the enclosure [370-8]. Metal enclosures **must** be bonded to the grounding conductor [370-4]. Use of wood or other flammable materials is discouraged. Conductors from different systems such as utility power, gas generator, hydro, or wind **shall** not be placed in the same enclosure, box, conduit, etc., as PV conductors unless the enclosure is partitioned [690-4b].

When designing a PV distribution system or panel board, an approved NEMA style box and approved disconnect devices and overcurrent devices should be used. The requirements for internal configuration of *NEC* Articles 370, 373, and 384 **must** be followed. Dead front-panel boards with no exposed current-carrying conductors, terminals, or contacts are generally required.

BATTERIES

Battery storage in stand-alone PV systems poses several safety hazards. In general, *NEC* Articles 480 and 690-71, 72, 73 should be followed. Several hazards are present:

- Hydrogen gas generation from charging lead-acid batteries
- High short-circuit currents
- Acid or caustic electrolyte
- Electric shock potential in higher voltage systems

Hydrogen Gas

When flooded, non-sealed, lead-acid batteries are charged at high rates or when the terminal voltage reaches about 2.4 volts per cell, the batteries produce hydrogen gas. Even sealed batteries may vent hydrogen gas under certain conditions. This gas, if confined and not properly vented, poses an explosive hazard if exposed to a spark from a charge controller relay or switch or battery-servicing tool. The amount of gas generated is a function of the battery temperature, the voltage, the charging current, and the battery bank size. Hydrogen is a light, small-molecule gas that is easily dissipated. Small battery banks (i.e., 1-8 220-amp-hour, 6-volt batteries) placed in a large room or a well-ventilated area do not pose a significant hazard. Larger numbers of batteries in smaller or tightly enclosed areas require venting. Venting manifolds may be attached to each cell and routed to an exterior location. A catalytic recombiner cap may be attached to each cell to recombine some of the hydrogen with oxygen in the air to produce water. If these combiner caps are used, they will require periodic maintenance. The batteries may be installed in a sealed box, (nearly impossible with hydrogen) and the box vented with a pipe to the outside in a safe location. It is rarely necessary to use power venting [480-8].

Certain charge controllers are designed to minimize the generation of hydrogen gas, but lead-acid batteries need periodic overcharging to equalize the cells. This produces gassing that should be dissipated.

In **no case** should charge controllers, switches, relays, or other devices capable of producing an electric spark be mounted in a battery enclosure or directly over a battery bank. Care **must** be exercised when routing conduit from a sealed battery box to a disconnect. Hydrogen gas may travel in the conduit to the arcing contacts of the switch.

High Short-Circuit Currents

Batteries are capable of generating tens of thousands of amps of current when shorted. A short circuit in a conductor not protected by overcurrent devices can melt wrenches or other tools, battery terminals and cables, and spray molten metal around the room. Exposed battery terminals and cable connections **must** be protected from accidental contact with metal objects. Live parts of batteries **must** be guarded. Battery voltages **must** be less than 50 volts in dwellings unless certain protective criteria are met [690-71]. This generally means that the batteries should be accessible only to a qualified person. A locked room, battery box, or other container and some method to prevent access by the untrained person should minimize the hazards from short circuits and electric shock. The danger may be reduced if insulated caps or tape are placed on each terminal and an insulated wrench is used for servicing, but in these circumstances, corrosion may go unnoticed on the terminals. The *NEC* requires certain spacings around battery enclosures and boxes to allow for unrestricted servicing. This is generally about three feet [110-16].

Acid or Caustic Electrolyte

A thin film of electrolyte can accumulate on the tops of the battery and on nearby surfaces. This material can cause flesh burns. It is also a conductor and in higher voltage battery banks poses a shock hazard. It should be removed periodically with an appropriate neutralizing solution. For lead-acid batteries, a dilute solution of baking soda and water works well. A mild vinegar solution works on nickel-cadmium batteries.

Charge controllers are available that minimize the dispersion of the electrolyte at the same time they minimize battery gassing. They do this by keeping the battery voltage from climbing into the vigorous gassing region where the high volume of gas causes electrolyte to bubble out of the cells.

Battery servicing hazards can be minimized by using protective clothing including face masks, gloves, and rubber aprons. Self-contained eyewash stations and neutralizing solution would be beneficial additions to any battery room. Water should be used to wash acid or alkaline electrolyte from the skin and eyes.

Electric Shock Potential

Storage batteries in dwellings **must** operate at less than 50 volts unless live parts are protected during routine servicing [690-71]. It is recommended that live parts of any battery bank should be guarded [690-71b(2)].

GENERATORS

Other electrical power generators such as wind, hydro, and gasoline/propane/diesel **must** comply with the requirements of the *NEC*. These requirements are specified in the following *NEC* articles:

Article 230	Services
Article 250	Grounding
Article 445	Generators
Article 700	Emergency Systems
Article 701	Legally Required Standby Systems
Article 702	Optional Standby Systems
Article 705	Interconnected Power Production Sources

When multiple sources of ac power are to be connected to the PV system, they **must** be connected with an appropriately rated and approved transfer switch. AC generators frequently are rated to supply larger amounts of power than that supplied by the PV/battery/inverter. The transfer switch **must** be able to safely accommodate this additional power.

CHARGE CONTROLLERS

A charge controller or self-regulating system **shall** be used in a stand-alone system with battery storage. The mechanism for adjusting state of charge **shall** be accessible only to qualified persons [690-72].

Presently, there are no charge controllers on the market that have been tested by *UL* or other recognized testing organizations. Some are scheduled to be tested in the near future.

Any charge controller should be mounted in a listed enclosure with provisions for ventilation. Surface mounting of devices with external terminals readily accessible to the unqualified person will not be accepted by the inspection authority. Dead-front panels with no exposed contacts are generally required for safety. A typical charge controller such as shown in Figure 18 should be mounted in a *UL*-listed enclosure so that none of the terminals are exposed.

DISTRIBUTION SYSTEMS

The *National Electrical Code* was formulated when there were abundant supplies of relatively cheap energy. As the code was expanded to include other power systems such as PV, many sections were not modified to reflect the recent push toward efficient use of electricity in the home. Stand-alone PV systems **may** be required to have services with 60- to 100-amp capacities to meet the code [230-79]. DC receptacles and lighting circuits **may** have to be as numerous as their ac counterparts [220, 422]. In a small one- to four-module system on a remote cabin or small home, these requirements are overstated, since the power source may be able to supply only a few hundred watts of power.

The local inspection authority has the final say on what is, or is not, required and what is, or is not, safe. Reasoned conversations may result in a liberal interpretation of the code. For a new dwelling, it seems appropriate to install a complete ac electrical system as required by the *NEC*. This will meet the requirements of the inspection authority, the mortgage company, and the insurance industry. Then the PV system and its dc distribution system can be added. If an inverter is

Figure 18. Typical Charge Controller.

used, it can be connected to the ac service entrance. DC branch circuits and outlets can be added where needed, and everyone will be happy. If or when grid power becomes available, it can be integrated into the system with minimum difficulty. If the building is sold at a later date, it will comply with the *NEC* and will have been inspected.

Square D has received a direct current (dc), *UL* listing for its standard QO residential **branch** circuit breakers. They can be used up to 48 volts (PV open-circuit voltage) and 70 amps dc. The AIC is 5,000 amps, so a current-limiting fuse (RK5 type) or APT fused switch **must** be used when they are connected on a battery system. The Square D QOM **main** breakers (used at the top of the load center) **do not** have this listing, so the load center **must** be obtained with main lugs and no main breakers.

In a small PV system, a two-pole Square D QO breaker could be used as the PV disconnect (one pole) and the battery disconnect (one pole). Also, a fused disconnect could be used in this configuration. This would give a little more flexibility since the fuses could have different current ratings. Figure 15 on page 41 shows both systems with only a single branch circuit.

In a system with several branch circuits, the Square D load center can be used. A standard, off-the-shelf Square D residential load center without a main breaker can be used for a dc distribution panel in 12- and 24-volt dc systems. The main disconnect would have to be a QO breaker "back fed" and connected in one of the normal branch circuit locations. Since the load center has two separate circuits, (one for each phase) they will have to be tied together to use the entire load center. Figure 19 illustrates this use of the Square D load center. "Backfed" circuit breakers must be retained by a special clip available from Square D [384-16f].

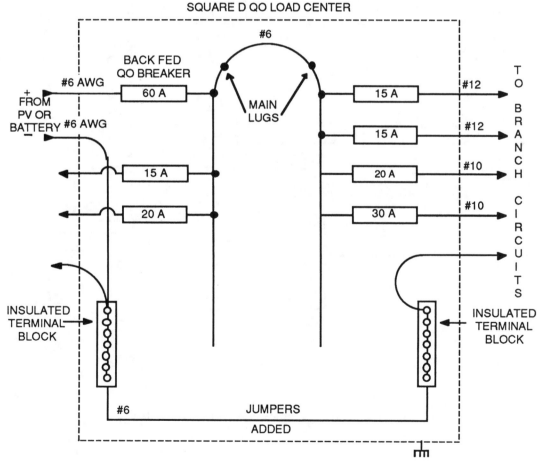

Figure 19. DC Load Center.

Another possibility is to use one of the phase circuits to combine separate PV source circuits, then go out of the load center through a breaker for the PV disconnect switch to the charge controller. Finally the conductors would be routed back to the other phase circuit in the load center for branch circuit distribution. Several options exist in using one and two-pole breakers for disconnects. Figure 20 demonstrates this system.

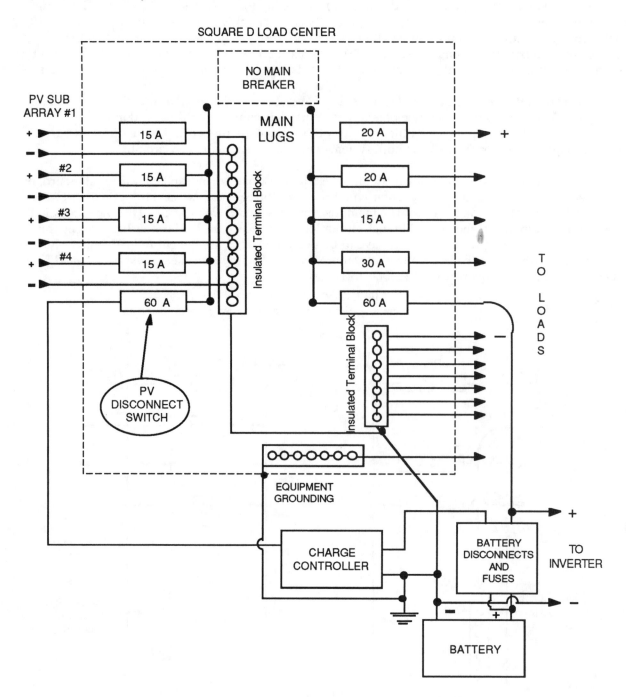

Figure 20. DC Combining Box and Load Center.

Interior Wiring and Receptacles

The interior wiring used in a PV system **must** comply with the *NEC*. Exposed single conductors are not permitted--they **must** be installed in conduit. Wires carrying the same current (i.e., positive and negative battery currents) **must** be installed in the same conduit or cable to prevent increased circuit inductances that would pose additional electrical stresses on disconnect and overcurrent devices [300-3(b)]. Nonmetallic sheathed cable may be used and it should be installed in the same manner as cable for ac branch circuits [300, 690-31a]. The bare grounding conductor in such cable **must** not be used to carry current and cannot be used as a common negative conductor for combination 12/24-volt systems [336-25].

The receptacles used for dc **must** be different than those used for any other service in the system [210-7f, 551-20f]. The receptacles should have a rating of not less than 15 amps and **must** be of the three-prong grounding type [210-7a, 720-6]. These requirements can be met in most locations by using the three-conductor 15-, 20-, or 30-amp 240-volt ac NEMA style 6-15, 6-20, 6-30 receptacles for the 12-volt dc outlets. If 24-volt dc is also used, the NEMA 125-volt locking connectors, style L5-15 or L5-20, are commonly available. The NEMA FSL-1 is a locking 30-amp 28-volt dc connector, but its availability cannot be determined. Numerous different styles of approved receptacles are available that meet this requirement. Cigarette lighter sockets and plugs frequently found on "PV" and "RV" appliances **do not** meet the requirements of the *National Electrical Code*.

It is not permissible to use the third or grounding conductor of a three-conductor plug or receptacle to carry common negative return currents on a combined 12/24-volt system. This terminal **must** be used for equipment grounding and may not carry current except in fault conditions [210-7].

Smoke Detectors

Many building codes require smoke and fire detectors to be wired directly into the ac power wiring of the dwelling. With a system that has no inverter, two solutions might be offered to the inspector. The first is to use the 9-volt or other primary-cell, battery-powered detector. The second is to use a voltage regulator to drop the PV system voltage to the 9-volt or other level required by the detector. The regulator **must** be able to withstand the PV open-circuit voltage and supply the current required by the detector alarm.

On inverter systems, the detector on some units may trigger the inverter into an "on" state, unnecessarily wasting power. In other units, the alarm may not draw enough current to turn the inverter on and thereby produce a reduced volume alarm or in some cases no alarm at all. Small dedicated power inverters might be used, but this seems a waste of power and reliability when the dc detectors are available.

Ground-Fault Circuit Interrupters

Some ac ground-fault circuit interrupters (GFCI) do not operate reliably on the output of some non sine wave inverters. If the GFCI does not function when tested, insure that the neutral (white-grounded) conductor of the inverter output is solidly grounded and connected to the grounding (green or bare) conductor of the inverter in the required manner. If this does not result in the GFCI testing properly, other options are possible. A direct measurement of an intentional ground fault may indicate that slightly more than the 5 milliamp internal test current is required to trip the GFCI. The inspector may accept this. Some inverters will work with a ferro resonant transformer to produce a waveform more satisfactory for use with GFCIs, but the no-load power consumption may be high enough to warrant a manual demand switch. A sine wave inverter could be used to power those circuits requiring GFCI protection.

Interior Switches

Switches rated for ac only **shall** not be used in dc circuits [380-14]. AC-DC general-use snap switches are available on special order from most electrical supply houses and they are similar in appearance to normal "quiet switches." *UL*-listed electronic switches with the proper dc voltage and current ratings might also be used, but the nonstandard appearance may require that the *UL*-listing specifications be provided to the inspectors.

There have been some failures of dc-rated snap switches when used as PV array and battery disconnect switches. If these switches are used on 12- and 24-volt systems and are not activated frequently, they may build up internal oxidation or corrosion and not function properly. Switches in these locations **must** be activated under load periodically to keep them clean.

SYSTEM LABELS AND WARNINGS
Photovoltaic Power Source

A permanent label **shall** be applied near the PV disconnect switch that contains the following information: [690-52]

- Operating Current (System maximum power current)
- Operating Voltage (System maximum power voltage)
- Open-Circuit Voltage
- Short-Circuit Current

Multiple Power Systems

Systems with multiple sources of power such as PV, gas generator, wind, hydro, etc., **shall** have diagrams and markings showing the interconnections [705-10].

Switch or Circuit Breaker

If a switch or circuit breaker might have its terminals energized in the open position, a label should be placed near it indicating: [690-17]

- WARNING - ELECTRIC SHOCK HAZARD - DO NOT TOUCH - TERMINALS ENERGIZED IN OPEN POSITION

General

Each piece of equipment that might be opened by unqualified persons should be marked with warning signs:

- WARNING - ELECTRIC SHOCK HAZARD - DANGEROUS VOLTAGES AND CURRENTS - NO USER SERVICEABLE PARTS INSIDE - CONTACT QUALIFIED SERVICE PERSONNEL FOR ASSISTANCE

Each battery container, box, or room should also have warning signs:

- WARNING - ELECTRIC SHOCK HAZARD - DANGEROUS VOLTAGES AND CURRENTS - EXPLOSIVE GAS - NO SPARKS OR FLAMES - NO SMOKING - ACID BURNS - WEAR PROTECTIVE CLOTHING WHEN SERVICING

APPENDIX A
Examples of Various PV Systems

These examples show some of the ways PV systems might be connected to meet the requirements of the *National Electrical Code*. The examples are presented for reference only and may require modification to meet site-specific and local jurisdiction requirements.

Figure A-1 shows a self-contained assembly with no external connections or user access to the electrical conductors. Such a unit might be an attic ventilation fan or a circulating pump on a domestic hot water system and might have *UL* certification as a unit. No disconnect switches or overcurrent devices would be required in a properly designed unit, although an on-off switch might be desired. This unit would also not have any exposed metal surfaces that might come in contact with internal live parts and therefore would not require a grounding conductor.

A system with external connections and wiring for a direct-drive load may or may not require disconnects and overcurrent protection. It depends on the design and the accessibility of live contacts.

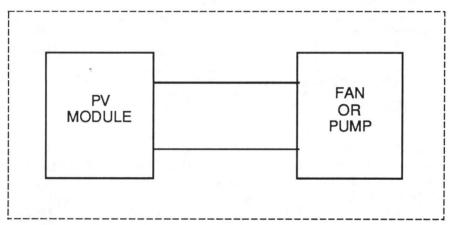

Figure A-1. Totally Self-Contained System.

It would be possible to design connectors that had no exposed live contacts for such a system, and the code-required disconnects could be managed by unplugging the connectors. If all components and conductors were sized to handle 125-150 percent of the PV short-circuit current, overcurrent devices might not be required. Such a system might look like Figure A-2.

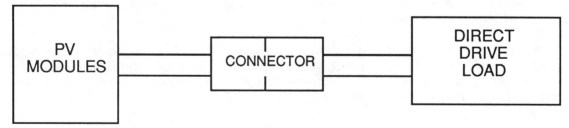

Figure A-2. Direct-Drive System--No Live Contacts.

As the system becomes more complex with battery storage, additional articles of the *NEC* will apply. Multiple component systems with modules, charge controllers, and batteries imply that external wiring will be needed and service will be required periodically. The *NEC* dictates the disconnects and various other protective devices. Figure A-3 illustrates some of the safety and performance requirements for a small stand-alone system.

In Figure B-3, the conductors are sized for the needed ampacity and to minimize voltage drop. The length of the wire and its resistance are calculated to keep potential short-circuit currents to a level that does not exceed the AIC of the circuit breakers. The battery bank will also be on the small side and most likely will be a sealed maintenance-free unit, further lowering the available short-circuit currents. Each circuit breaker might be replaced by a fuse and a switch, if care were

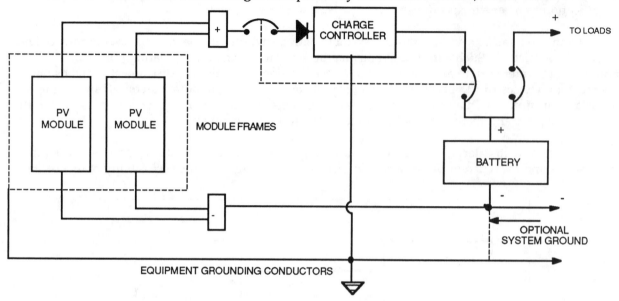

Figure A-3. Small System.

taken to insure that fuses could be serviced only when dead on both ends. Some charge controller designs may alleviate the need for the blocking diode. Note the method of connecting modules in parallel so one can be disconnected without disturbing the grounded conductor (if the system is grounded) of the other. Equipment grounds and three wire cables are needed to each load.

In medium-size systems where multiple strings of modules are connected in parallel, attention should be given to blocking and bypass diodes. Modules may or may not have internal bypass diodes to overcome problems caused by shaded cells, and the manufacturer's recommendations should be followed in this area. Blocking diodes not only prevent batteries from discharging into the array at night, but prevent parallel strings of modules from forcing current into shaded strings. Figure A-4 illustrates disconnects, overcurrent protection, short-circuit protection and diodes for a 1,500-watt stand-alone system. In very large arrays, blocking diodes might also be used at both ends of strings to prevent ground-fault currents from circulating.

Figure A-5 shows the use of blocking diodes on each end of long high-voltage strings of modules to prevent reverse biasing of the entire string when shaded. Normally only one diode would be used, but the use of one at each end of the string will serve to minimize the possibility of circulating ground fault currents should they occur in high-voltage arrays.

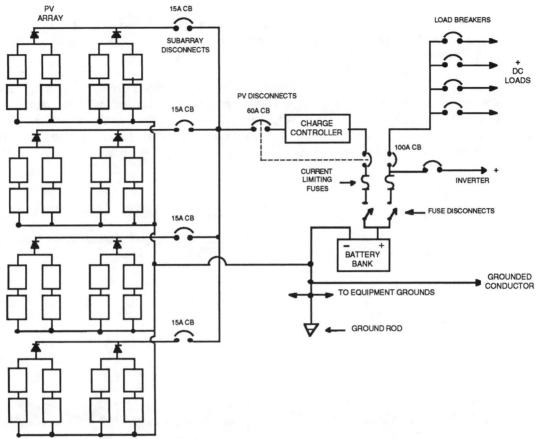

Figure A-4. Medium-Size Stand-Alone System.

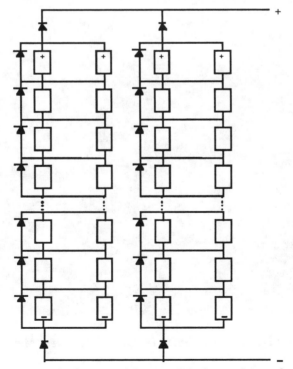

Figure A-5. Blocking and Bypass Diodes on Large Systems.

The cost of and the power lost in the blocking diodes **must** be weighed against the potential for damage if they are omitted. Bypass diodes protect individual modules or sets of parallel modules from the effects of shading.

On systems larger than about 2 kW and systems with array voltages greater than about 200 volts, careful attention **must** be given to system grounding, ground fault protection, and system disconnects. Tradeoffs **must** be made between cost, safety, component and system reliability, and component availability. These tradeoffs can only be made on a site by site basis and need to be made by an engineer experienced in dc power systems.

Southwest Technology Development Institute, Las Cruces, New Mexico

Lead-Acid Batteries for Home Power Storage

By Richard Perez, Home Power Magazine. (Note: All page references are to *Home Power Magazine*)

In 1970, we moved to the Mountains. The only desirable property we could afford was in the outback. Everything was many miles down a rough dirt road, far from civilized conveniences like electricity. We conquered the bad roads with a 4WD truck and countless hours of mechanical maintenance. The electrical power problem was not so easy to solve. We had to content ourselves with kerosene lighting and using hand tools. The best solution the marketplace could then offer was an engine driven generator. This required constant operation in order to supply power, in other words, expensive. It seemed that in America one either had power or one didn't.

We needed inexpensive home power. And we needed it to be there 24 hours a day without constantly running a noisy, gasoline eating, engine. At that time, NASA was about the only folks who could afford PVs. We started using lead-acid batteries to store the electricity produced by a small gas engine/generator. We'd withdraw energy from the batteries until they were empty and then refill them by running a lawnmower engine and car alternator. Since we stored enough energy to last about 4 days, we discharged and recharged the batteries about 100 times a year. Over years of this type service, we have learned much about lead-acid batteries— how they work and how to best use them. The following info has been hard won; we've made many expensive mistakes. We've also discovered how to efficiently and effectively coexist with the batteries that store our energy. Batteries are like many things in Life, mysterious until understood.

Before we can effectively communicate about batteries, we must share a common set of terms. Batteries and electricity, like many technical subjects, have their own particular jargon. Understanding these electrical terms is the first step to understanding your batteries.

Electrical Terms

Voltage. Voltage is electronic pressure. Electricity is electrons in motion. Voltage is the amount of pressure behind these electrons. Voltage is very similar to pressure in a water system. Consider a water hose. Water pressure forces the water through the hose. This situation is the same for an electron moving through a wire. A car uses 12 Volts, from a battery for starting. Commercial household power has a voltage of 120 volts. Batteries for renewable energy are usually assembled into packs of 12, 24, 32, or 48 volts.

Current. Current is the flow of electrons. The unit of electron flow in relation to time is called the Ampere. Consider the water hose analogy once again. If voltage is like water pressure, then current is like FLOW. Flow in water systems is measured in gallons per minute, while electron flow is measured in Amperes. A car tail light bulb consumes about 1 to 2 Amperes of electrical current. The headlights on a car consume about 8 Amperes each. The starter uses about 200 to 300 Amperes. Electrical current comes in two forms— direct current (DC) and alternating current (ac). In DC circuits the electrons flow in one direction ONLY. In ac circuits the electrons can flow in both directions. Regular household power is ac. Batteries store electrical power as direct current (DC).

Power. Power is the amount of energy that is being used or generated. The unit of power is the Watt. In the water hose analogy, power is can be compared to the total gallons of water transferred by the hose. Mathematically, power is the product of Voltage and Current. To find Power simply multiply Volts times Amperes. The amounts of power being used and generated determine the amount of energy that the battery must store.

Battery Terms

A Cell. The cell is the basic building block of all electrochemical batteries. The cell contains two active materials which react chemically to release free electrons (electrical energy). These active materials are usually solid and immersed in a liquid called the "electrolyte". The electrolyte is an electrically conductive liquid which acts as an electron transfer medium. In a lead acid cell, one of the active materials is lead dioxide (PbO_2) and forms the Positive pole (Anode) of the cell. The other active material is lead and forms the Negative pole (Cathode) of the cell. The lead acid cell uses an electrolyte composed of sulphuric acid (H_2SO_4).

During discharge, the cell's active materials undergo chemical reactions which release free electrons. These free electrons are available for our use at the cells electrical terminals or "poles". During discharge the actual chemical compositions of the active materials change. When all the active materials have undergone reaction, then the cell will produce no more free electrons. The cell is now completely discharged or in battery lingo, "dead".

Some cells, like the lead-acid cell, are rechargeable. This means that we can reverse the discharge chemical reaction by forcing electrons backwards through the cell. During the recharging process the active materials are gradually restored to their original, fully charged, chemical composition.

The voltage of an electrochemical cell is determined by the active materials used in its construction. The lead-acid cell develops a voltage of around 2 Volts DC. The voltage of a cell has no relationship to its physical size. All lead acid cells produce about 2 VDC regardless of size.

In the lead acid cell, the sulphuric acid electrolyte actually participates in the cell's electrochemical reaction. In most other battery technologies, like the nickel-cadmium cells, the electrolyte merely transfers electrons and does not change chemically as the cell discharges. In the lead-acid system, however, the electrolyte participates in the cell's reaction and the H_2SO_4 content of the electrolyte changes as the cell is discharged or charged. Typically the electrolyte in a fully charged cell is about 25% sulphuric acid with the remaining 75% being water. In the fully discharged lead-acid cell, the electrolyte is composed of less than 5% sulphuric acid with the remaining 95% being water. This happy fact allows us to determine how much energy a lead-acid cell contains by measuring the amount of acid remaining in its electrolyte.

A Battery. A battery is a group of electrochemical cells. Individual cells are collected into batteries to either increase the voltage or the electrical capacity of the resulting battery pack. For example, an automotive electrical system requires 12 VDC for operation. How is this accomplished with a basic 2 VDC lead-acid cell? The cells are wired together in series, this makes a battery that has the combined voltages of the cells. A 12 Volt lead-acid battery has six (6) cells, each wired anode to cathode (in series) to produce 12 VDC. Cells are combined in series for a voltage increase or in parallel for an electrical capacity increase. Figure 1 illustrates electrochemical cells assembled into batteries.

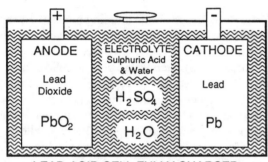

LEAD-ACID CELL FULLY CHARGED

$$PbO_2 + Pb + 2H_2SO_4 \xrightleftharpoons[\text{DISCHARGE}]{\text{CHARGE}} 2PbSO_4 + 2H_2O$$

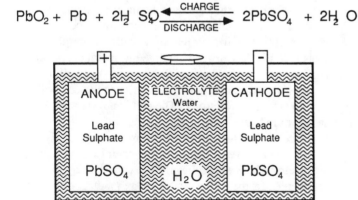

LEAD-ACID CELL FULLY DISCHARGED

FIGURE 1

Battery Capacity. Battery capacity is the amount of energy a battery contains. Battery capacity is usually rated in Ampere-hours (A-h) at a given voltage. Watt-hours (W-h) is another unit used to quantify battery capacity. While a single cell is limited in voltage by its materials, the electrical capacity of a cell is limited only by its size. The larger the cell, the more reactive materials contained within it, and the larger the electrical capacity of the cell in Ampere-hours.

A battery rated at 100 Ampere-hours will deliver 100 Amperes of current for 1 hour. It can also deliver 10 Amperes for 10 hours, or 1 Ampere for 100 hours. The average car battery has a capacity of about 60 Ampere-hours. Renewable energy battery packs contain from 350 to 4,900 Ampere-hours. The specified capacity of a battery pack is determined by two factors— how much energy is needed and how long the battery must supply this energy. Renewable energy systems work best with between 4 and 21 days of storage potential.

A battery is similar to a bucket. It will only contain so much electrical energy, just as the bucket will only contain so much water. The amount of capacity a battery has is roughly determined by its size and weight, just as a bucket's capacity is determined by its size. It is difficult to water a very large garden with one small bucket, it is also difficult to run a homestead on an undersized battery. If a battery based renewable energy system is to really work, it is essential that the battery have enough capacity to do the job. Undersized batteries are one of the major reasons that some folks are not happy with their renewable energy systems.

Battery capacity is a very important factor in sizing renewable energy systems. The size of the battery is determined by the amount of energy you need and how long you wish to go between battery rechargings. The capacity of the battery then determines the size of the charge source. Everything must be balanced if the system is to be efficient and long-lived.

State of Charge (SOC). A battery's state of charge is a percentage figure giving the amount of energy remaining in the battery. A 300 Ampere-hour battery at a 90% state of charge will contain 270 Amperes-hours of energy. At a 50% state of charge the same battery will contain 150 Ampere-hours. A battery which is discharged to a 20% or less state of charge is said to be "deep cycled". Shallow cycle service withdraws less than 10% of the battery's energy per cycle.

Lead-Acid Batteries. Lead-acid batteries are really the only type to consider for home energy storage at the present time. Other types of batteries, such as nickel-cadmium, are being made and sold, but they are simply too expensive to fit into low budget electrical schemes. We started out using car batteries.

Automotive Starting Batteries. The main thing we learned from using car batteries in deep cycle service is **DON'T.** Automotive starting batteries are not designed for deep cycle service; they don't last. Although they are cheap to buy, they are much more expensive to use over a period of several years. They wear out very quickly.

Car Battery Construction. The plates of a car battery are made from lead sponge. The idea is to expose the maximum plate surface area for chemical reaction. Using lead sponge makes the battery able to deliver high currents and still be as light and cheap as possible. The sponge type plates do not have the mechanical ruggedness necessary for repeated deep cycling over a period of many years. They simply crumble with age.

Car Battery Service. Car batteries are designed to provide up to 300 Amperes of current for very short periods of time (less than 10 seconds). After the car has started, the battery is then constantly trickle charged by the car's alternator. In car starting service, the battery is usually discharged less than 1% of its rated capacity. The car battery is designed for this very shallow cycle service.

Car Battery Life Expectancy & Cost. Our experience has shown us that automobile starting batteries last about 200 cycles in deep cycle service. This is a very short period of time, usually less than 2 years. Due to their short lifespan in home energy systems, they are more than 3 times as expensive to use as a true deep cycle battery. Car batteries cost around $60. for 100 Ampere-hours at 12 volts.

Beware of Ersatz "Deep Cycle" Batteries. After the failure of the car batteries we tried the so called "deep cycle" type offered to us by our local battery shop. These turned out to be warmed over car batteries and lasted about 400 cycles. They were slightly more expensive, $100. for 105 Ampere-hours at 12 volts. You can spot these imitation deep cycle batteries by their small size and light weight. They are cased with automotive type cases. Their plates are indeed more rugged than the car battery, but still not tough enough for the long haul.

True "Deep Cycle" Batteries. After many battery failures and much time in the dark, we finally tried a real deep cycle battery. These batteries were hard to find; we had to have them shipped in as they were not available locally. In fact, the local battery shops didn't seem to know they existed. Although deep cycle types use the same chemical reactions to store energy as the car battery, they are very differently made.

Deep Cycle Physical Construction. The plates of a real deep cycle battery are made of scored sheet lead. These plates are many times thicker than the plates in car batteries, and they are solid lead, not sponge lead. This lead is alloyed with up to 16% antimony to make the plates harder and more durable. The cell cases are large; a typical deep cycle battery is over 3 times the size of a car battery. Deep cycle batteries weigh between 120 and 400 pounds. We tried the Trojan L-16W. This is a 6 Volt 350 Ampere-hour battery, made by Trojan Batteries Inc., 1395 Evans Ave., San Francisco, CA (415) 826-2600. The L-16W weighs 125 pounds and contains over 9 quarts of sulphuric acid. The "W" designates a Wrapping of the plates with perforated nylon socks. Wrapping, in our experience, adds years to the battery's longevity. We wired 2 of the L-16Ws in series to give us 12 Volts at 350 Ampere-hours.

Deep Cycle Service. The deep cycle battery is designed to have 80% of its capacity withdrawn repeatedly over many cycles. They are optimized for longevity. If you are using battery stored energy for your home, this is the only type of lead-acid battery to use. Deep cycle batteries are also used for motive power. In fact, many more are used in forklifts than in renewable energy systems.

Deep Cycle Life Expectancy & Cost. A deep cycle battery will last at least 5 years. In many cases, batteries last over 10 years and give over 1,500 deep cycles. In order to get maximum longevity from the deep cycle battery, it must be cycled properly. All chemical batteries can be ruined very quickly if they are improperly used. A 12 Volt 350 Ampere-hour battery costs around $440. Shipping can be expensive on these batteries. They are corrosive and heavy, and must be shipped motor freight.

Deep Cycle Performance. The more we understood our batteries, the better use we made of them. This information applies to high antimony, lead-acid deep cycle batteries used in homestead renewable energy service. In order to relate to your system you will need a voltmeter. An accurate voltmeter is the best source of information about our battery's performance. It is essential for answering the two basic questions of battery operation— when to charge and when to stop charging.

Voltage vs. Current. The battery's voltage depends on many factors. One is the rate, in relation to the battery's capacity, that energy is either being withdrawn from or added to the battery. The faster we discharge the battery, the lower its voltage becomes. The faster we recharge it, the higher its voltage gets. Try an experiment- hook the voltmeter to a battery and measure its voltage. Turn on some lights or add other loads to the battery. You'll see the voltage of the battery is lowered by powering the loads. This is perfectly normal and is caused by the nature of the lead-sulphuric acid electrochemical reaction. In homestead service this factor means high powered loads need large batteries. Trying to run large loads on a small capacity battery will result in very low voltage. The low voltage can ruin motors and dim lights.

Voltage vs. State of Charge. The voltage of a lead-acid battery gives a readout of how much energy is available from the battery. Figure 2, on page 26, illustrates the relationship between the battery's state of charge and its voltage for various charge and discharge rates. This graph and its companion, Fig.3, are placed in the center of the magazine as a tearout so you can put them on your wall if you wish. This graph is based on a 12 Volt lead-acid battery at room temperature. Simply multiply the voltage figures by 2 for a 24 Volt system, and by 4 for a 48 Volt system. This graph assumes that the battery is at room temperature 78°F. Use the C/100 discharge rate curve for batteries at rest (i.e. not under charge or discharge).

Temperature. The lead-acid battery's chemical reaction is sensitive to temperature. See the graph, Figure 3 on page 25, which shows the same info as Figure 2, but for COLD lead-acid batteries. Note the voltage depression under discharge and the voltage elevation under charge. The chemical reaction is very sluggish at cold temperatures. Battery efficiency and usable capacity drop radically at temperatures below 40° F. At 40°F., a lead acid battery has effectively lost about 20% of its capacity at 78°F. At 0°F., the same battery will have effectively lost 45% of its room temperature capacity. We keep our batteries inside, where we can keep them warm in the winter. Batteries banished to the woodshed or unheated garage will not perform well in the

12 Volt Lead-Acid Battery Chart- 78°F.

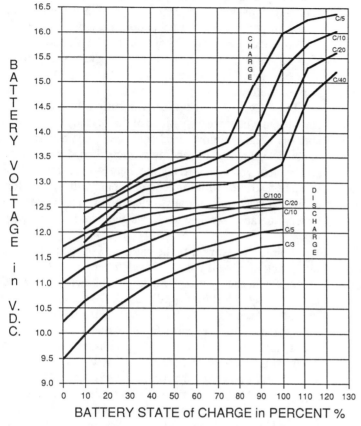

BATTERY VOLTAGE in V.D.C.

BATTERY STATE of CHARGE in PERCENT %

12 Volt Lead-Acid Battery Chart- 34°F.

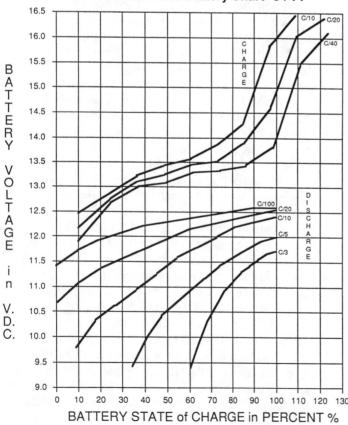

BATTERY VOLTAGE in V.D.C.

BATTERY STATE of CHARGE in PERCENT %

winter. They will be more expensive to use and will not last as long. The best operating temperature is around 78° F..

The situation with temperature is further complicated by the lead-acid system's electrolyte. As the battery discharges the electrolyte loses its sulphuric acid and becomes mostly (≈95%) water. IT WILL FREEZE. Freezing usually ruptures the cell's cases and destroys the plates. Lead-acid batteries at < 20% SOC will freeze at around 18°F. If you're running lead acid batteries at low temperatures, then keep them fully charged to prevent freezing on very cold nights.

Determining State of Charge with a Hydrometer. A hydrometer is a device that measures the density of a liquid in comparison with the density of water. The density of the sulphuric acid electrolyte in the battery is an accurate indicator of the battery's state of charge. The electrolyte has greater density at greater states of charge. We prefer to use the battery's voltage as an indicator rather than opening the cells and measuring the electrolyte's specific gravity. Every time a cell is opened there is a chance for contamination of the cell's inards. Lead- acid batteries are chemical machines. If their cells are contaminated with dirt, dust, or other foreign material, then the cell's life and efficiency is greatly reduced. If you insist on using a hydrometer, make sure it is spotlessly clean and temperature compensated. Wash it in distilled water before and after measurements.

Rates of Charge/Discharge. Rates of charge and discharge are figures that tell us how fast we are either adding or removing energy from the battery. In actual use, this rate is a current measured in Amperes. Say we wish to use 50 Amperes of current to run a motor. This is quite a large load for a small 100 Ampere-hour battery. If the battery had a capacity of 2,000 Ampere-hours, then the load of 50 Amperes is a small load. It is difficult to talk about currents through batteries in terms of absolute Amperes of current. Battery people talk about these currents in relation to the battery's capacity.

Rates of charge and discharge are expressed as ratios of the battery's capacity in relation to time. Rate (of charge or discharge) is equal to the battery's capacity in Ampere-hours divided by the time in hours it takes to cycle the battery. If a completely discharged battery is totally filled in a 10 hour period, this is called a C/10 rate. C is the capacity of the battery in Ampere-hours and 10 is the number of hours it took for the complete cycle. This capacity figure is left unspecified so that we can use the information with any size battery pack.

For example, consider a 350 Ampere-hour battery. A C/ 10 rate of charge or discharge is 35 Amperes. A C/20 rate of charge or discharge is 17.5 Amperes. And so on... Now consider a 1,400 Ampere-hour battery. A C/10 rate here is 140 Amperes, while a C/20 rate is 70 Amperes. Note that the C/10 rate is different for the two different batteries; this is due to their different capacities. Battery people do this not to be confusing, but so we can all talk in the same terms, regardless of the capacity (size) of the battery under discussion.

Let's look at the charge rate first. For a number of technical reasons, it is most efficient to charge deep cycle lead-acid batteries at rates between C/10 and C/20. This means that the fully discharged battery pack is totally recharged in a 10 to 20 hour period. If the battery is recharged faster, say in 5 hours (C/5), then much more electrical energy will be lost as heat. Heating the battery's plates during charging causes them to undergo mechanical stress. This stress breaks down the plates. Deep cycle lead-acid batteries which are continually recharged at rates faster than C/10 will have shortened lifetimes. The best overall charging rate for deep cycle lead-acid batteries is the C/20 rate. The C/20 charge rate assures good efficiency and longevity by reducing plate stress. A battery should be completely refilled each time it is cycled. This yields maximum battery life by making **all** the active materials participate in the chemical reaction.

We often wish to determine a battery's state of charge while it is actually under charge or discharge. Figure 2, on page 25, illustrates the battery's state of charge in relation to its voltage for several charge/discharge rates. This graph is based on a 12 Volt battery pack at room temperature. For instance, if we are charging at the C/20 rate, then the battery is full when it reaches 14.0 volts. Once again the digital voltmeter is used to determine state of charge without opening the cells and risking contamination. Figure 3, on page 24, offers the same information as Figure 2, but in Figure 3 the information pertains to a lead-acid battery at 34°F. Note the depression of voltage under discharge and the voltage elevation under charge. This reflects an actual change in the batteries internal resistance to electrical flow. The colder the battery becomes, the higher its internal resistance gets, and the more radical the voltage swings under charge and discharge become.

The Equalizing Charge. After several months, the individual cells that make up the battery may differ in their states of charge. Voltage differences greater than 0.05 volts between the cells indicate it is time to equalize the individual cells. In order to do this, the battery is given an equalizing charge. **An equalizing charge is a controlled overcharge of an already full battery. Simply continue the recharging process at the C/20 rate for 5 to 7 hours after the battery is full.** Batteries should be equalized every 5 cycles or every 3 months, whichever comes first. Equalization is the best way to increase deep cycle lead-acid battery life. Battery voltage during the equalizing charge may go as high as 16.5 volts, especially if the battery's temperature is < 40°F.. This voltage is too high for many 12 Volt electronic appliances. Be sure to turn off all voltage sensitive gear while running an equalizing charge.

Wind machines and solar cells are not able to recharge the batteries at will. They are dependent on Mama Nature for energy input. We have found that most renewable energy systems need some form of backup power. The engine/generator can provide energy when the renewable energy source is not operating. The engine/generator can also supply the steady energy necessary for complete battery recharging and equalizing charges. The addition of an engine/generator also reduces the amount of battery capacity needed. Wind and solar sources need larger battery capacity to offset their

intermittent nature. Home Power #2 discusses homebuilding a very efficient and supercheap 12 Volt DC source from a lawnmower motor and a car alternator.

Self-Discharge Rate vs. Temperature. All lead-acid batteries, regardless of type, will discharge themselves over a period of time. This energy is lost within the battery; it is not available for our use. The rate of self-discharge depends primarily on the battery's temperature. If the battery is stored at temperatures above 120° F., it will totally discharge itself in 4 weeks. At room temperatures, the battery will lose about 6% of its capacity weekly and be discharged in about 16 weeks. The rate of self-discharge increases with the battery's age. Due to self-discharge, it is not efficient to store energy in lead-acid batteries for periods longer than 3 weeks. Yes, it is possible to have too many batteries. If you're not cycling your batteries at least every 3 weeks, then you're wasting energy.If an active battery is to be stored, make sure it is first fully recharged and then put it in a cool place. Temperatures around 35° F. to 40° F. are ideal for inactive battery storage. The low temperature slows the rate of self-discharge. Be sure to warm the battery up and recharge it before using it.

Battery Capacity vs. Age. All batteries gradually lose some of their capacity as they age. When a battery manufacturer says his batteries are good for 5 years, he means that the battery will hold 80% of its original capacity after 5 years of proper service. Too rapid charging or discharging, cell contamination, and undercharging are examples of improper service which will greatly shorten any battery's life. Due to the delicate nature of chemical batteries most manufacturers do not guarantee them for long periods of time. On a brighter note, we have discovered that batteries which are treated with tender love and care can last twice as long as the manufacturer's claim they will. If you're using batteries, it really pays to know how to treat them.

Battery Cables. The size, length, and general condition of your battery cables are critical for proper performance. While the battery may have plenty of power to deliver, it can't deliver it effectively through undersized, too long or funky wiring. Battery (and especially inverter) cables should be made of large diameter copper cable with permanent soldered connectors. The acid environment surrounding lead-acid system plays hell with any and all connections. Connectors which are mechanically crimped to the wire are not acceptable for battery connection. The acid gradually works its way into the mechanical joint resulting in corrosion and high electrical resistance. See Home Power #7, page 36, for complete instructions on home made, low loss, soldered connectors and cables.

Battery Safety. Location plays a great part in battery safety. A battery room or shed, securely locked & properly ventilated, is a very good idea. Children, pets, and anyone not aware of the danger should never be allowed access to battery areas. Lead-acid batteries contain sulphuric acid, and lots of it. For example, a medium sized battery of 12 VDC at 1,400 A-h will contain some 18 gallons of nasty, corrosive, dangerous acid. Such a battery pack is capable of delivering over

4,000 Amperes of 12 VDC for short periods. Direct shorts across the battery can arc weld tools and instantly cause severe burns to anyone holding the tool. Be careful when handling wrenches or any metallic object around batteries. If tools make contact across the batteries electrical terminals, the results can be instantly disastrous.

When a lead-acid battery is almost full and undergoing recharging, the cell's produce gasses. These gasses are mostly oxygen and hydrogen- a potentially explosive mixture. Battery areas should be well ventilated during recharging and especially during equalizing charges to dissipate the gasses produced. If a blower is used in ventilation, make sure that it employs a "sparkless motor". See Home Power #6, page 31, for specific info on venting lead-acid batteries.

Battery Maintenance. There is more to battery care than keeping their tops clean. Maintenance begins with proper cycling. The two basic decisions are when to charge and when to stop charging. Begin to recharge the battery when it reaches a 20% state of charge or before. Recharge it until it is completely full. Both these decisions can be made via voltage measurement, amperage measurement and the information in Figures 2 and 3.

1. Don't discharge a deep cycle battery greater than 80% of its capacity.

2. When you recharge it, use a rate between C/10 and C/20.

3. When you recharge it, fill it all the way up.

4. Keep the battery as close to room temperature as possible.

5. Use only distilled water to replenish lost electrolyte.

6. Size the battery pack with enough capacity to last between 4 to 21 days. This assures proper rates of discharge.

7. Run an equalizing charge every 5 charges or every 3 months, whichever comes first.

8. Keep all batteries and their connections clean and corrision free.

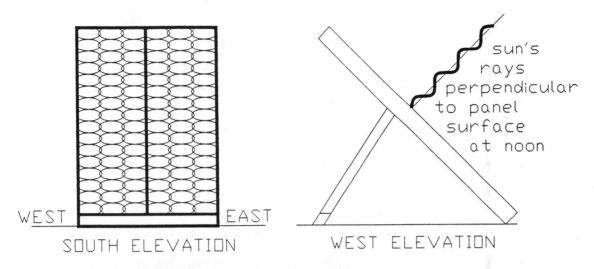

WEST EAST

SOUTH ELEVATION

WEST ELEVATION

sun's rays perpendicular to panel surface at noon

Solar Panel Alignment for the Northern Hemisphere
(Direction is reversed for the Southern Hemisphere)
A rough estimate would add 15° to your latitude for winter
and subtract 15° from your latitude for summer

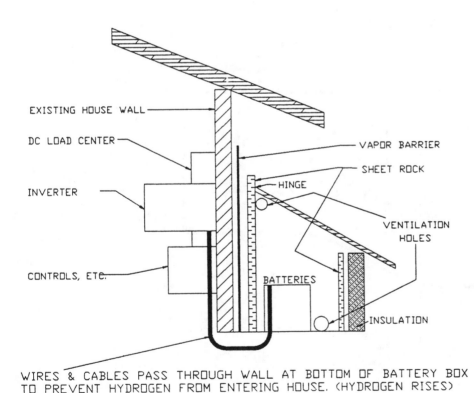

EXISTING HOUSE WALL

DC LOAD CENTER

INVERTER

CONTROLS, ETC.

VAPOR BARRIER

SHEET ROCK

HINGE

VENTILATION HOLES

BATTERIES

INSULATION

WIRES & CABLES PASS THROUGH WALL AT BOTTOM OF BATTERY BOX
TO PREVENT HYDROGEN FROM ENTERING HOUSE. (HYDROGEN RISES)

Recommended Battery Enclosure

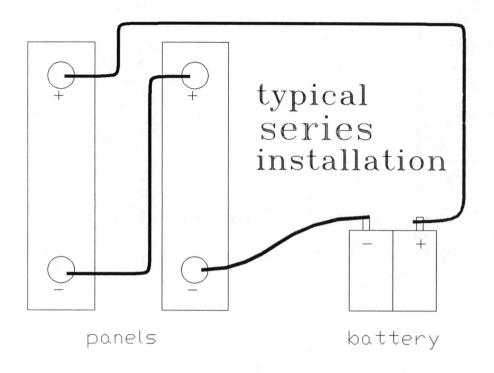

typical
series
installation

panels   battery

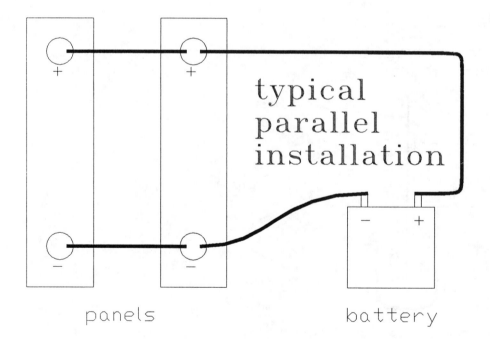

typical
parallel
installation

panels   battery

Maximum One-Way Distance
for Less than 2% Voltage Drop

12V Circuits, Distance x10 for 120V

Amps	Watts @ 12v	#14	#12	#10	#8	#6	#4	#2	1/0	2/0	3/0
1	12	45.0	70.0	115	180	290	456	720	-	-	-
2	24	22.5	35.0	57.5	90.0	145	228	360	580	720	912
4	48	10.0	17.5	27.5	45.0	72.5	114	180	290	360	456
6	72	07.5	12.0	17.5	30.0	47.5	75.0	120	193	243	305
8	96	05.5	08.5	11.5`	22.5	35.5	57.0	90.0	145	180	228
10	120	04.5	07.0	11.5	18.0	28.5	45.5	72.5	115	145	183
15	180	03.0	04.5	07.0	12.0	19.0	30.0	48.0	76.5	96.0	122
20	240	02.0	03.5	05.5	09.0	14.5	22.5	36.0	57.5	72.5	91
25	300	01.8	02.8	04.5	07.0	11.5	18.0	29.0	46.0	58.0	73
30	360	01.5	02.4	03.5	06.0	09.5	15.0	24.0	38.5	48.5	61
40	480	-	-	02.8	04.5	07.0	11.5	18.0	29.0	36.0	45
50	600	-	-	02.3	03.6	05.5	09.0	14.5	23.0	29.0	36
100		-	-	-	-	02.9	04.6	07.2	11.5	14.5	18
150		-	-	-	-	-	-	04.8	07.7	09.7	12
200		-	-	-	-	-	-	03.6	05.8	07.3	09

24V Circuits

Amps	Watts @ 24v	#14	#12	#10	#8	#6	#4	#2	1/0	2/0	3/0
1	24	90	142	226	360	573	911	-	-	-	-
2	48	45	71	113	180	286	455	724	-	-	-
4	96	22	36	57	90	143	228	362	576	726	915
6	144	15	24	38	60	95	152	241	384	484	610
8	192	11	18	28	45	72	114	181	288	363	458
10	240	9	14	23	36	57	91	145	230	290	366
15	360	6	10	15	24	38	61	97	154	194	244
20	480	-	7	11	18	29	46	72	115	145	183
25	600	-	-	9	14	23	36	58	92	116	146
30	720	-	-	8	12	19	30	48	77	97	122
40	960	-	-	-	9	14	23	36	58	73	92
50	1200	-	-	-	-	11	18	29	46	58	73
100	2400	-	-	-	-	5.7	9.1	14.5	23	29	36.6
150	3600	-	-	-	-	-	-	9.7	15.4	19.4	24.4
200	4800	-	-	-	-	-	-	7.2	11.5	14.5	18.3

Mean Daily Solar Radiation

An inspection of the U.S. solar map will show roughly how many hours of sun one can expect per day when averaged over the entire year. This can be a dangerous way to size a photovoltaic system, however, as the winter weather can radically skew the average, and result in a system grossly oversized for the summer yet still undersized for the winter.

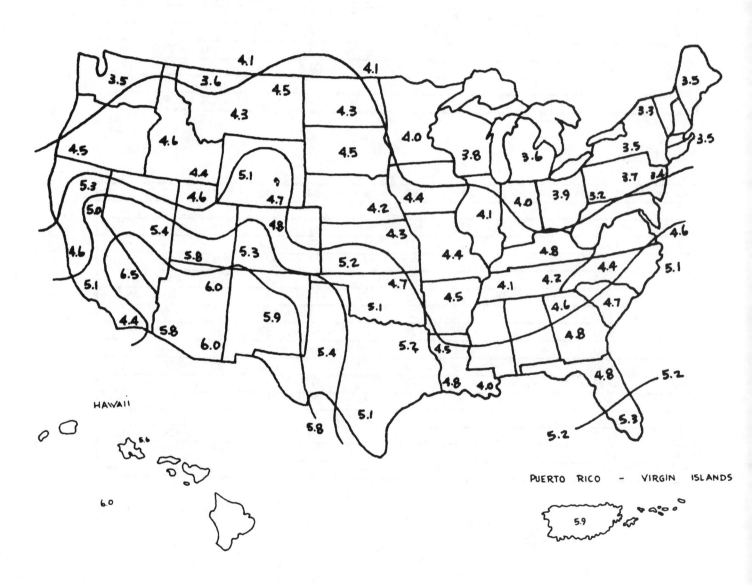

REAL GOODS

Magnetic Declinations in the United States

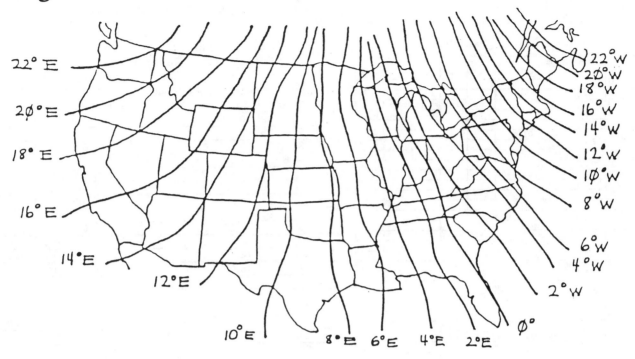

Figure indicates correction of compass reading to find true north. For example, in Washington state when your compass reads 22°E, it is pointing due north.

Maximum Number of Conductors for a Given Conduit Size

	Conduit size	½"	¾"	1"	1¼"	1½"	2"
Conductor size	#12	10	18	29	51	70	114
	#10	6	11	18	32	44	73
	#8	3	5	9	16	22	36
	#6	1	4	6	11	15	26
	#4	1	2	4	7	9	16
	#2	1	1	3	5	7	11
	#1		1	1	3	5	8
	#1/0		1	1	3	4	7
	#2/0		1	1	2	3	6
	#3/0		1	1	1	3	5
	#4/0		1	1	1	2	4

Friction Loss Charts for Water Pumping

Friction Loss- PVC Class 160 PSI Plastic Pipe
Pressure loss from friction in psi per 100 feet of pipe.

NOMINAL PIPE DIAMETER IN INCHES

Bold Numbers Indicate 5 Feet per Second Velocity

Flow GPM	1	1.25	1.5	2	2.5	3	4	5	6	8	10
1	0.02	0.01									
2	0.06	0.02	0.01								
3	0.14	0.04	0.02								
4	0.23	0.07	0.04	0.01							
5	0.35	0.11	0.05	0.02							
6	0.49	0.15	0.08	0.03	0.01						
7	0.66	0.20	0.10	0.03	0.01						
8	0.84	0.25	0.13	0.04	0.02						
9	1.05	0.31	0.16	0.05	0.02						
10	1.27	0.38	0.20	0.07	0.03	0.01					
11	1.52	0.45	0.23	0.08	0.03	0.01					
12	1.78	0.53	0.28	0.09	0.04	0.01					
14	2.37	0.71	0.37	0.12	0.05	0.02					
16	**3.04**	0.91	0.47	0.16	0.06	0.02					
18	3.78	1.13	0.58	0.20	0.08	0.03					
20	4.59	1.37	0.71	0.24	0.09	0.04	0.01				
22	5.48	1.64	0.85	0.29	0.11	0.04	0.01				
24	6.44	1.92	1.00	0.34	0.13	0.05	0.02				
26	7.47	2.23	1.15	0.39	0.15	0.06	0.02				
28	8.57	**2.56**	1.32	0.45	0.18	0.07	0.02				
30	9.74	2.91	1.50	0.51	0.20	0.08	0.02				
35		3.87	**2.00**	0.68	0.27	0.10	0.03				
40		4.95	2.56	0.86	0.34	0.13	0.04	0.01			
45		6.16	3.19	1.08	0.42	0.16	0.05	0.02			
50		7.49	3.88	1.31	0.52	0.20	0.06	0.02			
55		8.93	4.62	**1.56**	0.62	0.24	0.07	0.02			
60		10.49	5.43	1.83	0.72	0.28	0.08	0.03	0.01		
65			6.30	2.12	0.84	0.32	0.09	0.03	0.01		
70			7.23	2.44	0.96	0.37	0.11	0.04	0.02		
75			8.21	2.77	1.09	0.42	0.12	0.04	0.02		
80			9.25	3.12	1.23	0.47	0.14	0.05	0.02		
85			10.35	3.49	**1.38**	0.53	0.16	0.06	0.02		
90				3.88	1.53	0.59	0.17	0.06	0.03		
95				4.29	1.69	0.65	0.19	0.07	0.03		
100				4.72	1.86	**0.72**	0.21	0.08	0.03	0.01	
150				10.00	3.94	1.52	0.45	0.16	0.07	0.02	
200					6.72	2.59	**0.76**	0.27	0.12	0.03	0.01
250					10.16	3.91	1.15	0.41	0.18	0.05	0.02
300						5.49	1.61	**0.58**	0.25	0.07	0.02
350						7.30	2.15	0.77	0.33	0.09	0.03
400						9.35	2.75	0.98	0.42	0.12	0.04
450							3.42	1.22	**0.52**	0.14	0.05
500							4.15	1.48	0.63	0.18	0.06
550							4.96	1.77	0.76	0.21	0.07
600							5.82	2.08	0.89	0.25	0.08
650							6.75	2.41	1.03	0.29	0.10
700							7.75	2.77	1.18	0.33	0.11
750							8.80	3.14	1.34	**0.37**	0.13
800								3.54	1.51	0.42	0.14
850								3.96	1.69	0.47	0.16
900								4.41	1.88	0.52	0.18
950								4.87	2.08	0.58	0.20
1000								5.36	2.29	0.63	0.22
1500									4.84	1.34	0.46
2000										2.29	0.78
2500										3.46	1.18
3000											1.66

Friction Loss- Polyethylene (PE) SDR-Pressure Rated Pipe
Pressure loss from friction in psi per 100 feet of pipe.

NOMINAL PIPE DIAMETER IN INCHES

Numbers in Bold Indicate 5 Feet/Second Velocity

Flow GPM	0.5	0.75	1	1.25	1.5	2	2.5	3
1	0.49	0.12	0.04	0.01				
2	1.76	0.45	0.14	0.04	0.02			
3	3.73	0.95	0.29	0.08	0.04	0.01		
4	**6.35**	1.62	0.50	0.13	0.06	0.02		
5	9.60	2.44	0.76	0.20	0.09	0.03		
6	13.46	3.43	1.06	0.28	0.13	0.04	0.02	
7	17.91	4.56	1.41	0.37	0.18	0.05	0.02	
8	22.93	**5.84**	1.80	0.47	0.22	0.07	0.03	
9		7.26	2.24	0.59	0.28	0.08	0.03	
10		8.82	2.73	0.72	0.34	0.10	0.04	0.01
12		12.37	**3.82**	1.01	0.48	0.14	0.06	0.02
14		16.46	5.08	1.34	0.63	0.19	0.08	0.03
16			6.51	1.71	0.81	0.24	0.10	0.04
18			8.10	2.13	1.01	0.30	0.13	0.04
20			9.84	2.59	1.22	0.36	0.15	0.05
22			11.74	**3.09**	1.46	0.43	0.18	0.06
24			13.79	3.63	1.72	0.51	0.21	0.07
26			16.00	4.21	1.99	0.59	0.25	0.09
28				4.83	2.28	0.68	0.29	0.10
30				5.49	**2.59**	0.77	0.32	0.11
35				7.31	3.45	1.02	0.43	0.15
40				9.36	4.42	1.31	0.55	0.19
45				11.64	5.50	1.63	0.69	0.24
50				14.14	6.68	**1.98**	0.83	0.29
55					7.97	2.36	0.85	0.35
60					9.36	2.78	1.17	0.41
65					10.36	3.22	1.36	0.47
70					12.46	3.69	**1.56**	0.54
75					14.16	4.20	1.77	0.61
80						4.73	1.99	0.69
85						5.29	2.23	0.77
90						5.88	2.48	0.86
95						6.50	2.74	0.95
100						7.15	3.01	**1.05**
150						15.15	6.38	2.22
200							10.87	3.78
300								8.01

To understand how to use Table I, assume you have a family of four persons. There will be some lawn and garden watering. There is a small swimming pool. There are 30 head of dairy cattle, 40 hogs, and 1,000 laying hens. You would figure needs as follows:

TABLE I. APPROXIMATE DAILY WATER NEEDS FOR HOME AND FARM[19]

	Water Consumption Per Day (gallons)
HOME	
For kitchen and laundry use (including automatic equipment), bathing, sanitary use and other uses inside the home	100 per person
For swimming pool maintenance, per 100 sq. ft.	30
LAWN AND GARDEN	
For lawn sprinkling per 1,000 sq. ft., per sprinkling	600 (approx. 1 in.)
For garden sprinkling, per 1,000 sq. ft., per sprinkling	600 (approx. 1 in.)
FARM (maximum needs)	
Dairy cows (14-15,000 pounds milk) Average drinking rate	20 per head
Dry cows or heifers	15 per head
Calves	7 per head
Beef, yearlings, full feed, 90°F	20 per head
Beef, brood cows	12 per head
Sheep or goats	2 per head
Horses or mules	12 per head
Swine finishing	4 per head
Brood sows, nursing	6 per head
Laying hens (90°F)	9 per 100 birds
Broilers (over 100°F)	6 per 100 birds
Turkeys—15-19 weeks (over 100°F)	20-25 per 100 birds
Ducks*	22 per 100 birds
Dairy sanitation—milk room & milking parlor	500 per day
Flushing floors	10 per 100 sq. ft.
Sanitary hog wallow	100 per day

*Studies on water consumption of ducks were not available. The figure is based on rule-of-thumb method of multiplying amount of feed consumed per day by two. This method is sometimes used for other fowls.

Home

4 persons, 100 gal. per person	400 gal. per day
5,000 sq. ft. of lawn to be sprinkled (5 x 600 gal. per 1,000 sq. ft.)	3,000 gal. per day
1,000 sq. ft. of garden	600 gal. per day
200 sq. ft. of swimming pool (2 x 30)	60 gal. per day

Farm

30 dairy cattle, 20 gal. each	600 gal. per day
Dairy sanitation	500 gal. per day
40 hogs, 4 gal. each	160 gal. per day
1,000 laying hens, 9 gal. per 100 birds	90 gal. per day
Probable maximum water needs—total	5,410 gal. per day

To determine your own total daily water needs, use the table in the same way.

You can see from the example how lawn and garden sprinkling can greatly increase the total water needs for any one day. But, there is a good chance you will do your lawn and garden sprinkling over a two- or three-day period. If over a three-day period, this could lower the water needs per day for lawn and garden sprinkling from 3,000 gallons to 1,000 gallons. This would lower the overall daily needs from 5,410 gallons to 3,410 gallons.

With this information, you can now check your water source(s) to see if there is enough water to meet your maximum daily water needs.

Courtesy of Planning for an Individual Water System, 4th Edition See page 413 of our book chapter.

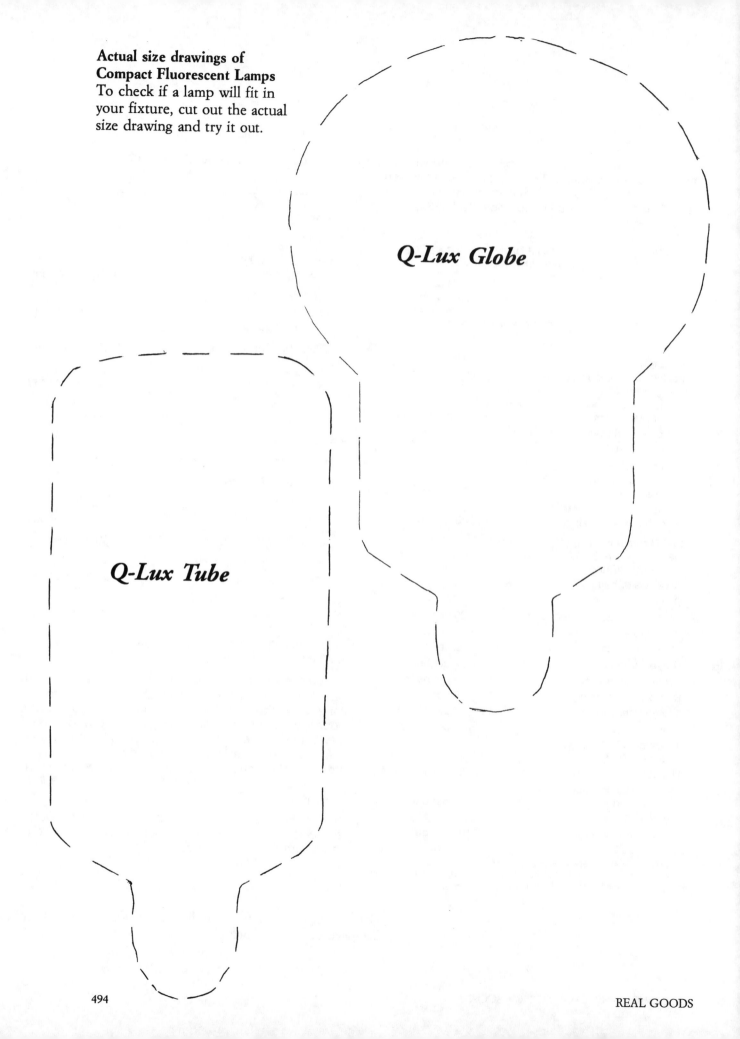

Actual size drawings of Compact Fluorescent Lamps
To check if a lamp will fit in your fixture, cut out the actual size drawing and try it out.

Q-Lux Globe

Q-Lux Tube

REAL GOODS

**Actual size drawings of
Compact Fluorescent Lamps**
To check if a lamp will fit in
your fixture, cut out the actual
size drawing and try it out.

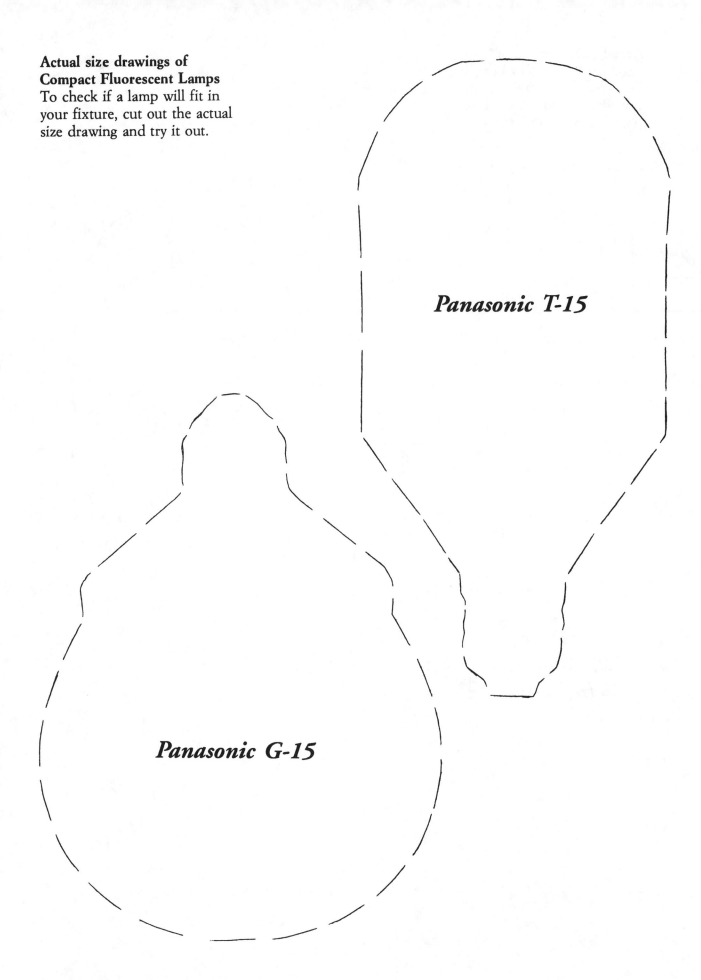

Panasonic T-15

Panasonic G-15

**Actual size drawings of
Compact Fluorescent Lamps**
To check if a lamp will fit in
your fixture, cut out the actual
size drawing and try it out.

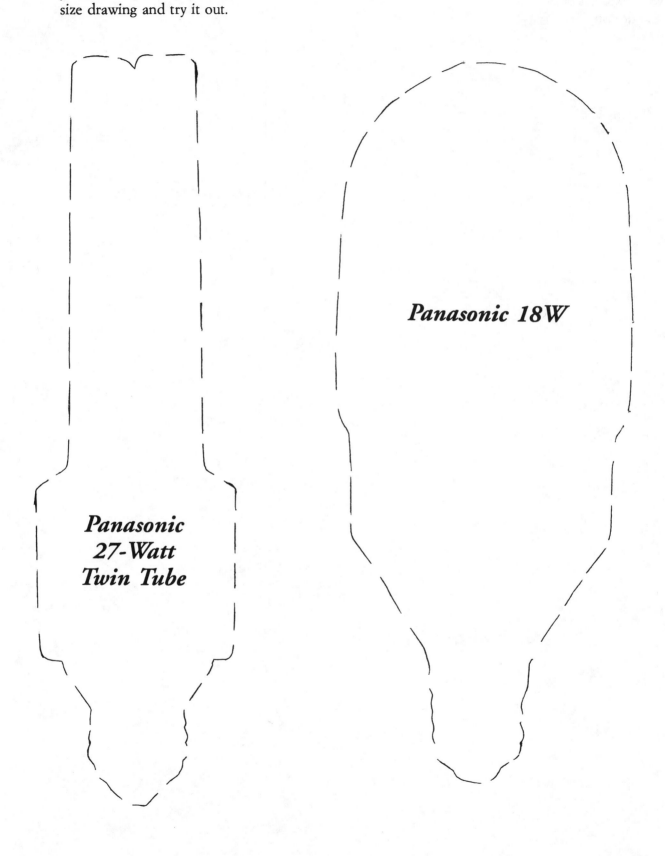

*Panasonic
27-Watt
Twin Tube*

Panasonic 18W

**Actual size drawings of
Compact Fluorescent Lamps**
To check if a lamp will fit in
your fixture, cut out the actual
size drawing and try it out.

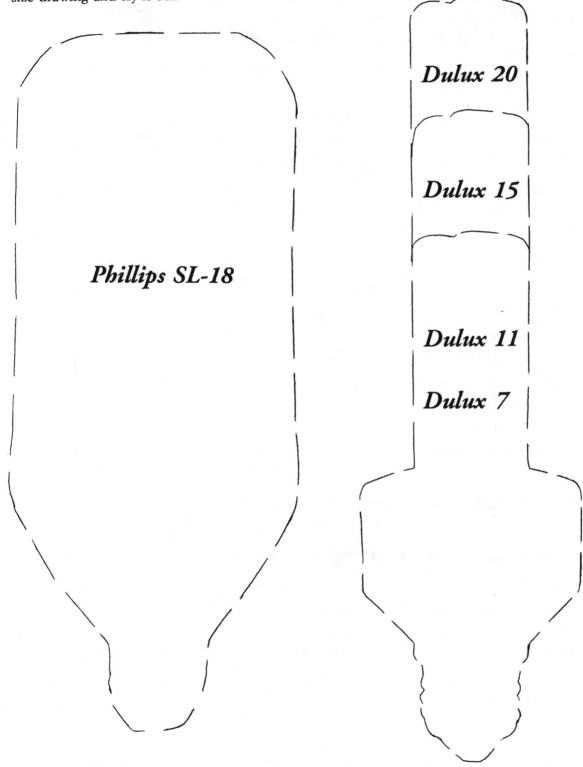

Phillips SL-18

Dulux 20

Dulux 15

Dulux 11

Dulux 7

System Demand Planning Chart

Name: _____ Site Location: _____

Appliance*	Qty		Wattage (volt x amp) Mult by 1.1 for AC		Hrs per Day		Days per Week				Avg Watt-Hrs/Day
		x		x		x		÷7		=	
		x		x		x		÷7		=	
		x		x		x		÷7		=	
		x		x		x		÷7		=	
		x		x		x		÷7		=	
		x		x		x		÷7		=	
		x		x		x		÷7		=	
		x		x		x		÷7		=	
		x		x		x		÷7		=	
		x		x		x		÷7		=	
		x		x		x		÷7		=	
		x		x		x		÷7		=	
		x		x		x		÷7		=	
		x		x		x		÷7		=	
		x		x		x		÷7		=	
		x		x		x		÷7		=	
		x		x		x		÷7		=	
		x		x		x		÷7		=	
		x		x		x		÷7		=	

*( ) Check all DC appliances

Step 5: Maximum AC wattage at one time

Total Watt-Hour Per Day Load

Total Watt-Hr Per Day Load		Battery Inefficiency Factor		Total Corrected Watt-Hours Per Day
	x	1.25	=	

Generator Direct Loads

Appliance	Qty	Wattage (Volts x Amps)	Hours Per Day	Days Per Week

REAL GOODS

Battery Sizing Worksheet

Total Corrected Watt-Hours per Day (from Demand Worksheet)		Wh/Day
System Voltage (usually 12 or 24)	÷	Volts
Load Amp-Hours per Day	=	Ah/Day
Days of Storage	x	Days
Amp-Hours Battery Storage Capacity	=	Amp-Hours
Maximum Depth of Discharge	÷	Percent
Total Battery Capacity	=	Ah at Volts

The following page describes calculations which need to be made to further clarify independent energy system design parameters. Understand that these are general calculations, and real life situations may vary from the results obtained. Please don't feel overwhelmed by any calculations on these worksheets—feel free to discuss any calculations with the Real Goods Technical Staff.

Battery charger sizing: We recommend Todd Engineering battery chargers because they have been proven to be efficient and reliable. When charging lead-acid batteries with a generator, we recommend the 15.5 volt model. The 14.0 volt chargers are suited for continuous charging of lead-acid batteries from a utility power source, whereas the 16.5 volt chargers are suited to higher voltage NiCad battery banks. Other chargers may be used, although the maximum charge rate, determined by dividing the battery capacity by ten, should not be exceeded by much. This is the optimal charge rate.

Generator Supply Sizing: The intent here is to determine how long you might need to recharge batteries based upon a fully discharged battery and a charger of a given amp output, as well as the minimum size generator necessary to operate the charger **alone**. If you intend to operate the charger concurrently with other generator loads (such as water pumps or washing machines) you need to size up the generator appropriately to handle these loads. The calculation of minimum generator size for charging assumes the use of the high efficiency Todd Engineering chargers. The use of less efficient chargers may require substantially larger generators. For example, the Trace 2012 inverter with Standby (battery charging) Option requires a minimum 6.5 KW generator to take full advantage of the 110 amp charger offered.

Photovoltaic Supply: This section determines the total number of modules you will need. The calculations are based upon a fixed array given the average daily hours of sunlight as defined by the Sun Map on page 490. You may choose to use a tracking device, which can boost yield by as much as 40%.

Wind and Hydro supply: Many independent energy producers take advantage of a variety of power sources and incorporate them into "hybrid" systems. The Real Goods staff can advise you as to the feasibility of these options as well as provide computer assisted hydro potential analysis services.

Power Supply Worksheets

Battery Charger

Total Battery Capacity in Amp-Hours		Amp-Hours
	÷ 10	
Maximum Battery Charger Amperage	=	Amps
Select a charger with the closest maximum charge rate		See the Todd Battery Chargers on Page 203
MODEL	AMPS	@ VOLTS

Generator Supply

Total Corrected Watt Hours per Day (from Demand Worksheet)		Wh/Day	Charger Amps		Amps
System Voltage	÷	Volts	Volts	x	Volts
Load Amp-Hours per Day	=	Ah/Day	Watts	=	Watts
Charger Amperage	÷	Amps	Gen. Derate (prevent overload)	x 1.25	
Approximate Hours of Generator Run Time	=	Hours	Minimum Generator Size For Battery Charging Only	=	Watts

Photovoltaic Supply

Total Corrected Watt-Hours per Day (from Demand Worksheet)		Wh/Day	
Hours of Sun per Day (Solar Installation)	÷	Hours	See Sun Map in Appendix on Page 490
Minimum Array Wattage Subtotal	=	Watts	
Photovoltaic Derate Factor	x 1.15		
Array Wattage Total	=	Watts	
Watts per Module	÷	Watts	See Module Specs Beginning on Page 149
Number of Modules Required	=		Round up (to even number if 24 volt system)

Wind Power Supply

Average Windspeed at Site (mph)		MPH	
Wattage at This Windspeed		Watts	Write or Call for Manufacturers Spec's
Hours Running	x 24		
Expected Watt-Hours per Day	=	Wh/Day	
Total Corrected Watt-Hours per Day (from Demand Worksheet)		Wh/Day	
Watt-Hours per Day from Wind	–	Wh/Day	
Watt-Hours To Be Satisfied By Other Sources	=	Watt-Hours	

Hydro Power Supply

Total Corrected Watt-Hours per Day (from Demand Worksheet)		Wh/Day	For More Detailed Information on Your Hydro-Electric Potential Order Our Hydro-Site Analysis
Watt-Hours Produced by Hydro	–	Watt-Hours	See Hydro Sizing Chart on Page 191 for Estimate
Watt-Hours To Be Satisfied By Other Sources	=	Watt-Hours	

Resource List

Real Goods has compiled the following list of resources, many with whom we have had personal experience. We do not necessarily endorse any of the actions of any of these organizations. Our apologies for those we have forgotten. To be considered for listing in the next edition of the Alternative Energy Sourcebook, send a brief description of your company or organization to Real Goods Eco-Desk, Attention: Resource List.

The Alliance to Save Energy
1725 K St. NW, Ste. 914
Washington, DC 20006-1401
(202) 857-0666
Promotes energy efficiency through research, demonstration projects, lobbying, and education.

The American Hydrogen Association
219 South Siesta Lane, Ste. 101
Tempe, Arizona, 85281
Publishes a newsletter about hydrogen fuel for cars, a new technology which could eventually solve a lot of problems

The American Solar Energy Society
2400 Central Avenue, Unit B-1
Boulder, CO 80301
(303) 443-3130
USA's member of the International Solar Energy Society — promotes the application of all renewable energies.

The American Wind Energy Association
777 N. Capitol St., Ste. 805
Washington, DC 20005

Appropriate Technology Transfer
for Rural Areas (ATTRA)
University of Arkansas
P.O. Box 3657
Fayetteville, AR 72702
(800) 346-9410

Audobon Society
950 Third Av.
New York, NY 10022
(212) 832-3200
Promotes rational strategies for energy development, and protects life from pollution, radiation, and toxic substances. Seeks solutions for our environmental problems.

Business Partnership for Peace
P.O. Box 658
Ithaca, NY 14851
(607) 273-1919
An organization which networks businesses and lobbies government.

The Consumer Federation of America
1424 16th St NW, Ste 604
Washington, DC 20036
(202) 387-6121
A lobbyist organization.

Defenders of Wildlife
1244 19th ST NW
Washington, DC 20036
(202) 659-9510
Non-profit organization whose purpose is to preserve, enhance, and protect, the natural abundance and diversity of wildlife.

Earth First!
106 W. Standley
Ukiah, CA 95482
(707) 468-1660
Eco-activist organization.

Earth Island Institute
300 Broadway
San Fransisco, CA 94133
(415) 788-3666
Environmental organizing and publishing network.

Ecology Action
John Jeavons
5798 Ridgewood Road
Willits, CA 95490
(707) 459-0150
Researches methods of growing food in small spares.

The Elmwood Institute
P.O. Box 5765
Berkeley, CA 94705
(510) 845-4595
A think tank that takes in and provides information on environmental issues.

Energy Conserving Passive Solar Houses
Drawing Room Graphic Services
Box 88627
North Vancouver, BC V7L 4L2
(604) 689-1841
Provides house plans for passive solar homes.

Environmental Careers Organization, Inc.
286 Congress St
Boston, MA 02210
(617) 426-4375
Offers information about careers and jobs that have to do with the environment.

Environmental Defense Fund
257 Park Av. South
New York, NY 10010
(212) 505-2100
An organization committed to combining the efforts of scientists, economists, and attorneys to devise solutions to environmental problems.

Environmental Hazards Management Institute
P.O. Box 932
Durham, NH 03824
(800) 446-5256

Florida Solar Energy Research Center
300 State Rd 401
Cape Canaveral, FL 32920-4099
(407) 783-0300
Publishes up to date information about solar energy and energy conservation.

Green Party of California
P.O. Box 20999
Oakland, CA 94620
(510) 649-9773

Green Party Information Clearinghouse
P.O. Box 30208
Kansas City, MO 64112
The Green Party is a worldwide political party, fairly new to the U.S. which focuses on environmental issues.

Greenpeace USA
1436 U St. NW
Washington, DC 20009
(202) 462-1177
An international non-profit organization dedicated to protecting the environment.

Lighthawk
PO Box 8163
Santa Fe, NM 87504
The environmental airforce.

Mendocino Environmental Center
106 W. Standley
Ukiah, CA 95482
(707) 468-1660
Grassroots environmental resource and support network.

National Appropriate Technology Assistance Service
PO Box 2525
Butte, MT 59702-2525
(800) 428-2525
Provides up to date information and assistance on energy conservation and renewable energy.

The National Renewable Energies Lab
1617 Cole Blvd.
Golden, CO 80401
(303) 231-1000
Conducts experiments and provides information about renewable energy.

Natural Resource Defense Council
40 West 20th St.
New York, NY 10011
Promotes the protection of natural resources by combining legal action, scientific research, and citizen action.

Northeastern Sustainable Energy Association (NESEA)
23 Ames St.
Greenfield, MA 01301
(413) 774-6051
Sponsors the American Tour de Sol and advocates electric vehicle research, development, and production.

The Owner Builder Center
1250 Addison St. Suite 209
Berkeley, CA 94702
(510) 848-6860

Passive Solar Environments
821 W. Main St.
Kent, OH 44240
(216) 673-7449
House design and planning.

The Rainforest Action Network
301 Broadway
San Fransisco, CA 94133
(415) 398-4404
Organization dedicated to saving our vital rainforests.

Redwood Alliance
Michael Welch
761 8th St. #4/Box 293
Arcata, CA 95521
(707) 822-7884

The Rocky Mountain Institute
1739 Snowmass Creek Road
Snowmass, CO 81654-9199
(303) 927-3128
Performs studies and accumulates information on energy efficiency.

Save America's Forests
4 Library Court, SE
Washington, DC 20003
(202) 544-9219
A nationwide coalition of grassroots environmental groups, public interest groups, responsible businesses, and individuals, working to pass laws to protect our forest ecosystems.

Sea Shepherd
1314 2nd St.
Santa Monica, CA 90404
(310) 394-3198
Militant defenders of the ocean environment.

Sierra Club
730 Polk St.
San Fransisco, CA 94109
(415) 776-2211
Promotes conservation of the natural environment by influencing public policy decisions, and promotes responsible use of the earth's resources and ecosystems.

Solar Energy Expo & Rally (SEER)
151 North Main Street
Willits, CA 95490
(707) 459-1256
Sponsors the Tour de Mendo Rally and an annual Solar Fair.

Solar Technology Institute
P.O. Box 1115
Carbondale, CO 81623-1115
(303) 963-0715
A place to learn about solar and other renewable energies. Hands on experience included.

Southface Energy Institute
P.O. Box 5506
Atlanta, GA 30307
Provides assistance for owner-builders interested in passive solar houses.

Union of Concerned Scientists
26 Church St.
Cambridge, MA 02238
(617) 547-5552
An organization of scientists and citizens concerned about the impact of advanced technology on society. Programs focuses on energy policy and national security.

The United States Department of Energy
100 Independence Av., SW
Washington, DC 20585
(202) 586-6210
A place to call if you have any questions, comments, or complaints about U. S. energy policy.

The United States
Environmental Protection Agency
401 M St., SW
Washington, 20460
(202) 382-4700
A place to call if you have any questions, comments, or complaints about the environment.

Volunteers in Technical Assistance (VITA)
Dania Granados
1815 N Lynn St. #200 Vita
Arlington, VA 22209
(703) 276-1800

Your Senator
Senate Office Building
Washington, DC 20510

Your Representative
House Office Building
Washington, DC 20515
Please take the initiative to write your representatives if you have something on your mind. Most of them want to know people's opinions and it will be read. Also, please vote!

The Capitol building switchboard
(202) 224-3121

Recommended Catalogs

Ag Access
Danielle Lindeman
603 4th St
Davis, CA 95616
(916) 753-9633
A great catalog for books on sustainable agriculture.

Dripworks
380 Maple St.
Willits, CA 95490
(707) 459-4710
Provides a good selection of drip and garden supplies. Free catalog.

Earth Care Paper Company
John & Carol Magee
Box 14140
Madison, WI 53714
Catalog of recycled paper for office and gifts.

Harmony Farm Supply
P.O. Box 460
Graton, CA 95444
(707) 823-9125
Also supplies drip irrigation and gardening supplies. Catalog available.

Gardner's Supply Company
128 Intervall Rd
Burlington, VT 05401
(802) 660-3506
Catalog of fine gardening supplies and accessories.

Music for Little People
Lieb Ostrow
Box 1460
Redway, CA 95560
(707) 923-3991
Catalog of great children's music for peace.

The Natural Choice
Rt 13 Lakeshore Plaza
Carterville, IL 62918
(618) 985-6224
Catalog for non-toxic personal products and cosmetics.

Electric Vehicle Clubs:

Lee Clouse
PO Box 11371
Phoenix, AZ 85061
(602) 943-7950

Mil Stults
2270 Minnie Street
Hayward, CA 94541
(510) 582-9713

George Schaeffer
211 Ballan Blvd.
San Rafael, CA 94901
(415) 456-9653

Jean Bardon
540 Moana Way
Pacifica, CA 94044
(415) 355-3060

Lee Hemstreet
787 Florales Drive
Palo Alto, CA 94306
(415) 493-5892

Don Gillis
5820 Herma
San Jose, CA 95123
(408) 225-5446

I.L. Weiss
2034 N. Brighton 'C'
Burbank, CA 91504
(818) 841-5994

Ken Koch
12531 Breezy Way
Orange, CA 92669
(714) 639-9799

Jim Cullen
Desert Research Institute
2505 Chandler Ave. Ste. 1
Las Vegas, NV 89120

Ken Bancroft
4301 Kingfisher
Houston, TX 77035
(713) 729-8668

Ray Nadreau
19547 23rd N.W.
Seattle, WA 98177
(206) 542-5612

Dave Pares
3251 S. Illinois
Milwaukee, WI 53207
(414) 481-9655

Kasimir Wysocki
293 Hudson St.
Hackensack, NJ 07601
(201) 342-3684

Bob Batson
1 Fletcher Street
Maynard, MA 01754
(508) 897-8288

Steve McCrea
1402 East Las Olas Blvd Ste 904
Fort Lauderdale, FL 33301

VEVA
543 Powell Street
Vancouver, BC, Canada
V6A-1G8
(604) 987-6188

American Solar Car Assn.
Robert Cotter
PO Box 158
Waldoboro, ME 04572

EVCO
Box 4044 Station 'E'
Ottawa, Ontario, Canada
K1S-5B1

DEVC
George Gless
Denver, CO
(303)442-6566

Fox Valley EVA
John Stockberger
2S 543 Nelson Lake Road
Batavia, IL 60510
(312) 879-0207

D. Goldstein
Maryland E.V.A.
9140 Centerway Rd.
Gaithersburg, MD 20879

PRODUCT INDEX
for an index of topics see the
following Topic Index

PRODUCT INDEX

for an index of topics see the
following Topic Index

PRODUCT INDEX
for an index of topics see the
following Topic Index

This is an index of topics in the Alternative Energy Sourcebook.
For an index of products, please see the preceding PRODUCT INDEX.

TOPIC INDE

This is an index of topics in the Alternative Energy Sourcebook.
For an index of products, please see the preceding PRODUCT INDEX.

This is an index of topics in the Alternative Energy Sourcebook.
For an index of products, please see the preceding PRODUCT INDEX.

TOPIC INDI

This is an index of topics in the Alternative Energy Sourcebook.
For an index of products, please see the preceding PRODUCT INDEX.

This is an index of topics in the Alternative Energy Sourcebook. For an index of products, please see the preceding PRODUCT INDEX.

TOPIC INDE

This is an index of topics in the Alternative Energy Sourcebook.
For an index of products, please see the preceding PRODUCT INDEX.

is is an index of topics in the Alternative Energy Sourcebook.
r an index of products, please see the preceding PRODUCT INDEX.

This is an index of topics in the Alternative Energy Sourcebook.
For an index of products, please see the preceding PRODUCT INDEX.

Daytime Phone ( )
Important for clarifying questions in orders

Ordered By:

Name

Address

City   *State/Zip*

Change Address If Necessary
Ship to: (If different from *Ordered By* address)

Name

Address

City   *State/Zip*

☐ Check here if you want a catalog sent with your order.

For Credit Card Orders Call Toll-Free
1-800-762-7325

Gift Orders: If you'd like to send a gift directly to a friend, complete the section below. We will provide a gift card. Use extra paper for more gifts or for any message that you'd like to include on the gift card.

ITEM #	DESCRIPTION	QTY	PRICE	TOTAL

Ship to:		**Total**	
Name		**Tax (CA)**	
Address		**Shipping**	
City/State/Zip		**Gift Total** (enter below)	

ITEM #	PAGE	QTY	DESCRIPTION	SIZE/COLOR	PRICE	TOTAL

Packing & Shipping Charges (for each delivery to each address)
For Total Amount of Order Add (do not include freight-collect items)

Amount of Order *Destination*	under $25	$25 – $74	$75 – $149	$150 – $299	$300 – $599	$600 – $999	Over $1,000
Rocky Mountains & West (CA, WA, OR, ID, NV, NM, AZ, CO, WY, MT, UT)	3.75	5.75	$8	$12	$15	$20	2%
East of the Rockies (The Rest of the 48 States) (Or U.S. Mail to HI & AK)	4.75	6.75	$11	$14	$21	$28	3%
Federal Express (2nd Day Air) AK & HI add $10 to these rates	$11	$13	$18	$28	$49	$60	5%
Canada (Air Parcel Post)	$8	$12	$18	$35	$60	$80	7%

Call for freight quote on foreign orders.
All items shipped FOB shipping point.

Payment Method:
☐ Money Order or Cashier's Check   ☐ Personal Check
(allow time to clear)
☐ Credit Card (MasterCard, Visa, Discover, or Amex)

Account Number: *(please include all numbers)*

☐☐☐☐☐☐☐☐☐☐☐☐☐☐☐☐☐

Exp. Date: ☐☐ – ☐☐

TOTAL OF GOODS	
Sales Tax 7¼% (CA Deliveries Only)	
Shipping (See Box)	
Gift Total (from above)	
Total Enclosed (U.S. Dollars Only!)	

☐ ***Rush Order Processing:*** Add $10 to Shipping Charges. Your order will be given high priority!

Signature _____ Date _____

Real Goods Guarantee • *Our merchandise represents the best value available anywhere.* Our trained staff will work with you to ensure that you select the best product for your individual needs. *We promise to treat you fairly.* You can return any stocked item for full cash if returned in original condition & packaging within 90 days of purchase for any reason whatsoever. No questions asked except for "How can we do better next time?"

Credit Card Orders By Telephone & Fax

We can process your order in the fastest possible way if you call us or fax us with your order, 707/468-0301. Please have your order form filled out completely and your credit card and expiration date ready. Always include your return phone number on faxed orders. Our regular business hours are 7 am to 7 pm (Pacific Time) Monday through Saturday. During off hours you can place your order with our answering machine.

Ordering By Mail

Be sure to supply all of the information requested on the order form. We can process your order right away if payment is made with a money order, cashier's check or credit card. Funds must be in US dollars drawn on a US bank. If you must use a personal check, allow ample time for it to clear our bank before we ship. *Note: All prices are subject to change without notice.* Where prices have increased, we will charge the balance to your credit card or contact you if the difference is significant.

Order Processing Time

We begin processing your order the moment we receive it. We ship the great majority of orders within three days of receipt. If you need your order right away, add $10 for **Rush Processing** and we'll get your order out that day if received by 2 pm (Pacific Time). The rush charge is for our processing time only — rush shipping can be provided by Federal Express. See the order form or call for FEDEX rates. You will be notified promptly if an item must be back ordered, and it will be marked B/O on

your invoice. While we stock nearly all the merchandise in our catalog, there are a few items that we ship directly from the manufacturer (they are indicated by a ■ next to the item number in the catalog). These items will be marked D/S on your invoice. Some large items like Sunfrost refrigerators and batteries are shipped freight collect from the manufacturer. Please allow extra time for receipt of these items.

Freight Damage, Returns & Adjustments

Inspect all shipments upon arrival. File a freight claim with the appropriate carrier if you discover damage. Be sure to save the damaged cartons in original condition until any claim is settled. If you return merchandise to us, be sure to include a copy of your invoice and a description of the problem. Items must be returned in *their original carton and in original condition to be eligible for replacement or refund.* You will be notified by mail when we've received your package. You may return any *normally stocked* item for a full refund within 90 days of the invoice date. You are responsible for the freight back to us. We may have to charge a restocking fee on items not normally stocked or shipped directly from manufacturers (these items are indicated by a ■ by the item code), or not in their original condition or container. *All TVs and video equipment must be returned directly to the manufacturer. We cannot accept returns on these items.*

Warranties

We guarantee that all our products meet the specifications listed in our catalog. If anything fails within its warranty period, contact the manufacturer for further instructions.

Take Control of Your Mailbox

Some people feel we are sending too much through the mails. We want *you* to determine how much you receive. (You choose when to go to the store; shouldn't you choose when to receive a catalog?) Send us a note with your order if you would like fewer catalogs.

Are You Getting Duplicate or Unwanted Catalogs?

Try as we may, our computer system isn't always perfect and we don't like wasting catalogs and postage any more than you do. If you receive multiple copies please send us all the mailing labels and then pass the extra along to a friend. If you don't want us to provide your name to other organizations that we consider environmentally and socially responsible, notify us in writing. If you want your name eliminated from all mailing list rentals, notify the *Direct Marketing Association (DMA) at P.O. Box 3861, New York, NY 10163.*

Customer Service

Customer service is available between 9 am and 5 pm, Pacific Time, Monday through Friday. Be sure to ask for customer serrvice so we can properly route your call. Call on the toll-free number or the regular number.

Technical Assistance

If you require technical assistance, system design assistance, installation help, etc., please call our technicians at 707/468-9214.

All prices subject to change without notice.

Check with our Color Catalogs and the Real Goods News for pricing updates & new products.